Environmental Science and Engineering

The ultimate goal of this series is to contribute to the protection of our environment, which calls for both profound research and the ongoing development of solutions and measurements by experts in the field. Accordingly, the series promotes not only a deeper understanding of environmental processes and the evaluation of management strategies, but also design and technology aimed at improving environmental quality. Books focusing on the former are published in the subseries Environmental Science, those focusing on the latter in the subseries Environmental Engineering.

Ananya Kuanar · Dattatreya Kar ·
Alok Prasad Das · Dipalee Bhanja
Editors

Recent Advances in Bioremediation and Phytoremediation

Springer

Editors
Ananya Kuanar
Centre for Biotechnology
Siksha O Anusandhan University
Bhubaneswar, Odisha, India

Alok Prasad Das
Department of Life Sciences
RDW University
Bhubaneswar, Odisha, India

Dattatreya Kar
Institute of Medical Sciences and SUM Hospital
Siksha O Anusandhan University
Bhubaneswar, Odisha, India

Dipalee Bhanja
All India Institute of Medical Sciences
Bhubaneswar, Odisha, India

ISSN 1863-5520 ISSN 1863-5539 (electronic)
Environmental Science and Engineering
ISBN 978-3-031-77883-4 ISBN 978-3-031-77884-1 (eBook)
https://doi.org/10.1007/978-3-031-77884-1

This Springer imprint is published by the registered company Springer Nature Switzerland AG
The registered company address is: Gewerbestrasse 11, 6330 Cham, Switzerland

Contents

Chapter 1
Rhizofiltration: A Sustainable Green Technology for Remediation of Heavy Metals from Aquatic Systems

Trinath Biswal

Abstract Rhizofiltration is the process of remediation of toxic contaminants and extra nutrients through adsorption by the plant root zone (rhizosphere) and is a green technology for removing contaminants from various water resources. Extensive industrial activity is the major cause of heavy metal contamination, which is now a major global issue. The addition of heavy metals to our clean water resources is gradually increasing. Therefore, there is an urgent requirement to develop an innovative approach to prevent the introduction of heavy metals into our environment. Rhizofiltration is a successful method for the removal of heavy metals from water resources through plant roots. This process uses the roots of some specific plant species, such as Indian mustard, sunflower, and numerous kinds of grasses, for effective removal of heavy metals, including Zn^{2+}, Cu^{2+}, Hg^{+}, Ni^{2+}, Ag^{+}, Cd^{2+}, and Pb^{2+},from various aqueous mediums. Roots of some other plant species, such as *Lemna* spp., *Pistoia stratiotes, Salvinia* spp., and *Eicchornia* spp., were also found to be effective for the removal of toxic heavy metals in sustainable pathways. Today, phytofiltration (rhizofiltration) has established a significant technique for the remediation of heavy metals through the roots of aquatic plants. This process uses the roots of the plants to absorb, accumulate, and precipitate heavy metals from surface water, groundwater sources, and contaminated industrial effluents. Hence, rhizofiltration is considered a novel green technology for the removal of heavy metals from aquatic systems.

Keywords Heavy metals · Plant roots · Bioremediation · Green pathway · Industrial wastes

---

T. Biswal (✉)
Department of Chemistry, Veer Surendra Sai University of Technology, Burla, Odisha, India
e-mail: biswal.trinath@gmail.com

A. Kuanar et al. (eds.), *Recent Advances in Bioremediation and Phytoremediation*, Environmental Science and Engineering,
https://doi.org/10.1007/978-3-031-77884-1_1

1.1 Introduction

The contamination of groundwater and other water bodies by heavy metals (HMs), and microplastics is now a promising global issue. Although many techniques have been developed for the remediation of HMs from water bodies, such as chemical processes, ion exchange, microbiological precipitation methods, etc., none of these have been satisfactory (Dey et al. 2023; Mishra et al. 2009, 2019). All these methodologies exhibit different effectiveness for different kinds of metals, and these methods are usually expensive if the standards of cleanup are high (Dushenkov et al. 1995; Rawat et al. 2012; Das et al. 2011). There are some higher plants in aquatic ecosystems whose root systems can be utilized for the removal of HMs and organic contaminants from water bodies. These plants include water hyacinth (Eichhornia crassipes), duckweed (Lemna minor), water velvet (Azollapinnutu), and pennywort (Hydrocotyle umbellata). But the efficacy of the removal of HMs in these plant species is quite slow due to the small size of the plants and the slower rate of growth of the plant roots. Indian mustard (Brassica junceu) is an important terrestrial plant whose plant roots grow rapidly and can remove HMs efficiently. Rhizofiltration is also analogous to phytoextraction; however, it is mainly associated with the remediation of contaminated groundwater but not soil. In this process, the remediation of contaminants can be done by absorption, precipitation, and concentration of contaminants through the roots or root systems of the plant species. The presence of heavy metals is highly dangerous to living organisms, including animals and humans, because of their properties of biomagnification, bioaccumulation, and persistence within the environment for no longer (Das and Ghosh 2023; Nag et al. 2023). Continuing exposure to heavy metals has a very serious impact on our respiratory system and causes cancer, cardiovascular disorders, etc. Even in low concentrations, heavy metals and their ions result in teratogens, toxins, carcinogens, and mutagens for the human body (Verma et al. 2006; Yadav et al. 2011; Raj and Das 2023). The technique of rhizofiltration is a sustainable pathway for the removal of organic contaminants and heavy metals (HMs) from groundwater, which therefore protects human health and the environment. But the traditional techniques of remediation of HMs from water are highly expensive, require toxic materials, generate hazardous by-products in the form of effluents, and are inadequate to remove most of the HMs from the groundwater. Rhizofiltration is a cost-effective, environmentally friendly green technology that remediates HMS from aquatic bodies. B. junceu was found to be an ideal plant species for rhizofiltration due to its capability of accumulating a high concentration of Pb and other HMs in its plant roots. Furthermore, several plant species in the Brassicaceae family possess the capability of removing HMs such as Pb, Cd, and Cu. Metals are natural components of the environment, and their presence in water bodies is highly hazardous to the health of human beings, animals, and aquatic biotic communities. The excessive addition of HMs to the water system is mainly due to anthropogenic activities like the discharge of industrial wastewater effluents, the combustion of fossil fuels, the discharge of urban effluents, and sewage wastewater (Mishra et al. 2022, 2023). The permissible values of HMs in natural water resources are <1.0 μgm/l. If the HMs present in the water bodies are exceeding the permissible limit, then they are highly toxic to human health. The degradation of biological ecosystems is primarily due to organic contaminants and, secondarily, to the persistence of HMs in the ecosystem. The existence of HMs in the ecosystem has numerous negative impacts on both the terrestrial and aquatic environments (Abubakar et al. 2014; Yadav et al. 2015). The HMs enter the food chain and food web of different ecosystems and cause hazards to human health and livestock, either indirectly or directly, and result in a great threat to the living communities around the globe. There are some natural phenomena, like volcanic eruptions, weathering of rocks, etc., that interring HMs into the groundwater via geochemical recycling. Phytoremediation is a sustainable technology that causes the removal of HMs from aquatic systems, which are of various kinds, such as phytoextraction, phytodegradation, and phytovolatilization. Phytostabilization, Rhizofiltration. But the process of rhizofiltration is most important because it retards the mobility of the pollutants, restricts their movement in the groundwater, and hence decreases their bioavailability for entering the food chain and food web (Awa and Hadibarata 2020).

1.2 Basic Concept Regarding Rhizofiltration

Phytoremediation is an energy-saving, environmentally friendly, cost-effective, and promising technology for the removal of HMs from aquatic systems and wastewater effluents. In the process of phytoremediation, one of the mechanisms is rhizofiltration, which uses the root zone (rhizosphere) of the plant species for the remediation of contamination through absorption, precipitation, and concentration. Rhizofiltration (hydraulic control) is a technique based on the capability of the plant root system to accumulate and absorb HM contaminants from water bodies. Root exudates can precipitate organic contaminants and HMs and be adsorbed by them, causing an improvement in the quality of water. There are many metals that promote the growth of plants, and key components of different kinds of proteins and enzymes are known as essential metals. However, the concentration of non-essential HM causes HM-promoted phytotoxicity, which replaces essential elements, inactivates enzymes, and blocks functional groups. The mechanism of rhizofiltration is to clean out HMs such as Cd, Ni, Cr, V, Cu, and Pb and some radionuclides such as Sr, U, and Cs. Trees with long roots can absorb huge amounts of water and HMs, along with other toxic organic contaminants. The increase in the number of oxygen radicals interferes with the activities of electron transport, which is one of the important HMs associated with phytotoxicity. The HMs can be detoxified through the thiol groups present in the protein of the plant roots. In-situ rhizofiltration is effective for polluted surface water, but ex-situ rhizofiltration is effective for polluted supply water and groundwater resources. The root extracts above-ground plant aquicultural systems are highly effective through the application of ex-situ rhizofiltration. Typha angustifolia is an important aquatic plant species that results in potential rhizofiltration for the removal of HMs. It absorbs Zn, Cd, and Pb effectively via its root system and reduces the metal concentration in water bodies. In addition to Typha angustifolia, there are many aquatic plant species such as water fern (Azolla), water hyacinth (Eichhornia), and water lettuce (Pistia) that also possess the capability of remediating HMs from water. Pistia has an excellent capability of executing phytoextraction, rhizofiltration, and phytostabilization of Pb and As, whereas Azolla and Eichhornia can absorb Cu and Ni efficiently from polluted water sources. The underground polluted plumes can be recovered and filtered by using rhizofiltration. It was found that the hydroponically cultivated roots of terrestrial plants possess a higher efficiency of remediating HMs than the root systems of usual plants. The method of rhizofiltration begins with the recognition and extraction of groundwater with a pump-and-treat (P&T) system. For effective rhizofiltration, specific plant species having a high adsorption capability of HM from contaminated water sources and wastewater effluents are hydroponically cultivated with the help of greenhouses. Therefore, plant species with high efficiency in HM remediation have been planted at the sources or sites of contaminated water resources. Subsequently, the plant species are removed from the contaminated areas when the roots of the plants are saturated with pollutants, and after that, these are properly disposed of. Table 1.1 illustrates the various plant species used in rhizofiltration for the removal of toxic HM contaminants (Hg, As, Cd, Fe, Cr, and Pb) and organic pollutants like poly aromatic hydrocarbons (PAHc), chlorinated biphenyl, etc. (Woraharn et al. 2021; Mithembu 2012; Mthembu 2016). The various plants used in rhizofiltration are shown in Table 1.1.

1.3 Data Needed for Rhizofiltration

- Depth of the source of contaminated water
- Kinds of HMs present in the contaminated water resources.
- Level or degree of contamination and its monitoring.
- The plant species must be emergent, aquatic, and sub-emergent.
- The hydraulic retention time and sorption by the root system of plants must be successfully designed and developed.
- Rhizofiltration is specifically effective in cases of low concentrations of HMs and huge volumes of water resources (Bakshe and Jugade 2023)

1.4 Mechanism of Rhizofiltration

Rhizofiltration is a kind of phytoremediation that involves the filtering of groundwater, surface water, and wastewater effluents through the root systems of plants to remediate toxic organic substances, HMs, and excess nutrients. The continuous buildup of potentially toxic elements (PTEs) such as cadmium (Cd), selenium (Se), mercury (Hg), lead (Pb), copper (Cu), chromium (Cr), nickel (Ni), and arsenic (As) is the cause of severe impacts on human health and challenges to global environmental

Table 1.1 Plants used for remediation of HMs and toxic organic materials through rhizofiltration

HMs to be remediated	Specific plant species	References
Cadmium (Cd)	Brassica juncea, Azolla filiculoides Lam, Setaria italica (L.) Beauv., Salvinia auriculata Aubl., Pistia stratiotes L., Salvinia minima Baker, and domingensis	Kristanti et al. (2021)
Arsenic (As)	Cynara cardunculus, Chinese brake fern (Pteris vittata), stratiotes, Spirodela polyrhiza, and Eichhornia crassipes	Akhtaret al. (2013)
Lead (Pb)	Pistia stratiotes L., Brassica juncea, Azolla filiculoides Lam, Oxycaryum cubense (Poep. & Kunth) Palla, Salvinia minima Baker, Azolla pinnata, Carex pendula, Salvinia auriculata Aubl., and Typha domingensis	Thayaparan et al. (2013)
Zinc (Zn)	Typha domingensis, Eichhornia crassipes (Mart.)Solms, Helianthus annuus, Jatropha curcas,Tagetes erecta, and Pongamia pinnata	Dürešová et al. (2014)
Aluminum (Al)	Pistia stratiotes L., Typha domingensis, Centella asiatica, Scirpus grossus, Thypa angustifolia.and Ipomoea aquatica	Kamusoko and Jingura (2017)
Nickel (Ni)	Eichhornia crassipes (Mart.) Solms, Spinach, sunflower, Jatropha curcas, Alocasia puber, and Pongamia pinnata	Leblebici et al. (2019)
Iron (Fe)	Pistia stratiotes L., Ipomoea aquatica Typha domingensis, and Centella asiatica	Mohanty et al. (2010)
Copper (Cu)	Eichhornia crassipes (Mart.) Solms, Limnocharis Flava, Brassica napus L., and Elodea	Pérez-Palacios et al. (2017)
Manganese (Mn)	Pistia stratiotes L., hyacinth, Cnidoscolus multilobus, duckweed, Platanus mexicana, azolla, Solanum diversifolium, Asclepius curassavica L., cattail, Pluchea sympitifolia, and poplar	Galal et al. (2018)
Antimony (Sb)	Brassica rapa, Cynodon dactylon, and Amaranthus mangostanus	Yazdanpanah and Rajaei (2014)
Mercury (Hg)	Salvinia natans, Lemna minor, Indian mustard, S. natans. Sansevieria trifasciata, Sunflower, Celosia plumosa.,tobacco, spinach, and rye	Raj and Maiti (2019)
Chromium (Cr)	Sunflower, rye, tobacco, Indian mustard, Typha angustifolia, Canna, Juncus acutus L., Leersia hexandra, and Spinach	Malaviya et al. (2020)
Synthetic dyes	Bryophyllum fedtschenkoi, Tagetes erecta, Arundo donax L, Catharanthus roseus, Trachyspermum ammi, Hibiscus rosa-sinensis, and Chrysanthemum indicum	El-Aassar et al. (2018)
Radionuclides	Sunflower, tobacco, Lactuca sativa, spinach, Raphanus sativus L., rye and Indian mustard, Brassica campestris L., and Oenanthe javanica	Tomé et al. (2008)

sustainability. These heavy metals persist for a long time in water bodies and cause adverse effects on human health and the surrounding environment (Belle et al. 2022).

- The mechanism of rhizofiltration for remediation of HMs from water is based on the impact of biochemical and physical factors associated with the root systems of plant species.
- The efficiency or effectiveness of the mechanism of rhizofiltration is determined by the effectiveness of root system creation, which causes the intake of HMs into the plant body through plant roots.
- Various factors, such as root exudates and pH changes in the rhizosphere, are the causes of the absorption and precipitation of HMs onto the surfaces of root systems.
- The root exudates or environment of the root system

The root systems or root exudates may create a biogeochemical environment that causes the precipitation of toxic pollutants into the water resources or roots of the plant species. When the roots or root systems are completely saturated by HMs, then the plants are picked for disposal. The exudates, like simple organic acids and phenolic compounds, are released from the entire cell at the time of the decay of the roots. These root exudates can also alter the speciation of the metals and the consumption of metal ions of various kinds with the instantaneous release of protons that cause acidification of the medium and facilitate the bioavailability and transport of metals. Mostly, some specific genes present in plant species result in the effective accumulation of metals. Basically, some organic acids and glutathione metabolism play a vital role in the tolerance of metal existence in plant roots. The native environmental conditions, including temperature, intensity of light, rainfall, pH, and wind velocity, also play a vital role in the acceptance of HMs by the root systems of the plants. This method involves the growth of plant species hydroponically and transplanting the HMs into the contaminated water, where the plant species used can concentrate, precipitate, and absorb the HMs in their roots, root systems, and shoots. The root system delivers a huge surface area, which facilitates the accumulation and absorption of nutrients and required quantities of water, along with other non-essential pollutants and HMs. To adapt the plant species, an adequate root system has to be created where plant species are planted in polluted water. The plant species are planted in contaminated zones of water bodies where the root systems of the plants consume water along with HMs, organic contaminants, and nutrients. Plants used for rhizofiltration in man-made wetlands and lagoons are termed in situ rhizofiltration, whereas in ex-situ rhizofiltration, the polluted water containing HMs is diverted into the system containing an array of plants that can purify the polluted water. The development of industrial sectors or the emission of toxic pollutants is not the actual problem, but the most key issue is the addition of toxic pollutants to a healthy and clean environment. Restraint in the transportation of pollutants is a suitable model strategy for limiting pollutants in a clean environment. Rhizofiltration and phytostabilization were found to be the most efficient sustainable green methodology for decreasing the movement of toxic pollutants, particularly HMs. Basically, HMs are discharged due to extensive industrial activity through industrial wastewater effluents and solid wastes. Therefore, it is vital to keep these industrial wastewater effluents and solid wastes in clean human-use water resources. The plantation of some specific plant species not only prevents the transport of pollutants but also supports improving the air quality and maintaining an adequate temperature nearer to the plant area. Likewise, rhizofiltration is the most efficient and cost-effective technique for limiting the toxic pollutants moving along the media of groundwater or surface water. The industrial location of most industries can be phytostabilized through the proper plantation of specific trees or plants according to the kind of pollutants, topography, and soil condition. The basic principle behind the processes of phytostabilization and rhizofiltration is the accumulation of pollutants in the plant tissues, which reduces the bioavailability and mobility of metals by accumulating within the root system or alleviating them inside the substrate molecules. The possible impact of the accumulation of excessive HMs is a reduction in the capability of their transportation through water. Both approaches have a very comparable primary mechanism in the method of complexation with root exudates or root systems and chelation due to the molecules that bind the molecules of HMs. Further, these mechanisms involve cell wall bonding and microbes that happen in phytostabilization; otherwise, the formation of antioxidant materials and the compartmentation of toxic pollutants occur in rhizofiltration. Again, the mobility of trace elements, organic wastes, texture, temperature, and redox potential are various key factors that influence the mechanism of rhizofiltration (Laghlimi et al. 2015; Castro-Castellon et al. 2021; Lee et al. 2010).

1.5 Limitations of Rhizofiltration

- The pH of the contaminated water was properly adjusted to constantly allow for the accumulation of optimum HMs in water bodies.
- The speciation of chemicals and the interaction of the HMs and other contaminants present in the water or wastewater must be properly understood for effective application.
- A properly engineered system is necessary to control and regulate the flow rate and concentration of wastewater effluent or contaminated water.
- Basically, plants in a terrestrial ecosystem are suitable for rhizofiltration and grow properly in a nursery or greenhouse system.
- Periodic removal of saturated plant species must be replaced and should be properly disposed of.
- Still, the study is limited to greenhouse systems and laboratories, and its application in the practical field is a challenge.
- The absorption of toxic contaminants generally occurs on or into the root systems of the surrounding water bodies in the rhizosphere (root zone).
- Specified plant species are planted or cultivated in polluted groundwater where the root system of the plant accumulates HMs in it via absorption. Rhizofiltration is effective if the concentration of HMs is low, and the volume of the water is large.

The plant species translocate HMs to the shoots that are never utilized in situations where the HMs and organic material contamination is comparatively higher.
- The products associated with bioavailability and toxicity after the biodegradation of the pollutants are not always recognized. The by-products obtained after degradation may be bioaccumulated in aquatic organisms and animals or can be mobilized in groundwater.
- It is necessary to know about the various materials produced at the time of degradation in the metabolic cycle of plant species that do not enter the food chain or food web.
- If the root system of the plants is not able to penetrate up to that depth of the water where the pollutants are persisting, then the water must be water out of the ground, and trees must be planted for remediation of the groundwater (Ghosh et al. 2005; Gaur et al. 2014)

1.6 Advantages of Rhizofiltration

- The method of rhizofiltration is used for remediation of toxic HMs and organic pollutants, which may be carried out either in situ, where specific plants or trees are grown directly in the polluted water body, or ex situ, where plants or trees are allowed to grow off-site and later on after proper growth is introduced to the polluted water sources.
- This method of water remediation is comparatively cost-effective and requires less operational and capital costs, depending on the kinds of contaminants present in the water bodies.
- In some cases, some specific pollutants present in the water bodies were reduced significantly within a very short interval of time, and in other cases, they required a longer time. For example, the root of a sunflower plant decreases the uranium level within approximately 24 h.
- The method of treatment of water is basically pleasing and reduces leaching and water infiltration of pollutants.
- After harvesting the saturated plant, the crop can be converted to biofuel briquettes, which are a proper substitute for fossil fuel (Ignatius et al. 2014; Koźmińska et al. 2014).

1.7 Genetic Engineering in Rhizofiltration

The contamination of HMs in the natural environment because of anthropogenic activities and excessive industrialization pose a problem that threatens the health of millions of people around the globe. Plant roots exposed to numerous kinds of microbes in the usual environment play a vital role in the preservation of soil structure, nutrient recycling, detoxification of hazardous chemicals, and management of pathogens and pests. The interaction between plant roots and microbes is either non-specific or specific. In cases of specific interactions, plant species deliver a source of carbon that boosts the bacterial ability to decrease the toxicity of the polluted water, and in cases of non-specific interactions, various kinds of metabolic reactions in plants facilitate the growth of microbes, which causes a decrease in the HMs of contaminants present in the water. To overcome the impact of phytotoxicity, the microorganisms found in the rhizosphere of plant species (bacteria and mycorrhizae) can actively contribute to the change in metal speciation and effective regeneration of used plant species. Various methods, including proper application of fertilizers, pH adjustment, and chelators, can be used to further enhance the capability of rhizofiltration. The conventional method of remediation of HMs from water is a non-specific and expensive process, whereas rhizofiltration is a successful method for decontaminating different kinds of HMs from water and wastewater. The enzymes, or microorganisms, play an integral role in the growth of plant species through interaction with various environmental factors. The use of genes encoding enzymes can change the oxidation state of bacteria. A gene can encode HgO reductase. It also changes the toxic metals into species of comparatively less toxic form; for example, enzymes or microorganisms can be able to methylate the metal Se into the less toxic form dimethylselenate. Multiple biotechnological approaches, such as genetic alteration of plant species, may be adopted to strengthen the capability of phytoremediation or rhizofiltration. Plant genetic engineering can be applied to rhizofiltration, which can be treated as a potential way of exploiting the appropriate active genes that are responsible for uptake of HMs, translocation, vacuolar sequestration, reduction, volatilization, and complexation. Genetic modification may be applied to express the enzymes or microbes that participated in the metabolic pathway of existing plant species or suggest new pathways for plant species. The various kinds of plant species and microorganisms possess a variety of mechanisms to protect against HM poisoning in water bodies. The genes introduced in rhizofiltration promote the detoxification of HM contaminants and subsequently improve the rate of degradation of the contaminants, including HMs and organic contaminants. Transgenic plants specifically containing bacterial genes can be able to convert Se and Hg into comparatively fewer toxic species. Some organic acids and glutathione (GHS) metabolism play a significant role in the metal tolerance level of plant species. Glutathione appears in plant species, basically as reduced GSH. Glutathione can be synthesized by glutathione synthetase and glutamylcysteine synthetase enzymes. The metabolism of glutathione is linked with sulphur and cysteine metabolism in plants.

The concentration of cysteine in plants restricts glutathione biosynthesis. Thiol peptide phytochelatins (PCs) of low molecular weight are always termed class III metallothionein and can be synthetized in plant species from glutathione induced by HM ions. Glutathione is the cause of the synthesis of peptides via the α-glutamylcysteine transferase enzyme, also termed phytochelatin synthase (PCS), which catalyses the transfer of the α-Glu-Cys group from the donor molecules of glutathione to glutathione, which is an acceptor molecule. PCS is a cytosolic, important enzyme that is activated through metal ions such as Cd^{2+}, Au^{2+}, Zn^{2+}, Pb^{2+}, Hg^{2+}, Ag^{1+}, Bi^{3+}, Cu^{2+}, and PCs, synthesizing chelate HMs and ultimately producing complexes. The complexes formed are then moved via the cytosol in an ATP-dependent pathway via the tonoplast into the vacuole. The impact of genetic engineering on phytoremediation is illustrated in Fig. 1.1. The increase in concentration of HMs in water bodies is the cause of the negative impact on all the living biota on our globe; therefore, methods must be developed to remediate pollutants in water bodies. The present technique of remediation for decontaminating HMs present in water bodies is environmentally hostile, highly expensive, and requires more labour. The use of genetic engineering in rhizofiltration for the removal of HMs from water and wastewater effluents makes the remediation method more environmentally friendly and cost-effective. This approach has attracted attention in the past few years. The use of plant species in genetic engineering provides a novel opportunity for rhizofiltration and phytoremediation of water contaminated by HMs. But this technique or method can be completely exploited only if the mechanisms of translocation, metal tolerance, and accumulation of HMs are completely known. In some cases, the decrease in pH value of the soil noticeably improves the phytoavailability of HMs by mixing bacteria like Alyssum murale, Sphingomonas macroscopica, M. arabinogalactanolyticum, etc., resulting in an improvement in phytostabilization. Brassica napus plants were injected to improve the tolerance level to different toxic HMs against varying kinds of bacterial species like microbacterium lactiu, pseudomonas chlororaphis, azotabacter vinelandii, Bacillus mucilaginosus, etc. (Eapen et al. 2005; kesan 2011; Raskin 1996).

1.8 Significance of Rhizofiltration

- Rhizofiltration is used for remediation or decontamination of groundwater, surface water, storm waters, industrial effluents, household effluents, downwashes released from power lines, effluents from acid mine drainage, water containing radionuclides, runoff from agricultural fields, and sludges in dilute forms. Ultimately, we can say that most of the categories of polluted water can be remediated through rhizofiltration.
- The root system of the plants used appropriately for application in rhizofiltration grows at a rapid rate, resulting in the effective remediation of toxic HMs.
- Although many aquatic plants were found to be suitable for rhizofiltration, different terrestrial plants have been recognised to be efficient for the remediation of toxic heavy metals like Cu^{2+}, Zn^{2+}, Cr^{6+}, Cd^{2+}, Ni^{2+}, and Pb^{2+} from water bodies or in aqueous solutions. It was observed that the presence of radioactive pollutants at low levels in water resources can be successfully remediated through rhizofiltration. The system to attain rhizofiltration contains a ‘feeder layer’ of soil, which is suspended overhead of the polluted stream in which plant species grow and their roots are entering the water bodies.
- The feeder layer permits the plant species to absorb fertilizers through their root systems with the simultaneous removal of HMs without contaminating the water bodies. The tree or plant species of various kinds utilised for the remediation of

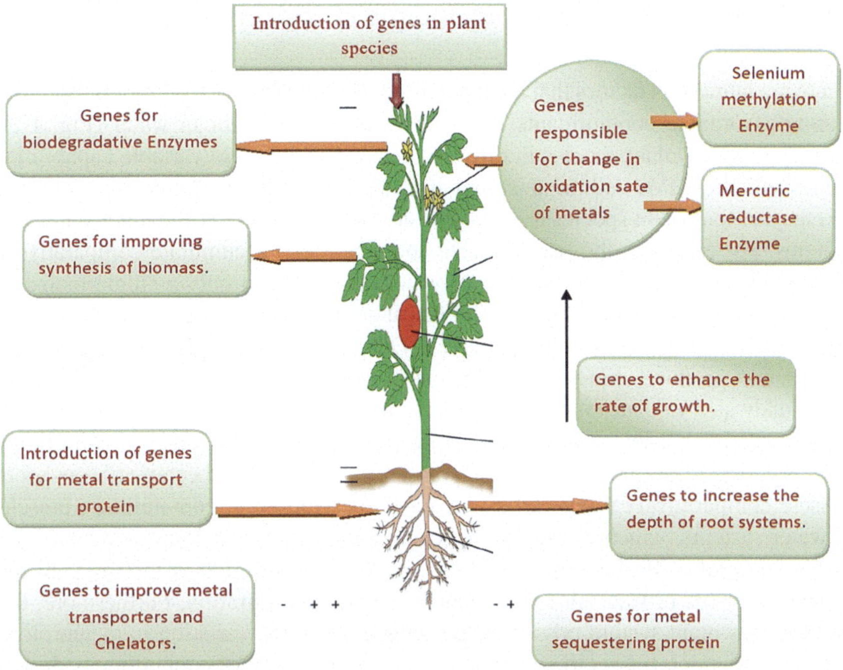

Fig. 1.1 Use of genetic engineering in Phytoremediation and rhizofiltration

water bodies must have low costs, long lifespans, and grow effectively in soil of marginal quality.

- The maintenance costs for rhizofiltration are very low, and the most popular plant species include poplars and willows that exhibit high flood tolerance capabilities. In cases of deep toxic pollutants, hybrid poplars with root systems extended up to 30 feet deep have been applied. Their root systems can enter microscopic pores in the matrix system of soil and can recycle 100 L of water per day per tree.
- These plant species behave like a pump for HMs from water bodies and cause the efficient remediation of contaminants in a green pathway from water bodies. Terrestrial plant species such as Brassica species and sunflower species possess the capability of being effective rhizofiltrators.
- It is a low-cost green technology with low labor costs, minimum maintenance costs, and operational costs. Hence, it is an appropriate alternative to the conventional approach for the remediation of wastewater and HM-contaminated water (Veselý et al. 2011; Banerjee et al. 2022).

1.9 Factors Affecting the Technique of Rhizofiltration.

There are numerous factors that are responsible for effective rhizofiltration for remediation of organic contaminants, HMs, and nutrients present in water (Fig. 1.2). Various kinds of plant species in different ecosystems possess variable capabilities for remediating HMs and organic contaminants from water systems; therefore, not all categories of plant species are appropriate for remediating pollutants from water systems. For example, the use of maize and wheat can accumulate comparatively more phthalic acid esters (PAEs) from water and soil than only a single maize or wheat species. The plants of maize and wheat are not suitable for remediating other pollutants like organochlorine pesticides (OCPs) such as HCHs, DDT, and their degradation products. Since rhizofiltration is specially needed for the remediation of groundwater, aquatic plants, particularly macrophytes, have been suitably used for remediating organic contaminants from groundwater. Plant species with fibrous roots or root systems have higher effectiveness in rhizofiltration because fibrous roots possess comparatively more surface area and are exposed to contaminants present in water systems, which enhances the accumulation of contaminants over the root systems of plants. Besides biological factors, the concentration of microorganisms present in plants or trees also significantly affects rhizofiltration. Furthermore, the multispecies plant system possesses less sensitivity to the variation in meteorological factors, which boosts the efficiency of the remediation of contaminants. It was observed that in the process of rhizofiltration, the mixed species of plants facilitate the treatment of wastewater or heavily contaminated water. The interaction between plant species and microbes influences the quality of water in the rhizosphere. In the process of rhizosphere, the region of the underground zone of the plant species is continuously exposed to water and is found to be an extremely active zone associated with diversified microorganisms. The microbes needed energy, carbon, and oxygen to boost the immunity of the plants and stimulate the exchange of minerals in plants. Therefore, nutrient consumption and protection from phytotoxicity are enhanced by the interaction between plant species and microorganisms (Sheoran et al. 2016; Malaviya et al. 2012) The various important factors that depend on the efficiency of the rhizosphere are as follows (Fig. 1.2).

1.10 Time of Rhizofiltration

The method of rhizofiltration is dependent on time because of the mechanisms of bioaccumulation and biosorption. A greater number of active sites on plant roots (the surface of the biosorbent) in the first hour during the biosorption resulted in a higher rate of adsorption of HMs from the water sources. The process of biosorption will be comparatively higher between one and three hours of the process, and after that, the number of active sites decreases and is exhausted because of the transfer of adsorbate (HMs) from the water bodies to the active sites.

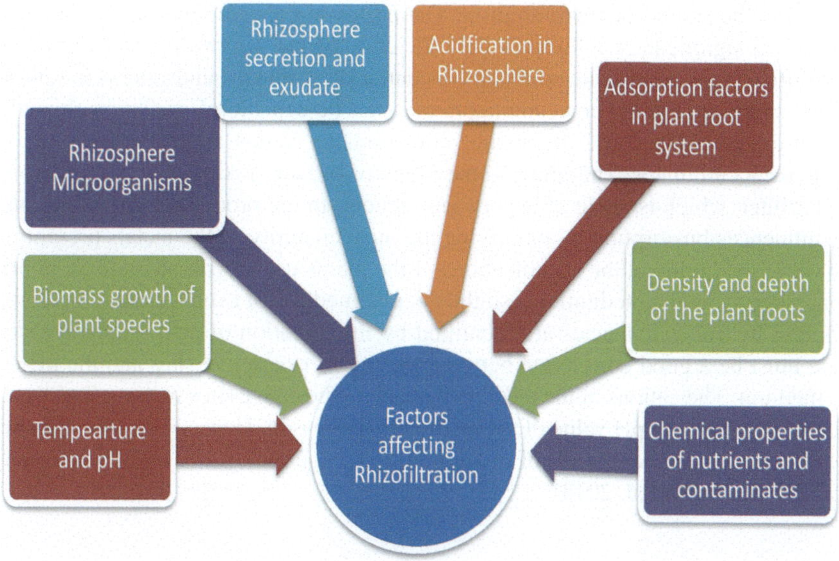

Fig. 1.2 Various factors affecting Rhizofiltration

Example: The plant species *Hydrilla vercillatta* eliminates Cu in 10 days (efficiency of 89.2%), whereas Fe elimination occurs with an efficiency of 91.2% in five days. Hence, the rate of elimination or remediation of HMs from a water sample is measured by knowing the variety of plant species used, the optimum values of contact times, and the kinds of metals that must be remediated (Odinga et al. 2019).

1.11 Temperatureand pH During Rhizofiltration

Temperature is one of the key factors that results in an increase in the efficiency of the HMs at the time of rhizofiltration. The rate of biosorption of HMs is highly dependent on the temperature maintained during that process. The rise in temperature results in an increase in contact time between the absorber and the HMs present in the water sample, which therefore increases the overall performance of absorption by the plant roots. But with a decrease in temperature, the number of active sites at the surface of the plant roots deactivates, resulting in a decrease in the rate of absorption of HMs. Furthermore, the rise in temperature reduces the time required to attain the state of equilibrium in the biosorption of HMs. The variation of temperature also causes a variation in the pH of groundwater, which influences the efficiency of rhizofiltration. At some specific pH, the growth of hyperaccumulators is not appropriate. If the growth of hyperaccumulators occurs in an unsuitable pH condition, then the uptake capability of organic pollutants by the hyperaccumulator will decrease or even die, triggering the failure of rhizofiltration. The pollutant concentration in the groundwater also significantly influences the efficiency of the removal of HMs through rhizofiltration. If the organic pollutant concentration within groundwater is low, then in that situation, the hyperaccumulators may not be capable of executing rhizofiltration effectively because the organic contaminants cannot approach the roots of the hyperaccumulator. Therefore, some kinds of organic toxins may travel across the biofilter, which is made of hyperaccumulators. In the process of rhizofiltration, pH influences biosorption by modifying the bioavailability of the metals present in the water medium and the specifications of the plants used. With an increase in pH from a highly acidic medium to a slightly acidic medium or even a neutral medium, different hydrolyzed species are identified by the variation of active sites. Hence, there must be a good correlation between the number of active sites and the pH of the medium. The cations of the metal will attach to the active sites in the cell walls of the plant roots if the pH value of the medium is increased. Hence, we cannot ignore the effect of pH in the process of rhizofiltration for the removal of HMs (Sikhosana et al. 2020; Singh et al. 2018).

1.12 Chemical Properties of the Pollutants and Nutrient

The nutrient concentration in the water bodies shows both a negative and positive effect on the biosorption of metals by the plant roots. If the availability of nutrients for plants is much higher, then the adsorbed metals can bond to the anions that are present in nutrients. Hence, the opposition between the cations of the nutrient and HMs results in the uptake of HMs. The growth of plants is stimulated by the presence of nutrients in the water bodies. Subsequently, more active sites for the absorption of metals are produced that assist the process of rhizofiltration, and its efficiency in the removal of HMs and other toxic contaminants increases. However, the acceptance of HMs and other organic contaminants depends on the response of the plant species to metals and the growth rate of the plant species. Some organic pollutants are hydrophobic in nature and can easily dissolve in groundwater or surface water. But still, the problem of absorbing organic pollutants from the root systems of plant species is challenging. In association with surfactants, the process of rhizofiltration efficiency for the removal of organic substances will be noticeably improved (Das and Paul 2023).

1.13 Density and Depth of the Plant Roots

The root density of the plant root depends on the depth of the water bodies, and the absorption of metal by the root systems of the plant is also dependent on the depth. A major part of the plant roots is found near the surface of the soil, and the root density declines with the increase in depth. Therefore, some specific plants whose roots are entered into adequate depth are only suitable for rhizofiltration. The method of acceptance of contamination from water by the root region of the plant species and subsequent translocation to the leaves and stems is termed transpiration. Many studies have reported that the translocation of metal from the roots through transpiration has been well recognized. The higher concentrations of metals absorbed by the roots cause water consumption by facilitating the migration of metals through mass flow to the surface of the root system, where precipitation occurs. The rate of transpiration also depends on the topographic or climatic condition of the environment. The plant roots are basically dormant during the winter season and consume very little water in the climate of the mid-latitude region, whereas the reverse is true in the summer, when more volume of water is facilitated to transpire for photosynthesis (Douglas et al. 2004).

1.14 Absorption Factor of Plantroot System

The total quantity of metals accumulated by the root system of the plant species normally equals the concentration of the metal in mg/l in its soluble state. The metal must be absorbed by the roots before translocation, either through the apoplastic or symplastic pathway, where a passive or active filter system may be created. The factors related to root absorption are influenced by rhizobiological activity, moisture, temperature, pH, concentration of competing ions, root exudate, and the concentration of metal in the soil solution (Bloom et al. 2002).

1.15 Acidification of Rhizosphere

The rhizosphere is considered the interface between the root system and the soil. In this region, numerous interactions between the solution of the soil and the root system of the plant species take place, which is the cause of the significant modification of the physicochemical characteristics of this region. Plants need huge quantities of nutrients for their optimum development and growth, and the absorption of such a huge amount of nutrients through the root system of plants may cause modification of the rhizosphere, specifically acidification, which is a continuous phenomenon in many soils throughout the globe. The mechanism behind it is the proton extrusion at the plasma membrane that improves cation influx, which is a key route utilized by the roots of plants to remove excess nutrients from the water or soil solution. The acidification of the rhizosphere plays a crucial role in the acquisition of nutrients by plant species and stimulates the activity of H^+ATPase present at the plasmalemma. The lower availability of nutrients proves the release of protons due to nutrient acquisition by the plant roots. Hence, a shortage of nutrients in the water bodies or less nutrients in the water or wastewater is an indication of high activity of H^+ATPase. Acidification in the rhizosphere is a significant natural process in which plant species consume a high amount of nutrients. Proton liberation usually promotes the consumption of one cation. Hence, the release of protons (H^+) is essential to facilitate nutrient transport into the plant roots. The secretion of H^+ ions by the root system of plants is the cause of acidification in the rhizosphere area and therefore improves the dissolution of metals. The secretion of H^+ ions from the roots of plants is regulated by the triphosphatase (H^+-ATPase) along with H^+ pumps (Carrillo et al. 2002; Petersen et al. 1991).

1.16 Rhizosphere Secretions and Exudates

The root systems of several plant species exudate organic acids, including malic, oxalic, and malonic acids, which is the cause of the decrease in the pH value of the rhizosphere that converts the cations of the metal bioavailable. This secretion of acids significantly impacts the rhizosphere and the biotic community living within it. The nature of the exudates and quantity of the material released usually depend on several factors, such as the architecture of the root system, the level of toxicity of metals, and the existence of harmful microbes. Histidine is an amino acid that is accountable for the hyperaccumulation and transport of Ni in Alyssum. There are numerous kinds of organic acids of low molecular weight, like succinic acid and acetic acid, which are found in the rhizosphere of a Cd-accumulating genotype such as Kyle but never in a non-accumulating genotype such as Arcola. In dicotyledons, the root reductases decrease the concertation of Cu^{2+} and $Fe3^+$ under less Cu and Fe to enhance the uptake of plant species of Fe, Mn, Cu, and Mg. The bioavailability of many metals may be increased through the secretion of phytosiderophores by plants into the rhizosphere. Siderophores are formed by a multitude of classes of microbes, from plant pathogens to animal pathogens and free-living organisms, which are symbiotic N-fixing microorganisms. But the formation of siderophore is the most stimulating plant growth rhizobacteria that exhibit their maximum growth and siderophore creation activity under drastic environmental conditions, such as lack of nutrients and the presence of high concentrations of HMs. The siderophores bacteria can bind a variety of metals, hence improving phytoavailability in the zone of contamination. Phytosiderophores, including avenic and mugineic acids, are released because of their action on the deficiency of nutrients and metals. This is the cause of improving the bioavailability of metals in the case of Fe, Cu, Pb, Cd, Zn, and Pt, which facilitates carrying these metals into the plant tissues. The complex based on metal chelate can be transported through the plasma membrane for iron-phytosiderophore in cereals. The soil immunisation with the help of Pseudomonas aeruginosa basically improves the bioavailable concentrations of Pb and Cr compared to uninoculated controls (Walker et al. 2003; Vives-Peris et al. 2020).

1.17 Rhizosphere Microorganisms

The zone of the rhizosphere of plant species is highly populated by a huge number of microorganisms, mostly mycorrhizal fungi and bacteria. The presence of these microorganisms is responsible for increasing the production of biomass and the tolerance level of plants to HMs and is considered a significant component in the pathway of phytoextraction. These root-colonizing mycorrhizae and bacteria remarkably enhance the bioavailability of different HM ions. The microbes within the rhizosphere function as catalysts for exudate organic material and redox transformations, which promote the bioavailability and root absorption of Cd^{2+}, Mn^{2+}, and some other metal ions. It was observed that microorganisms of the rhizosphere, such as Bacillus and Pseudomonas, enhance the quantity of Cd accumulation in Brassica juncea. The increase in phytoextraction of Zn in rackets with subsequent involutus such as Ectomycorrhizae The biosurfactant developed by Pseudomonas aeruginosa can be used for the remediation of HMs like Pb and Cd (salt et al. 1995; Dimitroula et al. 2015; Khan et al. 2022).

1.18 Biomass Growth of Plant Species

The biomass of the plant species above the ground is found to be one of the key factors, in addition to the bioavailability of the metal ions, which impacts the phytoextraction efficiency. The efficiency of phytoextraction must be improved if the biomass of the plant above the land is increased, which leads to an increase in the concentration of total metal in shoots. The contamination of HMs is the cause of retarding the growth of plant species because of the deficiency of organic substances and essential nutrients, including N, K, and P, on association with them. Irrigation and the use of fertilizers are two vital factors that can stimulate the growth of plant species by modifying the presence of essential nutrients and organic substances. Fertilizers not only promote the growth of plant species but also decrease the pH of the soil, therefore improving the bioavailability of metals present in the water or soil (Nedelkoska et al. 2000). Chou et al. (2005) studied the impact of K-fertilizer on plant species and found that rape (Brassica campestris) shows the maximum production of biomass above the ground parts and the maximum accumulation of contaminants in shoots. The impact of nitrate fertilizers on the development and growth of Brassica juncea (Indian mustard) on the accumulation of Zn was studied, and it was found that by applying the required quantity of nitrate fertilizers the formation of shoot biomass remarkably increases, which is the cause of more accumulation of Zn (Chou et al. 2005). If plant species take 90% of the NO_3^- and 10% of the total N as NH^+, then the capability of phytoextraction by Indian mustard becomes maximum. The increase in metal bioavailability and biomass of plant species is the cause of stimulating microbial-induced phytoextraction. We observed four kinds of bacterial strains from the rhizosphere contaminated by HMs that exhibit the activity of

ACC (1-aminocyclopropane-1-carboxylate) deaminase. Two kinds of strains, such as Pseudomonas fluorescens ACC and Pseudomonas tolaasii ACC, form acetic acid, and in that case, the activity of siderophores becomes more even under Cd stress (Arshad et al. 2007). These strains formed cannot only influence the Cd accumulation in the shoot and root systems of Bacillus napus but also cause an increase in the biomass formation of the plant species, leading to a higher accumulation of total Cd present in the water bodies and soil. It was reported that more Salix caprea biomass and bioavailability were achieved in the accumulation of Zn and Cd on the shoot when Streptomyces sp. AR17 and Agromyces sp. AR33 were inoculated (De Maria et al. 2011).

1.19 Conclusion

The rapid development of industrial and mining activity is the cause of the introduction of excessive amounts of toxic contaminants (HMs) into freshwater bodies and groundwater, which is the cause of significant health hazards for humans and degradation of the environment. Among the various toxic substances present in the water resources, HMs are found to be more dangerous and cause a severe threat to human beings, animals, and individuals. Hence, researchers throughout the globe emphasised developing an ecologically safe, low-cost, green technology that is viable for remediating toxic contaminants, especially HMs, from water bodies. Based on this contest, rhizofiltration and phytostabilization are the most commonly adopted viable plant-based methods that can remediate the toxin and HMs from the water bodies of industrial and mining zones. Furthermore, the benefits of rhizofiltration offer an excellent ecological approach for the removal of HMs from water systems in a cost-effective and eco-friendly way, reducing metal toxicity. The kind of plant species, the action plan of rhizofiltration, and innovative strategies for effective removal of toxic heavy metallic contaminants are the challenges for researchers. The method of rhizofiltration can use specific plants whose root systems are capable of accumulating HMs and is a good, cost-effective, eco-friendly, sustainable, and environmentally safe viable alternative approach for remediation of heavy metals from water systems. Furthermore, strategies must be developed to use microorganisms for developing HM bioavailability to facilitate more HM deposition in the root systems of plants and make the method more efficient. Rhizofiltration is an excellent environmental-friendly method for mitigating groundwater contamination. Extensive research work must be undertaken to develop plant species that can transfer rhizofiltration to commercialization.

References

Abubakar MM, Ahmad MM, Getso BU (2014) Rhizofiltration of heavy metals from eutrophic water using Pistia stratiotes in a controlled environment. IOSR J Environ Sci Toxicol Food Technol 8(6):27–34

Akhtar MS, Chali B, Azam T (2013) Bioremediation of arsenic and lead by plants and microbes from contaminated soil. Res Plant Sci 1(3):68–73

Arshad M, Saleem M, Hussain S (2007) Perspectives of bacterial ACC deaminase in phytoremediation. Trends Biotechnol 25(8):356–362

Awa SH, Hadibarata T (2020) Removal of heavy metals in contaminated soil by phytoremediation mechanism: a review. Water Air Soil Pollut 231(2):47–61

Bakshe P, Jugade R (2023) Phytostabilization and rhizofiltration of toxic heavy metals by heavy metal accumulator plants for sustainable management of contaminated industrial sites: a comprehensive review. J Hazard Mater Adv 10:1–16

Banerjee A, Roychoudhury A (2022) Rhizofiltration of combined arsenic-fluoride or lead-fluoride polluted water using common aquatic plants and use of the 'clean' water for alleviating combined xenobiotic toxicity in a sensitive rice variety. Environ Pollut 304:1–18

Belle GN, Oberholster PJ, Fossey A, Esterhuizen L, Moodley R (2022) Using pollution indices to develop a risk classification tool for gold mining contaminated soils. J Environ Sci Health, Part A 57(12):1047–1057

Bloom AJ, Meyerhoff PA, Taylor AR, Rost TL (2002) Root development and absorption of ammonium and nitrate from the rhizosphere. J Plant Growth Regul 21:416–431

Carrillo AE, Li CY, Bashan Y (2002) Increased acidification in the rhizosphere of cactus seedlings induced by Azospirillum brasilense. Naturwissenschaften 89:428–432

Castro-Castellon AT, Hughes JMR, Read DS, Azimi Y, Chipps MJ, Hankins NP (2021) The role of rhizofiltration and allelopathy on the removal of cyanobacteria in a continuous flow system. Environ Sci Pollut Res 28:27731–27741

Chou FI, Chung HP, Teng SP, Sheu ST (2005) Screening plant species native to Taiwan for remediation of 137Cs-contaminated soil and the effects of K addition and soil amendment on the transfer of 137Cs from soil to plants. J Environ Radioact 80(2):175–181

Das AP, Ghosh S (2023) Cost-effectiveness of the biomining process. Manganese Min Microorgan 143–151. https://doi.org/10.1016/B978-0-12-822147-1.00008-X

Das AP, Sukla LB, Pradhan N, Nayak S (2011) Manganese biomining: a review. Biores Technol 102(16):7381–7387. https://doi.org/10.1016/J.BIORTECH.2011.05.018

Das P, Paul KK (2023) Hydroponic rhizofiltration of dairy wastewater by Coleus Scutellarioides and Portulaca Oleracea. J Water Process Eng 52:103589

De Maria S, Rivelli AR, Kuffner M, Sessitsch A, Wenzel WW, Gorfer M, Puschenreiter M (2011) Interactions between accumulation of trace elements and macronutrients in Salix caprea after inoculation with rhizosphere microorganisms. Chemosphere 84(9):1256–1261

Dey S, Tripathy B, Kumar MS, Das AP (2023) Ecotoxicological consequences of manganese mining pollutants and their biological remediation. Environ Chem Ecotoxicol. https://doi.org/10.1016/J.ENCECO.2023.01.001

Dimitroula H, Syranidou E, Manousaki E, Nikolaidis NP, Karatzas GP, Kalogerakis N (2015) Mitigation measures for chromium-VI contaminated groundwater—the role of endophytic bacteria in rhizofiltration. J Hazard Mater 281:114–120

Douglas JA, Douglas MH, Lauren DR, Martin RJ, Deo B, Follett JM, Jensen DJ (2004) Effect of plant density and depth of harvest on the production and quality of licorice (Glycyrrhiza glabra) root harvested over 3 years. N Z J Crop Hortic Sci 32(4):363–373

Dürešová Z, Šuňovská A, Horník M, Pipíška M, Gubišová M, Gubiš J, Hostin S (2014) Rhizofiltration potential of Arundo donax for cadmium and zinc removal from contaminated wastewater. Chem Pap 68:1452–1462

Dushenkov V, Kumar PN, Motto H, Raskin I (1995) Rhizofiltration: the use of plants to remove heavy metals from aqueous streams. Environ Sci Technol 29(5):1239–1245

Eapen S, D'souza SF (2005) Prospects of genetic engineering of plants for phytoremediation of toxic metals. Biotechnol Adv 23(2):97–114

El-Aassar MR, Fakhry H, Elzain AA, Farouk H, Hafez EE (2018) The Rhizofiltration system consists of chitosan and natural Arundo donax L. for removal of basic red dye. Int J Biol Macromol 120:1508–1514

Galal TM, Eid EM, Dakhil MA, Hassan LM (2018) Bioaccumulation and rhizofiltration potential of Pistia stratiotes L. for mitigating water pollution in the Egyptian wetlands. Int J Phytoremed 20(5):440–447

Gaur N, Flora G, Yadav M, Tiwari A (2014) A review with recent advancements on bioremediation-based abolition of heavy metals. Environ Sci Process Impacts 16(2):180–193

Ghosh M, Singh SP (2005) A review on phytoremediation of heavy metals and utilization of it's by products. Asian J Energy Environ 6(4):214–23

Ignatius A, Arunbabu V, Neethu J, Ramasamy EV (2014) Rhizofiltration of lead using an aromatic medicinal plant Plectranthus amboinicus cultured in a hydroponic nutrient film technique (NFT) system. Environ Sci Pollut Res 21:13007–13016

Kamusoko R, Jingura RM (2017) Utility of Jatropha for phytoremediation of heavy metals and emerging contaminants of water resources: a review. CLEAN–Soil, Air, Water 45(11):1–17

Kensa VM (2011) Bioremediation-an overview. J Ind Pollut Control 27(2):161–168

Khan AU, Khan AN, Waris A, Ilyas M, Zamel D (2022) Phytoremediation of pollutants from wastewater: a concise review. Open Life Sci 17(1):488–496

Koźmińska A, Hanus-Fajerska E, Muszyńska E (2014) Possibilities of water purification using the rhizofiltration method. Woda Środowisko Obszary Wiejskie 14(47):89–98

Kristanti RA, Ngu WJ, Yuniarto A, Hadibarata T (2021) Rhizofiltration for removal of inorganic and organic pollutants in groundwater: a review. Biointerface Res Appl Chem 4:12326–12347

Laghlimi M, Baghdad B, El Hadi H, Bouabdli A (2015) Phytoremediation mechanisms of heavy metal contaminated soils: a review. Open J Ecol 5(08):375–388

Leblebici Z, Dalmiş E, Andeden EE (2019) Determination of the potential of Pistia stratiotes L. in removing nickel from the environment by utilizing its rhizofiltration capacity. Braz Arch Biol Technol 62:e19180487

Lee M, Yang M (2010) Rhizofiltration using sunflower (Helianthus annuus L.) and bean (Phaseolus vulgaris L. var. vulgaris) to remediate uranium contaminated groundwater. J Hazard Mater 173(1–3):589–596

Malaviya P, Singh A (2012) Phytoremediation strategies for remediation of uranium-contaminated environments: a review. Crit Rev Environ Sci Technol 42(24):2575–2647

Malaviya P, Singh A, Anderson TA (2020) Aquatic phytoremediation strategies for chromium removal. Rev Environ Sci Bio/Technol 19:897–944

Mithembu MS (2012) Nitrogen and phosphorus removal from agricultural wastewater using constructed rhizofiltration in Durban, South Africa. J Agric Sci Technol 2:1142–1148

Mishra S, Das AP, Seragadam P (2009) Microbial remediation of hexavalent chromium from chromite contaminated mines of Sukinda Valley, Orissa (India). J Environ Res Dev 1122–1127

Mishra S, Dash D, Das AP (2022) Detection, characterization and possible biofragmentation of synthetic microfibers released from domestic laundering wastewater as an emerging source of marine pollution. Mar Pollut Bull 185:114254. https://doi.org/10.1016/J.MARPOLBUL.2022.114254

Mishra S, Ghosh S, van Hullebusch ED, Singh S, Das AP (2023) A critical review on the recovery of base and critical elements from electronic waste-contaminated streams using microbial biotechnology. Appl Biochem Biotechnol 1–30. https://doi.org/10.1007/S12010-023-04440-X/METRICS

Mishra S, Charan Rath C, Das AP (2019) Marine microfiber pollution: a review on present status and future challenges. Mar Pollut Bull 140:188–197. https://doi.org/10.1016/J.MARPOLBUL.2019.01.039

Mohanty M, Dhal NK, Patra P, Das B, Reddy PSR (2010) Phytoremediation: a novel approach for utilization of iron-ore wastes. Rev Environ Contam Toxicol 206:29–47

Mthembu MS (2016) Removal of organic and inorganic nutrients in a constructed rhizofiltration system using macrophytes and microbial biofilms (Doctoral dissertation)
Nag M, Lahiri D, Ghosh S, Sarkar T, Pati S, Das AP, Ram DK, Bhattacharya D, Ray RR (2023) Application of microorganisms in biotransformation and bioremediation of environmental contaminant: a review. Geomicrobiol J. https://doi.org/10.1080/01490451.2023.2261443
Nedelkoska TV, Doran PM (2000) Characteristics of heavy metal uptake by plant species with potential for phytoremediation and phytomining. Miner Eng 13(5):549–561
Odinga CA, Kumar A, Mthembu MS, Bux F, Swalaha FM (2019) Rhizofiltration system consisting of Phragmites australis and Kyllinga nemoralis: evaluation of efficient removal of metals and pathogenic microorganisms. Desalin Water Treat 169:120–132
Pérez-Palacios P, Agostini E, Ibáñez SG, Talano MA, Rodríguez-Llorente ID, Caviedes MA, Pajuelo E (2017) Removal of copper from aqueous solutions by rhizofiltration using genetically modified hairy roots expressing a bacterial Cu-binding protein. Environ Technol 38(22):2877–2888
Petersen W, Böttger M (1991) Contribution of organic acids to the acidification of the rhizosphere of maize seedlings. Plant Soil 132:159–163
Raj D, Maiti SK (2019) Sources, toxicity, and remediation of mercury: an essence review. Environ Monit Assess 191:1–22
Raj K, Das AP (2023) Lead pollution: Impact on environment and human health and approach for a sustainable solution. Environ Chem Ecotoxicol 5:79–85. https://doi.org/10.1016/J.ENCECO.2023.02.001
Raskin I (1996) Plant genetic engineering may help with environmental cleanup. Proc Natl Acad Sci 93(8):3164–3166
Rawat K, Fulekar MH, Pathak B (2012) Rhizofiltration: a green technology for remediation of heavy metals. Intl J Inno Biosci 2(4):193–199
Salt DE, Blaylock M, Kumar NP, Dushenkov V, Ensley BD, Chet I, Raskin I (1995) Phytoremediation: a novel strategy for the removal of toxic metals from the environment using plants. Bio/technology 13(5):468–474
Sheoran V, Sheoran AS, Poonia P (2016) Factors affecting phytoextraction: a review. Pedosphere 26(2):148–166
Sikhosana MLM, Botha A, Mpenyane-Monyatsi L, Coetzee MA (2020) Evaluating the effect of seasonal temperature changes on the efficiency of a rhizofiltration system in nitrogen removal from urban runoff. J Environ Manage 274:1–16
Singh R, Ahirwar NK, Tiwari J, Pathak J (2018) Review of sources and effect of heavy metal in soil: Its bioremediation. Int J Res Appl Nat Soc Sci 2018:1–22
Thayaparan M, Iqbal SS, Chathuranga PKD, Iqbal MCM (2013) Rhizofiltration of Pb by Azolla pinnata. Int J Environ Sci 3(6):1811–1823
Tomé FV, Rodríguez PB, Lozano JC (2008) Elimination of natural uranium and 226Ra from contaminated waters by rhizofiltration using Helianthus annuus L. Sci Total Environ 393(2–3):351–357
Verma P, George KV, Singh HV, Singh SK, Juwarkar A, Singh RN (2006) Modeling rhizofiltration: heavy-metal uptake by plant roots. Environ Model Assess 11:387–394
Veselý T, Tlustoš P, Száková J (2011) The use of water lettuce (Pistia stratiotes L.) for rhizofiltration of a highly polluted solution by cadmium and lead. Int J Phytoremed 13(9):859–872
Vives-Peris V, De Ollas C, Gómez-Cadenas A, Pérez-Clemente RM (2020) Root exudates: from plant to rhizosphere and beyond. Plant Cell Rep 39:3–17
Walker TS, Bais HP, Grotewold E, Vivanco JM (2003) Root exudation and rhizosphere biology. Plant Physiol 132(1):44–51
Woraharn S, Meeinkuirt W, Phusantisampan T, Chayapan P (2021) Rhizofiltration of cadmium and zinc in hydroponic systems. Water Air Soil Pollut 232(5):204–220
Yadav AK, Pathak B, Fulekar MH (2015) Rhizofiltration of heavy metals (cadmium, lead and zinc) from fly ash leachates using water hyacinth (Eichhornia crassipes). Int J Environ 4(1):179–196
Yadav BK, Siebel MA, van Bruggen JJ (2011) Rhizofiltration of a heavy metal (lead) containing wastewater using the wetland plant Carex pendula. CLEAN–Soil, Air, Water 39(5):467–474
Yazdanpanah S, Rajaei P (2014) A mini review: metal remediation by microbes and plants. J Biodiver Environ Sci (JBES) 5(2):222–226

Chapter 2
Plant Wastes as Emerging and Efficient Scavengers for Inorganic and Organic Pollutants from Wastewater

Priyanka Priyadarsini Samal and Soumen Dey

Abstract Organic and inorganic pollutants are key sources of water contamination. Economic, eco-friendly, and effective approaches are essential for wastewater treatment. Biosorption using phyto-materials is gaining attention for the execution of effective pollutant removal due to their abundant availability, eco-friendliness, non-toxicity, and economical viability. This chapter highlights the application of plant-based materials such as leaves, flowers, peels, shells, seeds, stalks, and barks for pollutant removal from wastewater. Moreover, the advantages of modified phytosorbents over pristine variety are discussed. The removal efficiency of brinjal stalks, bamboo leaves, neem leaves, Kamla fruit peel, rice husk, sawdust, coco peat, mahua seeds, pomegranate peel, kashful stalks, cadamba flower, orange peel, papaya peel, date seed and related materials on organic and inorganic pollutant removal has been discussed. The adsorption capacity under optimized condition usually has been found to vary from 10 to 500 mg/g. Carboxyl, hydroxyl, and carbonyl functionalities are the key binding sites to remove toxins from the water body. The effect of physicochemical parameters such as contact time, input concentration, pH, and dose are clearly explained. Most of the photo-materials exhibit a pseudo-second-order kinetic model, with varying degrees of intra-particle diffusion. However, isotherms may vary depending on the multilayer and monolayer adsorption mechanism. The biosorption is spontaneous and endothermic but some cases show the exothermic. Regeneration and multicycle usage of the adsorbents have been explained to evaluate techno-economic potential. Such phytomaterials are also able to remove pollutants in the presence of industrial wastewater. Hydrogen bonding, electrostatic, and π–π interactions are the dominating driving forces behind adsorption. A comparative discussion of the adsorption capacities of different phytosorbents has been summarized to assess relative efficacy. The future possibility has been recommended. Overall, this chapter highlights the use of phyto-materials as a potential scavenger of toxic components from wastewater which will be beneficial for sustainable development.

P. P. Samal · S. Dey (✉)
Environment Protection Laboratory, Department of Chemistry, Central University of Jharkhand, Ranchi, India
e-mail: soumen.dey@cuj.ac.in

A. Kuanar et al. (eds.), *Recent Advances in Bioremediation and Phytoremediation*, Environmental Science and Engineering,
https://doi.org/10.1007/978-3-031-77884-1_2

Keywords Phytosorbent · Dye · Heavy metals · Adsorption

2.1 Introduction

The rapid expansion of populations in urban areas, coupled with industrialization and economic growth, has resulted in a concerning rise in the prevalence of microplastics, heavy metals, and dyes, which pose significant environmental challenges in today's world (Das et al. 2011; Mishra et al. 2019). The global concern over environmental pollution is rising daily to treat organic and inorganic pollutant-contaminated wastewater (Mansour et al. 2018; Mishra et al. 2022). Numerous conventional treatments have been prompted to mitigate the water pollution issues (Gokulan et al. 2019). Discharging inorganic pollutants (heavy metals) into water medium poses a major ecological challenge because of their toxicity, non-biodegradability, and carcinogenic nature. Notably, chromium, zinc, mercury, copper, cadmium, nickel, and lead are troublesome pollutants because of their detrimental impact (Paul et al. 2023; Raj and Das 2023). These heavy metals originate from diverse industrial sources such as refineries, coal-fired power plants, municipal wastewater, mining, tanneries, electronics, pharmaceuticals, batteries, fertilizers, electroplating, and the steel industrial sector (Lahiri et al. 2021; Mishra et al. 2023). In contrast to heavy metals, dyes are colored aromatic complex organic compounds. Due to their toxic and carcinogenic nature, they pose challenges to humans as well as flora and fauna when disposed into water streams. Various industries including but not limited to textile, leather, paper manufacturing, pharmaceutical, agriculture, and painting sectors discharge a wide array of dyes containing effluents into aqueous resources (Kocaman 2020; Das and Singh 2011). Due to the potentially harmful effect, untreated effluents create a substantial risk to ecosystems.

Various treatment methods such as membrane separation, ozonation, adsorption, precipitation, advanced oxidation, ion exchange, chemical oxidation, filtration, solvent extraction, coagulation, etc., have been focused on removing the pollutants from wastewater (Manzoor et al. 2022; Yadav et al. 2021). Among these, adsorption stands out as one of the most suitable techniques, offering simplicity, easy operation, and cost-effectiveness (Adegoke et al. 2023; Samal et al. 2023a, b). Activated carbon and different synthetic adsorbents have been employed for wastewater remediation. However, high operational costs, and regeneration issues, led to identify cost-effective alternatives (Kumari et al. 2023). This focus has resulted in significant advancements in the application of different eco-friendly, cost-effective, and non-conventional adsorbents derived from plant wastes. This chapter specifically delves into the effectiveness of various plant-based wastes for various pollutant sequestrations from wastewater. Utilizing these abundant plant wastes including leaves, stems, seeds, barks and peels as efficient phytosorbents, the wastewater remediation is possible in a sustainable and techno-economic manner.

The key focus of this chapter is to deliver a discussion on the application of various phytosorbents for heavy metals and dyes sequestration from aqueous solution. The type of phytosorbents and their effectiveness under various physicochemical parameters are presented. The chapter also includes a compilation of kinetics, isotherms, and thermodynamic models with a comparative discussion. The uptake capacities under different conditions with best fitted model are important contributions made in this chapter. The characterization of these materials such as FTIR, powder XRD, SEM-EDX, BET surface area analysis, and pH_{ZPC} have been discussed to know the material's physical and chemical characteristics and the mechanism of adsorption. The regeneration and multi-cycle usage of phytosorbents have been discussed. In addition to these, the practical application of these adsorbents has been provided with a clear discussion of the removal efficiency of plant-based material on real effluent treatment.

2.2 Toxicity of Dyes and Heavy Metals

The dyes are complex organic structures and they are categorized into vat, neutral, direct, azo, sulfur, reactive, anionic, disperse, mordant, and cationic, dyes (Bhattacharjee et al. 2021; Das and Swain 2013). The annual production of dye effluents is estimated to exceed 7×10^5 tons. When these untreated dyestuffs, originating from various industries, are introduced into aquatic environments, cause water pollution, posing threats to the environment. These effluents are harmful for their adverse effects such as toxicity, non-biodegradability, carcinogenicity, and mutagenicity (Hou et al. 2020). Additionally, their presence inhibits the development of aquatic plants and animals through blocking of sunlight, leading to a reduction in photosynthesis. This alters the environmental balance and sustains adverse conditions for aquatic life (Goswami et al. 2020; Silva et al. 2019).

Heavy metals are inorganic elements whose specific gravity are exceeding 5.0. Exposure to heavy metals poses significant health risks. Arsenic (As) has been linked to the development of cancers in the lungs, kidneys, and skin (Khan et al. 2021). Chromium is associated with a range of health issues, including kidney damage, chronic bronchitis, internal hemorrhage, emphysema, and stomach, skin, and lung cancers (Rambabu et al. 2020; Samal et al. 2024). Cadmium is known to cause skeletal damage and kidney disease (Shen et al. 2017), while lead is connected to nephritic syndrome, hepatitis, encephalopathy, anaemia, and complications in both the gastrointestinal and central nervous systems (Bouabidi et al. 2018). Copper exposure may result in food poisoning, nausea, liver toxicity, and gastrointestinal illness (Shahzad et al. 2017). Nickel has been linked to neurotoxicity, nephrotoxicity, hepatotoxicity, reproductive toxicity, gene toxicity, and an increased risk of cancer (Shi et al. 2018). Mercury exposure is associated with neurological damage, blindness, paralysis, chromosomal damage, chest pain, pulmonary problems, and dyspnea. So, heavy metal exposure causes diverse and serious health problems (Yadav et al. 2021). So, the eradication of these pollutants from the aqueous body is important.

2.3 Various Demonstrated Treatment Methods

Various physical treatments, chemical treatments, and biological treatments have been implemented for dyes and heavy metals eradication from industrial effluent. Each category presents its own set of advantages and limitations. Chemical methods encompass a range of techniques such as ozonation, ion exchange, coagulation, Fenton oxidation, electro-floatation, chemical oxidation, and co-precipitation. For instance, ozonation employs ozone gas to facilitate the interaction. Biological methods involve microorganisms such as bacteria, algae, and fungi for wastewater treatment but sludge formation in these methods is a major disadvantage (Das et al. 2024a, b). Physical methods include membrane filtration, adsorption, osmosis, and sedimentation. Notably, adsorption has added prominence in pollutant sequestration due to its cheap and simple operation (Kumari et al. 2023).

2.4 Adsorption Technique

The adsorption techniques have various advantages, encompassing a straightforward design, non-toxicity, cost-effectiveness, ease of implementation, adaptability, heightened efficiency, and an eco-friendly nature. The adsorption is attributed to the diverse binding forces such as pi–pi interaction, hydrogen bonds, and electrostatic attraction between an adsorbate and an adsorbent. Chemisorption is facilitated by electron exchange, while physisorption relies on π–π stacking, hydrogen bonds with Van der Waals forces (Yadav et al. 2021).

2.5 Natural Adsorbents

The efficiency of adsorption is generally influenced by the inherent characteristics of the material employed for adsorption. Key factors such as maximum absorption capacity, reusability, surface properties, cost-effectiveness, and impacts on the environment play crucial roles in determining the appropriate adsorbent for effective pollutant removal. Various types of plant-based materials, including both natural and modified forms, have been reported for the successful pollutant elimination from effluents. Within plant components, stems and stalks emerge as widely recognized adsorbents for eradicating the pollutants from water medium, due to their porous and amorphous nature. Notably, peels are particularly valuable due to their widespread availability as a waste material in bulk. The utilization of orange peel, and banana peel as pollutant scavengers exemplifies this approach. Additionally, plant fibers have proven highly efficient for pollutant adsorption. Seeds are favored for their cost-effectiveness, eco-friendliness, and abundant nature. Various plant seeds, such as apple seeds, papaya seeds, Arjuna seeds, apricot seeds, and pomegranate seed waste

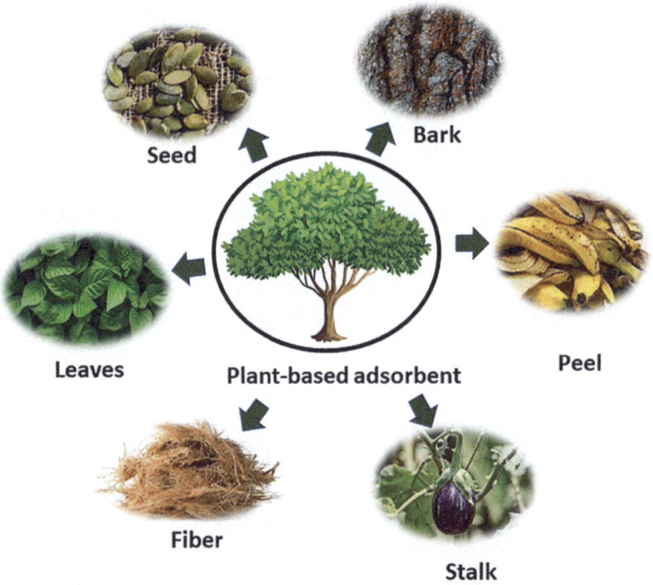

Scheme 2.1 Different plant-based adsorbents

have been extensively studied for dyes and heavy metals removal. The utilization of plant wastes is a promising avenue because it contains cellulose, hemicelluloses, and lignin that facilitate successful interactions for pollutant removal (Manzoor et al. 2022). The scheme is presented in Scheme 2.1. Table 2.1 presents a comparative discussion on the maximum uptake capacity of different plant-based adsorbents.

2.6 Stem and Stalk

To develop cost-effective and environmentally friendly adsorbents by utilizing plant-based adsorbents is important. Plant stems and stalks stand out as well-known materials and have been widely employed for dyes and heavy metals eradication from wastewater systems. The stalks and stems are amorphous and porous. Due to advantageous properties such as excellent uptake capability, eco-friendliness, porous structure, large surface area, and cost-effectiveness, stems and stalks are considered as excellent phytosorbents for pollutant elimination from wastewater. Numerous stem-based adsorbents, such as bamboo stem (Olayinka et al. 2020), *Centaurea* stem (Hassan et al. 2017), and *Haloxylon recurvum* stems (Naderi et al. 2018), and stalk-based adsorbents such as chili stalks (Panda et al. 2024), brinjal stalks (Rout et al. 2023), and wild sugarcane stalks (Bharadwaj et al. 2023) have proven effective for different dyes and heavy metals removal from an aqueous phase.

Table 2.1 Comparative study on maximum pollutant adsorption capacity of phytosorbents

Plant-based adsorbent	Target pollutant	Maximum adsorption capacity (mg/g)	Refs.
Cocos nucifera L. coir	Safranin-O dye	80.32	Ghosh et al. (2021)
Phosphoric acid-treated *Cocos nucifera* L. coir	Safranin-O dye	96.81	Ghosh et al. (2021)
Sulphuric acid-treated *Cocos nucifera* L. coir	Safranin-O dye	89.53	Ghosh et al. (2021)
Banana peel	Congo red	1.72	Mondal and Kar (2018)
Dragon fruit peel	Cu(II)	92.59	Phuengphai et al. (2021)
Rambutan peel	Cu(II)	192.30	Phuengphai et al. (2021)
Passion fruit peel	Cu(II)	121.95	Phuengphai et al. (2021)
Vicia faba peels	Methylene blue	140.00	Bayomie et al. (2020)
Litchi peel	Cd(II)	230.50	Chen et al. (2018)
Orange peel	Cd(II)	170.30	Chen et al. (2018)
Pomegranate peel	Cd(II)	132.50	Chen et al. (2018)
Banana peel	Cd(II)	98.40	Chen et al. (2018)
Pomegranate peel	Basic red 46	86.13	Akkari et al. (2023)
Annona muricata seed	Methylene blue	34.60	Sonde et al. (2023)
Treculia africana seed hull	Methylene blue	21.05	Sonde et al. (2023)
Citrullus colocynthis seeds	Methylene blue	18.83	Alghamdi and El Mannoubi (2021)
Citrullus colocynthis peel	Methylene blue	4.48	Alghamdi and El Mannoubi (2021)
Moringa seeds	Dispersed red 60	170.70	Soliman et al. (2019)
Moringa seeds	Congo red	196.80	Soliman et al. (2019)
Punica granatum L. seed	Crystal violet	434.80	Uddin and Nasar (2020)
Punica granatum L. seed	Methylene blue	99.10	Uddin and Nasar (2020)
B. vulgaris stem	Cd(II)	12.34	Olayinka et al. (2020)
B. vulgaris stem	Pb(II)	27.90	Olayinka et al. (2020)
Centaurea stem	Crystal violet	476.19	Hassan et al. (2017)

(continued)

Table 2.1 (continued)

Plant-based adsorbent	Target pollutant	Maximum adsorption capacity (mg/g)	Refs.
Alkali-treated *Trapa natans* L. shells	Malachite green	136.46	Qaiyum et al. (2022)
Acacia auriculiformis leaves	Crystal violet	67.25	Samal et al. (2023a, b)
Acacia auriculiformis leaves	Methylene blue	78.55	Samal et al. (2023a, b)
Solanum melongena stalk	Eriochrome Black-T	52.63	Rout et al. (2023)
Bambusa bambos leaves	Crystal violet	30.00	Mahato et al. (2023)
Neolamarckia cadamba pulp	Methylene blue	115.60	Das et al. (2024a, b)
Saccharum spontaneum stalk	Methylene blue	20.91	Bharadwaj et al. (2023)
Thiosulfate-impregnated waste tea leaves	Malachite green	126.80	Samal et al. (2023a, b)
Rauvolfia tetraphylla leaves	Crystal violet	109.99	Swain et al. (2024)

For the preparation of bamboo stem adsorbent, a bamboo stem (*B. vulgaris*) was collected and sundried. It was then cut into pieces and ground to get a powdery substance. After sieving, the stem powder was tested with Pb(II) and Cd(II) ions and 95.92% and 80.98% removal was possible in 90 min at pH 5, 298 K temperature, and concentration of 50 mg/L (Olayinka et al. 2020). In the same procedure, *Thapsia transtagana* stems adsorbent was exercised capacities for Cd(II) and Co(II) ions with uptake capacities of 35.35 and 14.84 mg/g (Elhalil et al. 2019). Centaurea stems and *Haloxylon recurvum* plant stems have been used for crystal violet dye and methylene blue dye removal with uptake capacities of 476.19 and 22.93 mg/g respectively (Hassan et al. 2017; Naderi et al. 2018). Stalk-based adsorbent chili stalk was used for the decontamination of two cationic dyes namely Crystal violet and Methylene blue with uptake capacity of 49.53 and 36.88 mg/g at neutral pH (Panda et al. 2024). Brinjal stalk and wild sugarcane stalk were exercised for Eriochrome Black T and Methylene blue dye eradication with maximum capacities of 52.63 and 20.91 mg/g respectively (Rout et al. 2023; Bharadwaj et al. 2023).

Chemical modification of phytosorbents has been done to enhance the uptake capacity of the phytosorbent. In a study, corn stalks were modified using citric acid and the modified corn stalk was able to remove methylene blue and malachite green with adsorption efficiency of 26.60 and 27.55 mg/g at 333 K (Soldatkina and Yanar 2021). Overall, the utilization of plant stems and stalks offers a sustainable and effective means for pollutant eradication from aqueous systems.

2.7 Leaves

Plant leaves are primarily composed of pectins, hemicellulose, lignin, and cellulose that contain hydroxyl, nitro, amino, carboxyl, and carbonyl functional groups. Those groups enable interactions between adsorbate and adsorbent (Bulgariu et al. 2019). Many studies have investigated the utilization of leaf-based biomass as an adsorbent for effectively removing pollutants. Phoenix leaves effectively adsorb MB dye with 89.70 mg/g adsorption capacity at 323 K. *Platanus orientalis* leaves powder has been also demonstrated for MB dye sorption with 114.94 mg/g sorption capability. Moreover, it exhibited good recycling ability for dye removal from textile effluent. The study on Gulmohar (*D. regia*) leaf powder for MB dye adsorption revealed a monolayer capacity of 186.22 mg/g (Bulgariu et al. 2019). Oak Leaves and *Posidonia oceanica* have been utilized for Crystal violet (CV) and Methylene blue (MB) adsorption. The oak tree leaves exhibited 31.65 and 33.5 mg/g adsorption capacity for CV and MB dye respectively where as *Posidonia oceanica* exhibited 22.93 and 27.78 mg/g adsorption capacity for CV and MB, respectively (Sulyman and Gierak 2020). Chinar leaf powder was utilized for Crystal violet dye removal with a removal capacity of 25.88 mg/g (Ahmad Khan et al. 2023). *Pteris vittata* plant leaves were chemically modified with HNO_3 and $CaCl_2$ and after chemical modification, the leaves were used for Cr(VI) remediation from wastewater with 66.60 mg/g uptake capacity (Khan et al. 2022). So, leaves-based adsorbents have been employed as cheap and promising adsorbents for the pollutants remediation from wastewater.

2.8 Peels

Peels have been widely used as adsorbent due to their abundant availability as a waste material. Annually tons of peels are produced worldwide. So, these peels can be utilized in the removal of hazardous pollutants as they are economically viable, non-toxic, and environmentally friendly in nature. The application of banana peels for Congo red adsorption has been demonstrated with an absorption capability of 1.72 mg/g at 10 pH (Mondal and Kar 2018). Another investigation has been employed on the utilization of pomegranate peel for adsorbing Basic Red 46, with an removal capability of 86.13 mg/g (Akkari et al. 2023). Fava bean peel powder was examined for methylene blue adsorption with an uptake capability of 140 mg/g (Bayomie et al. 2020). Utilization of litchi peel, orange peel, banana peel, and pomegranate peel on Cd(II) removal has been reported with the removal capacities of 230.50, 170.30, 98.40, and 132.50 respectively (Chen et al. 2018). Peel of dragon fruit, passion fruit, and rambutan has been utilized to adsorb Cu(II) with uptake capacities of 92.59, 121.95, and 192.30 mg/g respectively at 4 pH. The preparation involved washing with distilled water, drying, and grinding the peels (Phuengphai et al. 2021).

2.9 Fibers

In recent years, the focus on natural fibers has been intensified among researchers due to their sustainability and environmental friendly characteristics. For instance, a study has explored the use of coconut residual fiber for Malachite green (MG) and Methylene blue (MB) removal. The coconut fiber was treated with alkali to reduce the lignocellulosic moiety to make a water-stable adsorbent. The adsorption capacities were 110.74 and 133.11 mg/g for MG and MB respectively (Mohanta et al. 2021). Moreover, coconut coir has been treated with sulphuric acid and phosphoric acid for Safranin-O dye adsorption. The uptake capacity of untreated coir, sulphuric acid-treated coir, and phosphoric acid-treated coir were 74.07, 85.40, and 86.81 mg/g respectively (Ghosh et al. 2021).

2.10 Seeds

Typically, plant seeds have been most widely employed for treating pollutant-containing wastewater, owing to their abundant availability, cost-effectiveness, and environmentally friendly characteristics. Various types of plant seeds, including *Annona muricata* and *Treculia africana* seed (Sonde et al. 2023), *Citrullus colocynthis* seeds (Alghamdi and El Mannoubi 2021), pomelo rinds, *Carica papaya* seeds (Ho and Li 2019), *Punica granatum* L. seeds (Uddin and Nasar 2020), *Leucaena leucocephala* seeds (Kristanda et al. 2021) have been reported for organic and inorganic pollutant removal. In a systematic research study, Moringa seed was utilized for Congo Red decontamination from an aqueous medium, achieving an uptake capacity of 238.60 mg/g at neutral pH (Soliman et al. 2019). The seeds of *Anadenanthera macrocarpa* and *Cedrela fissilis* were utilized for Methylene blue dye removal. The uptake capacities were 228.00 and 230.00 mg/g respectively at 328 K (de Salomón et al. 2021).

2.11 Bark

Barks, primarily derived from woody trees, are acknowledged for their exceptional adsorption efficiency. These materials are eco-friendly, simple, and cost-effective. In studies focusing on the treatment of Congo red using *Pinus pinaster* tree bark, researchers observed the removal of Congo red ranging from 23.4 to 100%, showcasing the effectiveness of pineapple bark as an adsorbent (Manzoor et al. 2022).

2.12 Characterization Techniques

The chemical and physical characteristics of any phytosorbent are important. So, the characterization of adsorbents both before and after interactions is important to predict the mechanism involved. Key parameters contributing to these physical and chemical characteristics of adsorbent are surface area, functional group present, morphology, elements present, and zero-point surface charge. These parameters collectively impact on adsorption capability of the adsorbent. Thus, gaining an in-depth understanding of adsorbent's properties is crucial. To achieve this understanding, various analytical techniques are employed to characterize the adsorbents and to assess the efficiency of adsorbents on wastewater treatment (Hussain et al. 2018; Malik et al. 2019).

2.13 FESEM

Morphological structures of adsorbent are important because they offer valuable insights into the surface and shape of adsorbent. In FESEM analysis, the high-energy focused electron beam strikes the outer surface atoms of the adsorbent, producing various signals that provide a clear idea regarding the topography of the material. In this analysis, before adsorption presence of cavities, pores, and channels are detected while the morphological changes occur after adsorption.

In a study, the tube-like texture pores, and channels were detected on the surface of NaOH-treated acacia leaf powder while the surface underwent a significant transformation after Crystal violet (CV) and Methylene blue (MB) adsorption. The pores are filled with both the dye molecules with heterogeneous occupation (Samal et al. 2023a, b). In another study, oval-shaped cavities along with cracks, ridges, macropores, and channels were detected on the FESEM image of chili stalk powder, and after CV and MB sorption, the disappearance of oval-shaped cavities with compact surfaces were noticed. Moreover, In the EDX analysis of chili stalk, carbon and oxygen peaks were noticed while nitrogen and sulfur peaks were obtained after CV and MB adsorption (Panda et al. 2024). In a study, an octopus-like structure with pits, and ridges were found in the FESEM image of brinjal stalk. After Eriochrome black T (EBT) adsorption, the structure and architecture underwent noticeable changes (Rout et al. 2023). In the SEM image of activated soya bean seeds, pores, and cavities were seen. Uneven and irregular cavities facilitated lead and arsenic adsorption (Gaur et al. 2018).

2.14 FTIR

FTIR is usually employed to categorize functional groups present on adsorbent surfaces that play a significant role in governing interactions in adsorption processes. The spectra are usually recorded in a range of 400–4000 cm^{-1}. Phytosorbents contain lignocellulosic material which consists of carbonyl, carboxyl, phenolic, and hydroxyl, groups, as well as azo and nitro groups. These functional groups make the phytosorbents selective and highly effective in interacting with pollutants and play pivotal roles in the adsorption mechanism (Afroze and Sen 2018; Yagub et al. 2014).

From FTIR spectra of bamboo leaves, a broad band of –OH stretching at 3,342 cm^{-1} was noticed, indicating the presence of phenolic-OH moieties. Post-adsorption, a shift in this peak to 3,300 cm^{-1} was noted, attributed to potential hydrogen bonding. A distinct peak at 1,744 cm^{-1} suggested C=O functionality and shifted to 1,690 cm^{-1} after adsorption. An intense peak at 1,050 cm^{-1} indicated Si–O–Si stretching and after CV adsorption, the peak was shifted (Mahato et al. 2023).

In another stud, *Thapsia transtagana* stems were used for Cd(II) and Co(II) adsorption and the FTIR showed that the spectral bands between 3,600 and 3,200 cm^{-1} were responsible for OH and NH group. 1757 cm^{-1} peak was ascribed to the carboxyl groups, and 1,635 cm^{-1} band signified OH bending. The pronounced absorption band at 1,614 cm^{-1} peak was indicative of C=O stretching, characteristic of ketones, aldehydes, or carboxyl moieties. Correspondingly, the 1,414 cm^{-1} peak was for C–O stretching, and the 1,360 cm^{-1} for C–N groups, 1,160 –1,060 cm^{-1} peak was for P–O stretching, respectively. After adsorption, peak shifting might be attributed to the biosorption of metal ions (Cd(II), Co(II)) (Elhalil et al. 2019).

Moreover Pb(II) removal by saffron flower was confirmed by analyzing the FTIR spectra of pristine saffron flower and after Pb(II) uptake. The Saffron flower contains peaks at 3408, 2922, 1767, and 1028 cm^{-1} which correspond to –OH, –CH, C=O, and C–O stretching. These peaks were shifted after Pb(II) adsorption, and new bands appeared at 550 and 469 cm^{-1} responsible for Pb–O stretching (Khoshsang and Ghaffarinejad 2018). Moreover, Cr(VI) adsorption onto to walnut shell was also confirmed from FTIR analysis. The pristine material contained different –OH, –CH, –CO, and –C=C peaks. After adsorption, peak shifting was noticed and a new peak at 889 cm^{-1} was obtained because of Cr–O or Cr=O vibration. This observation proved chromium ions interaction on the walnut shell surface during the sorption process (Banerjee et al. 2018).

2.15 XRD

To ensure the efficiency of phytosorption, a comprehensive investigation of the crystalline characteristics of the adsorbent is necessary. Utilizing Powder X-ray diffraction (XRD), crystalline composition can be accessed. This technique facilitates the identification of crystalline phases and, the determination of lattice plane spacings. In the case of phytosorbents, a diffused band is observed at 2θ range of 25–35°, signifying lignocellulosic moieties presence. This diffused band signifies the amorphous nature of phytosorbents. In thiosulphate-treated tea leaves, a wide hump has been observed at 20–25°, strongly indicating the amorphous nature (Samal et al. 2023a, b). The amorphous broadband was also seen in *Sansevieria trifasciata* leaves. Secondary peaks were seen at 16 and 22° due presence of polysaccharide units (Grassi et al. 2023). *Dodonaea viscosa* bark also contains 2 peaks at 15° which is responsible for amorphous hemicellulose and at 22° which is responsible for cellulose. After Methyl red adsorption the crystallinity of the peaks was changed (Gul et al. 2022).

2.16 BET Surface Area

Adsorption is inherently a surface phenomenon. Specific surface area plays a pivotal role in adsorption, as a greater surface area ensures a large pore volume, fostering enhanced adsorption. The determination of specific surface area is commonly accomplished using the Brunauer–Emmett–Teller (BET) equation. The surface area of *Saccharum spontaneum* stalk was 6.20 m^2/g, with corresponding pore size of 6.196 nm and pore volumes of 9.604×10^{-3} cm^3/g. Post-adsorption, the surface area decreased to 4.11 m^2/g, accompanied by a decrease in pore size to 3.285 nm and volume of pore to 3.382×10^{-3} cm^3/g (Bharadwaj et al. 2023). Moreover, the surface area of alkali-treated water chestnut was 1.127 m^2/g, and the measured pore radius and volume were 1.09 nm and 6.68×10^{-3} cm^3/g, respectively (Qaiyum et al. 2022). The *Arachis hypogaea* shell's surface area was 32.96 m^2/g while after Cr(VI) adsorption, the surface area was decreased to 2.48 m^2/g. Moreover, the pore volume dropped from 0.017630 to 0.00012 cm^3/g after adsorption. This decrease occurred because the Cr(VI) ions were aggregated on the groundnut shell surface (Bayuo et al. 2019). Moreover, the surface area of the walnut shell was 10.16 m^2/g, which was decreased to 1.62 m^2/g after Cr(VI) adsorption due to pore blocking of adsorbent by Cr(VI) ions (Banerjee et al. 2018).

2.17 Zero Point Charge (ZPC)

The concept of ZPC holds fundamental significance in adsorption phenomena. It is the pH where the charge of a material's surface is zero. This parameter helps in comprehending the underlying mechanisms of the adsorption process across different pH conditions. Typically, the adsorption of cationic pollutants is more favorable at pH values exceeding pH_{ZPC}, whereas the adsorption of anionic pollutants is preferred when pH values fall below pH_{ZPC}. The pH_{ZPC} experiment was conducted using the conventional drift method. In this procedure, NaCl solutions were prepared, and the pH was adjusted from acidic to basic range using dilute acid and dilute alkali.

Subsequently, adsorbent was added to each solution and allowed to agitate for a particular time. The final pHs were examined and from the graph initial pH against the difference between the final and initial pH values, the pH_{ZPC} can be calculated (Samal et al. 2024). The pH_{ZPC} of *Punica granatum* L. was found to be 5.7 (Uddin and Nasar 2020). The pH_{ZPC} of banana peel was 4.51 (Mondal and Kar 2018) and *Haloxylon recurvum* stem was 6.3 (Hassan et al. 2017).

2.18 Factors Affecting Adsorption

Various factors affects the adsorption process, which includes parameters like solution pH, adsorbent dose, pollutant concentration, and contact time.

2.19 Solution pH

Solution pH significantly affects the uptake process. Specifically, the adsorbent exhibits a positive surface charge when the pH is less than pH_{ZPC} and a negative surface charge when the pH is greater than pH_{ZPC}. This dual nature of surface charge serves as the foundation for the preferential adsorption of different pollutants. The removal percentage of cationic pollutants is high at pH higher than pH_{ZPC}, whereas anionic pollutants are favorably adsorbed at pH lesser than pH_{ZPC} (Samal et al. 2023a, b).

Moreover, Hevira et al. (2021) conducted a study on the impact of pH on the uptake capacity of *Terminalia catappa* shells for Methylene blue. They observed a decline in uptake capacity when the pH decreased. Specifically, the maximum uptake capability was 2.45 mg/g at 5 pH (Hevira et al. 2021). As the pH_{ZPC} of *Terminalia catappa* shells was 2.01, it electrostatically attracted cationic methylene blue above 2.01 (Hevira et al. 2021). In another study, for Eosin dye removal, the maximum removal capacity of lemon peel was determined at 2 pH. This was attributed positive surface in low pH, thereby facilitating electrostatic attraction with anionic Eosin dye. Conversely, deprotonation at basic pH leading to electrostatic repulsion and decreased adsorption efficiency (Bukhari et al. 2022). Moreover, the sorption efficiency of *Allium Cepa* seed for Cd(II), Cr(VI), Cu(II), Zn(II), and Pb(II) removal was affected by pH. The maximum efficiencies were 80.00%, 15.00%, 87.32%, 33.00%, and 88.00%, respectively, at pH 7 (Sheikh et al. 2021).

2.20 Initial Pollutant Concentration

The concentration variation study underscores the significant influence of initial pollutant concentration on adsorption. This study reveals that a general decrease in removal percentage occurred with increasing concentration. The decrease in removal percentage is because of both the adsorbent active sites saturation and steric repulsion of pollutant species towards incoming pollutant species. Additionally, at lower concentrations, rapid filling of vacant active sites of adsorbent results in a higher adsorption percentage, whereas higher concentrations lead to inter-electronic repulsion which lower the adsorption percentage. The uptake of Methylene blue using a *Trichosanthes cucumerina* adsorbent was demonstrated where concentrations ranged from 10 to 80 mg/L. A decrease in adsorption efficiency from 95 to 66% was noticed when the concentration increased from 10 to 80 mg/L (Dey et al. 2022). The Cu(II) adsorption% was decreased from 60.00 to 31.20% with banana peel when the metal ions concentration increased 50–250 mg/L (Basirun et al. 2023). However, a rare increase in removal percentage is observed, attributed to the substantial driving force generated by higher pollutant concentrations, overcoming resistance during mass transfer between the solid and bulk phases. In a study it was found that the Cr(VI) uptake by *Stipa tenacissima* L. increased from 80 to 96% when the concentration of Cr(VI) was enhanced from 25 to 125 mg/L (Samir et al. 2019).

2.21 Contact Time

Contact time is crucial in adsorption. As the contact time increases, pollutant removal enhances in an upward trend. This is facilitated by the abundance of available binding sites that promote easy adsorption. However, as time progresses, a saturation point is reached on the adsorbent surface. Beyond this point, there is no significant increase in adsorption, marking the establishment of the equilibrium point. The determination of equilibrium time is important as it provides insights into the time when the adsorption process stabilizes. The attainment of equilibrium in the Pb(II) adsorption process was achieved within 150 min by using pomelo leaf adsorbent. Notably, a rapid uptake (64.00%) occurred within the initial 30 min. The fast uptake was attributed to the availability of empty sites on the pomelo leaf surface, allowing for the rapid filling of these sites by Pb(II) ions. However, as time progresses and the empty sites on the leaf surface are gradually occupied, the rate slows down and the establishment of equilibrium occurs (Chowdhury et al. 2022). In another report, it was found that Malachite green removal of 94.00% took place within the initial 20 min by using alkali-treated water chestnut, confirming efficient and quick adsorption within a short period (Qaiyum et al. 2022). In the same mechanism, Cr(VI) and Pb(II) adsorption onto *Juniperus procera* leaf-activated carbon occurred with 96% and 98% adsorptions were achieved at 120 and 100 min equilibrium time respectively (Ali et al. 2019).

2.22 Dose

The dosage of adsorbent is important in optimizing the economic efficiency of an adsorbent. Generally, the percentage pollutant removal tends to increase with higher material dose. As adsorbent dose is increased, the adsorbent's surface area is also increased. The expanded surface area leads to a higher number of binding sites of the adsorbent, improving the overall adsorption efficiency. In a specific study on the eradication of Cu(II), Pb(II), and Cd(II) using melon peel, it was revealed that increasing the biosorbent dosage from 0.30 to 3.00 g, the removal percentage for Cu(II) was increased from 94.00 to 99.10%, from 92.00 to 99.20% for Cd(II), and from 85.00 to 98.50% for Pb(II). This indicates a positive correlation between biosorbent dosage and removal efficiency, suggesting higher dosages provide more active binding sites and result an enhanced adsorption percentage (Ahmadi et al. 2022). Another data reveals that with an enhancement in the dose of *Cucumis sativus* peel from 0.4 to 5.0 g, an improvement in Crystal violet removal efficiency from 92.30 to 96.50% was noticed. This also indicates a positive correlation between the dosage and removal efficiency of peel.

2.23 Adsorption Kinetics

To predict the processes and mechanism of the pollutant adsorption, kinetics is crucial. Kinetic studies provide detailed information on the nature of adsorption, pollutant uptake rates, and rate-determining steps (Mohanta et al. 2021). Various kinetic studies commonly employ such pseudo-second-order, intraparticle-diffusion and pseudo-first-order, models.

2.24 Pseudo-First-Order (PFO)

Here, the adsorption process is conceptualized as the occupation rate of sorption sites is directly linked to the number of unoccupied sites. The linear equation is

$$\log(q_e - q_t) = \log q_e - (k_1/2.303) \cdot t \tag{2.1}$$

q_e = equilibrium uptake capacity, q_t = at a given time t uptake capacity, k_1 = rate constant of pseudo-first-order, t = time.

2.25 Pseudo-Second-Order (PSO)

Here, the adsorption rate order is dependent upon the interplay between the quantity of available empty sites and occupied sites on adsorbent surface. The linearized expression is

$$t/q_t = 1/\left(k_2 \cdot q_e^2\right) + t/q_e \tag{2.2}$$

k_2 = rate constant of pseudo-second-order.

2.26 Intra-particle Diffusion (IPD)

The adsorption process consists of three distinct stages: (1) migration of pollutant molecules from the bulk phase to the sorbent surface, (2) pore diffusion, and (3) equilibrium stage. This method was proposed by Weber and Morris, which reveals that the amount of adsorption is proportional to the square root of time ($t^{1/2}$). The linearized equation is

$$q_t = k_{id} t^{1/2} + C \tag{2.3}$$

k_{id} = rate constant of intra-particle-diffusion model and C = constant that suggests thickness of boundary layer.

The best-fitted kinetic model depends on the correlation coefficient (R^2). The model where R^2 values are close to 1, that model will be assumed as the best-fitted model. In many cases, PSO is found to be better fitted model. For instance, the kinetics mechanism of Malachite green onto base-modified *Artocarpus odoratissimus* was examined using PFO and PSO and IPD. The PSO model displayed a more linear trend with higher R^2 values of 0.970 than a PFO model with R^2 values of 0.910. To further assess the accuracy of model selection, the difference between theoretical or calculated adsorption capacity (q_{cal}) and experimental adsorption capacity (q_{exp}) was considered and the PSO model was closer and more accurate than the PFO, confirming that the uptake process followed a PSO model. PSO model suggests that the adsorption is primarily based on chemisorption, involving a chemical reaction between the adsorbent and adsorbate (Zaidi et al. 2019). In contrast, Congo red removal by Eucalyptus leaf was described by IPD, with an R^2 value of 0.996 (Kumari et al. 2020). Similarly, Wang et al. (2016), in their investigation of Methylene blue uptake onto citric acid-modified peanut shell, reported a very high R^2 values of 0.990 for the PSO model. These examples highlight the preference for the PSO model in describing adsorption kinetics in various systems, emphasizing the significance of R^2 values for model selection. In another study, it was noted that MB adsorption by *Citrullus colocynthis* seeds perfectly followed the PSO model (R^2 = 0.999) confirming chemisorption. Furthermore, the experimental capacity (18.830 mg/g) was near to the calculated capacity (19.531 mg/g) in the PSO model. Moreover, the PFO model was found to best-fit data ($R^2 = 0.994$) than the PSO model ($R^2 = 0.980$) for MB sorption by *Citrullus colocynthis* peel confirming physisorption (Alghamdi and El Mannoubi 2021).

2.27 Adsorption Isotherms

Isotherms play a crucial role in understanding the solute–solvent relationship at a specific temperature. In this context, this phenomenon also gives information about the adsorbate–adsorbent interaction type. The uptake capacities in the adsorption process vary as functions of temperature and concentration, and these variations can be effectively represented by isotherm models. Langmuir, Temkin, Dubinin-Radushkevich (D-R) and Freundlich models are among the most common studied isotherm models (Swain et al. 2024).

2.28 Langmuir Model

The Langmuir model is designed to elucidate the process of adsorption on a homogeneous surface (Das et al. 2023). The linear equation is

$$1/q_e = 1/q_m + 1/bq_mC_e \tag{2.4}$$

where q_e = equilibrium adsorption capacity, C_e = equilibrium concentration, q_m = maximum adsorption capacity, b = Langmuir constant.

Dimensionless parameter/separation factor is represented as

$$R_L = 1/(1 + bC_0) \tag{2.5}$$

R_L provides information about the adsorption favorability. The interaction is favorable at $0 < R_L < 1$, unfavourable at $R_L > 1$, linear at $R_L = 1$, and irreversible at $R_L = 0$.

2.29 Freundlich Model

The model involves multilayer, heterogeneous adsorption characterized by an unequal distribution of active sites, resulting in the coverage of the surface by a multilayer of solute molecules. A dimensionless parameter ‘n’ is employed to characterize the adsorption nature. n value 1–10 suggests favorability (Samal et al. 2023a, b).

The linear expressions are given by:

$$\ln q_e = \ln K_F + (1/n) \cdot \ln C_e \tag{2.6}$$

K_F = Freundlich constant.

2.30 Temkin Model

This model is grounded on the theory that the heat of adsorption for pollutants follows a linear decrease with surface coverage. This behavior is attributed to the interactions between the pollutant and adsorbent (Das et al. 2023). This model provides insights into the heat of adsorption and binding energy distribution during the adsorption process.

$$q_e = B_1 lnK_T + B_1 lnC_e \tag{2.7}$$

B_1 = Temkin constant.

2.31 Dubinin-Radushkevich (D-R) Model

The D-R model distinguishes between physisorption and chemisorption based on the calculated free energy. When the free energy value (E) is lower than 8 kJ/mol, it specifies physisorption, suggesting a physical interaction between the phytosorbent and pollutant. Conversely, if the free energy value exceeds 8 kJ/mol, chemisorption occurs, indicating a stronger chemical interaction between the adsorbent and adsorbate. This differentiation allows for a clearer understanding of the nature of adsorption in a given system (Panda et al. 2024). The linear expression is given as

$$\ln q_e = \ln q_s - k_{ad} \cdot \varepsilon^2 \tag{2.8}$$

where k_{ad} = activity coefficient, ε = Polanyi potential, E = free energy. The expression for ε and E is given as

$$\varepsilon = RT \ln(1 + 1/C_e) \tag{2.9}$$

$$E = 1/\sqrt{2k} \tag{2.10}$$

The uptake of Methylene blue onto *Neolamarckia cadamba* pulp surface occurred through Langmuir model with a high R^2 (correlation coefficient) of 0.990 and a removal capacity of 23.11 mg/g at 298 K. R_L values were ranging from 0.18 to

0.21 suggested adsorption favorability. Furthermore, the uptake capacity decreased from 23.11 to 18.62 mg/g with temperature rise from 298 to 328 K. The Freundlich n values were in the range of 1.47–1.90, confirming adsorption favorability. The Temkin and D-R models confirmed physical adsorption (Das et al. 2024a, b). In another study, *Mallotus philippensis* fruit peel was used for Methylene blue removal and the Freundlich model, with a high R^2 of 0.994, exhibited a better fit to the data compared to the other model. The n value in the Freundlich model was 1.08–2.03. Langmuir isotherm fits the experimental data quite well, evidenced by high correlation coefficients ($R^2 > 0.999$) (Qaiyum et al. 2023). In an isotherm study of Methylene blue adsorption onto activated carbon, the best-fitted isotherm model was Freundlich isotherm, indicating a multilayer and heterogeneous adsorption. Moreover, the Langmuir uptake capacity was 387.56 mg/g and from R_L and n values, favorable adsorption was found. From the Temkin model, physical adsorption was found (Fito et al. 2023).

The chromium adsorptions onto *Moringa stenopetala* seed, and banana peel powder and the obtained Langmuir capacities were 9.70 and 7.35 mg/g, respectively. R_L values were 0.325–0.275, indicating a favorable process of adsorption. Here, the Freundlich model was fitted better. Moreover, the Cr(VI) adsorption onto pomegranate peel followed Langmuir isotherm with a uptake capacity of 20.87 mg/g (Giri et al. 2021). Table 2.2 represents the best fitted kinetics and isotherm model of different adsorbents.

Table 2.2 Best-fitted kinetics and isotherm model of different adsorbents

Phytosorbent	Adsorbate	Best-fitted kinetics	Best-fitted isotherm	References
Cocos nucifera L. coir	Safranin-O dye	Pseudo-second-order	Langmuir and Temkin	Ghosh et al. (2021)
Platanus orientalis leaf	Crystal violet dye	Pseudo-second-order	Langmuir and Freundlich	Ahmad Khan et al. (2023)
Banana leaves	Copper (II)	Pseudo-second-order	Langmuir	Darweesh et al. (2022)
Activated carbon of *Rumex abyssinicus*	Methylene blue dye	Pseudo-second-order	Freundlich	Fito et al. (2023)
Pomegranate peel	Chromium (VI)	Pseudo-second-order	Langmuir	Giri et al. (2021)
Azadirachta indica (neem leaves)	Lead (II)	Pseudo-second-order	Freundlich	Elkhaleefa et al. (2021)
Neolamarckia cadamba pulp	Methylene blue dye	Pseudo-second-order	Langmuir	Das et al. (2024a, b)
Mallotus philippensis peel	Methylene blue dye	Pseudo-second-order	Freundlich	Qaiyum et al. (2023)

2.32 Adsorption Thermodynamics

Thermodynamic studies play a great role in examining the impact of temperature on adsorption capacity. The determination of essential parameters, including the enthalpy change (ΔH), the Gibb's free energy change (ΔG), and the entropy change (ΔS) held significance in assessing the feasibility of the process. Additionally, these parameters aid in evaluating the spontaneity and favorability nature of the adsorption system as well as determining exothermic or endothermic nature. Negative ΔG values specify spontaneous process of adsorption, a negative ΔH signifies exothermicity and a positive value signifies endothermicity. Positive ΔS values indicate an increase in randomness, whereas a negative value implies a decrease in randomness (Kumari et al. 2023). The expression for ΔG, ΔH, and ΔS are given below

$$\Delta G^{o} = \Delta H^{o} - T\Delta S^{o} \quad (2.11)$$

$$\Delta G^{o} = -RT\ln K \quad (2.12)$$

$$\ln q_e/C_e = \Delta S^{\circ}/R - \Delta H^{\circ}/RT \quad (2.13)$$

In a study on Basic red dye removal by using pomelo peel, the obtained negative value of ΔH° (–63.264 kJ mol^{-1}) indicated exothermic sorption and the corresponding ΔS° (–0.202 kJ mol^{-1} K^{-1}) indicated a reduction in disorder at the solid–liquid interface. ΔG° value ranged from 0 to –20 kJ mol^{-1} indicate C spontaneity. The endothermic nature was confirmed by of the positive ΔH° in the sorption process. Additionally, during Cr(VI) adsorption onto the banana peel, the obtained negative ΔG° values at 32° and 40 °C implied that the adsorption was spontaneous. The observed shift in ΔG° values from –0.625 to 1.473 kJ/mol, with increasing temperature indicated a more spontaneous uptake at lower temperatures.

From the thermodynamic study of Methyl orange dye adsorption onto Anchote peel, ΔG°, ΔS°, and ΔH° were found to be –8.53, –14.14, and –27.90 kJ/mol, respectively. The increase in ΔG° from –9.05 to –8.21 kJ/mol with rising temperature was observed. The negative ΔS° and ΔH° confirmed the exothermic process, highlighting decreased disorderliness at the interface (Hambisa et al. 2023).

The CV adsorption onto Cowpea husk was endothermic and feasible due to the positive ΔH (264.78 kJ/mol) value and negative ΔG (–2789.69 to –4459.32 kJ/mol) value. Negative ΔS (–0.014 J/mol K) indicated a reduction in randomness at interface (Ayuba and Idoko 2021). The adsorption process for Methylene blue removal by bilberry leaves was spontaneous (ΔG° = −18.91 to −23.1 kJ/mol), and endothermic (ΔH° = 2.02 kJ/mol). The ΔS° value (15.49 J/mol K) signified an enhancement of disorderness at the interface (Mosoarca et al. 2022). The Cr(VI) removal by mint waste was feasible, and the decrease in disorderness observed because ΔG°, ΔH°, and ΔS° values were −4.83, –28.93 kJ/mol, and −0.081 kJ/mol K respectively at 298 K.

2.33 Desorption

Desorption is crucial for various reasons, including, reducing process costs, enabling a constant supply of adsorbent through recycling, recovering contaminants, and minimizing secondary wastes. Indeed, the effectiveness of a particular phytosorbents dependent not only on its adsorptive capacity but also significantly on regeneration capacity and its reusability. The success of desorption depends on the proper selection of eluents which should be economically feasible, effective in regeneration, environmentally friendly, and retain the adsorbent properties. Various eluents, including HCl, EDTA, H_2SO_4, NaOH, methanol, and NaCl with different concentrations have been used for regeneration. Regeneration study was carried out by treating the Congo red dye-loaded Mahua seeds with 0.05 M NaOH, 0.05 M HCl, and 0.05 M NaCl solutions. Notably, in NaOH highest 88% regeneration was achieved, whereas approximately 12% was achieved in HCl and NaCl. The regenerated adsorbent was reused for three consecutive cycles (Kumari and Dey 2018). In another study, regeneration of Malachite green exhausted thiosulphate-treated tea leaves was carried out using hydrochloric acid (0.5 M), sodium hydroxide (0.5 M), sodium chloride (0.5 M), and a 1:1 ethanol–water. The highest regeneration efficiency (83%) was achieved in an acidic medium and reused for 4 cycles (Samal et al. 2023a, b). Moreover, 98% of the Pb(II) loaded mango seed was successfully recovered using 0.2 M HNO_3, citric acid, and EDTA (0.1 M). Similarly, 79–88% of the sequestered Cd(II) loaded mango seed was recovered by the same eluents (Wang et al. 2022). In another study, Auramine O dye-loaded *Ipomoea aquatica* roots were regenerated in an alkali medium and reused for up to 5 cycles (Lu et al. 2020).

2.34 Industrial Effluent Treatment

The investigation into the applicability of plant-based adsorbents for pollutant removal is currently in the developmental stage. Many researchers have conducted laboratory-scale experiments utilizing synthetic wastewater to assess the effectiveness of phytosorbent. Furthermore, certain studies have been expanded to assess the efficacy of plant-based adsorbents in real-world applications. To ensure practical applicability, bamboo leaves were demonstrated in real wastewater. The experiment involved collecting an industrial wastewater sample from a local textile industry followed by filtration, neutralization, and centrifugation. The material showed 36.8% efficiency in real wastewater (Mahato et al. 2023). In the same procedure, *Neolamarckia cadamba* was able to treat industrial effluent with 44.7%. In addition, spiked wastewater was prepared to enhance the optimization potential for field applications. Spike wastewater was prepared using distillation setups, Reverse Osmosis (RO) waste, tube well water, and tap water, and the efficiency of the material was tested. 90.83% adsorption rate was recorded in tube well water, while 81.86% adsorption was observed in RO waste (Das et al. 2024a, b).

Another study was conducted to measure the efficiency of white clover (WC) stem powder on Neutral red removal by using different spiked water such as tap water, river water, and distilled water. It was noticed that the uptake efficiency of the WC stem was higher in distilled water compared to river and tap water samples. This was due to the presence of impurities in the river and tap water, which occupied the active sites of the WC stem, thereby removal efficiency was decreased (Gul et al. 2023a, b). In another experiment, the adsorption efficiency of Banyan tree leaves for Brilliant green removal was also higher in distilled water than in tap and river water (Gul et al. 2023a, b). Moreover, Cr(VI) removal efficiency using spent tea waste was determined using tannery effluents, and 90% adsorption occurred in 60 min (Nur-E-Alam et al. 2018).

2.35 Mechanism

The adsorption mechanism gives an idea about the binding forces that facilitate the adsorption process. These forces encompass a range of interactions, including covalent forces, electrostatic attraction, ionic forces, dipole–dipole interaction, hydrogen bonding, π–π stacking, and van der Waals forces. These forces synergistically contribute to the overall adsorption process. The pH of medium and surface functional groups affect the adsorption forces. Lower pH conditions lead to electrostatic inhibition with cationic pollutants. Conversely, at higher pH levels, facilitates electrostatic attraction with cationic pollutants. Moreover, the interactions among the pollutant and the surface groups of phytosorbent are primarily occurred by H-bonding and non-ionic interactions. In a study, it was found that the carboxylate groups present in peel of dragon fruit, rambutan, and passion fruit helped to adsorb Cu^{2+} ions effectively. Typically, the adsorption mechanism involves complex formation, ion exchange, electrostatic interaction, and precipitation (Phuengphai et al. 2021). In another study, the CV dye removal using NaOH-treated *Rauvolfia tetraphylla* leaves (NRTL) occurred through π–π interaction, hydrogen bonding, and electrostatic attraction (Swain et al. 2024).

Uddin et al. reported that methylene blue removal by mango leaf powder (MLP) occurred through electrostatic interaction at higher pH. At higher pH, the ionization of carboxyl (C=O) and hydroxyl (-OH) groups occurred on the MLP surface which electrostatically interacted with the cationic Methylene blue dye. In another study, it was reported that Methylene blue dye adsorption onto fig leaves activated carbon involved hydrogen bonding, n–π-interaction, electrostatic forces, and π–π-interaction (Uddin et al. 2017). Nazir et al. reported that after citric acid modification on *Bougainvillea spectabilis*, the Pb(II) adsorption capacity increased. Initially, the pristine one contained numerous –OH groups, and after citric acid modification, oxygen-containing groups increased. As a result, the surface became negative because of the presence of electronegative atoms was responsible for enhancing lead adsorption capacity. Those functional groups provided active binding sites to interact strongly with Pb(II) ions (Nazir et al. 2019). In another study, it was reported that

Pb(II) and Cr(VI) removal by *Harpagophytum procumbent,* occurred through several interactions including ion exchange, hydrogen bonding, complexation, intraparticle diffusion, and electrostatic interactions (Shooto 2020).

2.36 Conclusions and Future Perspectives

This chapter provides an overview of significant organic and inorganic pollutants removal from wastewater, employing various phytosorbents. The exploration begins with the categorization and toxicity assessment of dyes and heavy metals, progressing to contemporary treatment methodologies for eliminating these pollutants from aqueous solutions, with a focus on adsorption techniques. A thorough discussion on different phytosorbent including stems, stalks, seeds, fibers, peels, and leaves-based adsorbents is presented. The surface characteristics of the phytosorbents, and factors affecting adsorption including adsorbent dose, contact time, pH, and concentration have been discussed. The discussion extends to the different pollutant uptake capacities of various phytosorbents that were compared. In-depth elucidation is provided on the mechanistic forces governing photosorption, including electrostatic attraction, dipole–dipole interaction, H-bonding, covalent forces, van der Waals forces, ionic forces, and π–π stacking interaction. The isotherm study highlights adsorption type along with the uptake capacities. The thermodynamic discussion unveils the nature, and feasibility of the uptake process. This regeneration study, along with reusing phytosorbents has been discussed. However, some concerns have been given that permit attention to the development of a more effective and economically viable process. These concerns include a thorough discussion on the adsorption capacity of material using simulated pollutant solutions, scaling up lab-scale production of phytosorbent, magnetic separation, and experiments in multi-component systems. In conclusion, the suggestion is made to thoroughly explore all available plant-based adsorbents in real sample analyses through column mode. A pilot scale operation may follow thereafter.

References

Adegoke KA, Akinnawo SO, Adebusuyi TA, Ajala OA, Adegoke RO, Maxakato NW, Bello OS (2023) Modified biomass adsorbents for removal of organic pollutants: a review of batch and optimization studies. Int J Environ Sci Technol 20(10):11615–11644. https://doi.org/10.1007/s13762-023-04872-2

Afroze S, Sen TK (2018) A review on heavy metal ions and dye adsorption from water by agricultural solid waste adsorbents. Water Air Soil Poll 229(7):225. https://doi.org/10.1007/s11270-018-3869-z

Ahmadi H, Hafiz SS, Sharifi H, Rene NN, Habibi SS, Hussain S (2022) Low cost biosorbent (Melon peel) for effective removal of Cu (II), Cd (II), and Pb (II) ions from aqueous solution. Case Stud Chem Environ Eng 6:100242. https://doi.org/10.1016/j.cscee.2022.100242

Ahmad Khan F, Ahad A, Shah SS, Farooqui M (2023) Adsorption of crystal violet dye using Platanus orientalis (Chinar tree) leaf powder and its biochar: equilibrium, kinetics and thermodynamics study. Int J Environ Anal Chem 103(16):4820–4840. https://doi.org/10.1080/03067319.2021.1931854

Akkari I, Graba Z, Bezzi N, Merzeg FA, Bait N, Ferhati A (2023) Raw pomegranate peel as promise efficient biosorbent for the removal of Basic Red 46 dye: equilibrium, kinetic, and thermodynamic studies. Biomass Convers Biorefinery 13(9):8047–8060. https://doi.org/10.1007/s13399-021-01620-9

Alghamdi WM, El Mannoubi I (2021) Investigation of seeds and peels of Citrullus colocynthis as efficient natural adsorbent for methylene blue dye. Processes 9(8):1279. https://doi.org/10.3390/pr9081279

Ali IH, Al Mesfer MK, Khan MI, Danish M, Alghamdi MM (2019) Exploring adsorption process of lead (II) and chromium (VI) ions from aqueous solutions on acid activated carbon prepared from Juniperus procera leaves. Processes 7(4):217. https://doi.org/10.3390/pr7040217

Ayuba AM, Idoko B (2021) Cowpea husk adsorbent for the removal of crystal violet dye from aqueous solution. Arab J Chem Environ Res 08:114–132. www.mocedes.org

Banerjee M, Basu RK, Das SK (2018) Cr(VI) adsorption by a green adsorbent walnut shell: adsorption studies, regeneration studies, scale-up design and economic feasibility. Process Saf Environ Prot 116:693–702. https://doi.org/10.1016/j.psep.2018.03.037

Basirun AA, Othman AR, Yasid NA, Shukor MYA, Khayat ME (2023) A green approach of utilising banana peel (Musa paradisiaca) as adsorbent precursor for an anionic dye removal: kinetic, isotherm and thermodynamics analysis. Processes 11(6):1611. https://doi.org/10.3390/pr11061611

Bayomie OS, Kandeel H, Shoeib T, Yang H, Youssef N, El-Sayed MMH (2020) Novel approach for effective removal of methylene blue dye from water using fava bean peel waste. Sci Rep 10(1):7824. https://doi.org/10.1038/s41598-020-64727-5

Bayuo J, Pelig-Ba KB, Abukari MA (2019) Adsorptive removal of chromium(VI) from aqueous solution unto groundnut shell. Appl Water Sci 9(4):107. https://doi.org/10.1007/s13201-019-0987-8

Bharadwaj B, Dutta S, Qaiyum MA, Samal PP, Dey B, Dey S (2023) Pristine wild sugarcane (Saccharum spontaneum) as a biosorbent for removal of methylene blue from wastewater: isotherm, kinetics and regeneration studies. Int J Phytorem. https://doi.org/10.1080/15226514.2023.2260002

Bhattacharjee J, Mishra S, Das AP (2021) Recent advances in sensor-based detection of toxic dyes for bioremediation application: a review. Appl Biochem Biotechnol 2021:1–20. https://doi.org/10.1007/S12010-021-03767-7

Bouabidi ZB, El-Naas MH, Cortes D, McKay G (2018) Steel-Making dust as a potential adsorbent for the removal of lead (II) from an aqueous solution. Chem Eng J 334:837–844. https://doi.org/10.1016/j.cej.2017.10.073

Bukhari A, Ijaz I, Zain H, Gilani E, Nazir A, Bukhari A, Raza S, Hussain S, Alarfaji SS, Naseer Y, Aftab R, Iram S (2022) Removal of Eosin dye from simulated media onto lemon peel-based low cost biosorbent. Arab J Chem 15(7):103873. https://doi.org/10.1016/j.arabjc.2022.103873

Bulgariu L, Escudero LB, Bello OS, Iqbal M, Nisar J, Adegoke KA, Alakhras F, Kornaros M, Anastopoulos I (2019) The utilization of leaf-based adsorbents for dyes removal: a review. J Mol Liq 276:728–747. https://doi.org/10.1016/j.molliq.2018.12.001

Chen Y, Wang H, Zhao W, Huang S (2018) Four different kinds of peels as adsorbents for the removal of Cd (II) from aqueous solution: kinetics, isotherm and mechanism. J Taiwan Inst Chem Eng 88:146–151. https://doi.org/10.1016/j.jtice.2018.03.046

Chowdhury IR, Chowdhury S, Mazumder MAJ, Al-Ahmed A (2022) Removal of lead ions (Pb^{2+}) from water and wastewater: a review on the low-cost adsorbents. Appl Water Sci 12(8):185. https://doi.org/10.1007/s13201-022-01703-6

Darweesh MA, Elgendy MY, Ayad MI, Ahmed AM, Elsayed NK, Hammad WA (2022) Adsorption isotherm, kinetic, and optimization studies for copper (II) removal from aqueous solutions by

banana leaves and derived activated carbon. S Afr J Chem Eng 40:10–20. https://doi.org/10.1016/j.sajce.2022.01.002
Das AP, Nayak B, Behera ID (2024) Applications of microorganisms in advanced geomicrobiological and biotechnological exploration. Geomicrobiol J 1–3. https://doi.org/10.1080/01490451.2024.2327059/ASSET//CMS/ASSET/979F6194-84CA-4A6E-8233-F10C1D4E33EF/01490451.2024.2327059.FP.PNG
Das AP, Singh S (2011) Occupational health assessment of chromite toxicity among Indian miners. Indian J Occup Environ Med 15(1):6. https://doi.org/10.4103/0019-5278.82998
Das AP, Sukla LB, Pradhan N, Nayak S (2011) Manganese biomining: a review. Biores Technol 102(16):7381–7387. https://doi.org/10.1016/J.BIORTECH.2011.05.018
Das AP, Swain S (2013) Algal biosorption of toxic dye Methylene blue. In: A potential source of food, feed, biochemicals, biofuels and biofertilizers, International conference on algal biorefinery, vol 13. Indian Institute of Technology
Das S, Samal PP, Qaiyum MA, Dutta S, Dey B, Dey S (2024) Neolamarckia cadamba (cadamba) waste pulp as a natural and techno-economic scavenger for methylene blue from aqueous solutions. Int J Phytorem 26(2):208–218. https://doi.org/10.1080/15226514.2023.2232861
de Salomón YLO, Georgin J, Franco DSP, Netto MS, Foletto EL, Allasia D, Dotto GL (2021) Application of seed residues from Anadenanthera macrocarpa and Cedrela fissilis as alternative adsorbents for remarkable removal of methylene blue dye in aqueous solutions. Environ Sci Pollut Res 28(2):2342–2354. https://doi.org/10.1007/s11356-020-10635-0
Dey S, Chakraborty R, Mohanta J, Dey B (2022) Tricosanthes cucumerina: a potential biomass for efficient removal of methylene blue from water. Bioremediat J. https://doi.org/10.1080/10889868.2022.2086530
Elhalil A, Tounsadi Sidi Mohamed Ben H, Machrouhi A, Farnane M, Abdennouri M, Tounsadi H, Barka N (2019) Heavy metals biosorption by Thapsia transtagana stems powder: kinetics, equilibrium and thermodynamics. Moroc J Chem 7. http://revues.imist.ma/?journal=morjchem&page=login
Elkhaleefa A, Ali IH, Brima EI, Shigidi I, Elhag AB, Karama B (2021) Evaluation of the adsorption efficiency on the removal of lead (II) ions from aqueous solutions using Azadirachta indica leaves as an adsorbent. Processes 9(3):559. https://doi.org/10.3390/pr9030559
Fito J, Abewaa M, Mengistu A, Angassa K, Ambaye AD, Moyo W, Nkambule T (2023) Adsorption of methylene blue from textile industrial wastewater using activated carbon developed from Rumex abyssinicus plant. Sci Rep 13(1):5427. https://doi.org/10.1038/s41598-023-32341-w
Gaur N, Kukreja A, Yadav M, Tiwari A (2018) Adsorptive removal of lead and arsenic from aqueous solution using soya bean as a novel biosorbent: equilibrium isotherm and thermal stability studies. Appl Water Sci 8(4):98. https://doi.org/10.1007/s13201-018-0743-5
Ghosh I, Kar S, Chatterjee T, Bar N, Das SK (2021) Adsorptive removal of Safranin-O dye from aqueous medium using coconut coir and its acid-treated forms: adsorption study, scale-up design, MPR and GA-ANN modeling. Sustain Chem Pharm 19:100374. https://doi.org/10.1016/j.scp.2021.100374
Giri R, Kumari N, Behera M, Sharma A, Kumar S, Kumar N, Singh R (2021) Adsorption of hexavalent chromium from aqueous solution using pomegranate peel as low-cost biosorbent. Environ Sustain 4(2):401–417. https://doi.org/10.1007/s42398-021-00192-8
Gokulan R, Prabhu GG, Jegan J (2019) Remediation of complex Remazol effluent using biochar derived from green seaweed biomass. Int J Phytorem 21(12):1179–1189. https://doi.org/10.1080/15226514.2019.1612845
Goswami M, Chaturvedi P, Kumar Sonwani R, Dutta Gupta A, Rani Singhania R, Shekher Giri B, Nath Rai B, Singh H, Yadav S, Sharan Singh R (2020) Application of Arjuna (Terminalia arjuna) seed biochar in hybrid treatment system for the bioremediation of Congo red dye. Biores Technol 307:123203. https://doi.org/10.1016/j.biortech.2020.123203
Grassi P, Netto MS, Jahn SL, Georgin J, Franco DSP, Sillanpää M, Meili L, Silva LFO (2023) Conversion of foliar residues of Sansevieria trifasciata into adsorbents: dye adsorption in continuous

and discontinuous systems. Environ Sci Pollut Res 30(4):9688–9698. https://doi.org/10.1007/s11356-022-22857-5

Gul S, Afsar S, Raza N, Gul H, Khezami L, Almufarij R, Hussain SN (2023) Efficient adsorptive removal of textile dye neutral red in environmental samples using white clover stem powder. Biomass Convers Biorefinery. https://doi.org/10.1007/s13399-023-04770-0

Gul S, Gul A, Gul H, Khattak R, Ismail M, Khan SU, Khan MS, Aouissi HA, Krauklis A (2023) Removal of brilliant green dye from water using Ficus benghalensis tree leaves as an efficient biosorbent. Materials 16(2). https://doi.org/10.3390/ma16020521

Gul S, Kanwal M, Qazi RA, Gul H, Khattak R, Khan MS, Khitab F, Krauklis AE (2022) Efficient removal of methyl red dye by using bark of hopbush. Water 14(18):2831. https://doi.org/10.3390/w14182831

Hambisa AA, Regasa MB, Ejigu HG, Senbeto CB (2023) Adsorption studies of methyl orange dye removal from aqueous solution using anchote peel-based agricultural waste adsorbent. Appl Water Sci 13(1):24. https://doi.org/10.1007/s13201-022-01832-y

Hassan W, Farooq U, Ahmad M, Athar M, Khan MA (2017) Potential biosorbent, Haloxylon recurvum plant stems, for the removal of methylene blue dye. Arab J Chem 10:S1512–S1522. https://doi.org/10.1016/j.arabjc.2013.05.002

Hevira L, Zilfa R, Ighalo JO, Aziz H, Zein R (2021) Terminalia catappa shell as low-cost biosorbent for the removal of methylene blue from aqueous solutions. J Ind Eng Chem 97:188–199. https://doi.org/10.1016/j.jiec.2021.01.028

Ho MQ, Li SFY (2019) Preparation, characterization and application of sorbent envelopes with Carica papaya seeds and Citrus grandis rind for cationic dyes removal. Green Chem Lett Rev 12(3):343–352. https://doi.org/10.1080/17518253.2019.1643927

Hou Y, Yan S, Huang G, Yang Q, Huang S, Cai J (2020) Fabrication of N-doped carbons from waste bamboo shoot shell with high removal efficiency of organic dyes from water. Biores Technol 303:122939. https://doi.org/10.1016/j.biortech.2020.122939

Hussain S, Anjali KP, Hassan ST, Dwivedi PB (2018) Waste tea as a novel adsorbent: a review. Appl Water Sci 8(6):165. https://doi.org/10.1007/s13201-018-0824-5

Khan Q, Zahoor M, Salman SM, Wahab M, Talha M, Kamran AW (2022) Removal of chromium (VI) from the steel mill effluents using the chemically modified leaves of Pteris vittata as adsorbent. Water 14(17):2599. https://doi.org/10.3390/w14172599

Khan SU, Zaidi R, Shaik F, Farooqi IH, Azam A, Abuhimd H, Ahmed F (2021) Evaluation of Fe-Mg binary oxide for As (III) adsorption—synthesis, characterization and kinetic modelling. Nanomaterials 11(3):1–16. https://doi.org/10.3390/nano11030805

Khoshsang H, Ghaffarinejad A (2018) Rapid removal of lead (II) ions from aqueous solutions by saffron flower waste as a green biosorbent. J Environ Chem Eng 6(5):6021–6027. https://doi.org/10.1016/j.jece.2018.09.020

Kocaman S (2020) Synthesis and cationic dye biosorption properties of a novel low-cost adsorbent: coconut waste modified with acrylic and polyacrylic acids. Int J Phytorem 22(5):551–566. https://doi.org/10.1080/15226514.2020.1741509

Kristanda J, Sintiago KS, Kristianto H, Prasetyo S, Sugih AK (2021) Optimization study of Leucaena leucocephala seed extract as natural coagulant on decolorization of aqueous Congo red solutions. Arab J Sci Eng 46(7):6275–6286. https://doi.org/10.1007/s13369-020-05008-1

Kumari R, Dey S (2018) Facile removal of Congo red using mahua (Madhuca longifolia) seeds, a low cost adsorbent. Int J Green Herb Chem 7(2):237–250. https://doi.org/10.24214/IJGHC/GC/7/2/23750

Kumari R, Mohanta J, Dey B, Dey S (2020) Eucalyptus leaf powder as an efficient scavenger for Congo red from water: comprehensive batch and column investigation. Sep Sci Technol 55(17):3047–3059. https://doi.org/10.1080/01496395.2019.1670208

Kumari R, Mohanta J, Sambasivaiah B, Qaiyum MA, Dey B, Samal PP, Dutta S, Dey S (2023) Dye sequestration from aqueous phase using natural and synthetic adsorbents in batch mode: present status and future perspectives. Int J Environ Sci Technol 20(12):14033–14052. https://doi.org/10.1007/s13762-023-04782-3

Lu YC, Priyantha N, Lim LBL (2020) Ipomoea aquatica roots as environmentally friendly and green adsorbent for efficient removal of Auramine O dye. Surf Interfaces 20:100543. https://doi.org/10.1016/j.surfin.2020.100543

Lahiri D, Nag M, Dey A, Sarkar T, Joshi S, Pandit S, Das AP, Pati S, Pattanaik S, Tilak VK, Ray RR (2021) Biofilm mediated degradation of petroleum products. Geomicrobiol J 1–10. https://doi.org/10.1080/01490451.2021.1968979

Mahato R, Qaiyum MA, Samal PP, Dutta S, Dey B, Dey S (2023) Exploring the promising potential of fallen bamboo leaves (Bambusa bambos) for efficient removal of crystal violet from wastewater. Int J Phytorem 25(8):1042–1051. https://doi.org/10.1080/15226514.2022.2125498

Malik LA, Bashir A, Qureashi A, Pandith AH (2019) Detection and removal of heavy metal ions: a review. Environ Chem Lett 17(4):1495–1521. https://doi.org/10.1007/s10311-019-00891-z

Mansour RA, Aboeleneen NM, AbdelMonem NM (2018) Adsorption of cationic dye from aqueous solutions by date pits: equilibrium, kinetic, thermodynamic studies, and batch adsorber design. Int J Phytorem 20(10):1062–1074. https://doi.org/10.1080/15226514.2018.1460306

Manzoor K, Batool M, Naz F, Nazar MF, Hameed BH, Zafar MN (2022) A comprehensive review on application of plant-based bioadsorbents for Congo red removal. Biomass Convers Biorefinery. https://doi.org/10.1007/s13399-022-02741-5

Mishra S, Dash D, Das AP (2022) Detection, characterization and possible biofragmentation of synthetic microfibers released from domestic laundering wastewater as an emerging source of marine pollution. Mar Pollut Bull 185:114254. https://doi.org/10.1016/J.MARPOLBUL.2022.114254

Mishra S, Ghosh S, van Hullebusch ED, Singh S, Das AP (2023) A critical review on the recovery of base and critical elements from electronic waste-contaminated streams using microbial biotechnology. Appl Biochem Biotechnol 1–30. https://doi.org/10.1007/S12010-023-04440-X/METRICS

Mishra S, charan Rath C, Das AP (2019) Marine microfiber pollution: a review on present status and future challenges. Mar Pollut Bull 140:188–197. https://doi.org/10.1016/J.MARPOLBUL.2019.01.039

Mohanta J, Kumari R, Qaiyum MA, Dey B, Dey S (2021) Alkali assisted hydrophobic reinforcement of coconut fiber for enhanced removal of cationic dyes: equilibrium, kinetics, and thermodynamic insight. Int J Phytorem 23(13):1423–1431. https://doi.org/10.1080/15226514.2021.1901850

Mondal NK, Kar S (2018) Potentiality of banana peel for removal of Congo red dye from aqueous solution: isotherm, kinetics and thermodynamics studies. Appl Water Sci 8(6):157. https://doi.org/10.1007/s13201-018-0811-x

Mosoarca G, Vancea C, Popa S, Dan M, Boran S (2022) The use of bilberry leaves (Vaccinium myrtillus L.) as an efficient adsorbent for cationic dye removal from aqueous solutions. Polymers 14(5):978. https://doi.org/10.3390/polym14050978

Naderi P, Shirani M, Semnani A, Goli A (2018) Efficient removal of crystal violet from aqueous solutions with Centaurea stem as a novel biodegradable bioadsorbent using response surface methodology and simulated annealing: kinetic, isotherm and thermodynamic studies. Ecotoxicol Environ Saf 163:372–381. https://doi.org/10.1016/j.ecoenv.2018.07.091

Nazir H, Muhammad S, Athar M, Farooq U, Wahab A, Akram M (2019) Citric acid functionalized bougainvillea spectabilis: a novel, sustainable, and cost-effective biosorbent for removal of heavy metal (Pb^{2+}) from waste water. Water Air Soil Pollut 230:303. https://doi.org/10.1007/s11270-019-04360-1

Nur-E-Alam M, Abu Sayid Mia M, Ahmad F, Mafizur Rahman M (2018) Adsorption of chromium (Cr) from tannery wastewater using low-cost spent tea leaves adsorbent. Appl Water Sci 8(5):129. https://doi.org/10.1007/s13201-018-0774-y

Olayinka JA, Tope BI, Olubunmi OO, Omoniyi MS (2020) Evaluation of kinetics and equilibrium studies of biosorption potentials of bamboo stem biomass for removal of lead (II) and cadmium (II) ions from aqueous solution. Afr J Pure Appl Chem 14(2):24–41. https://doi.org/10.5897/ajpac2019.0812

Panda A, Samal PP, Qaiyum MA, Dey B, Dey S (2024) Think before throw: waste chili stalk powder for facile scavenging of cationic dyes from water. Environ Monit Assess 196:118. https://doi.org/10.1007/s10661-023-12243-0

Paul A, Dey S, Ram DK, Das AP (2023) Hexavalent chromium pollution and its sustainable management through bioremediation. Geomicrobiol J 1–11. https://doi.org/10.1080/01490451.2023.2218377

Phuengphai P, Singjanusong T, Kheangkhun N, Wattanakornsiri A (2021) Removal of copper(II) from aqueous solution using chemically modified fruit peels as efficient low-cost biosorbents. Water Sci Eng 14(4):286–294. https://doi.org/10.1016/j.wse.2021.08.003

Qaiyum MA, Mohanta J, Kumari R, Samal PP, Dey B, Dey S (2022) Alkali treated water chestnut (Trapa natans L.) shells as a promising phytosorbent for malachite green removal from water. Int J Phytoremediation 24(8):822–830. https://doi.org/10.1080/15226514.2021.1977912

Qaiyum MA, Sahu PR, Samal PP, Dutta S, Dey B, Dey S (2023) Towards a win-win chemistry: extraction of C.I. orange from Kamala fruit (Mallotus philippensis), and simultaneous exercise of its peels for the removal of Methylene Blue from water. Int J Phytoremediation 25(7):907–916. https://doi.org/10.1080/15226514.2022.2119936

Raj K, Das AP (2023) Lead pollution: impact on environment and human health and approach for a sustainable solution. Environ Chem Ecotoxicol 5:79–85. https://doi.org/10.1016/J.ENCECO.2023.02.001

Rambabu K, Bharath G, Banat F, Show PL (2020) Biosorption performance of date palm empty fruit bunch wastes for toxic hexavalent chromium removal. Environ Res 187:109694. https://doi.org/10.1016/j.envres.2020.109694

Rout A, Qaiyum MA, Samal PP, Dutta S, Dey B, Dey S (2023) Brinjal (Solanum melongena) stalk waste as an effective scavenger for Eriochrome Black-T from water and wastewater: an approach towards waste to best. Int J Phytorem 25(8):956–964. https://doi.org/10.1080/15226514.2022.2123445

Samal PP, Kumari J, Qaiyum MA, Mohanta J, Kumari R, Dutta S, Dey B, Dey S (2023) Thiosulfate impregnated spent tea leaves for the remarkable uptake of malachite green. Int J Phytorem 25(11):1413–1422. https://doi.org/10.1080/15226514.2022.2161465

Samal PP, Qaiyum MA, Dutta S, Dey B, Dey S (2023) Augmented dye eradication from wastewater using alkali-aided, reinforced waste acacia (Acacia auriculiformis) leaves. Int J Phytorem 26(1):52–62. https://doi.org/10.1080/15226514.2023.2220404

Samal PP, Qaiyum MdA, Ghosh A, Kumari R, Mohanta J, Das S, Swain J, Dey B, Dey S (2024) Acacia auriculiformis leaf extract mediated green synthesis of goethite and boehmite embedded activated sawdust for Cr(VI) adsorption. J Hazard Mater Adv 13:100405. https://doi.org/10.1016/j.hazadv.2024.100405

Samir L, Samira A, hadj Mekatel E, Djamel N (2019) Adsorption of Cr(VI) on Stipa tenacissima L (Alfa): characteristics, kinetics and thermodynamic studies. Sep Sci Technol (Philadelphia) 54(6):876–887. https://doi.org/10.1080/01496395.2018.1521833

Shahzad A, Rasool K, Miran W, Nawaz M, Jang J, Mahmoud KA, Lee DS (2017) Two-dimensional $Ti_3C_2T_x$ MXene nanosheets for efficient copper removal from water. ACS Sustain Chem Eng 5(12):11481–11488. https://doi.org/10.1021/acssuschemeng.7b02695

Sheikh Z, Amin M, Khan N, Khan MN, Sami SK, Khan SB, Hafeez I, Khan SA, Bakhsh EM, Cheng CK (2021) Potential application of Allium Cepa seeds as a novel biosorbent for efficient biosorption of heavy metals ions from aqueous solution. Chemosphere 279:130545. https://doi.org/10.1016/j.chemosphere.2021.130545

Shen Y, Li H, Zhu W, Ho SH, Yuan W, Chen J, Xie Y (2017) Microalgal-biochar immobilized complex: a novel efficient biosorbent for cadmium removal from aqueous solution. Biores Technol 244:1031–1038. https://doi.org/10.1016/j.biortech.2017.08.085

Shi Y, Zhang T, Ren H, Kruse A, Cui R (2018) Polyethylene imine modified hydrochar adsorption for chromium (VI) and nickel (II) removal from aqueous solution. Biores Technol 247:370–379. https://doi.org/10.1016/j.biortech.2017.09.107

Shooto ND (2020) Removal of lead(II) and chromium(VI) ions from synthetic wastewater by the roots of Harpagophytum procumbens plant. J Environ Chem Eng 8(6):104541. https://doi.org/10.1016/j.jece.2020.104541

Silva F, Nascimento L, Brito M, da Silva K, Paschoal W, Fujiyama R (2019) Biosorption of methylene blue dye using natural biosorbents made from weeds. Materials 12(15):2486. https://doi.org/10.3390/ma12152486

Soldatkina L, Yanar M (2021) Equilibrium, kinetic, and thermodynamic studies of cationic dyes adsorption on corn stalks modified by citric acid. Colloids Interfaces 5(4):52. https://doi.org/10.3390/colloids5040052

Soliman NK, Moustafa AF, Aboud AA, Halim KSA (2019) Effective utilization of Moringa seeds waste as a new green environmental adsorbent for removal of industrial toxic dyes. J Market Res 8(2):1798–1808. https://doi.org/10.1016/j.jmrt.2018.12.010

Sonde CU, Nnaji N, Odoemelam SA, Ekuma FK, Ezeh GC, Okoli CP (2023) Methylene blue removal by agricultural wastes from Annona muricata and Treculia africana seeds. Biomass Convers Biorefinery 13(16):14901–14915. https://doi.org/10.1007/s13399-022-03419-8

Swain J, Samal PP, Qaiyum MA, Dey B, Dey S (2024) Biosorption of crystal violet, a cationic dye onto alkali treated *Rauvolfia tetraphylla* leaf: kinetics, isotherm and thermodynamics. Water Conserv Sci Eng. 9:1. https://doi.org/10.1007/s41101-023-00233-9

Sulyman M, Gierak A (2020) Green environmental approach for adsorption of hazardous dye from water using tree and sea plant leaves (Dead L.). Acta Sci Agric 4(2):01–10. https://doi.org/10.31080/asag.2020.04

Uddin MK, Nasar A (2020) Decolorization of basic dyes solution by utilizing fruit seed powder. KSCE J Civ Eng 24(2):345–355. https://doi.org/10.1007/s12205-020-0523-2

Uddin MT, Rahman MA, Rukanuzzaman M, Islam MA (2017) A potential low cost adsorbent for the removal of cationic dyes from aqueous solutions. Appl Water Sci 7(6):2831–2842. https://doi.org/10.1007/s13201-017-0542-4

Wang P, Ma Q, Hu D, Wang L (2016) Adsorption of methylene blue by a low-cost biosorbent: citric acid modified peanut shell. Desalin Water Treat 57(22):10261–10269. https://doi.org/10.1080/19443994.2015.1033651

Wang Q, Wang Y, Yang Z, Han W, Yuan L, Zhang L, Huang X (2022) Efficient removal of Pb(II) and Cd(II) from aqueous solutions by mango seed biosorbent. Chem Eng J Adv 11:100295. https://doi.org/10.1016/j.ceja.2022.100295

Yadav S, Yadav A, Bagotia N, Sharma AK, Kumar S (2021) Adsorptive potential of modified plant-based adsorbents for sequestration of dyes and heavy metals from wastewater—a review. J Water Process Eng 42:102148. https://doi.org/10.1016/j.jwpe.2021.102148

Yagub MT, Sen TK, Afroze S, Ang HM (2014) Dye and its removal from aqueous solution by adsorption: a review. Adv Colloid Interface Sci 209:172–184. https://doi.org/10.1016/j.cis.2014.04.002

Zaidi NAHM, Lim LBL, Usman A (2019) Enhancing adsorption of malachite green dye using base-modified *Artocarpus odoratissimus* leaves as adsorbents. Environ Technol Innov 13:211–223. https://doi.org/10.1016/j.eti.2018.12.002

Chapter 3
Studies on Thermal Curing and Thermal Degradation of Recyclable Thermoset Material

J. Dhanalakshmi and C. T. Vijayakumar

Abstract Hemiaminal was synthesized using 4,4'-diaminodiphenylsulphone (DDS). The structural characterization of the material was carried out using Fourier transform infrared spectrometer (FTIR). It is noted that the presence of absorption bands pertaining to NMP and the identification of -OH stretching confirms the proposed structure for the hemiaminal compound. The curing characteristic of the material was investigated using differential scanning calorimeter (DSC). From the DSC studies, it was found that the hemiaminal after 200 °C showed a strong exotherm which may be associated to the degradation of the hexahydrotriazine produced. This fact is supplemented using thermogravimetric (TG) study. The synthesized material was thermally cured and the polymer structure was investigated using FTIR. The absence of absorption band around 3300 cm^{-1} in the polyhexahydrotriazine compound indicates the conversion of the hemiaminal to the hexahydrotriazine form through cyclization reaction. The thermal degradation behaviors of PHT material was investigated by thermogravimetric analyzer at three different heating rates ($\beta = 10$, 20 and 30 °C min^{-1}). The most widely used integral methods like Flynn-Wall-Ozawa (FWO), Corrected Flynn-Wall-Ozawa (C-FWO), Kissinger-Akahira-Sunose (KAS), Corrected Kissinger-Akahira-Sunose (C-KAS), Vyazovkin (VYZ) and Advanced Vyazovkin (A-VYZ) methods are used to estimate the apparent activation energy for thermal degradation (Ea-D) of the material for various reaction extent (α). The results were presented and discussed.

Keywords Recycling thermoset plastic · Polyhexahydrotriazine · DSC · TG

J. Dhanalakshmi (✉)
Department of Chemistry, Kamaraj College of Engineering and Technology, Vellakulam, India
e-mail: j.dhanalakshmi81@gmail.com

C. T. Vijayakumar
Thangavel Match Industries, Thiruthangal, India

A. Kuanar et al. (eds.), *Recent Advances in Bioremediation and Phytoremediation*, Environmental Science and Engineering,
https://doi.org/10.1007/978-3-031-77884-1_3

3.1 Introduction

The synthesis and design of new high performance has witness little success in spite of challenges in production in the recent days owing to widespread commercial utility. The evolutionof novel synthetic platforms that imbibe structural modifications to evident diverse material qualities, such as high strength or self-healing is a great interest. Furthermore the thermosetting polymers are reminiscent for being non-recyclable as it cannot be remolded upon curing and being thermally decomposed upon heating to high temperatures (Lin and Pearce 1994).Also thermosetting polymers are non-recyclable due to which it will create environmental problems as well as wastage of valuable resources (Khan et al. 2021). As a result, a thermoset that can be depolymerized intrinsically would provide a viable pathway for thermoset recycling and redeployment (Chen et al. 2002). The creation of novel synthetic platforms that enable structural modification to impart a variety of desirable material qualities, such as high strength or self-healing, is thus of great interest. The idea of a rework-able thermoset is still relatively new.

Triazine is a class of nitrogen containing heterocycles. The triazines have planar six membered benzene like ring structure but three carbon atoms replaced by nitrogen. The best known triazines are derivatives of the 1,3,5-triazine derivatives melamine and cyanuric chloride. Polyhexahydrotriazines (PHTs) are polymers of hexahydro-1,3,5-triazines, a class of heterocyclic compounds with the formula $(CH_2NR)_3$. They are reduced derivatives of 1,3,5-triazine which have the formula $(CHN)_3$, a family of aromatic heterocycles. They are among the strongest known thermosetting plastics and are stable to solvents at $pH > 3$, but decompose to the monomers in acidic solutions (Hedrick et al. 2015; Jeannette et al. 2014). This property is unusual for thermosetting plastics and allows easy recycling.

Hence in the present investigation hemiaminal (HA-DDS) using 4,4'-diaminodiphenyl sulphone (DDS) was synthesized. The curing behaviour of the hemiaminal was studied using differential scanning calorimetry (DSC). The thermal stability of the material HA-DDS was studied using thermogravimetric analyzer (TGA). The synthesized hemiaminal was converted into polyhexahydrotriazine and thermal stability of the material (PHT-DDS) was studied using TGA and the results were presented and discussed below.

3.1.1 *Experimental*

4,4'-diaminodiphenylsulphone was obtained from HiMedia Laboratories Pvt. Ltd., Mumbai, India. Paraformaldehyde was purchased from SISCO Research Laboratory Pvt. Ltd., Mumbai, India. N-methylpyrrolidone was purchased from Sigma-Aldrich Pvt. Ltd., Hyderabad, India. Acetone purchased from Merck India Pvt. Ltd., Mumbai, India.

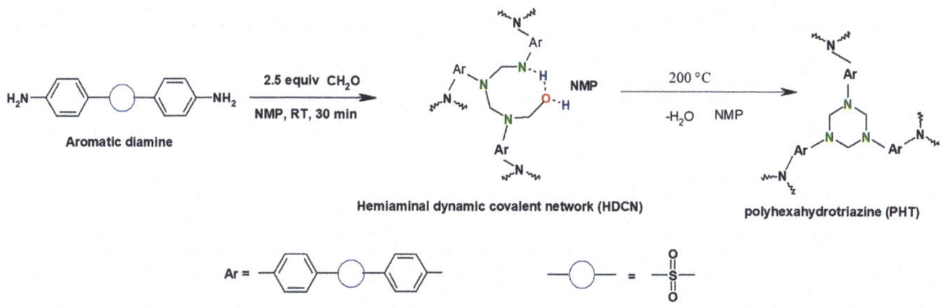

Fig. 3.1 Preparation of polyhexahydrotriazine (PHT) using hemiaminal (HA)

3.1.2 Preparation of Hemiaminals

Hemiaminal using aromatic diamine (DDS, 0.20 g, 1.0 mmol) and paraformaldehyde (PF, 0.15 g, 5.0 mmol) were weighed out in a 100 mL round bottom flask. N-methylpyrrolidone (NMP, 6.2 g, 6.0 mL) was added. The reaction mixture was heated in an oil bath at 50 °C for 24 h and hemiaminal was formed. After approximately 0.75 h, the hemiaminal precipitated in NMP. After the stipulated time, the hemiaminal (HA-DDS) was precipitated in acetone, filtered and dried in oven. The material was collected as an off-white solid (yield = 0.22 g, > 98%). The preparation procedure is shown in Fig. 3.1.

3.1.3 Thermal Curing

HA-DDS was taken in a micro test tube and flushed with inert gas (N_2) and then heated to 200 °C for 3 h. After, thermal polymerization the sample (PHT-DDS) was cooled to room temperature. The material wasremoved from test tube and ground to a granular powder and preserved for further analysis.

3.1.4 Methods

The FTIR spectrum of the cured material was run on a Fourier Transform Infrared—8400S Spectrophotometer, Shimadzu, Japan by using potassium bromide disc technique. Differential scanning calorimetric curves was recorded using TA Instruments DSC Q20. The material was heated from ambient to 350 °C at a heating rate of 10 °C min^{-1} in dry pure nitrogen atmosphere. The thermal degradation behaviour of the material was examined using TGA Model Q50 supplied by TA Instruments, Water India Pvt. Ltd., Bengaluru-560086, India and the heating rate of 10, 20 and

30 °C min^{-1}from ambient to 800 °C. The obtained DSC and TG curve was analysed using the universal analysis 2000 software provided by TA instruments.

3.1.5 *Kinetic Analysis*

3.1.5.1 Flynn-Wall-Ozawa (FWO) Method

The FWO (Yei et al. 2006) method is widely used for dynamic kinetic analysis and the equation used for this method is

$$E_a = \frac{-R\Delta ln\beta}{1.052\Delta\left(\frac{1}{T}\right)}$$

3.1.5.2 Corrected Flynn-Wall-Ozawa (C-FWO) Method

Farjas (Farjas and Roura 2011) introduced an iteration process that eliminates the necessity of the error table developed by Flynn. The equation used for this method is,

$$\ln\beta_i - \ln\xi_{\mathrm{FWO}} = \ln\left(\frac{AE}{RG(x)}\right) - 1.0518\frac{E}{RT_i} - 5.330$$

where

$$\xi_{FWO} = {}^{p(x_i)}\Big/_{e^{(-1.0518x-5.330)}}x_i = {}^{E}\Big/_{RT}$$

3.1.5.3 Kissinger-Akahira-Sunose (KAS) Method

The KAS (Kissinger 1957) method also uses the temperature integral toestimate the activation energy just similar to FWOmethod. But instead of using Doyle's approximation, KAS method employs the Coats-Redfern (Coats and Redfern 1964) approximation. The general equation can be transformed into:

$$\ln\frac{\beta_i}{T_i^2} = \ln\frac{AR}{g(\alpha)E} - \frac{E}{RT_i} + \ln\frac{E - 2RT_i}{E}$$

Based on the asymptotic approximation, the value of *RTi/E* (since the value is almost negligible) term can be entirely eliminated, thus giving rise to the KAS generalization.

$$\ln \frac{\beta_i}{T_i^2} = \ln \frac{AR}{g(\alpha)E} - \frac{E}{RT_i}$$

Now when by plotting ln(*β*i/Ti^2) against *1/RT*$_i$, the Ea can be determined. But very much similar to the organized errors in the FWO method, KAS too have a couple of systematic errors as the activation energy is not constant with respect to α in complex reactions and also the whole system is derived based on the errors of Coats and Redfern approximation.

3.1.5.4 Corrected Kissinger-Akahira-Sunose (C-KAS) Method

Similar to the way how the approximation was corrected in FWO method, Farjas (Farjas and Roura 2011) put forward the effective use of iterative procedure in KAS method as well.

$$\ln \frac{\beta_i}{T_i^2} - \ln \xi_{\mathrm{KAS}} = \ln \frac{AR}{g(\alpha)E} - \frac{E}{RT_i}$$

where

$$\ln \xi_{KAS} = {p(x_i)}/{p(x)}$$

p(x) is taken from the Coats and Redfern approximation. xi represents the E/RT at Ti the average temperature over the transformation.

3.1.5.5 Vyazovkin (VYZ) Method

VYZ method is the integral form of the Arrhenius equation. The activation energy can be determined at any particular degree of conversion by finding the value of for which the objective function is minimized (VyazovkinandDollimore1996).

$$\Omega = \sum_{i=1}^{n} \sum_{\substack{j=1 \\ j \neq i}}^{n} \frac{I(E_a, T_{\alpha i})\beta_j}{I(E_a, T_{\alpha j})\beta_i}$$

with

$$I(E_a, T_{\alpha i}) = \int_0^{T_{\alpha i}} \exp\left(\frac{-E_a}{RT}\right) dT$$

3.1.5.6 Advanced Vyazovkin(Ad-VYZ) Method

Vyazovkin (Vyazovkin 2001) proposed an advanced non-linear method which overcomes the minor errors in the activation energy obtained from Vyazovkin method. By using the following equation

$$\sum_{i=1}^{n}\sum_{j\neq 1}^{n}\frac{I[E_a, T_{\alpha i}]\beta_j}{I[E_a, T_{\alpha j}]\beta_i}$$

With

$$I[E_a, T_{\alpha i}] = \int_{\alpha - d\alpha}^{\alpha} \exp\left(\frac{-E_a}{RT}\right)dT$$

3.2 Results and Discussion

3.2.1 *FTIR Studies*

The functional groups and the corresponding wave number (cm^{-1}) for the compound HA-DDS was tabulated in Table 3.1. The presence of absorption bands at 674 and 686 cm^{-1} in the FTIR spectra of hemiaminal compound is attributed to the C-O-H bending vibration. The presence of absorption bands between 720 to 830 cm^{-1} are due to substituted benzene. The absorption band noted at the region 1040 to 1280 cm^{-1} is assigned to C-N stretching. The bands noted at around 1500 and 3200 cm^{-1} are associated to the -NH deformation. An absorption band at 1600 cm^{-1} is due to the N-methylpyrrolidone present in the material.

The appearance of a broad band at around 3300 cm^{-1} is due to the H-bonded -OH group stretching. The absorption bands at 1142 and 1328 cm^{-1} are assigned to $-SO_2-$

Table 3.1 FTIR studies: Functional groups identified and the corresponding wave numbers for the hemiaminal prepared using 4,4'-diaminodiphenylsulphone

Functional group	Wave number (cm^{-1})
	HA-DDS
C-O-H bending vibration	686
Disubstituted benzene	722, 827
Carbon ring in cyclic compound	997
C-N stretch	1101,1282
NH deformation	1513,1594
N-methylpyrrolidone	1635
H-Bonded -OH stretching	3100–3500

group present in HA-DDS. Hence the presence of absorption bands pertaining to NMP and the identification of -OH stretching confirms the proposed structure for the hemiaminal compound. The absence of absorption band around 3300 cm^{-1} in the FTIR spectra of PHT-DDS indicates the conversion of the hemiaminal to the hexahydrotriazine form through cyclization reaction. The absorption bands at 1142 and 1328 cm^{-1} are assigned to the sulphone group (-SO_2-) present in PHT-DDS.

3.2.2 *DSC and TG-DTG Studies of Hemiaminal(Monomer)*

The DSC curve recorded at a heating rate of 10 °C min^{-1} in an inert atmosphere (nitrogen, flow rate = 50 mL min^{-1}) for HA-DDS and TG-DTG (monomer) curves for the material was presented in Fig. 3.2.The hemiaminal synthesized using diaminodiphenylsulphone (HA-DDS) shows a broad complex exotherm (Fig. 3.2 DSC) in the temperature region 224 to 339 °C with several maxima at 255, 282 and 310 °C. The total enthalpy associated with the curing exotherm is 64 Jg^{-1}. From TG-DTG monomer curve, it is obvious that the small amount of material starts to lose mass from 99 to 258 °C and the amount of mass loss during this region is less than 10%. The major mass loss of 63% is shown in the temperature region around 258 to 507 °C.

The degradation is found to be maximum at 388 °C. The thermal degradation further proceeds very slowly up to 750 °C and the amount of mass lost during this final phase of degradation is 7%. The conversion of hemiaminal to hexahydrotriazine is proceeding with the elimination of NMP and water from hemiaminal. The mass loss noted below 200 °C may be safely attributed to the loss of NMP and water. Since the hemiaminal show major mass loss after 200 °C, the exotherm seen in the hemiaminal after 200 °C may be associated to the degradation of the hexahydrotriazine produced.

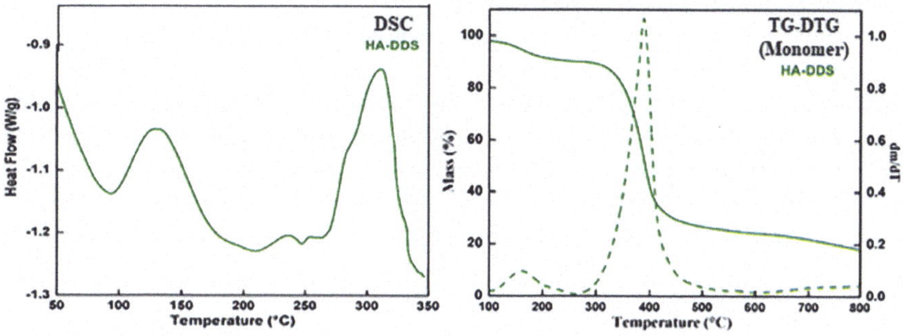

Fig. 3.2 DSC and TG-DTG (Monomer) traces of HA-DDS at a heating rate 10 °C min^{-1}

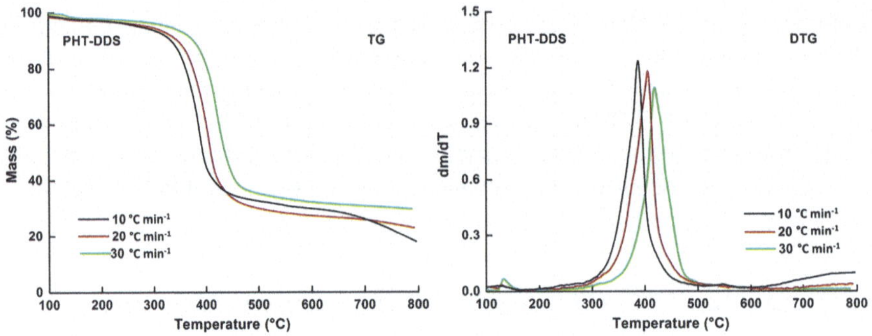

Fig. 3.3 TG and DTG curve for the sample PHT-DDS at different heating rates

Table 3.2 TG studies of different polyhexahydrotriazine at different heating rates ($\beta = 10$, 20 and 30 °C min^{-1})

Sample	Heating rates (°C min^{-1})	Onset Ts (°C)	Maximum Tmax(°C)			End set Te (°C)	Char residue (%) at 750 °C
			I	II	III		
PHT-DDS	10	275	–	388	–	485	22
	20	278	–	405	–	495	25
	30	282	–	419	–	515	30

3.2.3 *TG Studies of Polyhexahydrotriazine*

The TG and DTG curves recorded at different heating rates (β = 10, 20 and 30 °C min^{-1}) in nitrogen atmosphere for the material PHT-DDS shown in Fig. 3.3. The parameters such as onset of degradation (Ts), degradation maximum (Tmax), endset of degradation (Te) and char residue (%) at 750 °C obtained from the TG and DTG curves are tabulated in Table 3.2. The detailed observations of the thermogravimetric data obtained at a heating rate of 10 °C min^{-1} made for the sample PHT-DDS was discussed below. The PHT-DDS starts to degrade nearly at 275 °C and the degradation is over approximately 485 °C. The DTG curves for PHT-DDS show a prominent single degradation phenomenon. The amount of char residue noted at 750 °C for the material PHT-DDS was 22%. The bridging unit(-SO_2-)havethe electron withdrawing group in nature, so it is favor for the stabilization of the cross-linked network.

3.2.4 *Kinetics of Thermal Degradation*

The apparent activation energy for the cured material, PHT-DDS was calculated using seven isoconversional model free kinetic methods such as FWO, C-FWO, KAS, C-KAS, VYZ, and A-VYZ. The apparent activation energy values for the

Table 3.3 Kinetics of thermal degradation: The apparent activation energies for the thermal degradation of different polyhexahydrotriazines at different reaction extents

Sample	Methods	Ea-D value (kJ mol^{-1}) at different reaction extent (α)												
		0.2	0.25	0.3	0.35	0.4	0.45	0.5	0.55	0.6	0.65	0.7	0.75	0.8
PHT-DDS	FWO	87	89	88	92	95	98	100	101	102	104	118	142	175
	C-FWO	92	89	91	94	96	99	101	102	103	111	135	156	175
	KAS	81	82	84	86	89	92	94	95	96	98	112	138	173
	C-KAS	95	83	85	87	90	93	95	95	97	106	135	157	173
	VYZ	82	83	85	87	90	93	95	96	97	99	113	139	150
	A-VYZ	92	83	84	86	89	90	92	95	90	100	111	134	175

material obtained by these seven methods are tabulated in Table 3.3. Even though the apparent activation energy values calculated for the degradation of polymerized material by these methods are different, the fashions noted in the variation of apparent activation energy for the material obtained by the kinetic methods are same and the values are in good agreement. The slight deviation noted in the apparent activation energy values between these methods is due to the different approximation used in the kinetic methods. The apparent activation energies for thermal degradation (Ea-D kJ mol^{-1}) obtained using A-VYZ method for the material was considered for the discussion.

The material shows the constant increase in apparent activation energy (Ea-D) values is noted for the reaction extent levels 0.25 to 0.6 and then sudden increase is noted for further reaction extent levels. As evidenced by the activation energy values one can conclude that depending upon the nature of the swivel group present in between the aromatic nucleus decides the Ea-D value. Detailed use of hyphenated techniques like TG-FTIR and TG-MS are necessary to pinpoint the degradation products and associate it with the Ea-D values. From this further work it will be possible to provide detailed degradation mechanism for this family of compounds.

3.3 Conclusions

Using aromatic diamine and paraformaldehyde, the hemiaminal was synthesized. Synthesized material was thermally converted to polyhexahydrotriazine. Both the hemiaminal and the polyhexahydrotriazine were characterized for their structure using FTIR. The presence of the solvent N-methylpyrrolidone and the inter molecular hydrogen bonded structure is ascertained in the case of hemiaminal. The elimination of NMP and water from the hemiaminal is noted in the conversion from hemiaminal to polyhexahydrotriazine. The DSC and TG studies of the hemiaminal confirm the mass loss from room temperature till 200 °C. After 200 °C all the hemiaminal undergo exothermic degradation. The structure of the aromatic part of the hemiaminal dictates the amount of heat released during the thermal degradation of the hemiaminal.

Acknowledgements The authors would like to acknowledge the Management and the Principal of Kamaraj College of Engineering and Technology, S.P.G.C. Nagar, K. Vellakulam—625 701, India for providing all the facilities to do this work.

Declaration of Competing Interest The authors declare that they have no known competing financial interests or personal relationships that could have appeared to influence the work reported in this paper.

Funding This research did not receive any specific grant from fundingagencies in the public, commercial, or not-for-profit sectors.

References

Chen JS, Ober CK, Poliks MD (2002) Characterization of thermally reworkable thermosets: materials for environmentally friendly processing and reuse. Polymer 43:131–139

Coats AW, Redfern JP (1964) Kinetic parameters from thermogravimetric data. Nature 201:68–69

Farjas J, Roura P (2011) Isoconversional analysis of solid state transformations: A critical review. Part I. Single step transformations with constant activation energy. J Therm Anal Calorim 105:757–766

Hedrick JL, Horn HW, Jones GO, Brien JM, Virwani KR (2015) Methods of preparing polyhemiaminals and polyhexahydrotriazines. US Patent US 20150104579A1

Jeannette M, Gavin O, Virwani K, Mccloskey D, Boday J, Gijs M, Horn W, Coady J, Bintaleb M, Alabdulrahman MS, Alsewailem F, Almegren AA, Hedrick L (2014) Recyclable, strong thermosets and organogels via paraformaldehyde condensation with diamines. Science 344:732–735

ZI Khan, ZB Mohamad, Rahmat ARB, Abdullah NASB (2021) A novel recycled polyethylene terephthalate/polyamide 11 (rPET/PA11) thermoplastic blends. Prog Rubber Plast Recycl Technol 37:1–2

Kissinger HE (1957) Reaction kinetics in differential thermal analysis. Anal Chem 29:1702–1706

Lin SC, Pearce EM (1994) High performances thermosets: chemistry, properties, applications, 1st edn. Carl HanserVerlag, Munich Vienna New York

Vyazovkin S, Dollimore D (1996) Linear and nonlinear procedures in isoconversional computations of the activation energy of nonisothermal reactions in solids. J Chem Inf Comput Sci 36:42–45

Vyazovkin S (2001) Modification of the integral isoconversional method to account for variation in the activation energy. J Comput Chem 22:178–183

Yei DR, Fu HK, Chen WY, Chang FC (2006) Synthesis of a novel benzoxazine monomer-intercalated montmorillonite and the curing kinetics of polybenzoxazine/clay hybrid nanocomposites. J Polym Sci B Polym Phys 44:347–358

Chapter 4
Role of Single Cell Proteins in Bioremediation

Abdullah Al Noor and Anshika

Abstract Bioremediation is an eco-friendly and cost-effective process which utilizes microorganisms such as bacteria, fungi, and algae to convert pollutants into less toxic or non-toxic substances. This chapter provides an in-depth exploration of the role of single-cell proteins (SCPs) in bioremediation, focusing on their potential applications in the degradation and transformation of harmful pollutants. SCPs derived from various microorganisms have shown promise in bioremediation due to their ability to metabolize organic compounds and heavy metals. The book chapter discusses the different types of SCPs and their unique strengths and specializations in bioremediation. It also highlights the factors influencing the efficiency of bioremediation processes, including the type and concentration of pollutants, the presence of suitable microorganisms, and environmental conditions. The economic challenges associated with SCP-based bioremediation, such as production costs, scale-up challenges, waste management costs, and regulatory compliance, are also addressed. Technological advancements, including genetic engineering, omics technologies, nanotechnology, bioreactor technology, bioinformatics, and waste valorization technologies, are also explored for their potential to enhance the role of SCPs in bioremediation. The chapter further provides a comprehensive overview of the role of SCPs in bioremediation, emphasizing their potential as a sustainable solution for environmental cleanup, while also highlighting the future research directions and challenges in the field. Finally, it aims to contribute to the scientific understanding and advancement of bioremediation strategies using SCPs, providing valuable insights for researchers, practitioners, and policymakers in the field of environmental sciences and biotechnology.

Keywords Single cell proteins · SCPs · Bioremediation · Pollutants · Microorganisms · Genetic engineering · Omics technologies · Nanotechnology · Bioreactor technology · Waste valorization · Environmental impact

---

A. Al Noor (✉) · Anshika
Department of Molecular Biology and Genetic Engineering, School of Bioengineering and Biosciences, Lovely Professional University, Jalandhar, Punjab, India
e-mail: abdullahaljamiu3@gmail.com

A. Kuanar et al. (eds.), *Recent Advances in Bioremediation and Phytoremediation*, Environmental Science and Engineering,
https://doi.org/10.1007/978-3-031-77884-1_4

4.1 Introduction

4.1.1 Definition of Key Terms

- **Single-cell proteins (SCPs)**: Single cell protein (SCP) refers to an edible microbial biomass that is derived from single-celled organisms such as bacteria, yeasts, fungi, or algae. It serves as a protein-rich food source and can be produced using various methods, including fermentation or cultivation in bioreactors.
- **Bioremediation**: Bioremediation is a process that utilizes living organisms (such as bacteria, fungi, or plants) or their enzymes to degrade, transform, or remove pollutants from contaminated environments. It is an environmentally friendly approach used to mitigate the impact of pollutants in soil, water, and air. The organisms used in bioremediation can break down or detoxify substances such as petroleum hydrocarbons, heavy metals, pesticides, solvents, and other harmful chemicals. Bioremediation can occur naturally or be enhanced through the addition of specific microorganisms or nutrients to facilitate the degradation process. It is a cost-effective and sustainable method for cleaning up contaminated sites and restoring them to a more natural state.

4.1.2 Overview of Bioremediation

Bioremediation is a process that utilizes microorganisms, including bacteria, fungi, and algae, to degrade or transform harmful pollutants into less toxic or non-toxic substances. This eco-friendly and cost-effective method is widely used for the treatment of various environmental pollutants in soil, water, and air (Tyagi et al. 2011).

There are two primary types of bioremediation: in situ and ex situ. In situ bioremediation treats the contaminated material at the site, while ex situ involves the removal of the contaminated material to be treated elsewhere. The choice between the two depends on the nature and location of the contamination (Hassanshahian 2014).

Bioremediation strategies can be further classified into two categories: natural attenuation and bioaugmentation. Natural attenuation relies on the naturally occurring microorganisms at the site, while bioaugmentation involves the addition of specific strains of microorganisms that can degrade the pollutants (Megharaj et al. 2011).

The efficiency of bioremediation processes can be influenced by several factors, including the type and concentration of the pollutant, the presence of suitable microorganisms, and the environmental conditions (pH, temperature, oxygen availability, etc.) (Singh et al. 2017a, b) (Figs. 4.1 and 4.2).

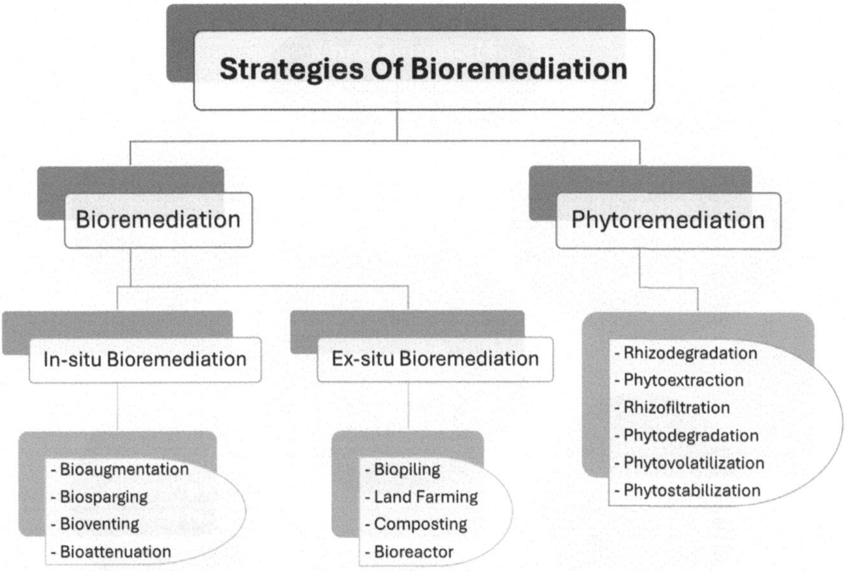

Fig. 4.1 Strategies of bioremediation

4.1.3 Importance of Single Cell Proteins

Single-cell proteins (SCPs) refer to proteins that are derived from microorganisms like bacteria, fungi, algae, and yeasts. These proteins have garnered attention from researchers and scientists in recent times due to their potential applications in different fields such as food and feed production, biofuel production, and environmental cleanup (Ritala et al. 2017). SCPs are not only rich in proteins in the dietary sense, but also contains carbohydrates, lipids, minerals and water, and even nucleic acids (Audu 2023).

In Bioremediation, the role of SCPs is also crucial because they possess the ability to break down or transform a range of pollutants or contaminants. These protein molecules can metabolize organic compounds like hydrocarbons polychlorinated biphenyls (PCBs) and heavy metals into less harmful substances. This makes SCPs an effective tool for treating forms of environmental pollution (Kumar et al. 2020a, b).

Furthermore, SCPs can be produced on a scale using waste materials making them a sustainable solution for environmental cleanup. Additionally, these protein molecules can be genetically engineered to enhance their capabilities in pollutants. This further increases their potential in bioremediation applications (Becker 2013).

Moreover, there is evidence supporting the advantages of SCPs over traditional remediation methods that position them as key players in bioremediation. These advantages include costs, minimal impact, on the environment, and the ability to treat diverse types of pollutants. Also, SCPs can be used in both in-situ and ex-situ

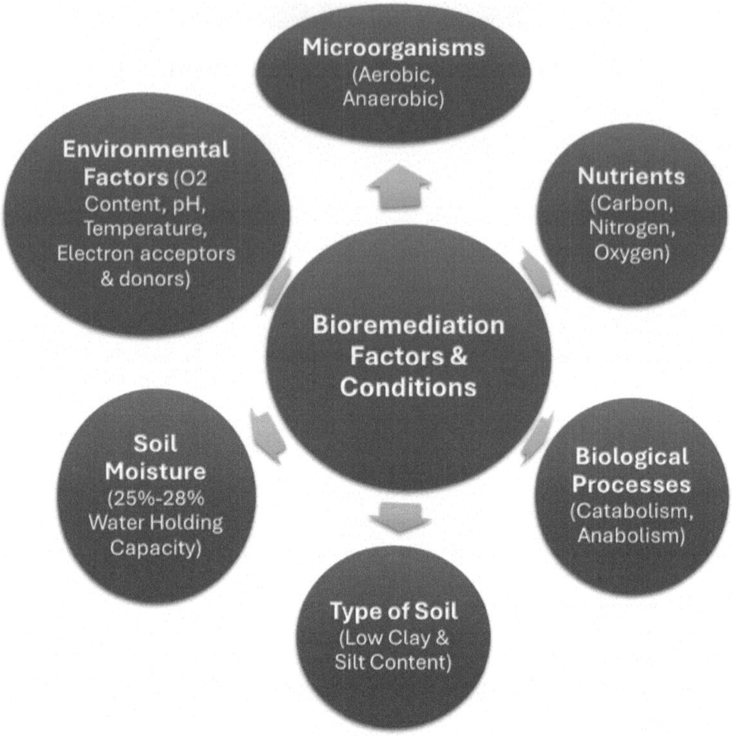

Fig. 4.2 Bioremediation factors and conditions

bioremediation strategies, providing flexibility in their application (Anastopoulos and Kyzas 2015).

4.2 Single Cell Proteins: An Overview

4.2.1 Definition and Characteristics

Single Cell Proteins (SCPs) are a fascinating group of proteins that are derived from microorganisms such as bacteria, fungi, algae, and yeasts. The term "single cell protein" might sound a bit technical, but it simply refers to the fact that these proteins are produced by individual, or single, cells of the microorganism (Ritala et al. 2017).

One of the most striking characteristics of SCPs is their high protein content. In fact, about 50–80% of their dry weight is pure protein. This, along with the presence of essential amino acids, vitamins, and minerals, makes SCPs a powerhouse of nutrition (Becker 2013).

Another fascinating aspect of SCPs is their adaptability. They can grow rapidly on a variety of substrates, including waste materials. This means they can turn waste into valuable protein, making them a sustainable and cost-effective source of nutrition. Moreover, unlike traditional agriculture, the production of SCPs doesn't compete for land or other resources, making them a friend to the environment (Kumar et al. 2020a, b).

In the context of bioremediation, SCPs are like little environmental superheroes. They have the ability to degrade or transform a wide range of pollutants. They can metabolize complex organic compounds, including hydrocarbons, polychlorinated biphenyls (PCBs), and heavy metals, into less harmful substances. This makes them an effective tool for cleaning up various types of environmental pollution (Anastopoulos and Kyzas 2015).

4.2.2 Production and Sources of Single Cell Proteins

Imagine a world where we could turn waste into valuable nutrition. That's exactly what happens when we produce Single Cell Proteins (SCPs). These proteins are derived from microorganisms like bacteria, fungi, algae, and yeasts, which are cultivated on various substrates, including agricultural waste, industrial waste, and even wastewater (Kumar et al. 2020a, b; Matassa et al. 2016) (Fig. 4.3).

The choice of substrate is like choosing the right soil for a plant. It can significantly influence the yield and quality of the SCPs. For example, substrates rich in carbohydrates can promote the rapid growth of microorganisms, leading to a higher yield of SCPs. On the other hand, substrates with a high content of specific nutrients

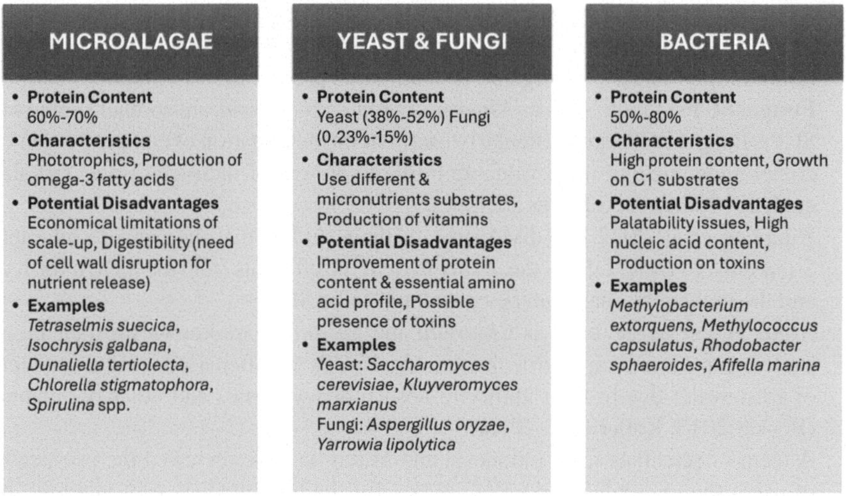

Fig. 4.3 Sources of SCPs

can enhance the nutritional value of the SCPs, much like a well-fertilized garden (Ritala et al. 2017; Saini and Keum 2018).

The production of SCPs can be carried out in different types of bioreactors, which are like the greenhouses of the microbial world. These include stirred tank reactors, airlift reactors, and fluidized bed reactors. The choice of bioreactor depends on several factors, including the type of microorganism, the nature of the substrate, and the desired characteristics of the SCPs (Becker 2013; Koller et al. 2016).

The sources of SCPs are the microorganisms themselves. Different types of microorganisms produce different types of SCPs. For instance, bacteria such as *Pseudomonas, Bacillus, and Escherichia coli* are known to produce SCPs with a high protein content. Fungi such as Aspergillus and yeast such as Saccharomyces cerevisiae are also commonly used for the production of SCPs, showcasing the diversity of nature's microscopic factories (Anastopoulos and Kyzas 2015; Ogunbanwo et al. 2019).

4.2.3 Types of Single Cell Proteins

Single Cell Proteins (SCPs) can be derived from a variety of microorganisms, each offering unique characteristics and potential applications. Here are some of the main types of SCPs:

1. **Bacterial SCPs:** Certain bacteria, such as *Pseudomonas*, *Bacillus*, and *Escherichia coli*, are known for their ability to produce SCPs with a high protein content. These bacterial SCPs are often used in bioremediation due to their ability to degrade a wide range of pollutants (Kumar et al. 2020a, b; Matassa et al. 2016). In the year 2022, Khumchai et al. had devised a novel strategy to treat the effluents produced by the paper and pulp industry. The 30 days long process makes use of the *Streptomyces tuirus* OS1, a multi-metal tolerant bacterial species, which fascinatingly reduces the organic contaminants present in PPI effluents.
2. **Fungal SCPs:** Fungi, such as *Aspergillus* and *Trichoderma*, are another source of SCPs. Fungal SCPs are particularly useful in the degradation of complex organic compounds, making them valuable in the treatment of industrial waste (Ritala et al. 2017; Saini and Keum 2018).
   Anupong et al. 2022 had discovered in their studies that the cyanide tolerant *Streptomyces tritici* D5 has the ability to treat the effluents from the Sago industry and the yielded biomass can be considered to be SCP.
3. **Algal SCPs:** Algae, such as *Chlorella* and *Spirulina*, are known for their high protein content and rapid growth rate. Algal SCPs are often used in the treatment of wastewater due to their ability to absorb heavy metals and other pollutants (Becker 2013; Koller et al. 2016).
   A team of scientists from Indonesia and Malaysia have reviewed the prospects of using the microalga *Chlorella vulgaris*for the bioremediation of mercury in wastewater. Not only that, but the derived biomass from the treatment can also

actually be converted into biodiesel, a sustainable alternative to chemical fuels (Kusuma et al. 2024).

4. **Yeast SCPs:** Yeasts, such as *Saccharomyces cerevisiae* and *Candida utilis*, are commonly used in the production of SCPs. Yeast SCPs are known for their high nutritional value and are often used in the food and feed industry (Anastopoulos and Kyzas 2015; Ogunbanwo et al. 2019).
   On that note, a group of scientists from Tunisia led by Afef Gamraoui, have documented a fascinating experiment in which they have coupled musical sounds with the bioremediation of wastewater from seafood canning using the yeast *Yarrowialipolytica.* Their findings suggest that musical sounds can actually stimulate the growth of Y. *lipolytica*, which in turn, enhances its efficiency in remedying the seafood canning wastewater (Gamraoui et al. 2023).

Each type of SCP has its unique advantages and potential applications. The choice of SCP often depends on the specific requirements of the application, including the type of pollutant, the environmental conditions, and the desired outcome of the bioremediation process (Ghosh and Das 2019).

4.3 Role of Single Cell Proteins in Bioremediation

4.3.1 Mechanisms of Action

Single Cell Proteins (SCPs) are like tiny environmental superheroes, each with their own special powers to combat pollution. Here's how they work:

1. **Biosorption:** SCPs can act like sponges, soaking up pollutants and holding them on their cell surface. It's like they're giving pollutants a big bear hug, keeping them from causing harm. This process, known as biosorption, is often used to remove heavy metals from wastewater (Anastopoulos and Kyzas 2015; Volesky 2010; Vijayaraghavan and Yun 2008).
   Environmental scientists have been considering biosorption techniques for years now in order to remedy the heavy metal pollution, an aspect where *Saccharomyces cerevisiae* has high potential. A team led by S Forough in 2022 utilized S. *cerevisiae*for zinc biosorption from industrial effluents and was quite successful (Forough et al. 2022).
2. **Biodegradation:** SCPs can also act like tiny recycling machines, breaking down pollutants and transforming them into less harmful substances. It's as if they're taking a complex puzzle and rearranging the pieces to create a beautiful picture. This process, known as biodegradation, is particularly useful in the treatment of organic pollutants such as hydrocarbons and polychlorinated biphenyls (PCBs) (Kumar et al. 2020a, b; Singh and Walker 2006; Haritash and Kaushik 2009).

The previously mentioned novel strategy of treating PPI effluents innovated by Khumchai et al. in 2022 is an example of the utilization of biodegradation technique using bacterial SCPs.

3. **Bioconversion:** Some SCPs can even turn pollutants into useful products. For instance, some SCPs can convert carbon dioxide into biomass, helping to mitigate climate change. It's like they're turning trash into treasure! (Ritala et al. 2017; Matassa et al. 2016; Acién et al. 2012).
   In a recent study by Rosario et al. the potential of enriched phototrophic purple bacteria (PPB), particularly *Rhodobacter*sp. and *Rhodopseudomonas* sp. was documentedfor the purpose of bioconversion of H_2 into single cell proteins (Rosario Rodero et al. 2024). The experiment yielded marvelous results.
4. **Bioaccumulation:** SCPs can act like tiny storage units, taking in pollutants and storing them within their cells, effectively removing them from the environment. It's as if they're tidying up the environment, one pollutant at a time. This process, known as bioaccumulation, is often used in the treatment of heavy metal pollution (Becker 2013; Koller et al. 2016; Gavrilescu 2004).
   To cite an example, BökeÖzkoç and his team published a study in 2023 on the bioremediation of copper using *Bacillus* sp. and *Streptomyces griseus.* The study found that the quantity of bioaccumulation increased with the increasing concentration of copper in case of both the species.
5. **Biomineralization:** Some SCPs can turn heavy metals into insoluble minerals, reducing their bioavailability and toxicity. It's like they're turning dangerous weapons into harmless stones. This process, known as biomineralization, is particularly useful in the treatment of heavy metal pollution (Saini and Keum 2018; Ghosh and Das 2019; Lloyd 2003).
   Recently, a team of scientists biomineralized contaminating materials into non-toxic inorganic substances in their novel method of bioremediation using nanoparticles (Yilmaz-Sercinoglu et al. 2023).
6. **Biostimulation:** SCPs can also act like cheerleaders, enhancing the bioremediation process by stimulating the growth and activity of other beneficial microorganisms. It's as if they're encouraging their teammates to work harder and perform better. This process, known as biostimulation, is often used in the treatment of oil spills (Ogunbanwo et al. 2019; Das and Chandran 2011; Atlas 2005).
   CD Rawat and his team, in their paper, discusses the use of this mechanism by SCPs in the remediation of contaminated environments (Rawat et al. 2023).

Each type of SCP has its unique advantages and potential applications. The choice of SCP often depends on the specific requirements of the bioremediation process, including the type of pollutant, the environmental conditions, and the desired outcome of the treatment (Volesky 2010; Singh and Walker 2006; Fomina and Gadd 2014).

4.3.2 *Specific Roles of Different Single Cell Proteins*

Different types of Single Cell Proteins (SCPs) play unique roles in bioremediation, each with their own strengths and specializations:

1. **Bacterial SCPs**: Bacterial-derived SCPs are regularly concerned with the biodegradation of pollution. This is because of the capacity of bacteria to supply enzymes that can break down complex natural compounds into less difficult and less toxic compounds. Additionally, they perform a role within the biosorption of heavy metals, where cell biomass binds and accumulates heavy metals from the environment, thereby contributing to the elimination of these pollutants in water or soil.
   In a perspective, bacteria like *Pseudomonas* and *Bacillus*may be termed as the superheroes of the SCP world since they can degrade a wide range of pollutants, including hydrocarbons and heavy metals. As such, they are often made use of to clean up industrial waste and oil spills (Kumar et al. 2020a, b; Das and Chandran 2011; Haritash and Kaushik 2009; Arulazhagan and Vasudevan 2011).
2. **Fungal SCPs**: SCPs in fungi and yeasts are powerful in breaking down complicated pollution such as hydrocarbons and xenobiotic compounds because of their complex enzymatic structures. In addition, fungal mycelium can bind heavy metals or radionuclides, contributing to their internalization and balance.
   Fungi such as *Aspergillus* and *Trichoderma* are the puzzle solvers of the SCP world. They excel at breaking down complex organic compounds, including lignin and cellulose, hence their popularity in the treatment of agricultural and PPI wastes (Ritala et al. 2017; Saini and Keum 2018; Singh and Walker 2006; Bugg et al. 2011).
3. **Algal SCPs**: Algal SCPs make contributions to phytoremediation methods where they are able to adsorb and acquire heavy metals and different pollutants, lowering their tiers inside the environment. Furthermore, they effectively capture the excess nutrients from water which prevents processes like eutrophication that results into harmful algal blooms.
   For example, algae like *Chlorella* and *Spirulina*can be called the sponges of the SCP world. They are known for their ability to absorb heavy metals and other pollutants. This is why, these algal species are often used to clean up wastewater and to restore aquatic environments (Becker 2013; Koller et al. 2016; Anastopoulos and Kyzas 2015; Mata et al. 2010).
4. **Yeast SCPs**: Yeast SCPs are experts in causing bioaccumulation of heavy metals from solutions, a property that can be well utilized in the bioremediation of contaminated water sources.
   Moreover, species such as *Saccharomyces cerevisiae* and *Candida utilis* act like the metabolizers of the SCP world. They can effectively break down a variety of organic compounds, including sugars and alcohols, which is why they are widely used to treat waste from the food and beverage industry (Ogunbanwo et al. 2019; Matassa et al. 2016; Ghosh and Das 2019; Papanikolaou and Aggelis 2011).

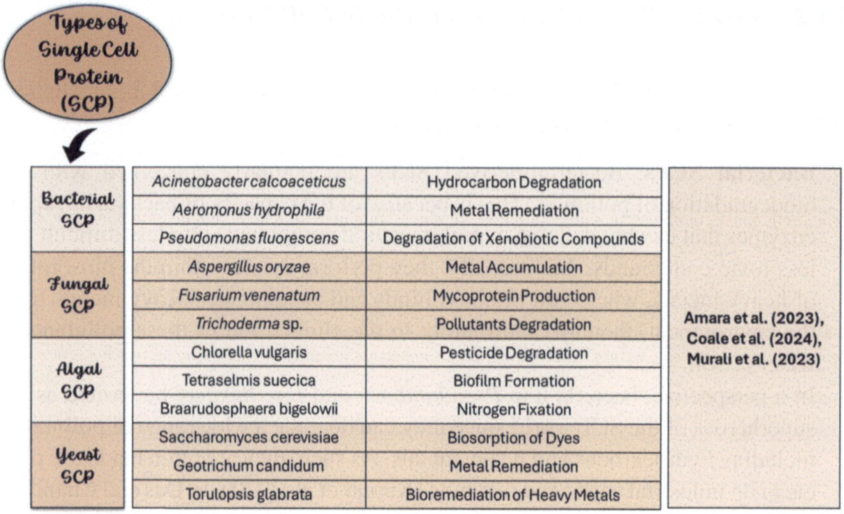

Fig. 4.4 Specific roles of different SCPs in bioremediation

The following chart illustrates a brief insight into the specific roles of different types of SCPs from particular species of organisms in bioremediation (Fig. 4.4).

Each type of SCP has its unique advantages and potential applications. The choice of SCP often depends on the specific requirements of the bioremediation process, including the type of pollutant, the environmental conditions, and the desired outcome of the treatment (Volesky 2010; Fomina and Gadd 2014; Vijayaraghavan and Yun 2008; Gadd 2010; Gavrilescu 2010).

4.3.3 Case Studies of Single Cell Proteins in Bioremediation

Single-cell proteins (SCPs) are emerging as a powerful tool in the field of bioremediation. These microscopic organisms, including bacteria, fungi, algae, and yeast, have shown remarkable abilities to degrade or transform various types of waste and pollutants (Kumar et al. 2020a, b; Ritala et al. 2017; Becker 2013; Ogunbanwo et al. 2019).

Several case studies highlight the role of Single Cell Proteins (SCPs) in bioremediation:

1. **Bacterial SCPs in Oil Spill Cleanup**: Bacteria such as *Alcanivoraxborkumensis* can degrade hydrocarbon molecules, which are major components of oil (Hassanshahian et al. 2020). As such, these bacteria can be used to break down the oil into more harmful components to bioremediate the affected area in case of an oil spill and SCPs can then be harvested from these cultures as a by-product of the filtration process.

A study by Al Disi et al. (2020) demonstrated the potential of bacterial SCPs in the context of oil cleanup where they *Pseudomonas aeruginosa*, a strain of bacteria that demonstrated a remarkable ability to break down crude oil components, thus aiding in the cleanup process.

2. **Fungal SCPs in Agricultural Waste Treatment**: Agricultural waste like lignocellulosic waste, poses a significant environmental challenge. This can be addressed with fungi such as *Aspergillus* and *Trichoderma* the reason being they are adept at decomposing complex plant materials found in agricultural waste. Not only that, but they can also convert lignocellulosic biomass into simpler sugars, which can then be fermented into fungal SCPs (Dhillon et al. 2013). This not only results in valuable protein production but also contributes to waste reduction.

   More recently in 2020, Koutrotsios and his associates explored fungal SCPs in dealing with agricultural waste. They employed *Pleurotus ostreatus*, a fungal strain capable of degrading this waste and, interestingly, producing edible mushrooms in the process.

3. **Algal SCPs in Wastewater Treatment:** Wastewater often contains heavy metals and other pollutants that are harmful to humans and the environment. It has been observed that this issue can be prevented by using algae such as *Chlorella* and *Spirulina*, as these bacteria can actually pull nitrogen, phosphorus and other nutrients from wastewater through a wellknown process called phycomediation (Olguín 2003). Along with it, this process generates algal biomass as a by-product rich in protein which can then be used as SCPs, and the clean water thus obtained can either be disposed of or recycled.

   Moreover, in a recent study by Wang et al. (2021), it was demonstrated algal SCPs, especially that of *Chlorella vulgaris* can actually be used in wastewater treatment since it was able to absorb the contaminants present in the wastewater, which effectively purified the water.

4. **Yeast SCPs in Food and Beverage Industry Waste Treatment**: The food and beverage industry produces a large amount of waste in each processing cycle. This is a concerning pollution but fascinatingly, such waste can be effectively treated using yeast SCPs.

   Recently, Pacheco et al. (2020) used a strain of *Saccharomyces cerevisiae*, which metabolized organic compounds in the waste and produced bioethanol, a renewable energy source.

   In addition, the yeast *Saccharomyces cerevisiae* is a traditionally used strain in the fermentation of sugars in food and beverage waste streams (Mussatto et al. 2007). This fermentation can generate yeast biomass (SCP) and also treat the waste by reducing its organic load. The biomass obtained can then be processed and used as a protein supplement in various industries.

5. **Bacterial SCPs in Plastic Waste Degradation**: Plastic waste, especially that of polyethylene, is a major environmental concern at the present age. In doing research to understand the possible solutions to this problem, certain bacteria have been found to utilize plastics as a carbon source, effectively breaking them down.

For instance, *Pseudomonas* species produce enzymes which are capable of degrading polyethylene terephthalate (PET) (Yoshida et al. 2016). This finding begets the idea if it could be cultivated on a mass scale, with SCPs extracted from the biomass.
More recently, Danso et al. (2019) demonstrated the use of bacterial SCPs, specifically that of *Pseudomonas citronellolis*, in degrading this type of plastic, offering a promising solution to plastic pollution.
6. **Fungal SCPs in Textile Dye Degradation**: Textile dyes, such as the azo dye Orange G, can be very harmful to the environment, and hence bioremediation of such dyes is required. This can be achieved by using certain species of fungi.
For instance, species like *Phanerochaetechrysosporium* can effectively break down the complex dye molecules which are found in textile effluents (Wesenberg et al. 2003). This biodegradation not only cleans the wastewater but also allows for the production of fungal SCPs from the resulting fungal growth.
In recent years, research works have been done extensively in order to incorporate fungal SCPs more efficiently in remedying such waste. In citing an example, Rawat et al. (2021) showed how fungal SCPs, specifically that of *Aspergillus niger*, can degrade these dyes, reducing their environmental impact.
7. **Algal SCPs in CO_2 Sequestration**: Algal SCPs can also play a role in mitigating climate change. For example, microalgae can photosynthesize CO_2 either from the atmosphere or from industrial emissions and incorporate it into their biomass (Singh et al. 2011). Thus, the cultivation of microalgae not only helps in CO_2 sequestration but also in producing nutrient-rich SCPs, which can either be used for consumption or as feed.
Furthermore, Liu et al. in the year of 2021, demonstrated successfully how a strain of *Chlorella vulgaris* could absorb CO_2, which is a major greenhouse gas, and produce biomass and lipid during the process. This was not only a desirable, but also a highly beneficial unexpected outcome.
8. **Yeast SCPs in Biofuel Production**: Finally, on a surprising note, yeast SCPs can actually contribute to renewable energy production.
To exemplify, *Saccharomyces* species and other yeasts can ferment sugars derived from agricultural wastes into a biofuel called bioethanol (Sánchez and Cardona 2008). The yeast biomass generated during this process can be harvested as SCPs. These SCPs, a sustainable source of protein also contribute to renewable energy production in this way.
Lian et al. (2019) further displayed how a strain of *Saccharomyces cerevisiae* could metabolize xylose, a sugar derived from plant biomass, in order to produce ethanol.

In conclusion, these case studies underscore the versatility and potential of SCPs in not only bioremediation, but also in meeting the sustainable development goals through alleviating our need for sustainable energy sources. The choice of SCP often hinges on the specific requirements of the bioremediation process, including the type of pollutant, the environmental conditions, and the desired outcome of the treatment (Kumar et al. 2020a, b; Ritala et al. 2017; Becker 2013; Ogunbanwo et al. 2019). As research continues, the role of SCPs in bioremediation is likely to expand, offering new solutions to our environmental challenges.

4.4 Applications of Single Cell Proteins in Bioremediation

4.4.1 Soil Bioremediation

One of the most significant roles SCPs play is in the treatment of soil contaminated with petroleum. Picture this: SCPs, derived from food waste, are introduced to soil tainted with polycyclic aromatic hydrocarbons (PAHs), a group of chemicals found in petroleum. The result? A significant reduction in PAH concentrations, as demonstrated by a study conducted by Wang and his team in 2020 (Fig. 4.5).

In a larger-scale study, Zhang and his colleagues in 2019 used SCPs to treat petroleum-contaminated soil. They observed a substantial decrease in petroleum hydrocarbon levels, further supporting the potential of SCPs in speeding up the degradation of petroleum contaminants.

Johnson and his team in 2020 also explored the use of SCPs for treating petroleum-contaminated soil. Their findings emphasized the potential of SCPs in reducing petroleum hydrocarbon levels, highlighting their role in tackling soil pollution caused by petroleum products.

But the capabilities of SCPs don't stop at petroleum. They also show promise in treating soil contaminated with heavy metals. Chen and his team in 2020 found that SCPs enhanced microbial activity and metal uptake, leading to a significant reduction in heavy metal concentrations in the soil.

Zhang and his team in 2017 delved deeper into the impact of SCPs on soil microbial communities when used for the bioremediation of petroleum-contaminated soil. Their findings suggested that SCPs not only improved the degradation of petroleum

SCP Type	Bioremediation Type
Algae (*Chlorella, Spirulina*)	Heavy Metal Contamination
Bacteria (*Pseudomonas, Bacillus, Clostridium*)	Organic Pollutants
Yeasts (*Saccharomyces cerevisiae, Candida tropicalis*)	Hydrocarbon Contamination
Fungi (*Pleurotus ostreatus, Trichoderma*)	Pesticide Contamination

Fig. 4.5 Applications of SCPs in bioremediation

contaminants but also influenced the composition and diversity of soil microbial communities.

Similar studies by Wang and his team in 2018 and Li and his team in 2017 further highlighted the potential of SCPs in enhancing the degradation of petroleum contaminants and improving the overall efficiency of remediation processes.

In conclusion, these studies collectively demonstrate the potential of SCPs in soil bioremediation, particularly in the treatment of petroleum hydrocarbons and heavy metal contaminants. The application of SCPs has shown promising results in enhancing the biodegradation capabilities of microorganisms and improving the overall efficiency of soil remediation processes. With the help of these mighty microscopic helpers, the future of bioremediation looks brighter.

4.4.2 Water Bioremediation

Single cell proteins (SCPs) have been recognized for their potential in bioremediation, particularly in the context of water treatment.

1. **Phycoremediation**: Phycoremediation is a process that uses algae to remove pollutants from wastewater. Algae, as a type of SCP, can absorb and metabolize various pollutants, including heavy metals, organic compounds, and nutrients like nitrogen and phosphorus. This process is cost-effective, eco-friendly, and can also lead to the production of biofuels, making it a sustainable option for water bioremediation (Bhatnagar et al. 2020).
2. **Bacterial SCPs in Bioremediation**: Bacterial SCPs, such as those from the Pseudomonas and Bacillus genera, have been used to degrade or detoxify pollutants in water. These bacteria can break down a wide range of pollutants, including petroleum hydrocarbons, pesticides, and heavy metals (Hassanshahian et al. 2020).
3. **Fungal SCPs in Bioremediation**: Fungi, another type of SCP, have been used in mycoremediation. Certain fungi, such as white-rot fungi, are capable of degrading complex organic pollutants, including polycyclic aromatic hydrocarbons (PAHs), endocrine-disrupting chemicals (EDCs), and pharmaceuticals (Kumar et al. 2020a, b).
4. **Yeast SCPs in Bioremediation**: Yeast, particularly Saccharomyces cerevisiae, has been used to absorb heavy metals from wastewater. The yeast cells can bind to heavy metals, reducing their concentration in the water (Ksheminska et al. 2020).
5. **Microalgae in Heavy Metal Removal**: Microalgae have been recognized for their ability to remove heavy metals from wastewater. They can accumulate these metals in their cells, reducing their concentration in the water (Rangsayatorn et al. 2020).

6. **Bacterial SCPs in Oil Spill Cleanup**: Bacterial SCPs have been used in the cleanup of oil spills. Certain bacteria can degrade the hydrocarbons in oil, reducing their environmental impact (Al-Sayegh et al. 2020).
7. **Fungal SCPs in Dye Removal**: Fungi have been used to remove dyes from wastewater. Certain fungi can degrade these dyes, reducing their concentration in the water (Rawat et al. 2016).
8. **Yeast SCPs in Radioactive Waste Treatment**: Yeast, particularly Saccharomyces cerevisiae, has been used to absorb radioactive isotopes from wastewater. The yeast cells can bind to these isotopes, reducing their concentration in the water (Brady et al. 2011).

In conclusion, SCPs offer a promising, sustainable, and cost-effective solution for water bioremediation. They can be used to treat a wide range of pollutants, and their use can also lead to the production of valuable by-products, such as biofuels.

4.4.3 Air Bioremediation

Single cell proteins (SCPs) have shown potential in air bioremediation, particularly in the degradation of airborne pollutants. Here are some applications:

1. **Microbial Degradation of Volatile Organic Compounds (VOCs)**
   Certain bacteria and fungi, as SCPs, can degrade VOCs, which are a major class of air pollutants. These microorganisms can metabolize VOCs, reducing their concentration in the air (Lebrero et al. 2016).
2. **Biofiltration of Air Pollutants**
   Biofiltration is a process that uses microorganisms to remove pollutants from air. Bacteria, fungi, and algae, as SCPs, can be used in biofilters to degrade or absorb pollutants, including VOCs, nitrogen oxides, and sulfur compounds (Estrada et al. 2013).
3. **Microbial Degradation of Greenhouse Gases**
   Certain bacteria and algae can degrade greenhouse gases, such as methane and carbon dioxide. These SCPs can metabolize these gases, reducing their concentration in the atmosphere and mitigating climate change (Strong et al. 2015).
4. **Microbial Degradation of Odorous Compounds**
   Certain bacteria and fungi can degrade odorous compounds, such as hydrogen sulfide and ammonia. These SCPs can metabolize these compounds, reducing their concentration in the air and improving air quality (Moe et al. 2012).
5. **Microbial Degradation of Particulate Matter**
   Certain bacteria and fungi can degrade particulate matter, a major air pollutant. These SCPs can metabolize or absorb particulate matter, reducing its concentration in the air (Wang et al. 2020).

6. **Microbial Degradation of Ozone**
   Certain bacteria and fungi can degrade ozone, a harmful air pollutant and greenhouse gas. These SCPs can metabolize ozone, reducing its concentration in the atmosphere (Zhou et al. 2018).
7. **Microbial Degradation of Nitrogen Oxides**
   Certain bacteria and fungi can degrade nitrogen oxides, a class of harmful air pollutants and greenhouse gases. These SCPs can metabolize nitrogen oxides, reducing their concentration in the atmosphere (Liu et al. 2017).
8. **Microbial Degradation of Sulfur Dioxide**
   Certain bacteria and fungi can degrade sulfur dioxide, a harmful air pollutant and greenhouse gas. These SCPs can metabolize sulfur dioxide, reducing its concentration in the atmosphere (Zhang et al. 2019).

4.4.4 Industrial Waste Bioremediation

Single cell proteins (SCPs) are microscopic life forms like bacteria, fungi, and algae that are packed with protein. They're not just good for nutrition, though. They're also incredibly useful for cleaning up our environment, especially when it comes to industrial waste. Here's how:

1. **Heavy Metal Heroes**
   Heavy metals are a big problem in industrial waste. But SCPs like certain bacteria, fungi, and algae can actually absorb and break down these metals, reducing their levels in the waste. It's like they're tiny, living, waste-treating factories (Rangsayatorn et al. 2020).
2. **Organic Pollutant Clean-up Crew**
   Organic pollutants like polycyclic aromatic hydrocarbons (PAHs) and volatile organic compounds (VOCs) are common in industrial waste. But certain bacteria and fungi can break these pollutants down, helping to clean up the waste (Lebrero et al. 2016).
3. **Radioactive Waste Wranglers**
   Some SCPs, like the bacteria Deinococcus radiodurans and the yeast Saccharomyces cerevisiae, can absorb radioactive isotopes from industrial waste. This helps to reduce the levels of these dangerous substances in the waste (Brady et al. 2011).
4. **Dye Degraders**
   Dyes are a major pollutant in waste from the textile industry. But certain bacteria and fungi can break these dyes down, helping to clean up the waste (Rawat et al. 2016).
5. **Tannery Waste Tacklers**
   Tannery waste is full of pollutants like chromium and organic compounds. But certain bacteria and fungi can break these pollutants down, helping to clean up the waste (Kumar et al. 2017).

6. **Pesticide Pulverizers**
   Pesticides are common pollutants in waste from the agricultural industry. But certain bacteria and fungi can break these pesticides down, helping to clean up the waste (Cycoń et al. 2019).
7. **Pharmaceutical Waste Fixers**
   Pharmaceutical compounds are common pollutants in waste from the pharmaceutical industry. But certain bacteria and fungi can break these compounds down, helping to clean up the waste (K'oreje et al. 2020).
8. **Oil Spill Saviors**
   Oil is a major pollutant in waste from the oil industry. But certain bacteria and fungi can break down the hydrocarbons in oil, helping to clean up the waste (Al-Sayegh et al. 2020).

The following graph depicts general contaminant removal (per day) data of each type of SCP in percentage (Fig. 4.6).

In conclusion, SCPs are like nature's little clean-up crew. They can help us tackle some of the biggest pollution problems in industrial waste, making them a key tool in our fight to protect the environment.

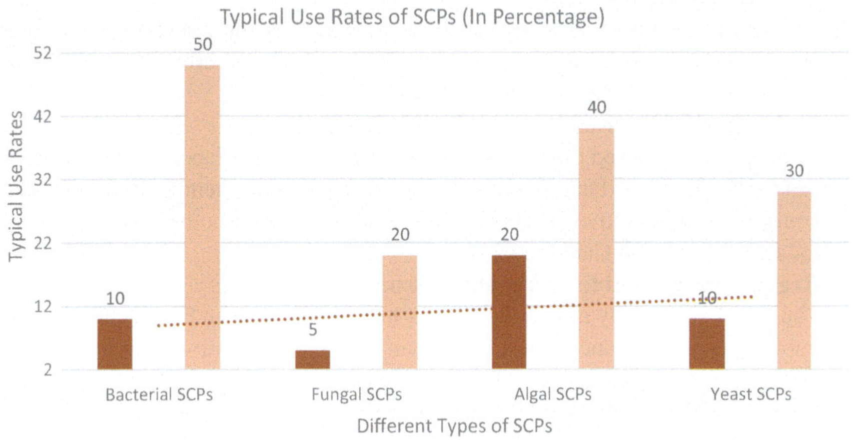

Fig. 4.6 Typical use rates of SCPs (The first column of each type of SCP indicates the lower and the second column the upper limit of the usage in percentage). This graph illustrates the efficiency of each type of SCP in removing contaminants on a daily basis

4.5 Challenges and Limitations in the Use of Single Cell Proteins for Bioremediation

4.5.1 Technical Challenges

While single cell proteins (SCPs) offer promising solutions for bioremediation, their application is not without challenges. Here are some of the technical hurdles and limitations:

1. **Survival and Activity of SCPs in Contaminated Environments**
   SCPs need to survive and remain active in contaminated environments to effectively degrade pollutants. However, high concentrations of pollutants can inhibit their growth and activity, limiting their effectiveness in bioremediation (Kuppusamy et al. 2016).
2. **Bioavailability of Pollutants**
   The bioavailability of pollutants can limit the effectiveness of SCPs in bioremediation. Some pollutants may be tightly bound to soil or sediment particles, making them less available for degradation by SCPs (Olaniran et al. 2013).
3. **Genetic Stability of SCPs**
   The genetic stability of SCPs can be a challenge in bioremediation. Genetic changes can occur during the growth of SCPs, which can affect their ability to degrade pollutants (Mrozik and Piotrowska-Seget 2018).
4. **Scale-up of SCP-based Bioremediation Processes**
   Scaling up SCP-based bioremediation processes from the laboratory to the field can be challenging. Factors such as nutrient availability, temperature, pH, and competition with native microorganisms can affect the effectiveness of SCPs in the field (Singh and Cameotra 2014).
5. **Regulatory and Public Acceptance Issues**
   The use of genetically modified SCPs for bioremediation can face regulatory and public acceptance issues. These issues can limit the application of these SCPs in the field (Bhattacharya et al. 2020).
6. **Cost of SCP Production and Application**
   The cost of producing and applying SCPs can be a challenge. While SCPs can be produced from waste materials, the cost of scaling up production and applying these SCPs in the field can be high (Rai et al. 2020).
7. **Incomplete Degradation of Pollutants**
   SCPs may not completely degrade certain pollutants, leading to the formation of potentially toxic metabolites. This can limit the effectiveness of SCPs in bioremediation (Kumar et al. 2021a, b).
8. **Lack of Knowledge on SCP-Pollutant Interactions**
   There is still a lack of knowledge on the interactions between SCPs and certain pollutants. This lack of knowledge can limit the development of effective SCP-based bioremediation strategies (Singh et al. 2017a, b).

4.5.2 Environmental Challenges

Single cell proteins (SCPs) are gaining recognition for their potential in bioremediation, a process that leverages organisms to remove or neutralize pollutants from contaminated sites. However, the application of SCPs in bioremediation is not without its challenges and limitations.

1. **Environmental Factors**: SCPs are sensitive to their surroundings. Factors such as temperature, pH, and nutrient availability can greatly influence their growth and metabolic activity. If these conditions are not optimal, the SCPs' ability to degrade pollutants can be compromised (Smith et al. 2020).
2. **Bioavailability of Pollutants**: Not all pollutants are readily accessible or available for degradation by SCPs. This bioavailability issue can limit the effectiveness of SCPs in bioremediation (Johnson and Kostal 2018).
3. **Toxicity of Pollutants**: Some pollutants can be harmful to SCPs, inhibiting their growth and metabolic activity. This toxicity can limit the use of SCPs in the bioremediation of sites contaminated with high concentrations of toxic pollutants (Chen et al. 2019).
4. **Genetic Stability**: The genetic stability of SCPs, especially genetically modified organisms used in bioremediation, is a concern. Genetic instability can lead to a loss of pollutant-degrading ability over time, reducing the effectiveness of SCPs in bioremediation (Gupta et al. 2021).
5. **Regulatory and Public Acceptance Issues**: The use of SCPs, particularly genetically modified organisms, in bioremediation can face regulatory hurdles and public acceptance issues. These challenges can delay or prevent the implementation of SCP-based bioremediation strategies (Bennett et al. 2020).
6. **Economic Feasibility**: The cost of producing and applying SCPs can be high, especially for large-scale bioremediation projects. This can limit the economic feasibility of using SCPs in bioremediation (Singh and Cameotra 2020).
7. **Management of Bioremediation**: The management of bioremediation processes, including the use of SCPs, can be complex and challenging. This includes monitoring the effectiveness of bioremediation, managing potential risks, and ensuring compliance with environmental regulations (Kumar et al. 2021a, b).
8. **Production of SCPs from Waste Resources**: The production of SCPs from waste resources can present challenges, including the need for efficient waste processing technologies and the management of potential environmental impacts (Liang et al. 2020).

4.5.3 Economic Challenges

Single cell proteins (SCPs) have shown great potential in the field of bioremediation, but their application is not without economic challenges and limitations.

1. **Production Costs**: Producing SCPs can be a costly affair. Expenses for raw materials, energy, and labor can add up, making it particularly challenging for large-scale bioremediation projects (Singh and Cameotra 2020).
2. **Scale-Up Challenges**: Taking the production of SCPs from a small laboratory scale to an industrial scale can be a daunting and expensive task. It requires larger bioreactors, more sophisticated control systems, and increased waste management (Liang et al. 2020).
3. **Waste Management Costs**: Using SCPs in bioremediation can generate waste products that need to be managed and disposed of. The costs associated with waste management can add to the overall cost of SCP-based bioremediation (Chen et al. 2019).
4. **Regulatory Compliance Costs**: Using SCPs, especially genetically modified organisms, in bioremediation can involve costs related to regulatory compliance. These can include costs for environmental impact assessments, permits, and monitoring (Bennett et al. 2020).
5. **Market Acceptance and Pricing**: Gaining market acceptance for SCPs and pricing SCP-based products can be challenging. This can limit the economic viability of SCP-based bioremediation (Kumar et al. 2021a, b).
6. **Research and Development Costs**: Developing new SCPs and bioremediation technologies can involve significant research and development costs. These costs can be a barrier to the commercialization of SCP-based bioremediation (Gupta et al. 2021).
7. **Infrastructure Costs**: The infrastructure required for the production and application of SCPs can be costly. This includes the need for production facilities, storage facilities, and application equipment (Johnson and Kostal 2018).
8. **Cost of Monitoring and Evaluation**: The effectiveness of SCP-based bioremediation needs to be monitored and evaluated, which can involve additional costs (Smith et al. 2020).

4.6 Future Perspectives and Potential Solutions

4.6.1 Technological Advancements and Their Impact

Technological advancements have significantly impacted the role of single cell proteins (SCPs) in bioremediation, enhancing their effectiveness and broadening their applications.

1. **Genetic Engineering**: Genetic engineering has enabled the modification of SCPs to enhance their pollutant-degrading abilities. This has expanded the range of pollutants that SCPs can degrade and improved the efficiency of SCP-based bioremediation (Gupta et al. 2021).

2. **Omics Technologies**: Omics technologies, including genomics, proteomics, and metabolomics, have provided insights into the metabolic pathways and mechanisms of pollutant degradation by SCPs. This has facilitated the optimization of SCP-based bioremediation strategies (Smith et al. 2020).
3. **Nanotechnology**: Nanotechnology has been used to enhance the delivery and effectiveness of SCPs in bioremediation. Nanoparticles can protect SCPs from adverse environmental conditions and improve their interaction with pollutants (Singh and Cameotra 2020).
4. **Bioreactor Technology**: Advances in bioreactor technology have improved the production and application of SCPs in bioremediation. Bioreactors can provide controlled conditions for the growth of SCPs and their interaction with pollutants (Liang et al. 2020).
5. **Bioinformatics**: Bioinformatics tools have facilitated the analysis of omics data and the prediction of SCP behavior in bioremediation. This has helped in the design of more effective SCP-based bioremediation strategies (Johnson and Kostal 2018).
6. **Waste Valorization Technologies**: Technologies for the valorization of waste resources have enabled the production of SCPs from waste materials. This has reduced the cost of SCP production and contributed to waste management (Chen et al. 2019).
7. **Sensing and Monitoring Technologies**: Advances in sensing and monitoring technologies have improved the monitoring of SCP-based bioremediation. These technologies can provide real-time data on the progress of bioremediation and the behavior of SCPs (Bennett et al. 2020).
8. **Artificial Intelligence (AI)**: AI and machine learning algorithms have been used to model and predict the performance of SCP-based bioremediation. This can facilitate the optimization of bioremediation strategies and the management of bioremediation projects (Kumar et al. 2021a, b).

4.6.2 Policy and Regulatory Considerations

The use of single cell proteins (SCPs) in bioremediation is subject to various policy and regulatory considerations. These considerations can influence the development, implementation, and effectiveness of SCP-based bioremediation strategies.

1. **Regulation of Genetically Modified Organisms**: The use of genetically modified SCPs in bioremediation is subject to regulations that vary by country. These regulations can influence the development and application of genetically modified SCPs in bioremediation (Gupta et al. 2021).
2. **Environmental Impact Assessments**: The use of SCPs in bioremediation may require environmental impact assessments to evaluate potential environmental risks. These assessments can influence the approval and implementation of SCP-based bioremediation projects (Bennett et al. 2020).

3. **Waste Management Regulations**: The production and application of SCPs in bioremediation can generate waste products that are subject to waste management regulations. These regulations can influence the cost and feasibility of SCP-based bioremediation (Chen et al. 2019).
4. **Regulation of Bioremediation Technologies**: Bioremediation technologies, including those involving SCPs, are subject to regulations that can influence their development and application. These regulations can include requirements for efficacy testing, safety testing, and monitoring (Johnson and Kostal 2018).
5. **Intellectual Property Rights**: The development of new SCPs and bioremediation technologies can involve intellectual property rights. These rights can influence the commercialization and dissemination of SCP-based bioremediation technologies (Singh and Cameotra 2020).
6. **Public Acceptance and Social Considerations**: The use of SCPs, particularly genetically modified organisms, in bioremediation can be influenced by public acceptance and social considerations. These can influence policy decisions and regulatory approvals (Smith et al. 2020).
7. **International Treaties and Agreements**: The use of SCPs in bioremediation can be influenced by international treaties and agreements, such as the Convention on Biological Diversity and the Cartagena Protocol on Biosafety. These can influence the international transfer and application of SCP-based bioremediation technologies (Kumar et al. 2021a, b).
8. **Regulation of SCP Production**: The production of SCPs from waste resources can be subject to regulations, including those related to waste valorization and the use of waste resources in production processes (Liang et al. 2020).

4.6.3 Future Research Directions

The field of single cell proteins (SCPs) in bioremediation is rapidly evolving, and there are several exciting areas that researchers could explore in the future.

1. **Boosting SCP Efficiency**: Researchers could look into ways to make SCPs even better at breaking down pollutants. This might involve tweaking their genes, adjusting their metabolic processes, or optimizing the conditions in which they grow (Gupta et al. 2021).
2. **Broadening the Scope of Pollutants**: There's potential to use SCPs to tackle a wider range of pollutants. This could involve modifying existing SCPs or discovering new ones with unique abilities to degrade pollutants (Smith et al. 2020).
3. **Improving SCP Delivery**: Future research could focus on getting SCPs to polluted sites more effectively. This might involve using nanotechnology, encapsulation technologies, or developing better methods of application (Singh and Cameotra 2020).
4. **Scaling Up SCP Production**: Researchers could work on ways to produce SCPs on a larger scale. This could involve improving bioreactor designs, developing

more efficient production processes, or using waste resources to produce SCPs (Liang et al. 2020).

5. **Understanding SCPs in the Environment**: Future research could aim to better understand how SCPs behave in the environment. This could involve using omics technologies, bioinformatics, and modeling approaches (Johnson and Kostal 2018).
6. **Developing Integrated Bioremediation Strategies**: Researchers could look at combining SCPs with other bioremediation agents or technologies. This could make bioremediation more effective and broaden its applications (Chen et al. 2019).
7. **Assessing SCP Environmental Impact**: Future research could focus on assessing the environmental impact of SCPs. This could involve looking at potential risks associated with using SCPs, particularly genetically modified ones, in bioremediation (Bennett et al. 2020).
8. **Exploring Socioeconomic Aspects of SCP-Based Bioremediation**: Researchers could explore the socioeconomic aspects of SCP-based bioremediation. This could involve looking at economic feasibility, market acceptance, and policy implications (Kumar et al. 2021a, b).

4.7 Conclusion

4.7.1 Summary of Key Findings

The role of single cell proteins (SCPs) in bioremediation has been a focus of significant research in recent years. Here, we summarize the key findings from this body of work.

1. **Efficiency of SCPs**: Research has shown that SCPs can be highly efficient in degrading pollutants, with their efficiency potentially enhanced through genetic and metabolic engineering (Gupta et al. 2021).
2. **Range of Pollutants**: SCPs have been found to degrade a wide range of pollutants, with the potential to expand this range through modification or the discovery of new SCPs (Smith et al. 2020).
3. **Delivery of SCPs**: Advances in nanotechnology and encapsulation technologies have improved the delivery of SCPs to polluted sites, enhancing their effectiveness in bioremediation (Singh and Cameotra 2020).
4. **Scaling Up SCP Production**: Research has demonstrated the feasibility of scaling up SCP production from laboratory to industrial scale, with potential improvements through the optimization of bioreactor designs and production processes (Liang et al. 2020).

5. **Behavior of SCPs in the Environment**: Omics technologies and bioinformatics have provided insights into the behavior of SCPs in the environment, facilitating the optimization of SCP-based bioremediation strategies (Johnson and Kostal 2018).
6. **Integrated Bioremediation Strategies**: The integration of SCPs with other bioremediation agents or technologies has been shown to enhance the effectiveness of bioremediation and broaden its applications (Chen et al. 2019).
7. **Environmental Impact of SCPs**: Research has highlighted the need for environmental impact assessments to evaluate potential risks associated with the use of SCPs, particularly genetically modified organisms, in bioremediation (Bennett et al. 2020).
8. **Socioeconomic Aspects of SCP-Based Bioremediation**: Studies have explored the socioeconomic aspects of SCP-based bioremediation, including economic feasibility, market acceptance, and policy implications (Kumar et al. 2021a, b).

4.7.2 Implications for the Field of Bioremediation

The research on single cell proteins (SCPs) in bioremediation has significant implications for the field. Here, we summarize these implications based on key findings from recent studies.

1. **Enhanced Bioremediation Efficiency**: The use of SCPs, particularly those enhanced through genetic and metabolic engineering, could significantly improve the efficiency of bioremediation processes (Gupta et al. 2021).
2. **Expanded Range of Treatable Pollutants**: The ability of SCPs to degrade a wide range of pollutants could expand the scope of bioremediation, allowing it to address a broader array of environmental contaminants (Smith et al. 2020).
3. **Improved Delivery Mechanisms**: Advances in the delivery of SCPs to polluted sites, such as the use of nanotechnology and encapsulation technologies, could enhance the effectiveness of bioremediation strategies (Singh and Cameotra 2020).
4. **Scalable Production**: The potential to scale up SCP production could make SCP-based bioremediation strategies more feasible and cost-effective, particularly with the optimization of bioreactor designs and production processes (Liang et al. 2020).
5. **Better Understanding of SCP Behavior**: Insights into the behavior of SCPs in the environment could facilitate the design of more effective and predictable bioremediation strategies (Johnson and Kostal 2018).
6. **Integrated Bioremediation Strategies**: The integration of SCPs with other bioremediation agents or technologies could lead to more comprehensive and effective bioremediation solutions (Chen et al. 2019).
7. **Informed Environmental Impact Assessments**: The focus on assessing the environmental impact of SCPs could lead to safer and more responsible use of

these organisms in bioremediation, particularly when it comes to genetically modified organisms (Bennett et al. 2020).

8. **Consideration of Socioeconomic Aspects**: The exploration of socioeconomic aspects of SCP-based bioremediation could lead to strategies that are not only effective but also economically feasible, socially accepted, and policy-compliant (Kumar et al. 2021a, b).

References

Acién FG, Fernández JM, Magán JJ, Molina E (2012) Production cost of a real microalgae production plant and strategies to reduce it. Biotechnol Adv 30(6):1344–1353

Al Disi Z, Jaoua S, Al-Thani D, Al-Meer S, Zouari N (2020) Efficient biodegradation of total petroleum hydrocarbons from the oily sludge of re-refinery industries by a newly isolated consortium RLZ-1. J Environ Manage 253:109741

Al-Sayegh A, Al-Wahaibi Y, Al-Bahry S, Elshafie A, Al-Bemani A, Joshi S (2020) Microbial bioremediation of petroleum hydrocarbon–contaminated marine environments. In: Recent insights in petroleum science and engineering. InTech, pp 425–448

Amara AA, El-Baky NA (2023) Fungi as a source of edible proteins and animal feed. J Fungi 9(1):73. MDPI AG. https://doi.org/10.3390/jof9010073

Amara NI, Chukwuemeka ES, Obiajulu NO, Chukwuma OJ (2023) Yeast-driven valorization of agro-industrial wastewater: an overview. Environ Monit Assess 195(10):1252

Anastopoulos I, Kyzas GZ (2015) Progress in batch biosorption of heavy metal ions onto algae. J Mol Liq 209:77–86

Anastopoulos I, Kyzas GZ (2020) Progress in batch biosorption of heavy metal ions onto algae. J Mol Liq 297:112038

Anupong W, Jutamas K, On-Uma R, Sabour A, Alshiekheid M, Karuppusamy I, Pugazhendhi A et al (2022) Sustainable bioremediation approach to treat the sago industry effluents and evaluate the possibility of yielded biomass as a single cell protein (SCP) using cyanide tolerant Streptomyces tritici D5. Chemosphere 304:135248

Arulazhagan P, Vasudevan N (2011) Role of a moderately halophilic bacterial consortium in the biodegradation of polyaromatic hydrocarbons. Mar Pollut Bull 62(3):741–747

Atlas RM (2005) Bioremediation. Appl Microbiol Biotechnol 66(5):551–564

Audu Y (2023) Production and nutrient analysis of single cell protein from agros-waste using saccharomycescerevisiae (Doctoral dissertation)

Becker EW (2013) Micro-algae as a source of protein. Biotechnol Adv 25(2):207–210

Bennett R, Phipps R, Strange A, Grey P (2020) Environmental and human health impacts of spreading oil and gas wastewater on roads. Environ Sci Technol 54(13):7905–7914

Bhatnagar A, Kumari P, Bhatnagar M (2020) Microalgal technology for sustainable wastewater treatment and bioenergy production: recent advances and future perspectives. Bioresour Technol Rep 9:100385

Bhattacharya S, Dubey A, Mehra S (2020) Microbial biosurfactants: a review on current research and future prospects. Curr Genomics 21(6):432–448

BökeÖzkoç H, Aliustaoğlu MT, Şentürk İ (2023) Bioremediation of copper with endophytic bacteria *Bacillus sp.* and *Streptomyces griseus*. J Environ Eng 149(11):04023073

Brady D, Duncan JR, Van Heerden E (2011) Bioremediation of a radionuclide (Uranium) in the presence of a heavy metal (Lead) by Saccharomyces cerevisiae. Biotechnol Bioprocess Eng 16(5):937–944

Bugg TD, Ahmad M, Hardiman EM, Singh R (2011) The emerging role for bacteria in lignin degradation and bio-product formation. Curr Opin Biotechnol 22(3):394–400

Chen M, Xu P, Zeng G, Yang C, Huang D, Zhang J (2019) Bioremediation of soils contaminated with polycyclic aromatic hydrocarbons, petroleum, pesticides, chlorophenols and heavy metals by composting: applications, microbes and future research needs. Biotechnol Adv 37(2):107–122

Coale TH, Loconte V, Turk-Kubo KA, Vanslembrouck B, Mak WKE, Cheung S, Ekman A, Chen JH, Hagino K, Takano Y, Nishimura T, Adachi M, Le Gros M, Larabell C, Zehr JP (2024) Nitrogen-fixing organelle in a marine alga. Science (New York, N.Y.) 384(6692):217–222. https://doi.org/10.1126/science.adk1075

Cycoń M, Mrozik A, Piotrowska-Seget Z (2019) Bioaugmentation as a strategy for the remediation of pesticide-polluted soil: a review. Chemosphere 220:880–894

Danso D, Chow J, Streita WR (2019) Plastics: environmental and biotechnological perspectives on microbial degradation. Appl Environ Microbiol 85(19):e01095-19

Das N, Chandran P (2011) Microbial degradation of petroleum hydrocarbon contaminants: an overview. Biotechnol Res Int

del Carmen Carranza-Méndez R, Sepúlveda-Torre L, Ramos-González R, Verma DK, Baranwal D, Aguilar CN, Chávez-González ML (2022) Agro-industrial wastes for production of single-cell protein (SCP). In: Innovations in fermentation and phytopharmaceutical technologies. Elsevier, pp 381–396. https://doi.org/10.1016/b978-0-12-821877-8.00012-9

del Rosario Rodero M, Magdalena JA, Steyer JP, Escudié R, Capson-Tojo G (2024) Potential of enriched phototrophic purple bacteria for H2 bioconversion into single cell protein. Sci Total Environ 908:168471

Dhillon GS, Brar SK, Verma M, Tyagi RD (2013) Green approach for nanoparticle biosynthesis by fungi: current trends and applications. Crit Rev Biotechnol 33(4):349–363

Estrada JM, Kraakman NJR, Lebrero R, Muñoz R (2013) A sensitivity analysis of process design parameters, commodity prices and robustness on the economics of odour abatement technologies. Biotechnol Adv 31(8):1524–1530

Fomina M, Gadd GM (2014) Biosorption: current perspectives on concept, definition and application. Biores Technol 160:3–14

Forough S, Kumarss A, Azam H, Mohaddeseh L (2022) Application of saccharomyces cerevisiae isolated from industrial effluent for zinc biosorption and zinc-enriched SCP production for human and animal. Food Sci Technol 42:e82021

Gadd GM (2010) Metals, minerals and microbes: geomicrobiology and bioremediation. Microbiology 156(3):609–643

Gamraoui A, Hamimed S, Landoulsi A, Chatti A (2023) Musico-bioremediation of seafood canning wastewater by *Yarrowialipolytica*. World J Microbiol Biotechnol 39(11):303

Gavrilescu M (2004) Removal of heavy metals from the environment by biosorption. Eng Life Sci 4(3):219–232

Ghosh A, Das A (2019) Single cell protein production and its waste water treatment by using purple non-sulphur bacteria *Rhodopseudomonas palustris*. Heliyon 5(8):e02279

Gupta A, Joia J, Sood A, Sood R, Sidhu C, Kaur G (2021) Microbes as potential tool for remediation of heavy metals: a review. J Microb Biochem Technol 13(1):1–10

Haritash AK, Kaushik CP (2009) Biodegradation aspects of polycyclic aromatic hydrocarbons (PAHs): a review. J Hazard Mater 169(1–3):1–15

Hassanshahian M (2014) The effects of crude oil on marine microbial communities in sediments from the Persian Gulf and the Caspian sea: a microcosm experiment. Mar Pollut Bull 86(1–2):62–69

Hassanshahian M, Emtiazi G, Caruso G (2020) Alcanivorax: a potential bacteria for bioremediation of oil spills. Mar Pollut Bull 161:111773

Hussain K, Haris M, Qamar H, Hussain T, Ahmad G, Ansari MS, Khan AA (2021) Bioremediation of waste gases and polluted soils. In: Microbial rejuvenation of polluted environment. Springer, Singapore, pp 111–137. https://doi.org/10.1007/978-981-15-7455-9_5

Johnson A, Kostal J (2018) Bioavailability: a key factor in contaminant risk. Environ Sci Technol 52(12):6632–6633

Khumchai J, Wongchai A, On-Uma R, Sabour A, Alshiekheid M, Narayanan M, Chi NTL et al (2022) A viable bioremediation strategy for treating paper and pulp industry effluents and assessing the prospect of resulted bacterial biomass as single cell protein (SCP) using indigenous bacterial species. Chemosphere 304:135246

Koller M, Muhr A, Braunegg G (2016) Microalgae as versatile cellular factories for valued products. Algal Res 18:296–306

K'oreje KO, Vergeynst L, Ombaka D, De Wispelaere P, Okoth M, Van Langenhove H, Demeestere K (2020) Occurrence patterns of pharmaceutical residues in wastewater, surface water and groundwater of Nairobi and Kisumu city, Kenya. Chemosphere 244:125492

Koutrotsios G, Larou E, Mountzouris KC, Zervakis GI (2020) Use of spent mushroom substrate for production of pleurotus ostreatus and pleurotus cystidiosus—biodegradation and upgrading of a major agro-industrial waste. J Clean Prod 242:118477

Ksheminska H, Fedorovych D, Babyak L, Yanovych D, Kaszycki P, Koloczek H (2020) Yeast and yeast-like microorganisms as active agents in environmental detoxification from heavy metal ions. In: Yeasts in biotechnology and human health. Springer, Cham, pp 1–30

Kumar A, Bisht BS, Joshi VD, Dhewa T (2021) Review on bioremediation of polluted environment: a management tool. Int J Environ Sci 1(1):107–123

Kumar A, Kumar A, Sharma J (2020) Fungal bioremediation of polycyclic aromatic hydrocarbons (PAHs): metabolic pathways and regulatory mechanisms. Environ Sci Pollut Res 27(19):23841–23856

Kumar A, Kumar A, Sharma J (2021) Fungal bioremediation of polycyclic aromatic hydrocarbons (PAHs): metabolic pathways and regulatory mechanisms. Environ Sci Pollut Res 28(1):1–16

Kumar A, Singh LK, Ghosh P, Verma M (2020) Single cell protein production: a review. Int J Curr Microbiol App Sci 9(1):293–303

Kumar P, Chandra R, Chowdhary P (2017) Decolourization and detoxification of synthetic molasses melanoidins by individual and mixed cultures of *Bacillus* spp. Biores Technol 235:176–184

Kuppusamy S, Thavamani P, Megharaj M, Venkateswarlu K, Lee YB, Naidu R (2016) Pyrosequencing analysis of bacterial diversity in soils contaminated long-term with PAHs and heavy metals: implications to bioremediation. J Hazard Mater 317:169–179

Kusuma HS, Illiyanasafa N, Jaya DEC, Darmokoesoemo H, Putra NR (2024) Utilization of the microalga chlorella vulgaris for mercury bioremediation from wastewater and biomass production. Sustain Chem Pharm 37:101346

Lebrero R, Rodríguez E, Martin M, García-Encina PA, Muñoz R (2016) Volatile fatty acids production from food wastes for biorefinery platforms: a review. J Chem Technol Biotechnol 91(2):487–502

Lian J, Garcia-Perez M, Coates R, Wu H, Chen S (2019) Yeast fermentation of carbohydrates in biomass hydrolysates for producing cellulosic ethanol. Biores Technol 279:197–204

Liang X, Gadd GM, Chen B (2020) Single cell protein production from waste resources: a review. Crit Rev Biotechnol 40(1):26–41

Liu J, Huang J, Jiang Y, Chen F (2021) CO_2 biofixation and lipid production with microalga chlorella sp. under different CO_2 concentrations. J Appl Phycol 33:279–288

Liu Y, Pan Y, Li J, Wang L, Tittel FK (2017) Atmospheric ammonia measurement with quartz-enhanced photoacoustic spectroscopy using a distributed feedback quantum cascade laser. Appl Phys B 123(4):144

Lloyd JR (2003) Microbial reduction of metals and radionuclides. FEMS Microbiol Rev 27(2–3):411–425

Mata TM, Martins AA, Caetano NS (2010) Microalgae for biodiesel production and other applications: a review. Renew Sustain Energy Rev 14(1):217–232

Matassa S, Boon N, Pikaar I, Verstraete W (2016) Microbial protein: future sustainable food supply route with low environmental footprint. Microb Biotechnol 9(5):568–575

Matassa S, Boon N, Pikaar I, Verstraete W (2020) Can direct conversion of used nitrogen to new feed and protein help feed the world? Environ Sci Technol 54(7):4114–4122

Megharaj M, Ramakrishnan B, Venkateswarlu K, Sethunathan N, Naidu R (2011) Bioremediation approaches for organic pollutants: a critical perspective. Environ Int 37(8):1362–1375
Moe WM, Yan J, Nobu MK, You H, Ginige MP (2012) Microbial community structure and dynamics in a pilot-scale submerged membrane bioreactor treating domestic wastewater under real-time, external, selective pressure. Water Res 46(6):1873–1887
Mrozik A, Piotrowska-Seget Z (2018) Genetic diversity and functional analysis of sigma54-dependent genes in Pseudomonas putida KT2440. Microbiol Res 207:192–202
Murali Sankar P, Karthiba L, Shreedevasena S, Anantha Raju P, Vanitha S, Salama EAA, Kamalakannan A, Jeyakumar P (2023) Bacterial single cell protein: applications, productions, and commercialization: opportunities and challenges. In: Food microbiology based entrepreneurship. Springer Nature, Singapore, pp 153–172. https://doi.org/10.1007/978-981-19-5041-4_9
Mussatto SI, Fernandes M, Mancilha IM, Roberto IC (2007) Effects of medium supplementation and pH control on lactic acid production from brewer's spent grain. Biochem Eng J 34(3):179–186
Nandy S, Andraskar J, Lanjewar K, Kapley A (2021) Challenges in bioremediation: from lab to land. In: Bioremediation for environmental sustainability. Elsevier, pp 561–583. https://doi.org/10.1016/b978-0-12-820524-2.00023-7
Ogunbanwo ST, Sanni AI, Onilude AA (2019) Characterization of bacteriocin produced by lactobacillus plantarum F1 and lactobacillus brevis OG1. Afr J Biotech 2(8):219–227
Olaniran AO, Balgobind A, Pillay B (2013) Bioavailability of heavy metals in soil: impact on microbial biodegradation of organic compounds and possible improvement strategies. Int J Mol Sci 14(5):10197–10228
Olguín EJ (2003) Phycoremediation: key issues for cost-effective nutrient removal processes. Biotechnol Adv 22(1–2):81–91
Pacheco AM, Gondim DR, Gonçalves LRB (2020) CBioethanol production by saccharomyces cerevisiae, pichia stipitis and zymomonas mobilis from sago starch. Ind Crops Prod 145:112082
Papanikolaou S, Aggelis G (2011) Lipids of oleaginous yeasts part II: technology and potential applications. Eur J Lipid Sci Technol 113(8):1052–1073
Pereira AG, Fraga-Corral M, Garcia-Oliveira P, Otero P, Soria-Lopez A, Cassani L, Cao H, Xiao J, Prieto MA, Simal-Gandara J (2022) Single-cell proteins obtained by circular economy intended as a feed ingredient in aquaculture. Foods 11(18):2831. MDPI AG. https://doi.org/10.3390/foods11182831
Rai R, Keshavarz T, Roether JA (2020) Microbial production of surfactants and their commercial potential. Microbiol Mol Biol Rev 84(1):e00010-19
Rangsayatorn N, Pokethitiyook P, Upatham ES, Lanza GR (2020) Cadmium and zinc removal from water by the marine alga Caulerpa lentillifera. Water Air Soil Pollut 131(1–4):273–281
Rawat CD, Phian S, Gupta R, Verma H, Kumar M, Kaur J, Rawat VS (2023) Microbial bioprocesses in remediation of contaminated environments and resource recovery. In: Microbial bioprocesses. Academic Press, pp 225–274
Rawat D, Mishra V, Sharma RS (2016) Detoxification of azo dyes in the context of environmental processes. Chemosphere 155:591–605
Ritala A, Häkkinen ST, Toivari M, Wiebe MG (2017) Single cell protein—state-of-the-art, industrial landscape and patents 2001–2016. Front Microbiol 8:2009
Saini R, Keum YS (2018) Microbial platforms to produce commercially vital carotenoids at industrial scale: an updated review of critical issues. J Ind Microbiol Biotechnol 45(5):307–321
Saini R, Keum YS (2021) Microbial platforms to produce commercially vital carotenoids at industrial scale: an updated review of critical issues. J Ind Microbiol Biotechnol 48(1):45–64
Sánchez ÓJ, Cardona CA (2008) Trends in biotechnological production of fuel ethanol from different feedstocks. Biores Technol 99(13):5270–5295
Singh A, Cameotra SS (2014) Efficiency of lipopeptide biosurfactants in removal of petroleum hydrocarbons and heavy metals from contaminated soil. Environ Sci Pollut Res 20(10):7367–7376

Singh A, Cameotra SS (2020) Environmental applications of nanoparticles. Chemosphere 78(9):1136–1150

Singh A, Walker A (2006) Microbial degradation of organophosphorus compounds. FEMS Microbiol Rev 30(3):428–471

Singh A, Nigam PS, Murphy JD (2011) Renewable fuels from algae: an answer to debatable land based fuels. Biores Technol 102(1):10–16

Singh R, Paul D, Jain RK (2017) Biofilms: implications in bioremediation. Trends Microbiol 14(9):389–397

Singh R, Wu B, Tang X, Wang H, Gao Y (2017) Bioremediation: environmental clean-up through pathway engineering. Front Microbiol 8:2300

Smith E, Scott J, Niu X, Webster G, Singleton I, Richardson D (2020) Microbial community responses to organophosphate substrate additions in contaminated subsurface sediments. PLoS ONE 15(6):e0234896

Strong PJ, Xie S, Clarke WP (2015) Methane as a resource: can the methanotrophs add value? Environ Sci Technol 49(7):4001–4018

Tyagi M, da Fonseca MMR, de Carvalho CCCR (2011) Bioaugmentation and biostimulation strategies to improve the effectiveness of bioremediation processes. Biodegradation 22(4):231–241

Vijayaraghavan K, Yun YS (2008) Bacterial biosorbents and biosorption. Biotechnol Adv 26(3):266–291

Volesky B (2010) Biosorption and me. Water Res 44(9):2777–2788

Volesky B, Holan ZR (2020) Biosorption of heavy metals. Biotechnol Prog 11(3):235–250

Wang L, Li Y, Chen P, Min M, Chen Y, Zhu J, Ruan RR (2021) Anaerobic digested dairy manure as a nutrient supplement for cultivation of oil-rich green microalgae Chlorella sp. Biores Technol 108:124–130

Wang Q, Xie J, Zhang B, Ni L, Yang Y, Wang X, Zhang Y et al (2020) The diversity and role of bacterial community in the corrosion of cast iron pipes in a drinking water distribution system. Environ Sci: Water Res Technol 6(1):58–70

Wesenberg D, Kyriakides I, Agathos SN (2003) White-rot fungi and their enzymes for the treatment of industrial dye effluents. Biotechnol Adv 22(1–2):161–187

Yilmaz-Sercinoglu Z, Durmaz-Sam S, Ulucan-Karnak F, Kuru CI (2023) Smart nanomaterials for bioremediation and biodegradation: laboratory and on-site applications and future trends. In: Sustainable nanomaterials for biosystems engineering: trends in renewable energy, environment, and agriculture, p 341

Yoshida S, Hiraga K, Takehana T, Taniguchi I, Yamaji H, Maeda Y, Oda K et al (2016) A bacterium that degrades and assimilates poly(ethylene terephthalate). Science 351(6278):1196–1199

Zhang Y, Wang Y, Cheng H, Yao Z, He K (2019) Air pollution reduction in China: recent success but great challenge for the future. Sci Total Environ 663:329–337

Zhou X, Zhang Y, Ji M, Li X, He J (2018) Ozone pollution around a heavily trafficked suburban highway. Atmos Environ 177:70–77

Chapter 5
Green Technologies for Sustainable Wastewater Management

Sandeep Bharti, Ganesh Dattatraya Saratale, Luiz Fernando R. Ferreira, Ranyere Lucena de Souza, Sikandar I. Mulla, Abhay Raj, and Ram Naresh Bharagava

Abstract The increase in freshwater contamination due to global population growth, industrialization, emerging anthropogenic pollutants, heavy metals, increased growth of pathogenic microbes and consumption of contaminated drinking water represents a critical problem. Because of these problems, significant emphasis should be placed on the recycling of wastewater. Conventional wastewater management methods require high economic input, provide poor treatment efficiency and cause environmental disbalances by emitting different toxic pollutants. To highlight and alter these types of issues, it is necessary to continuously evaluate the economically feasible technologies and environment-friendly biological methods that complement and improvise the conventional methods. This chapter represents the classifications,

S. Bharti · R. N. Bharagava (✉)
Laboratory of Bioremediation and Metagenomics Research (LBMR), Department of Environmental Microbiology (DEM), Babasaheb Bhimrao Ambedkar University, Vidya Vihar, Raebareli Road, Lucknow, Uttar Pradesh, India
e-mail: bharagavarnbbau11@gmail.com; ramnaresh_dem@bbau.ac.in

G. D. Saratale
Department of Food Science and Biotechnology, Dongguk University, Seoul, Ilsandong-gu, Goyang-si, Gyeonggi-do, Republic of Korea

L. F. R. Ferreira
Graduate Program in Genomic Sciences and Biotechnology, Catholic University of Brasília, Brasília, Brazil

R. L. de Souza
Graduate Program in Process Engineering, Tiradentes University, Av. Murilo Dantas, Farolândia, Aracaju, SE, Brazil

Institute of Technology and Research, Av. Murilo Dantas, Farolândia, Aracaju, SE, Brazil

S. I. Mulla
Department of Biochemistry, School of Applied Sciences, REVA University, Bangalore, India

A. Raj
Environmental Microbiology Division, Council for Scientific and Industrial Research (CSIR), Indian Institute of Toxicology Research (IITR), VishVigyan Bhavan, Mahatma Gandhi Marg, Lucknow, Uttar Pradesh, India

A. Kuanar et al. (eds.), *Recent Advances in Bioremediation and Phytoremediation*, Environmental Science and Engineering,
https://doi.org/10.1007/978-3-031-77884-1_5

sources, characteristics of wastewater, and best-known green technologies available for the treatment of wastewater, as knowing more about the content and sources that discharge the hazardous wastewater may aid in developing more efficient and sustainable wastewater treatment technologies. This chapter also highlighted the different conventional wastewater treatment technologies and different green technologies, including physical, chemical, and biological treatment methods and these methods have been critically reviewed. The state-of-the-art discussed the advantages and limitations of different conventional and modern wastewater treatment techniques. Finally, it overviewed the advantages of the various green technologies used for wastewater treatment and discussed the upcoming, advanced and more promising environment-friendly technologies used for wastewater treatment.

Keywords Wastewater classification · Wastewater characteristics · Anthropogenic pollutants · Conventional technologies · Green technologies · Environment-friendly technologies

5.1 Introduction

The availability of fresh water on the surface of the earth is one of the vital problems and approximately, only one thirdof water can be utilized for drinking purposes this water is obtained from various sources like dams, rivers, lakes, ponds and, canals. Hence, these sources also have been utilized as sites to discharge polluted domestic wastewater and various industrial wastewaters (Edokpayi et al. 2017). This is because the rapid increase in population in urban areas has caused urbanization in the last few decades which has resulted in the production of a higher quantity of wastewater from both industrial and domestic wastage. The proper treatment of both municipal wastewater and industrial wastewater is necessary because the pollutants present in wastewater have various harmful effects on the health of the environment and the ecosystem (Singh 2021). Management of wastewater is an essential approach to conserving the resources of water and it is defined as the reduction, reuse and treatment of wastewater. Wastewater collection strategies and management should be important by providing facility-like infrastructure, higher treatment efficiency and economic feasibility. Treatment of wastewater by conventional methods has different approaches like physical and chemical methods which are utilizing conventional treatment technologies and directly or indirectly harm the environment. Therefore, the prediction of wastewater treatment factors in treatment plants is essential (Roozbahani 2021).

The effluent released from different sources like different industries and domestic uses normally contains much dissolved solids, suspended solids, organic molecules and various metals and non-metals impurities. Different conventional and advanced green treatment processes are utilized for the removal of colloidal particles and different anthropogenic pollutants from wastewater like membrane filtration, coagulation, precipitation, flocculation, flotation, adsorption, and various green biological technologies (Lee et al. 2014). The essential aim of wastewater treatment is to conserve the original characteristics of water and to prevent environmental and human health by disposing of wastewater after treatment. The important aim of wastewater treatment is to recycle water for irrigation and drinking purposes and to protect the world's limited water supplies in arid and semiarid areas. Wastewater was not given any special treatment in the past. Rather, wastewater was diverted from houses into waterways via gutters and canals, where it was eventually disposed of and used by humans as rivers, streams, lakes, and seas. The natural treatment due to microbial and environmental factors leads to the elimination and removal of various contaminations early age and because of the lower population, the pollutants were limited to be degraded by the available bacteria as compared to present bacteria (Edokpayi et al. 2017).

There are several sources that pollute the available water e.g. industrial waste like mining industries, pharmaceutical industries, agricultural wastewater and wastewater released from domestic uses. Factually it is considered that if water is in use it gets polluted by different sources and activities whether it is domestic or industrial. However, the pollutants produced by industries like mining, pesticide factories, and several radioactive industries are more toxic therefore, on-the-spot treatment of effluent produced by industries is important and a constant effort should be made to conserve the original characteristics of water (Crini and Lichtfouse 2019).

5.2 Wastewater

The definition of wastewater can be considered as the combination of both liquid and solid material produced by various industries, residences, institutions, and different commercial establishments together merged with groundwater and stormwater to contaminate the water resources. The effluent released in the form of wastewater contains high doses of pathogenic and disease-causing microbial agents, toxic compounds, high inorganic and organic load, excess nutrients, excess biological and chemical oxygen demands and several minerals and sediments (Rodelas 2021). Some various contaminating sources of water resources are domestic waste effluent, mines, and most toxic pollutants-producing sources are from industries. Generally, four different classifications are considered as wastewater which is wastewater originating from domestic utilities, rainwater called storm wastewater, industrial wastewater and agricultural wastewater. Industrial wastewater can further be subdivided into various types of industries. In General, the treatment of wastewater poses the greatest problems this is because controlling and modulating the different parameters and characteristics of wastewater is a tough job (Crini and Lichtfouse 2019).

5.2.1 *Classification of Wastewater*

The history of wastewater management was quite irresponsible and where fairly new to practices although the drainage system was invented before the nineteen centuries (Aderomose and Oyebode 2022). Classification of wastewaters based on different sources as described in Fig. 5.1, which release or produce wastewater is as follows:

1. Domestic wastewater: Discharge wastewater from different commercial institutes, households and residencies
2. Industrial wastewater: industries like tannery, dairy, electronics and several other manufacturing industries utilise raw materials to manufacture products and release different types of hazardous waste merged with water to produce industrial wastewater.
3. Strom wastewater: a rainwater runoff resulting from floods due to heavy rainfall.
4. Agricultural wastewater: water containing different pesticides, chemical fertilizers and anonymously stimulating chemical compounds.
5. Thermal wastewater: Any waste treatment system that processes the waste feedstock at high temperatures is referred to as thermal treatment.

Different wastewater has its unique specific characteristics and also has different environmental impacts. The effluents produced from different sources are difficult to manage and treat. Therefore, different and unique tools and techniques are utilized for managing different wastewater. Proper treatment and disposal of wastewater is necessary in order to protect or conserve the health of the environment and ecosystem.

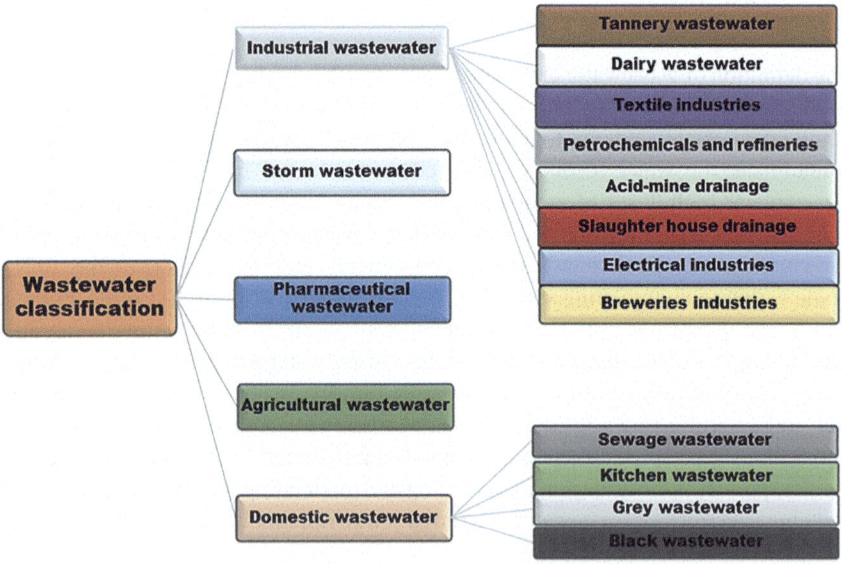

Fig. 5.1 Classification of wastewater

Standard protocols and regulations are implemented in different countries in order to manage the safer handling and treatment of wastewater effluent from domestic waste and different types of industries (Chowdhary et al. 2020).

5.2.2 *Characteristics of Wastewaters*

Wastewater has various contaminants and its characteristics will be defined on the basis of its contents and the compounds it carries as described in (Table 5.1). It is very difficult to analyse the characteristics of wastewater. However, domestic wastewater is utilized as a basis to analyse the characteristics of water and it is important that the characteristic of wastewater drastically changes with changes in sources. Wastewater characteristics are divided into biological, chemical and physical characteristics (Ali et al. 2021).Depending on the sources, wastewater characteristics may vary from industry to industry and their management techniques also vary with varieties in their contaminants. colour, taste, odour, temperature, turbidity, solids, and electrical conductivity are among the physical characteristics. Conversely, chemical parameters can include hardness, dissolved oxygen, pH, acidity, alkalinity, chlorine, and biological oxygen demand (Kordbacheh and Heidari 2023).

The pollutants can originate both from artificial as well as natural in our environment. Some of the pollutants like nitrogen, hydrocarbon, heavy metals, and some radioactive compounds are naturally occurring pollutants and chemically originated anthropogenically like various pesticides, surfactants, petrochemicals and plastics are entirely man-made (Kordbacheh and Heidari 2023). Usually, the source of water pollution is either anthropogenic or natural like storm flow or industrial and domestic wastewater. Some naturally water-polluting sources are storm water, volcanoes, dust, underground rocks etc. (Singh et al. 2020).

5.3 Objectives of Wastewater Treatment

Considering the impact on health of the environment, it is necessary to assess the objective of the treatment of wastewater. Wastewater treatment should have several objectives as it should have economic feasibility and eco-friendly treatment efficiency should be a must in every treatment process. With knowing the exact objective of wastewater, it will be easy to develop a treatment process which will fulfil the minimum requirement of treatment. Several objectives are there like operative cost, eco-friendly approaches, technical (microbiology-related TSS separation issues); or legal (fulfilment of effluent requirements with respect to varying pollution loads). The most important objective of wastewater treatment is to remove all unwanted and toxic contaminants from water and make water reusable resources (Flores-Alsina et al. 2010). Lots of chemicals like metal salts and flocculating compounds are used to increase the precipitation of phosphorous, and the addition of external carbon

Table 5.1 Table for the characteristics of wastewater

S. No	Physical characteristics	Chemical characteristics		Biological characteristic
		Inorganic compound	Organic compound	
1.	Total solids (TS)	Free ammonia (NH_4^+)	(CBOD5) Five-day carbonaceous biochemical oxygen demand	Coliform organisms (MNPs)
2.	Total volatile solids (TVS)	Organic nitrogen (Org N)		
3.	Total fixed solids (TFS)	Total Kjeldahl nitrogen (TKN)		
4.	Total suspended solids (TSS)	Nitrites (NO_2^-)	Ultimate carbonaceous biochemical oxygen demand (UBOD)	Toxicity (TUa and TUc) (Toxic unit acute, Toxic unit chronic)
5.	Volatile suspended solids (VSS)	Nitrates (NO_3^-)		
6.	Fixed suspended solids (FSS)	Total nitrogen (TN)		
7.	Total dissolved solids (TDS)	Inorganic phosphorous	Nitrogenous oxygen demand (BOD)	Specific microorganisms or pathogenic microbes (bacteria, algae, protozoa, helminthes, viruses)
8.	Volatile dissolved solids (VDS)	Organic phosphorous (Org P)		
9.	Total fixed dissolved solids (FDS)	Total phosphorous (TP)	Chemical oxygen demand (COD)	
10.	Particle size distribution (PSD)	pH	Total organic carbon (TOC)	
11.	Turbidity (NTU)	Alkalinity	(Specific organic compounds and classes of compounds)	
12.	Colour	Chloride (Cl^-)		
13.	Transmittance (% T)	Sulfate (SO_4^{2-})		
14.	Odour	Heavy Metals		
15.	Temperature	Specific inorganic compounds		
16.	Density	Various gases		
17.	Conductivity (EC)			

sources increases the efficiency of nitrogen removal which makes the process of wastewater treatment costly and economically ineffective.

Various physical methods like aeration, pumping, temperature controlling, grinding or mixing require power and energy which also require operating costs (Rashid et al. 2020). Problems like depletion of fossil fuels, pollution, contamination of water and water resources consumption lead to the shortage of healthy water resources. Therefore, it is necessary to develop a sustainable wastewater treatment in order to reach the required energy resources and to maintain a healthy environment. Indeed, the removal of contamination and unhealthy compounds is a crucial part of the treatment. Unfortunately, every treatment process requires high expenses, high energy intensives and toxic technologies. During the treatment process, the emission of several GHGs like methane, CO_2, nitrous oxides and several other volatile compounds are released into the atmosphere. Further, an extensive amount of important resources is found in wastewater, such as phosphate (PO_4^{3-}), ammonium (NH_4^+), and some metals which are not removed or utilized for any sustainable use (Li et al. 2014). Therefore, it is necessary to develop eco-friendly, economically feasible, and High-treatment efficiency technologies for the betterment of the future world (Rashid et al. 2020).

5.4 Impact of Wastewater on the Environment and Health

Water is necessary for humankind, wildlife and the ecosystem to maintain the equilibrium of the life cycle and the functioning of the balanced environment depends on the availability of clean water resources without anthropogenic activity and emerging contaminants. In recent years, water has been affected by several contaminants that have the potential to disturb the equilibrium of the ecosystem and have an adverse effect on the health of living beings (Varsha et al. 2022).

Wastewater-originated industries and domestic uses have the potential to pollute groundwater, soil property and even cause air pollution as shown in Fig. 5.2. In developing countries, numerous textiles industries, dairy industries and some other manufacturing factories have zero technologies to treat their wastewater before it is released into the environment. These industries contain high amounts of hazardous and carcinogenic compounds in their wastewater and several compounds released by these industries are non-biodegradable. It has been reported that around 2,000 types of different chemical compounds are found alone in textile industries and almost all of them have a serious impact on the health of the environment. These compounds enter water resources through the source of wastewater and solid wastes (Wang et al. 2022). Consequently, the chemicals present in dye are toxic, mutagenic, carcinogenic and teratogenic to aquatic beings and cause the extinction of various species of fish. The pollutants of textile and other industries are diversified containing large volumes of organic and inorganicpollutants like acids, hypochlorite of sodium, hydrogen peroxide and dying bases (Mohan and Shukla 2022).

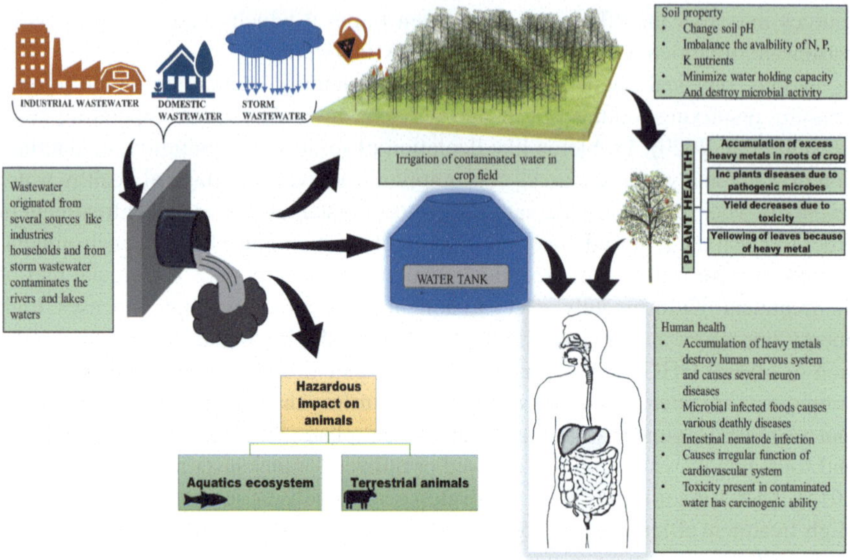

Fig. 5.2 Impact of wastewater on the environment and health

The availability of different heavy metals, biological compounds, and other ion particles likes nitrates (NO_3^-), ammonia, ammonium ions and phosphorous ions in water through discharge have been described for several decades and their impact on the health of humans and the environment are also known. But nowadays emerging contaminants like phthalates, pharmaceutical compounds, PAHs, PCBs, and bisphenol A are little known to human beings and their impact on humankind and the environment is brutal. The treatment of these compounds requires green treatment technologies and also requires specialised treatment technologies based on the contaminants. These compounds like analgesics, antibiotics, beta-blockers, anticonvulsants, anticancer agents, and several other pharmaceutical compounds are unknown and their impacts on the environment are hazardous and life threatening. The problem with this emerging contaminant is the lack of knowledge of its effect on living beings, the environment and aquatic life (Deblonde et al. 2011).

Heavy metal-laden wastewater contains abundant toxicity originating from different industries and their toxicity has various impacts on human and aquatic life. These toxic heavy metals accumulate in small aquatic animals and organisms like fishes, turtles and microbes. Moreover, the non-biodegradable nature of heavy metals makes them more hazardous because they get biomagnified and travel to higher and larger organisms including humans. The unrestricted release of wastewater from industries and mines containing hazardous heavy metals can pollute the surface water and groundwater which leads to serious damage to the environment for example water resources contaminated by iron precipitate from mine drainage leads to turbidity and minimizes the penetration of light and oxygen in the water. Iron precipitate available in irrigated water also causes damage to crops. Human exposure to heavy metals can cause serious diseases like kidney damage, liver diseases, nausea, vomiting, abdominal pain, loose motion, respiration problems, cancer, and bronchitis (Sinharoy et al. 2023). PTEs (potentially toxic elements) like cadmium, chromium, copper, nickel, arsenic, lead, zinc, etc. are known to be one of the greatest hazards to human health when it comes to environmental contamination. Parent weathering of rocks is one of the reasons that PTEs can arise naturally in the environment.

However, PTE concentrations have risen significantly as a result of human activities, including mining, industrial work, and chemical fertiliser use. These activities have brought attention to PTEs because of their continued presence in the environment, tendency to bioaccumulate in the food chain, and toxicity to humans and other organisms. The origin of the metal and its ability to react with its environment determine the geochemical interactions that a PTE establishes with the soil, which in turn determines the metal's impact on the environment (Almeida et al. 2023).

Different toxic compounds from pharmaceutical industry wastewater like estrogen and other hormonal contamination lead to abnormal development in aquatic animals and land animals like male fishes develop into females due to the presence of estrogen, which changes male traits into females causes shift in traits. Moreover, contamination from pharmaceutical waste in drinking water has the potential to mutate or cause genetic changes in new born babies and adults. The contamination of estrogen in drinking water can slacken men's fertility and also have the ability to cause serious problems or damage to organs like kidneys and liver. In women, it increases the chance of breast cancer and organ cancer (Samal et al. 2022).

Dairy industries also contain various contaminating pollutants which have an impact on living beings the effluent from industries contains casein, when casein is put through for degradation it gets converted into highly unpleasant smelling black colour waste. The black-coloured waste is sludge which is very harmful to water animals and plants when these sludges are released into water resources they cause the death of aquatic life and accumulate in water resources. The wastewater from the dairy industries blends with water and is released into freshwater resources which cause serious hormonal disorders in aquatic life and human beings. Dairy wastewater also contains a high amount of N, P, and K contamination which causes eutrophication in water (Joshiba et al. 2019).

5.5 Technologies Utilized for Wastewater Treatment

5.5.1 Conventional Treatment Methods

Removing contaminants from wastewater by utilizing conventional technologies including chemical and physical, which mainly focus on solid, organic and nutrient waste. The physical treatment of wastewater includes nanofiltration, reverse osmosis, electrodialysis (ED), forward osmosis, microfiltration, and dead-end filtration. These methods or technologies are briefly elaborated and disadvantages and advantages are also mentioned below. The chemical treatment process of wastewater includes different chemical precipitation techniques like coagulation and flocculation methods, ion exchange methods, neutralization methods, adsorption, and disinfection methods like chlorination/dichlorination, ultraviolet light and ozone (Rendón-Castrillón et al. 2023). Some chemical conventional technologies are beneficial, advantageous and very compact but very few of them are eco-friendly in nature and chemical treatment methods have several other disadvantages including cost-effectiveness, requiring excess operating energy, chemical consumption, and physicochemical monitoring of different parameters of wastewater. All these concerns have led the researchers to study and think of the most reliable technologies to overcome these types of limitations and implement more sustainable technologies (Bairagi and Ali 2020).

5.5.1.1 Physical Treatment Methods

Nanofiltration

The nanofiltration term was first utilized during the 1980s to describe the ultrafiltration methods through membranes and reverse osmosis. Nanofiltration has various approaches and different factors and parameters in wastewater treatment. The nanofiltration technology is proficient in removing colour, decreasing the volume of wastewater generated, and having the ability to recycle and recover it. The mechanism of nanofiltration depends mainly on the pore size and the rate of diffusion flow. Its membrane pores size is nominally $\sim$1 nm which makes them selectively permeate to pass even small particles through it, even tiny solutes are rejected but the electrostatic properties of the surface allow singly charged ions to be transmitted easily and with polyvalent ions are unable to transmit. These characteristics of nanofiltration technologies are really useful in the separation and selective elimination of unwanted solutes from complex process streams. It is one of the novel methods to treat wastewater containing organic and inorganic pollutants (Rendón-Castrillón et al. 2023).

Advantages and Disadvantages of Nanofiltration Technologies

1. The process is cost-effective as compared to reverse osmosis and requires less energy
2. The selective permeability of this process makes it a desirable technique to separate the required particles and compounds and to remove unwanted waste particles.
3. The most rated disadvantage of nanofiltration is that it is prone to membrane fouling because of the deposition of organic contamination on the surface of the membrane or because of biofouling.

Reverse Osmosis

Reverse osmosis is identified as hyperfiltration because it allows selective passage to liquid and most of the ionic and solute compounds are retained and it is non-porous in nature. Reverse osmosis operates in high pressure of around 20 to 100 bar and it is classified into two different types: aromatic polyamide and cellulose ester. Reverse osmosis comprises plate, frame, roll, tube and hollow fibre types. It utilizes high pressure to eliminate a variety of organic matter present in wastewater, inorganic salts and impurities.it is also utilized for the separation of organic compounds like oxygen compounds, nitrogen compounds, carbon compounds, colloidal particles with <3.10 μm, pesticides and ions (Liu et al. 2021) (Table 5.2).

Advantages and Disadvantages of Reverse Osmosis Technologies

1. Reverse osmosis is also prone to membrane fouling and has an effective cost for the treatment of wastewater.
2. This process requires excess energy and operating cost.

Forward Osmosis

It's a membrane driven process in which natural osmotic pressure is applied to separate feed solutions or different concentrated solutions through a semi-permeable membrane. It doesn't require any external hydraulic pressure and only a low-pressure apparatus is required for application which is installable and cost efficient. These membranes are made of cellulose triacetate membrane and are produced by diffusing cellulose triacetate with polyester mesh to upgrade water permeability without sacrificing the integrity of the membrane. Forward osmosis has various utility in desalination and secondary and tertiary sewage wastewater treatment. This technology is potentially effective and can be utilized as an alternative to expensive physical treatment technologies.

Advantages and disadvantages of forward osmosis methods

1. Forward osmosis is cost-efficient and easily installable. It requires low pressure to separate unwanted ions from wastewater.
2. It nearly separates all solute solids as well as suspended solids at a particular temperature.
3. The main demerit of this process is that with the reduction in concentration gradient across the membrane rejection layer, the performance of concentration polarization (CP is also reduced.
4. Small-scale dissolved solute always breaks into the feed solution because of reverse salt diffusion and it is also prone to membrane fouling like organic fouling, inorganic fouling, biofouling, and colloidal fouling.

Table 5.2 Table for the advantages and disadvantages of different types of wastewater

S. No	Technologies for wastewater-treatment	Characteristic of process	Advantages of process	Disadvantages of process
1	Physical methods	• Non-destructive separation method and semipermeable barrier to separate • Membrane technology in which osmotic pressure or applied force is applied to separate feed solutions	• Low capital cost, integrated physicochemical process, and straightforward process design • Excellent features for sludge settling and dewatering, highly effective for SS and colloidal particles • Significant decreases in the chemical and biological oxygen demands; a noteworthy decrease in the amount of total organic carbon and adsorbable organic halogen (from the pulp and paper sector); the capacity to inactivate bacteria; and a quick and effective removal process for insoluble pollutants (such pigments, etc.)	• Expensive energy demand and frequently prohibitively expensive investment expenditures for small and medium-sized businesses • There are wide variations in membrane filtration system designs, and high-concentration fouling, or quick membrane clogging, is a concern • Beads that are easily contaminated by particles and organic matter, such as oils and organics; these pollutants must be removed by a physicochemical pre-treatment, such as sand filtering or carbon adsorption • Saturation of the cationic exchanger before to the anionic resin (precipitate of metals and reactor blocking); big volume necessitates large columns and quick saturation and clogging of the reactors

(continued)

Table 5.2 (continued)

S. No	Technologies for wastewater-treatment	Characteristic of process	Advantages of process	Disadvantages of process
2	Chemical methods	• Utilization of chemicals and oxidants (e.g., O_3, Cl_2, ClO_2, H_2O_2, $KMnO_4$) • Chemically flocculation and precipitation to extract and separation of the pollutants	• A simple, quick, and effective procedure that produces ozone on-site without posing any risks related to storage • The outflow's quality (pollutants effectively destroyed and colour effectively reduced) • Effective colour and odour removal (ozonation procedure) • Effective cyanide and sulfuride elimination treatment • Initiates and increases the breaking of azo bonds (hypochlorite therapy) • Improves the product's biodegradability • High productivity, no sludge generation, potential for water recycling, and disinfection (viruses and bacteria)	• The initial expense for the effluents' transportation and storage is high expenses for operation • Dioxin and other pollutant (metals, etc.) formation • The local population has traditionally objected to the existence of plants that burn in the area • Saturation of the cationic exchanger before anionic resin (metallic precipitation and reactor blockage) • Beads that are easily contaminated by particles and organic matter, such as oils and organics; these pollutants must be removed by a physicochemical pre-treatment, such as sand filtering or carbon adsorption

(continued)

Table 5.2 (continued)

S. No	Technologies for wastewater-treatment	Characteristic of process	Advantages of process	Disadvantages of process
3	Green technology for wastewater treatment	• Use of biological (pure or mixed) organisms or the biologically manufactured compound for the management • combination utilisation of both electro-kinetics and bio-remediation processes	• Using microbes to biodegrade organic pollutants is a straightforward, financially advantageous, and widely accepted method • A wide variety of species utilised in pure cultures (white-rot fungus) or mixed cultures (consortiums) • Effectively removes biodegradable organic matter, NH3, NH4 + , and iron attenuates colour well • High removal of biochemical oxygen demand and suspended particles • Use of a range of extracellular enzymes with high biodegradability capacity and treatment efficiency • Microbiological processes will have a deciding influence on future technology employed to remove emerging pollutants from waterways	• Once a person is exposed to nanoparticles, their characteristics can occasionally change without warning • Additional specifications for water polishing technologies due to the quality of the water • Modest parameter adjustments affect the treatment process, which is a specialist procedure for pollutants • Contamination from wastewater accumulates in edible portions of crops and vegetables. Metals and other toxins are accumulated using specific and selective biological components. Handling of the accumulated material is a major issue

Electrodialysis (ED)

It is a combination of electrolytic and dialysis processes, in which wastewater is treated by membrane separation technique. The process separates anions and cations from dissolved salts in wastewater by utilising electric fields that move toward the cathode and anode. This leads to the minimization of anions and cations in the intermediate compartment and results in the separation and recovery of compounds. Anions and cations are passed through the membrane having positive and negative charges respectively. The efficiency of wastewater treatment through electrodialysis depends on the density of current, structure of the cell, flow rate, characteristic of the membrane used for ion exchange, feed water ionic concentration and pH (John et al. 2023).

Advantages and Disadvantages of Electrodialysis Treatment Methods

1. Electrodialysis consumes less energy and causes less environmental pollution
2. Operating and automation of electrodialysis is simple but electrodialysis only separates salts from wastewater and has lower desalination efficiency
3. Electrodialysis is also prone to membrane fouling

Microfiltration

This is a process of separation based on size, in this process, one size of a particle is separated from other sizes which come in a range of 0.01 μm to 20 μm. This process operates by utilizing pressure to remove suspended solids from wastewater. The Microfiltration media are obtainable with a variety of manufacturers and materials. This process depends upon the percentages of particles captured of identical size or it depends on the media's retention rate. Depending on the size of the pore, the membrane separation processes are differentiated into microfiltration, ultrafiltration (UF), conventional filtration (CF), and reverse osmosis (Gul et al. 2021).However, polymeric membranes are not suggested for use in large-scale wastewater treatment due to their limited lifespan and fouling issues.

Dead-End Filtration

Dead-end filtration is the most basic form of filtration through the membrane. The completefeed flow is pushed through the membrane which causes filtered matter to segregate on the surface of the membrane. Dead-end filtration is a batch process in which the process of filtration decreases with the accumulation of matter on the surface of the membrane. This process has effective treatment efficiency for concentrating compounds (Royce et al. 2018).

Crossflow Filtration

In this process, the feed flow is pushed through a tangential membrane. The membranes used in this process are tubular where the inside wall of a tube is the layer with membranes. The feed flows with high velocity through the membrane tubes in order to create turbulence and pressure to be utilized as a driving force to separate filtrate. Because the flow of feed flow and filtration flow are perpendicular to each other, therefore, the process is known as cross flow. It is the effect treatment technology to separate or filter liquid containing a high content of filterable matter (Ripperger and Altmann 2002).

Advantages and Disadvantages of Crossflow Filtration

Using a ceramic filter with a pore size of 0.8 μm for filtration works well to eliminate turbidity caused by suspended particles. Most of the living microbes are eliminated and recovery of enzymes is done through this process. This process is effective in producing low turbidity in water. Disadvantages of the crossflow filtration method are that it requires a high degree of filtration pressure and it is also prone to membrane fouling (Ripperger and Altmann 2002).

5.5.2 Chemical Treatment Methods

5.5.2.1 Flocculation and Coagulation Methods

Generally, the flocculation and coagulation methods are used as pre and post-methods to treat wastewater. Conventionally, aluminium, silicon, calcium and iron-based compounds were utilized for the treatment of wastewater as inorganic flocculants and coagulants to decrease the COD, total solvent, and the various contents of wastewater before further treatment. It is interesting that the process like electrocoagulation used for treating wastewater also utilizes the aluminium and iron electrodes. Water splitting, anode dissolution, and allowance of the origination of hydroxide coagulants with the application of an electric field are some general mechanisms of the electrocoagulation process. Electrodialysis (ED) could be used to precipitate and separate ions from wastewater towards anodic and cathodic zones, such as NH_4^+, SO_4^{2-}, Cr (III), etc., when ion exchange membranes were placed. For the treatment of Wastewater with better results, a combination of electrocoagulation process and electrodialysis process would be anticipated. Despite their apparent simplicity and ability to eliminate both organic and inorganic contaminants, flocculation and coagulation are unsuitable for use as the primary methods for treating wastewater. The harsh condition of wastewater and its differential properties have a large impact on the treatment process. Therefore, large quantities of chemical coagulants and flocculants are required to be added to the wastewater in order to get desirable results. However, the utilization of increased amounts of chemical coagulants and flocculants for treatment also increases the cost of treatment. The utilization of chemical coagulants and flocculants increases the toxicity of sludge produced from treatment which causes new pollutants in the environment (Zhao et al. 2022).

5.5.2.2 Ozonation Process

In ozonation method, the introduction of ozone gas into the wastewater in order to oxidation of organic compounds present in wastewater. Ozone gas has availability to oxide by bubbling ozone gas from the bottom of the tank. Ozone gas has a characteristic which makes it react directly or indirectly with organic pollutants through oxidative chain reaction to produce free radicles as mentioned in (Table 5.2). However, in the presence of water, ozone gas reacts with accessiblehydroxide ions to produce less selective and stronger hydroxyl radicles (Gligorovski et al. 2015). Recent research showed that the ozonation of pharmaceutical aromatic compounds are accessible because of the richness of electrons in compounds like ciprofloxacin, diclofenac, sulfamethoxazole, azithromycin, clarithromycin, carbamazepine, erythromycin and metoprolol (Ohoro et al. 2019). Conversely, $HO^{\bullet}$ radicals oxidise a variety of pharmaceutical compounds, even some that are resistant to ozone in alkaline conditions (such as cephalexin, primidone, loperamide, and penicillin), because of their expeditious action and non-selective nature (Ohoro et al. 2019).

Ozone gases have recently been used as a chemical treatment process to oxidise compounds present in wastewater. However, obstacles like the presence of organic carbon compounds and other oxidizable compounds make it hard to implement the ozonation process and require a greater amount of ozone gases to fulfil for ozonation processes (Tripathi and Hussain 2022).

Since ozone gas is a challenging chemical to handle during treatment process, practically all WWTPs must manufacture ozone on the treatment site. Ozone gas can be produced via a variety of methods. Because, the production of ozone gas needs high voltage electricity, transformers and power supply. Furthermore, because ozone is a toxic gas, any excess ozone that remains in the water after treatment must be eliminated. Ozone generators operating intermittently may also lessen the amount of leftover ozone while simultaneously improving the ozonation process's energy efficiency. However, an intermittent ozonation approach may result in lower rates of medicinal ingredient elimination (Ghazal et al. 2022).

5.5.2.3 Photocatalysis

Photocatalysis is an advanced oxidative process. This process shows similarities with the ozonation process in the production of hydroxyl radicles but in this process, hydroxyl radicles are produced from the interactions of water and with the surface of the florescent catalyst. The catalyst utilized in the treatment process is TiO_2 because it is chemically and biologically inert in nature and shows effective photocatalysis treated with lesser cost-effectiveness. Photocatalysis by TiO_2 is much safer to the human being as compared as compare to the ozonation process. Photocatalysis of wastewater from TiO_2 primarily focused on the formation of hydroxyl radicles (strong oxidant and non-selective radicals) which oxidized almost all contaminants present in wastewater unbiasedly. In order for photocatalysis to be competitive with other techniques, like ozonation, the other pollutants would need to be eliminated by upstream conventional wastewater treatment processes (such as the activated sludge process). This would leave photocatalysis procedures free to target the more persistent pollutants that largely escape conventional processes (Carbuloni et al. 2020). The efficiency of heterogeneous photocatalysis is dependent on factors such as irradiation, the ratio of catalyst amount to the initial concentration of the target compounds, and ambient variables like pH, in addition to matrix effects that may restrict the aspects for photocatalysis use. A catalyst load that is less than ideal may result in a restricted production of hydroxyl radicals and, as a result, insufficient elimination of the target chemicals. Conversely, a higher quantity might increase solution turbidity and hence reduce radiation transmittance. A solar-driven photocatalysis application is a more economical and energy-efficient substitute; nonetheless, the primary challenge of this process is the small overlap between the absorption spectrums of natural sunlight and the catalyst (Ghazal et al. 2022).

5.5.3 Green Technologies for Wastewater Treatment

See Fig. 5.3.

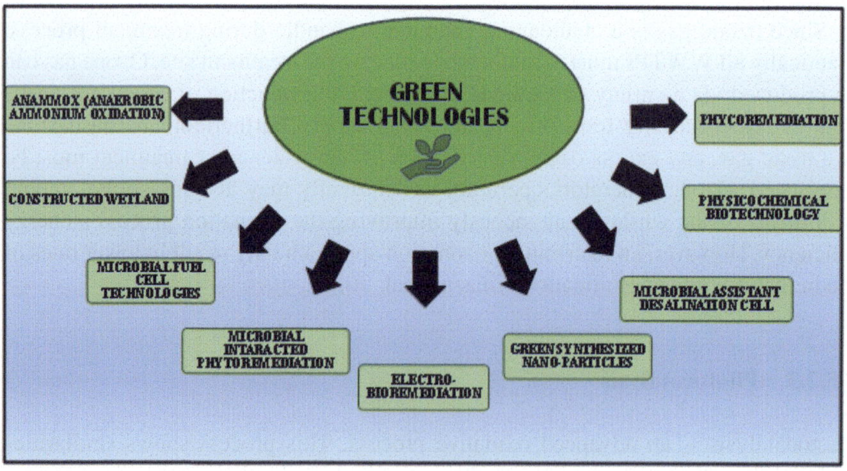

Fig. 5.3 Green technologies and their different types

5.5.3.1 Anammox (Anaerobic Ammonium Oxidation)

Decades ago, the oxidization of NH_4^+ anaerobically had several indications. In 1941 anaerobic oxidation of NH_4^+ was identified as a source of N_2 in the sea. Later, from the observation of the anoxic water it was identified that NH_4^+ present in anoxic water disappears in anoxic conditions because of the oxidation of NH_4^+ and NO_3^- present in anoxic water and produces N_2 gas and a similar phenomenon was also observed in anoxic sulphide free water from the black sea (Dalsgaard et al. 2005). In annamox, the anaerobic ammonium-oxidizing bacteria convert the present NH_4^+ to nitrogen gas with nitrite as an intermediate compound in the presence of oxygen. Nitritation–anammox process is an eco-friendly and cost-effective process and can be utilized as an alternative to other conventional physicochemical technology. Currently, this anammox treatment is only limited to the removal of ammonium compounds from wastewater. However, recent improvements in anammox technologies can be utilized in the treatment of municipal wastewater, removal of methane contamination by utilizing nitrite-reducing methane-oxidizing bacteria and utilization in the oxidation of organic compounds and other nitrogen compounds (Hu et al. 2013).

5.5.3.2 Constructed Wetlands

Constructed wetlands treatment of wastewater has been utilized as a promising technology in domestic, agriculture and industry wastewater contamination removal. The constructed wetlands focus on the leaching of contamination from landfills and storm runoff water. Contaminated wastewater is purified through wetlands naturally but in constructed wetlands, the parameters and conditions are properly regulated to maximize the efficiency of contamination removal from wastewater. Organic and suspended waste elimination increases the efficiency of treatment in constructed wetlands and by applying special media in constructed wetlands, the removal or absorption of nitrogen and phosphorous increases. The operation is cost-effective and requires less maintenance. The design of the wetland is framed in such a way that it maintains the condition of the environment in harsh conditions and even in floods. It also provides shelter and environment to wildlife animals and plants (Dinesh et al. 2020).

5.5.3.3 Microbial Fuel Cell (MFCs)

Microbial Fuel cells are an evolving biological treatment process and can be utilized in the removal of chemical oxygen demands (COD). It's a sustainable utility in eco-friendly electricity generation simultaneously with the treatment of wastewater. MFC technologies use oxidizing properties to oxidize simple carbonates into carbon dioxide and biochemically reduce them to transfer electrons. This process is similar to batteries in which anode and cathodes are differentiated and the substrates are located toward the anode side for bioconversion of the substrate. MFCs generate electrons and protons by converting substrates on the anode sides and transport these electrons and protons through resistor and proton-exchange membranes respectively where they react with oxygen and generate electricity. Sludge volume reduction, high-value product recovery, and high conversion efficiency from substrates to energy are some of the key benefits of MFC technology. MFCs still have to deal with certain challenges, such as weak power output and costly electrode and catalyst prices. Therefore, from an industrial perspective, it is imperative to increase efficiency and provide new catalysts for MFC technology (Munoz-Cupa et al. 2021).

5.5.3.4 MicrobeAssisted Phytoremediation

Microbial interacted phytoremediation is the utilization of plants with the interaction of microbial communities to clean anthropogenic pollutants from wastewater. This technique has become a promising method as it is eco-friendly and cost-effective. The efficiency of microbial-interacted phytoremediation depends on the establishment of microbial interaction with the plants. Plants support hydrocarbon-degrading microbes' communities in their rhizosphere area by providing nutrients that assist in phytoremediation in the root zone. For example, the activity of microbial communities capable of degradation of pollutants in rhizosphere areas of perennial ryegrass and alfalfa increased. In turn, the soil fertility and nutrients of soil availability to plants are enhanced. However, the toxicity present in wastewater is harmful to both plant and rhizosphere microbes although microbial interaction with plants is beneficial for the removal of contamination but the knowledge about plants and microbes is very limited and how interactions of plants and microbes are influenced by toxic pollutants which are present in different wastewater. Further research and development are required to understand the symbiotic relationship of microorganisms and plants to improve the sustainability and practicability of microbial interacted phytoremediation (Nie et al. 2011).

Microbial assist phytoremediation through plants has been examined as a medium for the removal of many anthropogenic pollutants and several other organic and inorganic pollutants including agrochemicals; polycyclic aromatic hydrocarbons, chlorinated solvents, polychlorinated biphenyls, radionuclides, munitions and heavy metals. These organic and inorganic pollutants are transported into the plants through the rhizosphere by the mass flow or diffusion methods which make this process the most amenable to phytoremediation technologies (Yang et al. 2020).

Disadvantages of Microbe Assisted Phytoremediation

1. The huge disadvantage of microbial-interacted phytoremediation is that unknown modulation and disabilities will divert the factor for plant species causing disadvantages to phytoremediation.
2. Accumulation of contaminant from wastewater in edible parts of crops and vegetables. So hyperaccumulation of toxicity in plants makes plant life shorter and biomagnification of toxicity in a food chain.

3. Due to the extreme toxicity of wastewater less biomass is obtained so, to solve this issue several plantation and harvesting is required.
4. Specific and selective plants are used for the plantation to accumulate metallic elements and disposal and handling of accumulated plants through the process of phytoremediation are some major challenges faced.
5. Chelate–enhanced phytoremediation causes environmental pollution and requires seasonal effective treatment.
6. Radionuclide mobilisation in plants via translocation is some disadvantage of these methods.

5.5.3.5 Electro-Kinetic-Bioremediation

Electro-kinetic-bioremediation is a popular hybrid technology in recent days which uses a combination of both electro-kinetics and bioremediation processes to eliminate toxicity from wastewater. Electro-kinetic phenomenon required for orientation and the acceleration of environmental pollution and microbial communities for bioremediation. This process utilises diffusion, electrolysis, electrophoresis, electroosmosis, electromigration and electric current around 0.2–2 Vcm^{-1}. Several other processes and technologies utilise similar electro-kinetic methods like electro-bio-attenuation, electro-bio-stimulation, electro-bioaugmentation, and electro-phytoremediation but hydraulic transport problems or the contaminant-microbe-(electron acceptor)/(electron donor) contact and mixing may restrict each. Electro-kinetics can be used to create alternate flow fields that will enhance the interaction between contaminants and microorganisms or to provide a necessary modification to a zone that is restricted by the surrounding environment (Saxena et al. 2020).

Disadvantages of Electro-Kinetic-Bioremediation

The oxidation process of hydrocarbon during aerobic digestion is faster as compared to the biodegradation process in the absence of oxygen. The option for the requirement of appropriate and potential material, or the significance of radius influence and limitations during advancement and scaling up, shouldn't be overlooked while conducting electro-bioremediation workability. One of the important bottlenecks of this process is that the presence of hydrocarbons is essential in the initiation of a degrading pathway (Choudhury et al. 2021).

5.5.3.6 Green Synthesized Nanoparticle Bio-Remediation

Strategies that are unable to be followed through bioremediation and physicochemical which make treatment ineffective but utilizing metals-based Nanoparticles for the remediation of wastewater, solves minute leakage of pollutants from treatment. Copper, zinc, silver, gold, etc., are used as nanoparticles in various nano-formulations and nano-composites, and are used more and more as they show effective resistance toward the uptake of toxic compounds by facilitating immobilisation in plants.

According to recent research, the biological synthesis of nanoparticles for the remediation of pollutants and toxicity from wastewater is an eco-friendly approach. It is efficient in eliminating carcinogens, pollution from heavy metals, hazardous dye, pesticide-laced water, and wastewater tainted with high levels of phosphorus, nitrogen, and hazardous algae. Plants, algae, fungi and bacteria are examples of biogenic resources that are widely exploited in the production of nanoparticles (NPs), especially metallic NPs (Singh et al. 2023).

Disadvantages of Green Synthesized Nanoparticle Bio-Remediation

Sometimes, the properties of green synthesizednanoparticles are unpredictable once they are introduced to the human. Treatment of wastewater through the utilization of green synthesizednanoparticles required to monitored or controlled by regulating various parameters like temperature, pH, reactant concentration, reaction time, metallic precursor and the extract utilization type for treatment. Preparation of green synthesizednanoparticle is costly therefore; the utilization of green synthesizednanoparticles technology is a not economical method.

5.5.3.7 Microbial Assistant Desalinisation Cell (MDC)

Microbial assistant desalination cell is a promising approach to clean up or capture chemical energy present in wastewater. it has the potential to fulfil the requirement of energy for the desalination industries and treatment of contaminated wastewater. The anode and cathode electrodes, as well as the ion-exchange membranes that divide them and cause cations to move towards the cathode and anions to migrate towards the anode, are the two essential parts of the MDC (Ping et al. 2013). We can adjust the arrangement to achieve the best desalination performance. One crucial problem in desalination work is boron removal, which can be efficiently accomplished by combining the device with Donnan Dialysis or Microbial Fuel Cell (Ping 2016).

Disadvantages of Microbial Assistant Desalinisation Cell

Further requirements of water polishing technologies as because of the water quality in terms of salinity and ion concentration it achieves irrigation standards but needs to assimilate the other essential toxic component. Another issue with the desalinated water product is its high turbidity. The back diffusion of contaminant organics, along with fouling and scaling on the membrane, collectively deteriorate the water quality, even though the anion exchange membrane, which serves as the primary barrier between wastewater and the desalinated water product, can effectively block all microorganisms in the wastewater. However, one possible method for getting rid of the organic matter traces in desalinated wastewater is filtration (Ping 2016).

5.5.3.8 Electro-Kinetic Phytoremediation

Electro-kinetic with combination phytoremediation is a sustainable and eco-friendly method to treat wastewater by utilizing electro-kinetic energy and plant accumulating nature. The extraction of contaminant from any wastewater, phytoextraction is the most beneficial and effective strategy to accumulate toxic compound. Nevertheless, these metals (loid)s may exhibit phytotoxicity and stress indicators at elevated concentrations, such as altered antioxidant enzyme and antioxidant levels, growth rate, biomass, photosynthetic pigments, and seed germination. Furthermore, the efficacy of phytoextraction is heavily dependent on the phytoavailability of heavy metals (loid), and amendments are given to enhance metal dissolution in soil (Gavrilescu 2022). As compared to other physicochemical treatment processes the process utilized for the treatment of wastewater is eco-friendly and beneficial and the management of the process is cost-effective. However, it is difficult to solve problems like hyperaccumulation with suitable accumulation of toxicity. It is tough to increase the photo-availability of soil metal to plants. Therefore, electro-kinetic phytoremediation is required to increase the availability of soil heavy metals to plants easily (Mao et al. 2015).

Electro-kinetic phytoremediation generally requires a very low amount of electric field in order to grow plants in polluted soil. the electric field increases the bioavailability of heavy metals present in the soil to plants and plants uptakes or accumulate by desorption and transportation process. The intensity of the electric field, voltage mode, electrode characteristic and electrode configurations have a conclusive influence on the effectiveness of electro-kinetic enhanced phytoremediation (Luo et al. 2018). Since significant changes in soil pH and contaminants' speciation also occur during remediation, chelating agents, complexing agents, organic agents, and fertiliser are applied in addition to other assisted amendments to improve the bioavailability of metal(loid)s, control soil pH within a favourable range, and promote plant growth (Mao et al. 2015).

5.5.3.9 Phycoremediation

Bioremediation is the process of using biological agents to restore natural assets. Strategies for phycoremediation, a type of bioremediation, address the potential use of a variety of algae to remove dangerous pollutants (such as pesticides, hydrocarbons, radioactive matter, and heavy metals) and to value-add products (such as fertilisers and biofuel) from the treated biomass. A sustainable method seems to be phycoremediation, which is the use of algae (micro or macroalgae) to remove or biotransform pollutants such as heavy metals, nutrients, and other toxins (including xenobiotics) from wastewater. Algae effectively lower the carbon footprint (the amount of carbon dioxide in the atmosphere that causes pollution, global warming, and greenhouse effects) and are a great carbon dioxide sink. As a result of their widespread distribution and excellent adaptation to a variety of environments, they are generally categorised as marine algae, microalgae, and macroalgae, all of which are commonly known as seaweeds. With an estimated 200,000–800,000 species globally, microalgae are rich in biodiversity and highly adaptable, enabling them to thrive in a variety of settings. For these reasons, they are excellent candidates for both wastewater treatment and the generation of biofuel and biofertilizers. In addition to removing nutrients, heavy metals, and toxins, biological treatment provides oxygen to heterotrophic aerobic bacteria, which helps them mineralize organic pollutants. Since the efficiency of phyco-remediation is directly related to the growth of microalgae, algal biomass can be increased by promoting microalgal development. Algal biomass is beneficial for human nutrition, aquaculture, the manufacture of biofuel, and medicines (Koul et al. 2022).

5.6 Advantages of Green Technology in Wastewater Treatment Over Conventional Methods

It is clearly known that the utilization of green technologies for the treatment of wastewater shows great potential in the removal of toxicity without producing addition pollution. Whereas, the content of wastewater is rich with nutritive compounds like ammonium, nitrate nitrogen, phosphorus, organic or inorganic carbon and several trace elements which support biological growth and biological wastewater treatment. The biological treatment of wastewater eliminates various carcinogenic compounds, heavy metals, and pesticides laced water with extreme availability of nitrogen, and phosphorous nutrients which are utilized by microbes to eliminate other toxicity from water. Algae, bacteria and fungi are some common examples which utilize inorganic and organic chemicals for their growth and remove unwanted toxic contamination, these organisms are also utilised in the production of materials like bio-nano-particles and other toxicity-removing agents (Sharma et al. 2023). Green technology utilized for wastewater treatment has a promising effect on decreasing cost and also lowering the technology requirements for the disposal of toxic sludges. The chemical and physical composition of wastewater has made to make afford by several researchers to develop sustainable wastewater treatment technology to trap the potential of wastewater in the production of eco-friendly energy, bioproducts and utilized in several other applications (Chen et al. 2016).

Green technology, also known as clean technology, refers to the development of technologies, methods, and materials that are used to protect the environment and natural resources while reducing the negative impact on human activity. When green technology meets the requirements, it does three things firstly, it reduces environmental degradation secondly, it ensures that greenhouse gas emissions are zero and that their use is safe and improves the environment to a state that is healthy and better for all living things. Lastly, it's a conserved energy and natural resource and makes better use of renewable energy sources (Yadav 2021).

5.7 Utility of Different Green Wastewater Treatment Technologies in the Production of Different Types of Bioproducts

Biologically produced products from wastewater (both municipal and industrial) have a wide range of applications. Treatment of wastewater and the value-adding of bioproducts from the concept of bioremediation, or the green treatment of wastewater, plays a significant role in the transition to the circular economy through the expansion of integrated and multifunctional methods for biomass or the manufacturing of marketable intermediates and end products. Researchers and the general public have shown an increasing interest in the revival of more value-added products and the manufacture of biologically derived chemicals (Sharma et al. 2022). Many compounds such as vitamins, polyphenols, carotenoids, flavonoids and antioxidants are present in the content of wastewater in different form and can be extracted by utilizing biological components like bacteria and algae (Safafar et al. 2015). Various bio-pesticides and bio-stimulants are produced from the biomass produced from the green treatment technologies. And activated sludge produced from dairy wastewater can be utilized in the production of protein and amino acid sources. Since the waste obtained from the treatment of wastewater by utilizing green technologies contains high organic compounds in a complex form like lipids, polysaccharides and proteins. Among all, the content of protein is around 50% of the dry weight of bacterial cells. Therefore, the sludge produced by green technologies has a significant capability to resolve the problem of the production of feed for cattle and domestic animals but toxicity like heavy metals and hazardous aromatic compounds should be prohibited and removal should be necessarily carried out (Zhang et al. 2018).

Algae and cyanobacteria-produced biomass from the treatment of wastewater can be utilized as a biofertilizer because it has the ability and content to fulfil the requirement of crop and have the ability to maintain the moisture in the soil. Algae and bacterial biomass release controlled amounts of nitrogen and phosphorous into the soil which results in the decrease in CO_2 liberation from the soil (Ammar et al. 2022).

Microbial biomass grown in wastewater treatment plants also proved to be a pertinent substrate for biogas production, this biomass can be either processed for lipid extraction (biofuel) or it can undergo anaerobic digestion to produce biogas. For biogas production, algae and oleaginous bacteria are preferred for these methods because algae and oleaginous bacteria show greater production efficiency than plants and no constraint of land or freshwater is required for the cultivation of algae and oleaginous bacteria (Cho et al. 2018).

Biological sludge from wastewater also can be utilised in the production of methane gas. Several researches have been conducted on methane yield from wastewater sludge and algal biomass (Ajeej et al. 2015). Sludge management is currently the biggest and most challenging issue in wastewater treatment facilities. Composting, incineration, and traditional stacking all have drawbacks, including high costs and poor efficiency. Studies have shown that components of polyhydroxyalkanoates

(PHAs) are present in sludge that is produced during wastewater treatment. Bacterial consortia and volatile fatty acids (VFAs) derived from surplus sludge can manufacture PHAs. Sludge could be used to produce PHAs, converting waste into goods with additional value. After that, the sludge will be cleaned and recycled, lowering the cost of producing bioplastic. Activated sludge can swiftly convert the readily degradable substrate into PHAs in the green biological treatment of sewage. The research on the production of PHAs from activated sludge has produced some results because of this property (Liu et al. 2019).

5.8 Conclusion

To date, discussions on the environmental and human risk of exposure to contaminants from the release of wastewater are still in serious debate. It is safe to say that although many studies developing toxicity tests have proven that industrial and municipal wastewater in high and low concentrations impose a risk to human health, in long-term exposure the effective risk of toxicity from waste might be life-threatening. Finally, it becomes understandable that certain measurements and precautions should be considered.

Biologically green wastewater treatment processes are a long way, and many newer and novel approaches have been developed in this connection in recent years. The selection of the green treatment process depends on a variety of factors, which include compliance with environmental discharge quality regulations. These factors help in selecting the suitable type of treatment process, whether aerobic or anaerobic. It is also observed that clubbing of artificial intelligence, typically machine learning with wastewater treatment processes would also be highly promising, and attempts in this connection must also be very much encouraged. Green Technologies forwastewater treatment can be used to achieve any level of treatment desired. Green treatment is necessary in treatment systems to remove nutrients from wastewater. Green wastewater treatment utilizes less sophisticated processes and equipment. They are relatively less expensive to run and have lesser operating costs as well as effluent quality as compared to other conventional physical and chemical treatment methods. Wastewater treatment processes require careful management to ensure the protection of the water body that receives the discharge. Trained and certified treatment plants operator should measure and monitor the incoming wastewater, the treatment process and the final effluent. The ultimate goal of wastewater treatment should be managing wastewater effectively, economically, and ecologically.

Acknowledgements The author Mr. Sandeep Bharti is highly thankful to the CSIR-UGC for providing the financial support for his research work.

References

Aderomose KS, Oyebode OJ (2022) Effective wastewater management for health and sustainable environment in Afe Babalola university (ABUAD)

Ajeej A, Thanikal JV, Narayanan CM, Kumar RS (2015) An overview of bio augmentation of methane by anaerobic co-digestion of municipal sludge along with microalgae and waste paper. Renew Sustain Energy Rev, 50, 270–276

Ali I, Naz I, Peng C, Abd-Elsalam KA, Khan ZM, Islam T, Sehar S (2021) Sources, classifications, constituents, and available treatment technologies for various types of wastewater: An overview. Aquananotechnology, 11–46

Almeida Â, Cotas J, Pereira L, Carvalho P (2023) Potential role of spirogyra sp. and chlorella sp. in bioremediation of mine drainage: a review. Phycology, 3(1), 186–201

Ammar EE, Aioub AA, Elesawy AE, Karkour AM, Mouhamed MS, Amer AA, El-Shershaby NA (2022) Algae as Bio-fertilizers: Between current situation and future prospective. Saudi J Biol Sci 29(5):3083–3096

Bairagi S, Ali SW (2020) Conventional and advanced technologies for wastewater treatment. Environ Nanotechnol Water Purif, 33–56

Carbuloni CF, Savoia JE, Santos JSP, Pereira CAA, Marques RG, Ribeiro VAS, Ferrari AM (2020) Degradation of metformin in water by TiO2–ZrO2 photocatalysis. J Environ Manag. 262, 110347. https://doi.org/10.1016/j.jenvman.2020.110347

Chen P, Xie Q, Addy M, Zhou W, Liu Y, Wang Y, Ruan (2016) Utilization of municipal solid and liquid wastes for bioenergy and bioproducts production. Biores Technol 215:163–172

Cho HU, Park JM (2018) Biodiesel production by various oleaginous microorganisms from organic wastes. Biores Technol 256:502–508

Choudhury SP, Kalamdhad AS (2021) Optimization of electrokinetic pretreatment for enhanced methane production and toxicity reduction from petroleum refinery sludge. J Environ Manage 298:113469

Chowdhary P, Bharagava RN, Mishra S, Khan N (2020) Role of industries in water scarcity and its adverse effects on environment and human health. Environ Concerns Sustain Dev: Vol 1: Air, Water Energy Resour, 235–256

Crini G, Lichtfouse E (2019) Advantages and disadvantages of techniques used for wastewater treatment. Environ Chem Lett 17:145–155

Dalsgaard T, Thamdrup B, Canfield DE (2005) Anaerobic ammonium oxidation (anammox) in the marine environment. Res Microbiol 156(4):457–464

Deblonde T, Cossu-Leguille C, Hartemann P (2011) Emerging pollutants in wastewater: a review of the literature. Int J Hyg Environ Health 214(6):442–448

Dinesh Kumar M, Gopikumar S, Uan DK, Adishkumar S, Rajesh Banu J (2020) Constructed wetlands: an emerging green technology for the treatment of industrial wastewaters. Emerg Eco-Friendly Green Technol Wastewater Treat, 21–44

Edokpayi JN, Odiyo JO, Durowoju OS (2017) Impact of wastewater on surface water quality in developing countries: a case study of South Africa. Water Quality 10(66561):10–5772

Flores-Alsina X, Gallego A, Feijoo G, Rodriguez-Roda I (2010) Multiple-objective evaluation of wastewater treatment plant control alternatives. J Environ Manage 91(5):1193–1201

Gavrilescu M (2022) Enhancing phytoremediation of soils polluted with heavy metals. Curr Opin Biotechnol 74:21–31

Ghazal H, Koumaki E, Hoslett J, Malamis S, Katsou E, Barcelo D, Jouhara H (2022) Insights into current physical, chemical and hybrid technologies used for the treatment of wastewater contaminated with pharmaceuticals. J Clean Prod 361:132079

Gligorovski S, Strekowski R, Barbati S, Vione D (2015) Environmental implications of hydroxyl radicals (• OH). Chem Rev 115(24):13051–13092

Gul A, Hruza J, Yalcinkaya F (2021) Fouling and chemical cleaning of microfiltration membranes: A mini-review. Polymers 13(6):846

Hu Z, Lotti T, van Loosdrecht M, Kartal B (2013) Nitrogen removal with the anaerobic ammonium oxidation process. Biotech Lett 35:1145–1154

John J, Gopikumar A, Gayathry P, Vincent A, Kaleekkal NJ (2023) Enhanced recovery using membrane-based zero liquid discharge systems. In Concept Zero Liq Disch (pp 129–161). Elsevier

Joshiba GJ, Kumar PS, Femina CC, Jayashree E, Racchana R, Sivanesan S (2019) Critical review on biological treatment strategies of dairy wastewater. Desalin Water Treat 160:94–109

Kordbacheh F, Heidari G (2023) Water pollutants and approaches for their removal. Mater Chem Horizs 2(2):139–153

Koul B, Sharma K, Shah MP (2022) Phycoremediation: A sustainable alternative in wastewater treatment (WWT) regime. Environ Technol Innov 25:102040

Lee CS, Robinson J, Chong MF (2014) A review on application of flocculants in wastewater treatment. Process Saf Environ Prot, 92(6), 489–508

Li WW, Yu HQ, He Z (2014) Towards sustainable wastewater treatment by using microbial fuel cells-centered technologies. Energy & Environ Sci, 7(3), 911–924)

Liu C, Takagi R, Saeki D, Cheng L, Shintani T, Yasui T, Matsuyama H (2021) Highly improved organic solvent reverse osmosis (OSRO) membrane for organic liquid mixture separation by simple heat treatment. J Membr Sci 618:118710

Liu F, Li J, Zhang XL (2019) Bioplastic production from wastewater sludge and application. In IOP conference series: earth and environmental science. 344(1), p 012071. IOP Publishing

Luo J, Ye L, Qi S, Wu J, Sophie Gu XW (2018) Effect of electrode configurations on phytoremediation efficiency and environmental risk. Plant and Soil, 424, 607–617

Mao X, Han FX, Shao X, Guo K, McComb J, Njemanze S (2015) Electro-kinetic-enhanced phytoremediation for the restoration of multi-metal (loid) contaminated soils. Int J Energy, Environ Econ 23(3):301

Mohan D, Shukla SP (2022) Hazardous consequences of textile mill effluents on soil and their remediation approaches. Clean Eng Technol 7:100434

Munoz-Cupa C, Hu Y, Xu C, Bassi A (2021) An overview of microbial fuel cell usage in wastewater treatment, resource recovery and energy production. Sci Total Environ, 754, 142429

Nie M, Wang Y, Yu J, Xiao M, Jiang L, Yang J, ...& Li, B. (2011) Understanding plant-microbe interactions for phytoremediation of petroleum-polluted soil. PLoS ONE 6(3):e17961

Ohoro CR, Adeniji AO, Okoh AI, Okoh OO (2019) Distribution and chemical analysis of pharmaceuticals and personal care products (PPCPs) in the environmental systems: A review. Int J Environ Res Public Health 16(17):3026

Ping Q (2016). Advancing microbial desalination cell towards practical applications (Doctoral dissertation, Virginia Tech

Ping Q, Cohen B, Dosoretz C, He Z (2013) Long-term investigation of fouling of cation and anion exchange membranes in microbial desalination cells. Desalination 325:48–55

Rashid SS, Liu YQ, Zhang C (2020) Upgrading a large and centralised municipal wastewater treatment plant with sequencing batch reactor technology for integrated nutrient removal and phosphorus recovery: Environmental and economic life cycle performance. Sci Total Environ 749:141465

Rendón-Castrillón L, Ramírez-Carmona M, Ocampo-López C, González-López F, Cuartas-Uribe B, Mendoza-Roca JA (2023) Treatment of water from the textile industry contaminated with indigo dye: A hybrid approach combining bioremediation and nanofiltration for sustainable reuse. Case Stud Chem Environ Eng, 100498

Ripperger S, Altmann J (2002) Crossflow microfiltration–state of the art. Sep Purif Technol 26(1):19–31

Rodelas B (2021) Nitrogen cycle in wastewater treatment plants. In Nitrogen Cycle: Ecol, Biotechnol Appl Environ Impacts (pp 144–202). CRC Press

Roozbahani A (2021) Application of bayesian networks modelling in wastewater management. In Soft Comput Tech Solid Waste Wastewater Manag (pp 111–130). Elsevier

Royce J, Liderfelt J, Robinson C (2018) Filtration methods for use in recovery processes. In Biopharm Process (pp 295–315) Elsevier

Safafar H, Van Wagenen J, Møller P, Jacobsen C (2015) Carotenoids, phenolic compounds and tocopherols contribute to the antioxidative properties of some microalgae species grown on industrial wastewater. Mar Drugs 13(12):7339–7356

Samal K, Mahapatra S, Ali MH (2022) Pharmaceutical wastewater as Emerging Contaminants (EC): Treatment technologies, impact on the environment and human health. Energy Nexus 6:100076

Saxena G, Thakur IS, Kumar V, Shah MP (2020) Electro-bioremediation of contaminants: concepts, mechanisms, applications and challenges. Combined application of physico-chemical & microbiological processes for industrial effluent treatment plant, 291–313

Sharma M, Agarwal S, Agarwal Malik R, Kumar G, Pal DB, Mandal M, …& Gupta, V. K. (2023) Recent advances in microbial engineering approaches for wastewater treatment: a review. Bioeng 14(1):2184518

Sharma R, Mishra A, Pant D, Malaviya P (2022) Recent advances in microalgae-based remediation of industrial and non-industrial wastewaters with simultaneous recovery of value-added products. Biores Technol 344:126129

Singh A (2021) A review of wastewater irrigation: Environmental implications. Resour Conserv Recycl 168:105454

Singh A, Tyagi P, Ranjan R, Sushkova SN, Minkina T, Burachevskaya M, Rajput VD (2023) Bioremediation of hazardous wastes using green synthesis of nanoparticles. Process 11(1):141

Singh J, Yadav P, Pal AK, Mishra V (2020) Water pollutants: Origin and status. SensS Water Pollut Monit: Role Mater, 5–20

Sinharoy A, Kumar M, Pakshirajan K (2023) Engineered yeast as a hyperaccumulator for heavy metal removal and recycling from waste streams. In Adv Yeast Biotechnol Biofuels Sustain (pp 503–520). Elsevier

Tripathi S, Hussain T (2022) Water and wastewater treatment through ozone-based technologies. In Dev Wastewater Treat Res Process (pp. 139–172). Elsevier.

Varsha M, Kumar PS, Rathi BS (2022) A review on recent trends in the removal of emerging contaminants from aquatic environment using low-cost adsorbents. Chemosphere, 287, 132270

Wang X, Jiang J, Gao W (2022) Reviewing textile wastewater produced by industries: Characteristics, environmental impacts, and treatment strategies. Water Sci Technol 85(7):2076–2096

Yadav A (2021) A review of green and innovative technology for a sustainable environment. Sustain Environ Clean-Up, 71–83

Yang Y, Liu Y, Li Z, Wang Z, Li C, Wei H (2020) Significance of soil microbe in microbial-assisted phytoremediation: an effective way to enhance phytoremediation of contaminated soil. Int J Environ Sci Technol 17:2477–2484

Zhang W, Alvarez-Gaitan JP, Dastyar W, Saint CP, Zhao M, Short MD (2018) Value-added products derived from waste activated sludge: a biorefinery perspective. Water 10(5):545

Zhao J, Wu Q, Tang Y, Zhou J, Guo H (2022) Tannery wastewater treatment: conventional and promising processes, an updated 20-year review. J Leather Sci Eng 4(1):10

Chapter 6
Bioremediation Approaches for Soil and Aquatic Environmental Restoration to Revive Ecosystems

Twinkle Rout, Soumyashree Rout, Audifax Kpeno, Bhagyashree Nanda, Dattatreya Kar, and Ananya kuanar

Abstract Bioremediation is a sustainable and eco-friendly approach which has gained prominence as a powerful tool in mitigating the adverse environmental impacts of soil and aquatic contamination. In contaminated soil, bioremediation techniques harness the remarkable metabolic capabilities of microorganisms to degrade, transform, or immobilize pollutants. Microbes, including bacteria, fungi, and archaea play pivotal roles in breaking down organic contaminants, such as hydrocarbons and pesticides, and in facilitating the removal of heavy metals through processes like biomineralization. Plant-assisted phytoremediation and engineered systems, like bio-augmentation and biostimulation, offer tailored solutions to address diverse soil contaminants. Recent research has explored the potential of synthetic biology to design custom microbes with enhanced remediation capabilities. In aquatic environments, bioremediation strategies are equally essential in combatting water pollution. Microbial communities, both free-living and associated with plants, contribute

---

T. Rout
Department of Surgical Oncology, IMS & SUM Hospital, Siksha 'O' Anusandhan (Deemed to Be University), Bhubaneswar, Odisha, India

S. Rout
Department of Neuroscience, IMS & SUM Hospital, Siksha 'O' Anusandhan (Deemed to Be University), Bhubaneswar, Odisha, India

A. Kpeno · B. Nanda
Department of Psychiatry, IMS & SUM Hospital, Siksha 'O' Anusandhan (Deemed to Be University), Bhubaneswar, Odisha, India

D. Kar
Department of Medical Research, IMS & SUM Hospital, Siksha 'O' Anusandhan (Deemed to Be University), Bhubaneswar, Odisha, India

A. kuanar (✉)
Centre for Biotechnology, School of Pharmaceutical Sciences, Siksha 'O' Anusandhan (Deemed to Be University), Bhubaneswar, Odisha, India
e-mail: ananyakuanar@gmail.com

A. Kuanar et al. (eds.), *Recent Advances in Bioremediation and Phytoremediation*, Environmental Science and Engineering,
https://doi.org/10.1007/978-3-031-77884-1_6

significantly to removing contaminants, including nutrients, petroleum hydrocarbons, and toxic metals. Natural systems such as wetlands and constructed treatment wetlands have proven effective in water purification. Additionally, biofilm-based approaches and genetically modified microorganisms promise to enhance the degradation of recalcitrant pollutants. This study provides an overview of the fundamental principles, strategies, and recent advancements in bioremediation; focusing on its application in both soil and aquatic environments with the importance of considering environmental conditions, microbial diversity, and biogeochemical processes. Moreover, it underscores the need for interdisciplinary collaboration among microbiologists, environmental engineers, ecologists, and geneticists to develop innovative and sustainable solutions. As society grapples with increasing pollution and its ecological consequences, bioremediation offers hope for restoring contaminated ecosystems, ultimately contributing to a more sustainable and healthier planet. The continuous evolution and refinement of bioremediation technologies are poised to play a pivotal role in addressing the pressing environmental challenges of our time.

6.1 Introduction

In the modern era, the pervasive issue of environmental pollution has emerged as a global concern, casting a shadow over the delicate balance of our planet's ecosystems. From industrial emissions to improper waste disposal, human activities have significantly altered the natural environment, leading to widespread contamination of air, water, and soil. This escalating problem poses threats not only to the health of the planet but also to the well-being of all living organisms inhabiting it (Skenderovic et al. 2015). Among the critical components of the Earth's ecosystems, soil and aquatic environments play pivotal roles in supporting biodiversity, sustaining life, and providing essential ecosystem services. Soil, the foundation of terrestrial ecosystems, nurtures plant life and facilitates nutrient cycling. Meanwhile, aquatic ecosystems, encompassing rivers, lakes, and oceans, are crucial habitats for a myriad of species and are vital for regulating global climate patterns. The health of these ecosystems is intricately linked to the overall well-being of our planet (Raj et al. 2019; Funabashi 2018). As environmental pollution continues to escalate, the need for comprehensive environmental restoration has become more apparent than ever. Degraded ecosystems face a multitude of challenges, including loss of biodiversity, compromised water quality, and diminished soil fertility. Urgent action is required to reverse these trends and restore the natural balance that sustains life on Earth. Recognizing the interconnectedness of ecosystems and the well-being of all living organisms, environmental restoration emerges as a fundamental imperative for a sustainable future. In the pursuit of effective environmental restoration, bioremediation stands out as a promising and environmentally friendly approach. This process harnesses the power of microorganisms to naturally degrade, transform, and remove pollutants from the environment (Adla et al. 2022). By utilizing the metabolic capabilities of bacteria, fungi, and archaea, bioremediation offers a targeted and sustainable method for cleaning up contaminated soil and water. The potential of bioremediation lies not only in its ability to remediate polluted environments but also in its capacity to promote the revival of ecosystems, allowing them to recover and thrive once again (Sharma 2020).

As we delve into the web of environmental challenges, understanding the importance of soil and aquatic ecosystems, acknowledging the urgent need for restoration, and recognizing the transformative role of bioremediation become imperative steps towards fostering a harmonious coexistence with our planet. In this context, this exploration seeks to delve into the depths of bioremediation as a beacon of hope for ecosystem revival in the face of environmental adversity.

Bioremediation Basics: Bioremediation is a biotechnological approach where various microorganisms or their enzymes are used to decontaminate the surrounding environment from different soil and water pollutants. Both plants and microorganisms like bacteria, fungi etc. are used in this process to break down harmful chemicals into lesser toxic forms by changing chemicals through different reactions that occur as a portion of their metabolism and catabolism activities. The microorganisms may be isolated from another location and transported to the polluted site, or they may be native to the contaminated region. Bioremediation will be successful and efficacious when the environment possesses the appropriate temperature, nutrients, optimal pH, etc. for better and faster microbial growth and degradation. Different microorganisms whose DNA can be genetically altered by various techniques have been shown to decontaminate ecosystems from petroleum, toluene, benzene, and various harmful chemicals. In addition to this, living microbes can eliminate different contaminants from acidic, hyperthermal, basic hypersaline, etc. wastes from the ecosystem (Gomathi et al. 2020; Bala et al. 2022; Harekrushna and Das 2012; Ayilara and Babalola 2023). It is well known that archaea contribute to bioremediation in different bacteria-based applications (Karzmarzick et al. 2018). Bioremediation has also limited approaches for certain pollutants like aromatic hydrocarbons or chlorinated organics which possess restricted resistance to microbial assault, resulting in lower or maybe not all degradation. Therefore, bioremediation possesses various types of degradation to decontaminate various pollutants.

6.2 Different Types of Bioremediation

6.2.1 Microbial Bioremediation

It is one of the biodegradation mechanisms used by removing and transporting various wastes through two processes (i) in situ and (ii) ex-situ. The ex-situ method implies eliminating contaminated soil from the pollution site to somewhere else where it will go through a microbiological process to immobilize various metal ions that can cause pollution. Various polluted surfaces, soil, groundwater, etc. were decontaminated by organisms directly at the same location by utilizing the in-situ method. Furthermore, in-situ is safer than the ex-situ process by transferring chemotactic microorganisms to the pollution site to break down harmful substances. (Ayilara and Babalola 2023; Abo-Alkasem et al. 2023) According to research, in-situ bioremediation applied carbon nutrients for a better acceleration of breaking down various pollutants through enhancing microbial activities (Bala et al. 2022). Even if it is used for the degradation of inorganic and toxic metals in the ecosystem. For ex. A variety of heavy metals like Co (III), Cr(VI), U(VI), etc. can degraded by microbes that possess Fe^{3+} and sulfate-reducing enzymatic capacities (Abo-Alkasem et al. 2023). Due to the small size and complicated structure of the microbes, they efficiently arrange a big surface area as well as enhance the biosorption of heavy metals which makes this process more effective to decontaminate the environment (Tarfeen et al. 2022). Microorganisms develop a hydrophobic solvent efflux pump that covers the outer membrane for protection from harmful pollutants (Ayilara and Babalola 2023). The most important matter is to choose the appropriate microorganism or mass of organisms that can sterilize the destructive and harmful metals. According to another research, this process also decreases up to 50–60% price thus making it a valid and the most preferable process for the pollution site (Tarfeen et al. 2022).

6.2.2 Microbial Bioremediation

Microbial bioremediation is a sustainable and environmentally friendly approach to address pollution and contamination issues in soil and water (Nag et al. 2023a). It harnesses the metabolic capabilities of microorganisms to degrade or transform various pollutants into less harmful substances. This natural process has gained significant attention as an alternative to traditional remediation methods due to its cost-effectiveness and minimal environmental impact. By understanding the principles of microbial degradation and utilizing specific microorganisms, we can design effective and sustainable strategies to mitigate the impact of various contaminants on the environment. The mentioned case studies highlight successful applications of microbial bioremediation, showcasing its potential as a viable solution for environmental remediation challenges (Mishra and Das 2023).

1. **Principles of Microbial Bioremediation**: Microbial degradation involves the utilization of microorganisms such as bacteria, fungi, and archaea to break down or transform contaminants into less toxic forms. These microorganisms possess enzymes that can target and metabolize specific pollutants, turning them into harmless byproducts. The success of microbial bioremediation depends on environmental factors such as temperature, pH, nutrient availability, and the presence of other substances. Optimizing these conditions is crucial for enhancing microbial activity and promoting efficient pollutant degradation (Nag et al. 2023b)

6.2.3 Specific Microorganisms Involved

1. **Bacteria:** (a) **Pseudomonas spp.**: Known for their versatility, Pseudomonas bacteria play a vital role in degrading hydrocarbons and pesticides.
   (b) **Bacillus spp.**: These bacteria are effective in breaking down organic pollutants, including oil and heavy metals.
2. **Fungi: White Rot Fungi (e.g., Phanerochaete chrysosporium)**: Particularly adept at degrading complex organic compounds, including certain persistent pollutants like polycyclic aromatic hydrocarbons (PAHs).
3. **Archaea**: **Methanogenic Archaea**: Involved in the anaerobic degradation of organic matter, especially in contaminated sediments and wastewater.

Phytoremediation: There is another type of remediation called phytoremediation which occurs by utilizing plant species at physical, chemical and biological levels to decrease pollution. This process can used to remove contaminants from toxic heavy metals through extraction, rhizofiltration, volatilization and stabilization methods. The process by which toxic pollutants are taken up by plant roots and then accumulate in the aerial portion of the plants is known as phytoextraction. When pollutants are absorbed by plant roots from the groundwater, then it is called rhizofiltration method. However, during the phytovolatilization process, the accumulation of pollutants is discarded into the atmosphere after transformation into minor toxic pollutants. But phytostabilization is another mechanism where root zone enzymes are excreted. By using these enzymatic activities, toxic metals are converted to lesser harmful forms as well and stop their accession into the food chain. Polluted water can be decontaminated through this method and by using plant species and various bacteria (Bala et al. 2022; Tan et al. 2023).

Biostimulation and Bioaugmentation: In the biostimulation process, particularly specified nutrients like phosphorous, nitrogen, oxygen, etc. are added to the pollution site for an acceleration of different metabolic activities of microorganisms to decontamination. For example, a petroleum-polluted area always requires the addition of a variety of nutrients as well as biosolids which are derived from fertilizers, and domestic sewage can degrade different pollutants. Different environmental factors like temperature, pH, etc. decide the rate and frequency of the biostimulation process. According to research, the addition of high levels of various nutrients can possess adverse events like a decrease in oxygen amount in the water which can cause mortality in aquatic life. Therefore, it is required to control the additional amount of nutrients in the polluted site which will help to degrade various contaminants properly (Goswami et al. 2018).

The bioaugmentation process refers to the addition of a variety of microorganisms in the contaminated area to degrade the hazardous substances as well as increase the polluted site's genetic potential. This process also requires a balanced environmental condition like temperature, moisture, organic and nutrient content, type of soil etc. For example, nitrophenolic pesticides can degraded at a temperature of around 30 degrees and a pH of slightly alkaline by Burkholderia species FDS-1, effectively.

The amount of organic matter in the soil is one of the key factors in describing bioaugmentation work capacity. Mostly, this process is applied at oil polluted areas (Goswami et al. 2018).

Mycoremediation:-The term "myco-remediation," which was coined by Paul Stamets, means the utilization of various fungal species—eukaryotic microbes representing a range of environments—to detoxify areas of pollution (Purohit et al. 2018). It is a bioremediation procedure involving stimulation and biological enhancement methods to detoxify environmental pollutants. While biological enhancement involves adding microbes or trees to contaminated areas to speed up degradation, biostimulation means increasing already-existing pollutant-degrading microorganisms (EPA 2014). The microbial remediation of metal is influenced by various factors including metallic ions' nature, cell wall structure, physiology, pH, temperature, time, ionic strength, and metal levels (Mishra et al. 2012). Metals can be bound by microorganisms from an aqueous solution. Mycosorption, a unique metal biosorption process in yeast and fungi, is a sustainable method for degrading environmental pollutants and transforming industrial and agro-industrial waste into products.

Fungal bioremediation, a unique process with diverse enzymes and growth rate, is suitable for environmentally stressed conditions, especially in bacterial-limited bioactivity-scarce conditions, where complexation is the main mechanism (Gadd 2010). Fungal bioremediation is a promising method for removing pollutants from soil, but success depends on factors like temperature, moisture, pH, water availability, and oxygen levels. Candida and Trichoderma, oil-degrading aquatic yeast species, can degrade agrochemicals and enhance plant growth (Ghahfarokhy et al. 2011). The eating oyster mushroom P. ostreatus is a stable mycoremedient for chemical formulation and a biological barrier for anti-nematode disinfestations because it can decompose 80% of all pesticides in soil in thirty-five days. Trichoderma harzianum, a robust agricultural pesticide, has potential as a remedial agent due to its ability to reduce cadmium levels, impacting biomass production (De Freitas et al. 2011). Fungi, like Suillus luteus, can enhance environmental benefits by promoting phytoextraction and reducing metal absorption, influenced by factors like metal tolerance, mycorrhizal status, and soil nutrition. The use of fungus, a wide variety of microorganisms that have high ecosystems, in bioremediation—a vital method for reducing toxins and waste—has advanced rapidly.

A. Fungi species involved in mycoremediation

Biological approaches in industrial and environmental biotechnology focus on developing "clean technologies" that minimize waste generation, and treat and convert waste into useful forms. One such method is mycoremediation, which uses fungi and mushrooms for waste degradation (Purnomo et al. 2013). Fungi play a crucial role in the recycling of nutrients and in the metabolism of contaminants into carbon and energy. Mushrooms are effective decomposers due to their mycelium and extracellular enzymes, which effectively eliminate hazardous pollutants through various methods. The mycoremediation mechanisms of fungi, such as bioaccumulation, biosorption, and bioconversion, are studied in this book as are the impact of various environmental conditions on this process (Yadav et al. 2021). Mushrooms are increasingly used for bioremediation due to their ability to produce food protein from waste through the degradation of waste through hydrolyzing and oxidizing enzymes (Zhu et al. 2013). Mushrooms play a crucial role in bioremediation of waste through biodegradation, biosorption, and bioconversion (Lamrood and Ralegankar 2013). However, safety aspects and the potential for food utilization are not well-reported. There is a lack of information on the pros and cons of mushroom cultivation on waste and its potential as a product. Mushrooms utilize various methods such as biodegradation, biosorption, and bioconversion to decontaminate and stimulate the environment.

6.2.4 Ecological Importance

Fungi are versatile organisms that inhabit various habitats, including terrestrial and aquatic ones. They have a heterotrophic mode of nutrition and can be classified into saprobic, ectosymbiotic, endosymbiotic, and parasitic. Saprobic fungi are essential for nutrient recycling and ecological balance, inhabiting both terrestrial and aquatic environments. They are also found in extreme environments characterized by extreme temperatures, pressure, salinity, acidity, alkalinity, radiation, toxins, and harmful chemicals. Extremophile fungi exhibit extraordinary metabolic profiles, utilizing unusual energy sources and organic substrates as carbon sources. These fungi are successful in survival at polluted sites and efficient in reducing pollutants, making them suitable for bioremediation (Frac et al. 2018). Fungal species isolated from extreme environments can be used to bioremediate polluted sites, such as industrial processes at high temperatures or acidic or alkaline conditions (Tkavc et al. 2017).

6.2.5 Fungi's Morphological Features Help Bioremediation

Fungi have various morphologies, from microfungi to macrofungi, and can be unicellular or multicellular. filamentous fungi have a mycelium, a porous network of elongated cells called hypha. These hyphae can penetrate solid substrates for nutrition and their topology depends on their branching pattern and density (Walker and White 2017). Fungal species can effectively absorb pollutants through biosorption, accumulating them in their fruiting bodies. Fungal mycelia have been used for biosorption of organic and inorganic pollutants, such as 2,4-dichlorophenol and 4-chlorophenol. Fungal cells use enzymes to degrade pollutants and transport them inside through their cell wall and membrane. The cell wall, composed of polysaccharides, proteins, lipids, and pigments, facilitates biosorption and uptake of pollutants. Metal-binding functional groups sequester heavy metal ions, resulting in active processes and detoxification.

6.3 Diversity of Fungi

Around 2.2–3.8 million fungal species worldwide play crucial roles as decomposers, mutualists, and pathogens in various environments. They exhibit high adaptability, with many strains from Basidiomycota and Ascomycota demonstrating efficient pollutant degradation (Hawksworth and Lücking 2017).

1. Basidiomycetes, including white rot, brown rot, and leaf litter fungi, have versatile physiological abilities to survive and thrive in extreme environments, effectively degrading pollutants like polyaromatic hydrocarbons through lignin modifying enzymes, but may be limited by other factors (Marco-Urrea et al. 2015). Basidiomycetes have bioremediation capabilities, allowing them to degrade pollutants in vitro. They produce degrading enzymes like laccase and phenol oxidase (Martinkova et al. 2016). Some fungal species, including saprotrophic and ectomycorrhizal basidiomycetes, play a significant role in this process. These species have abundant and diverse Cyt P450 genes, with CYP53 being a common CYP clan associated with different pollutant degradation categories (Olicon-Hernadez et al. 2017).
2. Zygomycota, also known as Mucoromycotina, is a non-ligninolytic lower fungal group that has been explored for its potential to degrade xenobiotics and hydrocarbons. Biomass of Cunninghamella bertholletiae is effective for landfill leachate biosorption, while Rhizopus oligosporus and Rhizopus oryzae have been found to bioremediate copper, zinc, chromium, arsenic, lead (Prabhakar et al. 2019) and cobalt (Tigini and Varese 2018). Fungal phyla like chytridiomycota, neocallimastigomycota, and blastocladiomycota may evolve genes encoding degradative enzymes, potentially enhancing mycoremediation potential. Blastocladiomycota, Glomeromycota, Microsporidia, Cryptomycota, and Neocallimastigomycota have been analyzed for plant biomass-degrading enzymes. Fungal species from Microsporidia and Cryptomycota lost many genes, while Neocallimastigomycota exhibited potential (Kameshwar and Qin 2019).
3. Ascomycetes fungi, including Fusarium and Aspergillus, can effectively metabolize organic compounds like monosaccharides, polysaccharides, proteins, lipids, and hydrocarbons due to intracellular CYPs (Medaura et al. 2021). Fungal biomass from contaminated wastewater was explored for bioremediation and waste water treatment. Penicillium brevicompactum and Pseudallescheria boydii strains showed significant toxicity abatement and decolorization potential (Spina et al. 2018).

6.3.1 Applications and Limitations in Soil and Water Restoration

Degradation of land and water resources impacts biodiversity and ecosystem services. However, degraded environments offer opportunities for environmental restoration, with over two billion hectares globally reclaimed. Intergovernmental commitments like the Aichi Targets and Bonn Challenge highlight these opportunities (Minnemeyer et al. 2014).

Nearly 1% of the globe's surface is being utilized by human activity such as mineral extraction, road construction, and waste disposal caused by global development. It is essential to use restoration methods like soil spreading and geographic modeling. Technosols, which are products of human activity, are beneficial. The use of organic waste for Technosol construction accelerates biological colonization and maintains soil humus and nutrient balance (Blanco et al. 2010). Pedogenic processes in Technosols are similar to natural soils, but the ingredients used influence their evolution, soil properties, and ecosystem services. Technosols have a fast evolution compared to natural soils, including biological activity, which is crucial for ecosystem services and soil pedogenesis p (Watkinson et al. 2017).

Restoration methods using Passive restoration, which favor natural regeneration, are less expensive than active methods, making them more cost-effective. However, studies on passive restoration are underrepresented, highlighting the need for more comparisons and affordable options (CalFed 2008). Human activities have significantly impacted watersheds worldwide, causing pollution and habitat degradation in rivers and coastal oceans. Despite efforts to manage and restore watersheds, significant investments have been made in diverse waterways, such as the California Delta region, the Baltic region, Moreton Bay sub catchments, and the Mekong River Basin (Duarte and Conley 2008). The author advocates for a significant reform in watershed restoration due to insufficient application of ecological science, inadequate testing, and slow production of use-inspired knowledge. They propose a rigorous approach that requires scientists to identify under-utilization, test practices, and recognize practitioners' needs for new science.

Rivers and coastal waters continue to degrade despite large investments due to factors such as development, land use change, population growth, over-harvesting of living resources, narrow management approaches, and scientific community oversimplification. Current approaches to restoration are insufficient and require more input from scientists. The scientific community should also be blamed for oversimplifying ecosystem dynamics. The need for scientific input is crucial for addressing this issue (Chazdon and Uriarte 2016).

Applications in Need of Science: Restoration Approaches Needs Additional Research.

Test ideas deemed closest to restoration activity: The scientific community recognizes the importance of societal research but often fails to address real problems due to a gap between ideas and implementation. Prioritizing research with immediate implementation results is crucial, especially in areas like urban water quality improvement. Researchers should explore available restoration tools and how they can enhance ecological restoration, aiming to accelerate the pace of restoration science. The sub-basin area is flat to gently undulating, bounded by steep escarpments. Altitude ranges from 1,680 m to 2,700 m. Most slopes are flat, with moderately sloping and steep land. Vegetation grows 120–180 days in the west and 180–300 days in the eastern highlands. Four major soil types are identified.

Soil carbon sequestration can improve soil quality, water quality, biodiversity, agronomic productivity, and food security. However, converting degraded soils to restorative land use and adopting renewable management practices (RMPs) on agricultural soils requires policies, commodification, and carbon credit trading. The current value of soil carbon is too low for meaningful financial rewards.

6.4 Challenges and Future Directions

6.4.1 Discuss Challenges in Bioremediation

The major challenges in bioremediation are:

The first challenge is the bioremediation timing because it takes much longer time may take 7 months to 1 year to completely degrade various contaminants. Not only duration but also limitations in appropriate environmental conditions, availability of nutrients, pollution site, and kind and amount of pollutants are required for successful bioremediation. Another challenge of bioremediation in the soil is the aging of pollutants which changes with time in the soil. The aging of contaminants like PAH (polycyclic aromatic hydrocarbons) and other harmful substances in the soil causes the presence of a lower amount of microorganisms in the soil. Therefore, the presence of maximal concentration of pollutants is necessary. Decomposing of a variety of wastes causes an alteration of the pH of the soil, resulting in instability of the contaminants and becoming an obstacle for the accumulation of microbes. Poor compliance of exogenous microorganisms with the contaminated soils slows down their adaptability to act on a particular site which causes bioaugmentation more ineffective (Nag et al. 2023c). During bioremediation, metabolic pathways between pollutants, heavy metals, and microbes are not still understandable. Even, the catalytic action of different enzymes is responsible for the degradation of pollutants which actually is hidden and therefore extensive research is required. Several challenges such as long duration, low-biomass production of hyperaccumulator plants, climatic interference in the accumulation capacity of plants, and invasiveness of the hyperaccumulator plant species affect the indigenous flora diversity. The above challenges need to be overcome to increase the use of phytoremediation for waste treatment (Vishwakarma et al. 2020; Sharma et al. 2021; Banik et al. 2013).

6.4.2 Emerging Trends and Technologies

There are different developmental technologies are present and necessary to utilize for successful bioremediation strategies. Modern advances in the field of molecular and omic technologies including ultrafast pyrosequencing, microarrays, micro profiling, proteomic, metabolomic, transcriptomics, metatranscriptomics, etc. have the potential to know about various mechanisms of microorganisms used for decontamination of environmental pollution. Different genomic methods including PCR (polymerase chain reaction), exometabolomics, DNA hybridization, and isotopic distribution are required to study the microbial biodegradation process. Various PCR techniques like AFLP (amplified fragment length polymorphism), ARISA (automated ribosomal intergenic spacer analysis), ARDRA (amplified ribosomal DNA restriction analysis) etc. are used for microbial fingerprinting methods to identify a variety of bacteria and microbes. To better study microbial m-RNA expressions, a DNA microarray (transcriptomic tool) is used. Through this process, scientists will easily determine the various microorganism's metabolism and adaptability changes according to environmental pollution by studying their cellular phenotype, genome, and proteome (Bala et al. 2022; Dutta et al. 2021).

Electromagnetic and nanotechnology are useful for purifying water. Sewer water can be sanitized by utilizing effective microorganisms through the application of effective microbes technology and purified water is subsequently used for irrigation. Different pollutants are treated with different enzymes after being isolated from various species. DE (direct evolution of enzymes) is a new way for scientists to redesign the biocatalyst which accelerates chemical reactions between pollutants and enzymes. Even enzyme structures can be evaluated and improved by studying and analyzing PDB (protein database) which is later used for bioremediation. The application of genetically modified organisms is vast for the bioremediation of toxic compounds, xenobiotics, etc. from the environment (Bala et al. 2022; Dutta et al. 2021).

6.4.3 Potential Future Developments in the Field

Numerous studies on the application of microbial enzymes for waste material bioremediation have been conducted; yet, process optimization is crucial to provide and preserve environmental resources. It is necessary to research new microbes that can degrade various pollutants very easily. It is also critical to do more research to develop different quick identification and biodegradation techniques that can clearly explain the underlying mechanisms between pollutants and microbes. Maximum microorganisms are utilized to deteriorate only organic substrates as compared to inorganic ones. Therefore, scientists will try to develop various microbes that are biodegradable for inorganic substrates and lessen their effect. Usually, the degradation of pollutants and survival of microbes are temperature-dependent, and their dependency level on temperature is also different. Thus, scientists should increase their research in this field and find out the ideal temperature for contaminants and microbes for biodegradation as well as discover their stability point in between them. The extent of future changes in environmental resources after bioremediation is also another factor for research by studying new microbes that are used for bioresource cleaning (Ayilara and Babalola 2023). Another research key factor is to identify, develop, and differentiate gene and protein sequences of various microbes to eliminate contaminants through genomics, metabolomics, and proteomics processes (Bala et al. 2022). Nanoparticles are also used for bioremediation but the introduction of nanoparticles in the environment and food chain develops a question about environmental safety (Singh and Saxena 2022).

6.5 Case Studies Demonstrating Successful Microbial Bioremediation in Soil and Water

1. **Exxon Valdez Oil Spill (1989)**: Microbial bioremediation, particularly the use of Pseudomonas bacteria, played a crucial role in the cleanup of the oil-contaminated beaches. The application of these microorganisms helped break down the hydrocarbons, facilitating the natural recovery of the affected ecosystems.
2. **BTEX (Benzene, Toluene, Ethylbenzene, Xylene) Contamination in Groundwater**: Bacillus spp. have been successfully applied to remediate groundwater contaminated with BTEX compounds. These bacteria utilize the pollutants as a carbon source, converting them into less harmful substances through metabolic processes.
3. **Mycoremediation of Heavy Metal Contamination**: Fungi, such as Aspergillus and Penicillium species, have demonstrated effectiveness in removing heavy metals from contaminated soils. Their ability to absorb and accumulate metals makes them valuable in reducing the environmental impact of mining and industrial activities.

V: **Phytoremediation: Harnessing Plant Power for Environmental Clean-up**: Phytoremediation is a sustainable and innovative approach that utilizes the natural abilities of plants to remediate contaminated environments. This eco-friendly technique has gained prominence as an effective and cost-efficient method for mitigating soil, water, and air pollution. Through various physiological and biochemical processes, plants can absorb, transform, and stabilize pollutants, contributing to the restoration of ecosystems impacted by anthropogenic activities.

Phytoextraction or phytoaccumulation is the process wherein plant roots absorb toxic metals, which are subsequently transported to the shoots and deposited in metabolically inactive parts such as vacuoles, cell walls, cell membranes, and other plant tissues. Hyperaccumulator plants, recognized for their ability to mitigate heavy metals (HMs), gather a heightened concentration of toxic metals in both their root and shoot tissues. The fundamental mechanisms involved in the accumulation of these toxic heavy metals includes the absorption of metal cations, subsequently leading to the formation of metal-phytochelatin complexes (M-PC) or metal–ligand complexes within the plant cell (Garcia-martin et al. 2020). Following the formation of the metal-phytochelatin complex (M-PC), these complex molecules are transported to the plant's vacuole and stored (Asgher et al. 2017). The extraction capacity of plant species is predominantly influenced by plant biomass and the concentration of heavy metals (HMs) in above ground plant tissues (Asgari et al. 2019). Consequently, an ideal phytoremediation species must not only exhibit tolerance and efficient absorption of HMs but also display fast growth, high biomass production, and offer economic benefits (Li et al. 2021). As a result, some plant species are more suited to utilize phytoextraction process during phytoremediation of certain metals as indicated in Table 6.1.

Rhizofiltration: In rhizofiltration, plants with extensive root systems, such as water hyacinth and sunflowers, are utilized to extract pollutants from water sources. The plants absorb contaminants through their roots, providing a natural filtration system. This method is especially effective in addressing water pollution caused by heavy

Table 6.1 Some Examples of plants that use phytoextraction during phytoremediation (source, Kafle et al. 2022)

Common name	Scientific name	Contaminants	Remarks
Mouse ear cress	Arabdidopsis helleri	Cd, Zn	Cd transported by other pathways
Tobacco	Nicotiana tabacum	Cd	Cd higher in stems and leaves
Common Cocklebur	Xanthium strumarium	Cd, Pb, Ni, Zn	Has highest BCF for Ni-1.651, Cd-1.574 and Pb-1.048
Geranium	Pelargonium hortorum	Pb	Bacteria tolerant to Pb used
Perennial rye grass	Lolium perenne	Ni, Co, and Fe	Greater survival time
Rattlebush	Sesbania drummondii	Pb	EDTA enhanced uptake of Pb and accumulation
Chinese Brake ferns	Pteris vittate	Arsenic (As)	126-fold Arsenic in the shoots versus soil, without the need of chelating agents or other soil amendments
Corn	Zea mays	Pb, Ti	Chelators induced the phytoextraction of Pb and Ti
Garden Geranium	Pelargonium hortorum	Pb, Cd	EDTA enhanced Pb and Cd phytoextraction
Alpine pennygrass	Thlaspi caerulescens	Cd, Zn	Cd transported by Zn and Ca pathway

metals and organic pollutants (Ramana et al. 2021). Aquatic plants typically participate in phytoremediation by employing rhizofiltration as shown in Table 6.2. This approach is particularly efficient for addressing the remediation of soil and water that is heavily contaminated with nutrients such as nitrogen and phosphorus (Odinga et al. 2019). Plant roots can facilitate the removal of ammonia in aqueous environments through rhizosphere nitrification, followed by subsequent denitrification (Sikhosana et al. 2020). In the rhizofiltration of water contaminated with heavy metals (HMs), root-mediated precipitation occurs, leading to the fixation of HMs inside the root in an insoluble form. For instance, sunflower and mustard have the capacity to filter water contaminated with lead (Pb) by precipitating Pb-phosphate within the root (Chaney et al. 2020).

Phytostabilization: Phytostabilization involves the use of plants to immobilize and stabilize contaminants in the soil, preventing their migration and entry into the food chain. Certain plant species have the ability to modify the chemical and physical properties of the soil, reducing the bioavailability of pollutants and minimizing their ecological impact (Sarwar et al. 2017). Certain plants have the ability to create bound residues of toxic contaminants, rendering them either unavailable in a poisonous form or unable to be released from the solid matrix post-accumulation. For instance, the Ryegrass plant can absorb the herbicide trifluralin and convert it into bound residue (Zhang et al. 2022). The contaminants can be immobilized in the rhizosphere with the

Table 6.2 Examples of plants that uses Rhizofiltration during phytoremediation (source, Kafle et al. 2022)

Common name	Scientific name	Contaminants	Remarks
Water starwort	Callitriche lusitanica	As	BCF: 0.002346
Water moss	Fontinalis antipyretica	U	BCF: 0.00023479
Duckweed	Lemna minor	U	BCF: 0.0000529
Threadleaf crowfoot	Ranunculus trichophyllus	As	BCF: 0.000354
Water hyacinth	Eichhornia crassipes	Fe, Cr, Cu, Cd, Zn, Ni, As	Translocation factor: Fe: 0.69, Cr:0.56, Cu:0.79, Cd:0.78, Zn:0.67, Ni:0.55, As:0.45
Carolina mosquito fern/ Water Velvet	Azolla caroliniana	As	BCF: 0.000397
Threadleaf crowfoot	Ranunculus trichophyllus	As	BCF: 0.000354
Cattail	Typha angustifolia	Cu, Pb, Ni, Fe, Mn, and Zn	Industrial waste containing phenol (100–800 mg l − 1) and melanoidin (2500–8500 Co–Pt)

assistance of phytochemical exudates, trapped on the root surface through transport proteins, or sequestered within the vacuole of root cells via cellular processes (Nag 2022). Heavy metals (HMs) and certain organic contaminants can undergo conversion into non-toxic compounds by forming conjugates with sugar, protein, and amino acid derivatives, or by creating complexes within the rhizosphere. As an illustration, the plant root epidermis immobilizes arsenic (As) by binding it to ferric sulfate within vacuoles, forming a trivalent As-tris-thiolate complex in the rhizosphere (Hammond et al. 2018) (Table 6.3).

Phytodegradation: Phytodegradation relies on the capacity of plants to break down or transform organic pollutants through metabolic processes. It is described as the process where plants metabolize and destroy contaminants within plant tissues (Kafle et al. 2022). This mechanism is particularly effective in addressing soil and water contamination caused by pesticides, petroleum hydrocarbons, and other organic compounds. Certain plant enzymes play a crucial role in the degradation of these pollutants (Spaczynski et al. 2012). During the direct process, plants absorb contaminants, metabolize or break them down into smaller, less harmful compounds, and distribute them within plant tissues. Due to the absence of active transporters in plants, these organic contaminants are absorbed through passive uptake. When the degradation of contaminants takes place in the rhizosphere, the process is referred to as rhizodegradation (Kafle et al. 2022) Table 6.4 shows the plants associated with utilsing phytodegradation.

Table 6.3 Examples of plants that uses phytostabilization or phytoimmobilizationduring phytoremediation (source, Kafle et al. 2022)

Common name	Scientific name	Contaminants	Remarks
Black nightshade	Solanum nigrum	Cu, Zn, Cd	10% biochar/attapulgite addition recommended
Cowpea	Vigna unguiculata	Pb, Zn	Mine waste phytostabilization
Coastal Mallow/ Seashore Mallow	Kosteletzkya pentacarpos	Cd, Zn	Salinity favors plants against metal toxicity. Cytokinin promotes Zn resistance
Sunflower	Helianthus annuus	Cu, Zn, Pb, Hg, As, Cd, Ni	Vermicompost as an supplement, usually for low level of metal contaminants
Holy rope/ hemp-agrimony	Eupatorium	As	Citric acid (CA) addition 20 mg/ L addition favored phtostabilization
Willow	Salix sps	Cd	Non-flooding Salix has higher BCF than flooding species

Table 6.4 Examples of plants that uses Phytodegradation for phytoremediation (Kafle et al. 2022)

Common name	Scientific name	Contaminants	Remarks
Blumea	Blumea malcolmii	Malachite Green (93.41% decolorization within 24 h	Phytodegradation of industrial waste
Elodea	Elodea Canadensis	DDT	Degradation catalyzed by enzyme cofactors or biomolecules
Perennial reed grass	Phragmites australis	Ibuprofen	P. australis could be a good plant in construction on wetland
Kudzu	Pueraria thunbergiana	DDT	Reductive dehalogenation of DDT
Ipil-ipil	Leucaena leucocephala	Ethylene dibromide (EDB; dibromoethane) and trichloroethylene (TCE)	First evidence of tropical tree to remediate EBD and TCE like organic pollutants
Water hyacinth	Pontederia crassipes	Sodium dodecyl sulfate (SDS)	Removal efficiency as well as degradation of SDS were enhanced by ascorbate peroxidase (APX) activity in the presence of Chromolaenaodorata L. extract
Unicellular green algae	Chlorella pyrenoidosa	Pentachlorophenol (PCP)	Algae activity may have been reduced by the cycling of light exposure
Pickly burr	Datura innoxia	Petroleum	Petroleum degradation was higher in vegetated contaminated soil than nonvegetated soil

Soil and Water Decontamination: Phytoremediation effectively removes contaminants from soil and water. Plants, through their roots, uptake and concentrate pollutants such as heavy metals, pesticides, and organic compounds, helping to detoxify the environment (Ramana et al. 2021; Odinga et al. 2019). This process reduces the risk of soil and water pollution, safeguarding ecosystems and human health.

Cost-Effective Remediation: Compared to traditional remediation methods, phytoremediation is often more cost-effective. The technique utilizes natural processes and requires minimal infrastructure, making it an economically viable alternative for environmental cleanup. Additionally, it can be applied in conjunction with other remediation methods to enhance overall effectiveness.

Sustainable Land Use: Phytoremediation promotes sustainable land use by restoring contaminated sites for future purposes. The process not only remediates polluted areas but also improves soil quality and structure, making the land suitable for agriculture, forestry, or urban development (Cunningham and Berti 2020).

Air Quality Improvement: Certain plants employed in phytoremediation contribute to air quality improvement. Trees and vegetation absorb pollutants, including volatile organic compounds (VOCs) and heavy metals, thereby reducing air pollution levels and enhancing overall environmental health (Meena et al. 2021).

Biodiversity Conservation: Phytoremediation supports biodiversity conservation by creating habitats for various organisms. The introduction of plant species with phytoremediation capabilities fosters ecological diversity and restores balance to ecosystems affected by pollution (Ugya et al. 2016).

6.6 Showcase Successful Bioremediation Projects

Phytoremediation is a method to restore ecosystem services in rural and urban areas, focusing on soil–plant interactions. It is a plant-based technology for restoring contaminated land and water resources, focusing on physical, chemical, and biological mechanisms to reduce pollutants' transport through enhanced infiltration/evapotranspiration (Placek et al. 2016). Microbial symbionts can enhance adaptation and survival. Field-scale testing validates genotypes' performance, and soil functions related to environmental pollution should be considered. Site selection for restoration projects in the USA is often opportunity-based, neglecting factors like population structure, ecological subsidies, and waterway health. Only a small fraction of projects are initiated due to a broader management plan. Most projects are selected based on land availability, which may lead to suboptimal sites. Dispersal is important, but survival in the local environmental context is also crucial.

Remediation technologies, such as grasslands, urban afforestation, riparian buffer systems, SRWCs, and woody species, have proven successful in reducing soil phosphorus, enhancing ecosystem services, and reclaiming mines.

6.7 Highlight Different Approaches Used in Various Ecosystems

Human preference for coastal environments is linked to economic and social wealth, with 90% of world trade reliant on oceans and two-thirds of the planet's mega-cities on coastlines (Reported by Green economy in a blue world: synthesis repoet). Agricultural systems cover 40% of the Earth's surface, requiring ecological integration for sustainability. Managing ecosystem services (ES) is a powerful approach, but it introduces new ecological and social dimensions into the design and management of agricultural systems. Agricultural systems must provide a web of supporting and regulating services, such as soil fertility, pest control, and pollination, to support and regulate biophysical systems. Simultaneous ES-based management is challenging due to multiple positive, negative, and non-linear relationships between services

(Robertson and Swinton2005; Zhang et al. 2007). Management greatly affects the synergies and tradeoffs between services in agro-ecosystems. Additionally, practices acting on unfunctionally linked services may create apparent relationships between these services, such as fertilizer use (Bennett et al. 2009). Oceans and nearshore environments provide significant ecosystem benefits to human society, with 59% and 38% of these benefits respectively (Deidun 2011). Sustainable management of marine resources is crucial due to pressures on biodiversity and loss of ecosystem functioning, which aggravates global poverty and undermines economic development. Urban areas consume three-quarters of global resources and are expected to double by 2050 (Dominguez-Tejo et al. 2016). The Ecosystem-based Approach (EBA) is an emerging paradigm in sustainable ocean management, but challenges include incorporating non-environmental factors.

According to research the US National Science Foundation has increased funding for Long-Term Ecological Research (LTER) sites to study urban ecological systems, leading to the flourishing of urban ecology in the USA. The research focuses on understanding how urbanization patterns alter ecological conditions and how these consequences impact the social system. The research focuses on identifying ecological and social patterns in individual land-use patch types, mosaics of patches, Phoenix metro, CAP ecosystem in a regional context, and urbanizing central Arizona in a global context. It explores the potential for lessons learned from CAP to be applied to other dryland regions and the global importance of urban areas. The study also explores the role of a global landscape of cities.

The Ecosystem-based Approach (EBA): an overarching planning framework.

Ecosystem-based management (EBA) is a concept that combines Integrated Coastal Management (ICM) and Large Marine Ecosystem Management (LME) to address the failure of single-species approaches in fisheries and land-based impacts (Griffis and Kimball 1996). EBA focuses on the health and resilience of marine ecosystems, adopting the precautionary principle for adaptive management. It acknowledges the interdependence between natural systems and ecological, social, economic, and institutional systems (Curtin and Prellezo 2010). EBA implementation challenges include lack of implementation guidelines and planning tools. The Intergovernmental Oceanographic Commission (IOC) supports EBA through its MSP Initiative, providing documentation, training opportunities, and implementation guidelines (David et al. 2004).

6.8 Adopting the Precautionary Approach (Statement 6.3) and the Adaptive Management Approach

The Environmental Business Administration (EBA) recommends the Precautionary Approach for best-available science-based management decisions, while Adaptive Management is suggested for complex ecosystem changes. However, case studies show a lack of key elements for implementing these approaches. For example, Rhode Island excluded the Precautionary Approach, Irish Sea recommended it, and Australia implied it through a legal provision (Tejo et al. 2016). The Adaptive Management Approach was considered important but had slight details (Fernandes et al. 2005). Most case studies stated the adoption but failed to explain how it was to be delivered. The gap between adoption and actual implementation could be due to lack of methodological understanding, misunderstandings from stakeholders, or reluctance to develop additional measures from management monitoring programs.

6.9 Future and Ongoing Environmental Assessments and Monitoring Programs

The study analyzed environmental monitoring programs in marine-related industries, focusing on compliance with the Strategic Environmental Assessment (SEA) Directive 2001/42/EC. Most European case studies reported compliance with this directive, promoting sustainable development. The Irish Sea case highlighted the benefits of integrating spatial planning and SEA at a regional scale. Other European examples included Belgium, Germany, Shetland Islands, Portugal, and Spain. Most case studies reported weaknesses in their monitoring programs, acknowledging key information gaps. Various approaches were used to design monitoring programs, with some focusing on monitoring the state, functioning, dynamics, ecological integrity, and resilience of ecosystems (Fee et al. 2009). Some European case studies linked their monitoring efforts to regional legislation, such as the SEA Directive and the MSFD. Few case studies proposed monitoring programs reflecting adaptive management, such as the Waden Sea program (Waylen et al. 2014).

6.10 Overall Performance of Selected Case Studies

The integration of Malawi Principles within MSP case studies was evaluated, with the Shetland Islands case study scoring highest (21.8 out of 25) and Spain, Canada, and Rhode Island following closely behind. The German case study scored lowest (16.4), with an average score of 20.27. The analysis highlighted challenges in implementing the principles and their effects on management solutions. Despite these challenges, evidence of the integration of EBA was found across the case studies (Franzmann et al. 2002).

6.10.1 Lessons Learned and Best Practices

Over the past 40 years, research on the biodegradation of polluting chemicals in groundwater has advanced significantly. Microorganisms in aquifers can constrain and remove contaminant plumes, particularly for petroleum hydrocarbons, chlorinated hydrocarbons, nutrients, and some metals (Johnston et al. 2013). However, research on PFAS bioremediation is only recent. Engineered bioremediation strategies can be effective in degrading significant contaminant mass and lowering concentrations of groundwater plumes (Janssen and Stucki 2020). Key lessons learned include understanding the geochemical, redox, and electron donor/acceptor conditions controlling specific contaminant biodegradation across various contaminant types. Accurate conceptualization and characterization of groundwater plumes are critical, as biodegradation rates are not constant over the entire plume. Amendment delivery mechanisms are effective in stimulating bioremediation in groundwater if they align with groundwater physics and chemistry controls. Bioaugmentation of isolated indigenous organisms is likely effective in selected cases (Janssen and Stucki 2020).

The use of above ground infrastructure for controlled treatment of larger aquifer areas, potentially paired with extraction wells, depends on the radius of influence. The method uses easily controlled above ground infrastructure to treat larger aquifer areas, targeting primary and secondary plume sources based on the radius of influence and injection depths. The treatment of groundwater flowing past the installed location, suitable for site boundaries or compliance points, may potentially lower maintenance and operational costs.

Gas injection infrastructure is needed for oxygen delivery, making it cost-effective and potentially removing volatile mass beyond biodegradation. Configuration as a PRB curtain is also possible. Bio-stimulants can be released over longer periods with targeted emplacement, potentially reducing costs if reapplication is infrequent or replacement time is prolonged.

6.11 Conclusion

Microbial bioremediation has provided insights into a sustainable and effective approach for addressing environmental pollution in both soil and water. The principles of microbial degradation, involving the metabolic capabilities of bacteria, fungi, and archaea, offer a natural solution to transform contaminants into less harmful forms. Environmental factors play a crucial role in optimizing microbial activity for efficient pollutant degradation. There is a pressing need for continued research and implementation of bioremediation strategies. Researchers and practitioners should explore novel microbial strains, optimize environmental conditions, and enhance our understanding of microbial interactions to maximize the efficiency of bioremediation processes. A call to action extends beyond the laboratory and into policy-making and industry practices. Governments, businesses, and communities should actively consider integrating bioremediation into environmental management plans. Investing in research, developing innovative technologies, and implementing bioremediation strategies on a broader scale can contribute significantly to the restoration and preservation of our ecosystems. Microbial bioremediation stands as a beacon of hope in our collective efforts to address environmental challenges. By acknowledging its importance, fostering collaboration, and committing to further research and implementation, we can build a more sustainable future where our ecosystems is restored through the power of microbial communities.

References

Abo-Alkasem MI, Hassan NH, Elsoud MMA (2023) Mirobial bioremediation as a tool for the removal of heavy metals. Bull Natl Res Cent 47:31

Adla K, Dejan K, Neira D, Dragana Š. Degradation of ecosystems and loss of ecosystem services. InOne Health 2022 Jan 1 (pp 281–327). Agricultural soil: Bioprospection for sustainable development. In Mycoremediation andenvironmental sustainability (pp 91–120). Cham: Springer. and landscape restoration in the tropics. Biotropica 48:709–715. Applications. 3rd ed., Chapter 1, Wiley, USA, p1–35

Asgari Lajayer B, Khadem Moghadam N, Maghsoodi MR, Ghorbanpour M, Kariman K (2019) Phytoextraction of heavy metals from contaminated soil, water and atmosphere using ornamental plants: mechanisms and efficiency improvement strategies. Environ Sci Pollut Res 1(26):8468–84

Asgher M, Per TS, Anjum S, Khan MI, Masood A, Verma S, Khan NA (2017) Contribution of glutathione in heavy metal stress tolerance in plants. Reactive oxygen species and antioxidant systems in plants: role and regulation under abiotic stress 297–313

Ayilara MS, Babalola OO (2023) Bioremediation of environmental wastes: the role of microorganisms. Front Agron 5:1183691

Bala S, Garg D, Thirumalesh BV, Sharma M, Sridhar K, Inbaraj BS, Tripathi M (2022) Recent strategies for Bioremediation of emerging pollutants: a review for a Green and sustainable Environment, Toxics 10(8):484

Banik S, Das KC, Islam MS, Salimullah M (2013) Recent advancements and challenges in microbial bioremediation of heavy metals contamination. JSM Biotechnol Bioeng 2(1):1035

Bennett EM, Peterson GD, Gordon LJ (2009) Understanding relationships among multiple ecosystem services. Ecol Lett 12(12):1394–404. Berlin, Germany, p 256. https://doi.org/10.1007/978-1-4020-8709-7; Bioaugmentation of native fungi, an efficient strategy for the bioremediation of an aged biomasses for bioremediation and detoxification of landfill leachate. Environ Earth Sci 77(9):1–10. Bioremediation of chlorinated and polycyclic aromatic hydrocarbons. Nat Biotechnol 32(6):620–628

Blanco HY, Lal R (2010) Principles of soil conservation and management. Springer

CalFed (2008) The california bay-delta ecosystem restoration program. http://www.delta.dfg.ca.gov/erp/. Accessed 12 Feb 08

Chaney RL, Li YM, Brown SL, Homer FA, Malik M, Angle JS, Baker AJ, Reeves RD, Chin M (2020) Improving metal hyperaccumulator wild plants to develop commercial phytoextraction systems: approaches and progress. In: Phytoremediation of contaminated soil and water 2020 Nov 25, pp 129–158. CRC Press

Chazdon RL, Uriarte M (2016) Natural regeneration in the context of large-scale forest

Cunningham SD, Berti WR. Phytoextraction and phytostabilization: technical, economic, and regulatory considerations of the soil-lead issue. In: Phytoremediation of contaminated soil and water 2020 Nov 25, pp 359–376. CRC Press
Curtin R, Prellezo R (2010) Understanding marine ecosystem based management: a literature review. Mar Policy 34:821–830. https://doi.org/10.1016/j.marpol.2010.01.003
David Tyldesley and Associates, Irish Sea pilot project coastal and marine spatial planning framework, 2004, p 35
De Freitas AL, de Moura GF, de Lima MAB, de Souza PM, da Silva CAA, de Campos GMT, do Nascimento AE (2011) Role of the morphology and polyphosphate in Trichoderma harzianum related to cadmium removal. Molecules 16:2486–2500. Decolorizing ability from an edible mushroom Russula virescens. Int Biodeterior Biodegrad 82:33–39.https://doi.org/10.1016/j.ibiod.2013.02.010
Deidun. A, Borg. S, Micallef. A (2011) Making the case for marine spatial planning in the Maltese Islands. Ocean Dev Int Law 42:136–154, https://doi.org/10.1080/00908320.2011.542108
Domínguez-Tejo E, Metternicht G, Johnston E, Hedge L (2016) Marine spatial planning advancing the ecosystem-based approach to coastal zone management: a review. Mar Policy 1(72):115–30
Duarte C, Conley DJ (2008) Return to neverland: shifting baselines affect eutrophication restoration targets. Estuaries Coasts. https://doi.org/10.1007/s12237-008-9111-2
Dutta K, Shityakova S, Khalifa I (2021) New trends in Bioremediation technologies toward environment-friendly society: a mini-review. Front Bioeng Biotechnol 9:666858
EPA (Environmental Protection Agency) (2014) Bioremediation overview. Office of
Fee E, Gerber K, Rust J, Haggenmueller K, Korn H, Ibisch P (2009) Stuck in the clouds: bringing the CBD's ecosystem approach for conservation management down to earth in Canada and Germany. J Nat Conserv 17:212–227. https://doi.org/10.1016/j.jnc.2009.04.005
Fernandes L, Day JON, Lewis A, Slegers S, Kerrigan B, Breen DAN et al (2005) Establishing representative No-Take areas in the Great Barrier Reef: largescale implementation of theory on marine protected areas. Conserv Biol 19:1733–1744, https://doi.org/10.1111/j.1523-1739.2005.00302.x
Frąc M, Hannula SE, Bełka M, Jędryczka M (2018) Fungal biodiversity and their role in soil health. Front Microbiol 9:707
Franzmann PD, Robertson WJ, Zappia LR, Davis GB (2002) The role of microbial populations in the containment of aromatichydrocarbons in the subsurface. Biodegradation 13(1):65–78
Funabashi M (2018) Water and ecosystem cycles mediated by plant genetic resources for food and agriculture. In: Genetic diversity in plant species-characterization and conservation 2018 Nov 5. IntechOpen
Gadd GM (2010) Metals, minerals and microbes: geomicrobiology and bioremediation. Microbiology 156:609–643
García Martín JF, González Caro MD, López Barrera MD, Torres García M, Barbin D, Mateos PÁ (2020) Metal accumulation by Jatropha curcas L. adult plants grown on heavy metal-contaminated soil. Plants 9(4):418
Ghahfarokhy MR, Goltapeh EM, Purjam E, Pakdaman BS, Modarres Sanavy SAM, Varma A (2011) Potential of mycorrhiza-like fungi and Trichoderma species in biocontrol of take-all disease of wheat under greenhouse condition. J Agric Technol 7:185–195
Gomathi T, Madhavan S, Radha E, Kumar V et al (2020) Bioremediation, Encyclopedia of Marine biotechnology, 3139–3172, https://doi.org/10.1002/9781119143802.ch140
Goswami M, Chakraborty P, Mukherjee K, Mitra G, Bhattacharya P, Dey S, Tribedi P (2018) Bioaugmentation and biostimulation: a potential strategy for environmental pollution. J Microbiol Exp 6(5):223–231
Green Economy in a Blue World: Synthesis Report Accessed on [25th Dec 2023] https://www.unep.org/resources/report/green-economy-blue-world-synthesis-report.
Griffis RB, Kimball KW (1996) Ecosystem approaches to coastal and ocean stewardship. Ecol Appl 6:708–712, http://www.jstor.org/stable/2269467

Hammond CM, Root RA, Maier RM, Chorover J (2018) Mechanisms of arsenic sequestration by Prosopis juliflora during the phytostabilization of metalliferous mine tailings. Environ Sci Technol 52(3):1156–64

Harekrushna S, Das CK, A review on: bioremediation. Int J Res Chem 2(1):13–21

Hawksworth DL, Lücking R (2017) Fungal diversity revisited: 2.2 to 3.8 million heptachlor epoxide by white-rot fungus Pleurotus ostreatus. Int Biodeterior Biodegrad 82:40–44, https://doi.org/10.1016/j.ibiod.2013.02.013. http://www.cluin.org/techfocus/default.focus/sec/Bioremediation/cat/Overview/. Industrially polluted soil with heavy hydrocarbons. Front Microbiol 12:713

Janssen DB, Stucki G (2020) Perspectives of genetically engineered microbes for groundwater bioremediation. Environ Sci Process Impacts 22(3):487–9

Johnston CD, Davis GB, Bastow TP, Annable MD, Trefry MG, Furness A, Geste Y, Woodbury RJ, Rao PS, Rhodes S (2013) The use of mass depletion–mass flux reduction relationships during pumping to determine source zone mass of a reactive brominated-solvent DNAPL. J Contam Hydrol 144(1):122–37

Kafle A, Timilsina A, Gautam A, Adhikari K, Bhattarai A, Aryal N (2022) Phytoremediation: mechanisms, plant selection and enhancement by natural and synthetic agents. Environ Adv 1(8):100203

Kameshwar AKS, Qin W (2019) Systematic review of publicly available non-Dikarya fungal proteomes for understanding their plant biomass-degrading and bioremediation

Karzmarzick MJ, Taylor DK, McCuthan AL (2018) Diversity and Niche of Archaeain Bioremediation, Archaea, 3194108

Lamrood PY, Ralegankar SD (2013) Biosorption of Cu, Zn, Fe, Cd, Pb and Ni by non-treated biomass of some edible mushrooms. Asian J Exp Biol Sci 4:190–195. Landscape Restoration Opportunities. World Resources Institute

Li X, Xiao J, Salam MM, Ma C, Chen G (2021) Impacts of bamboo biochar on the phytoremediation potential of Salix psammophila grown in multi-metals contaminated soil. Int J Phytorem 23(4):387–99

Marco-Urrea E, García-Romera I, Aranda E (2015) Potential of non-ligninolytic fungi in

Martínková L, Kotik M, Marková E, Homolka L (2016) Biodegradation of phenolic compounds by Basidiomycota and its phenol oxidases: a review. Chemosphere 149:373–382

Medaura MC, Guivernau M, Moreno-Ventas X, Prenafeta-Boldú FX, Viñas M (2021)

Meena M, Sonigra P, Yadav G (2021) Biological-based methods for the removal of volatile organic compounds (VOCs) and heavy metals. Environ Sci Pollut Res 28:2485–508

Minnemeyer S, Laestadius L, Sizer N, Saint-Laurent C (2014) Atlas of Forest and

Mishra S, Das AP (2023) Microbial remediation of synthetic microfiber contaminated wastewater. Microbial Technol Indus Wastewater Treatment 337–350. https://doi.org/10.1007/978-981-99-2435-6_17/COVER

Mishra V, Majumder CB, Agarwal VK (2012) Sorption of Zn (II) ion onto the surface of activated carbon derived from eucalyptus bark saw dust from industrial wastewater: isotherm, kinetics, mechanistic modeling, and thermodynamics. Desalin Water Treat 46(1–3):332–351

Nag A (2022) Textbook of agricultural biotechnology. PHI Learning Pvt. Ltd

Nag M, Lahiri D, Ghosh S, Sarkar T, Pati S, Das AP, Ram DK, Bhattacharya D, Ray RR (2023) Application of microorganisms in biotransformation and bioremediation of environmental contaminant: a review. Geomicrobiol J 1–18. https://doi.org/10.1080/01490451.2023.2261443

Nag M, Lahiri D, Ghosh S, Sarkar T, Pati S, Das AP, Ram DK, Bhattacharya D, Ray RR (2023) Application of microorganisms in biotransformation and bioremediation of environmental contaminant: a review. Geomicrobiol J. https://doi.org/10.1080/01490451.2023.2261443

Odinga CA, Kumar A, Mthembu MS, Bux F, Swalaha FM (2019) Rhizofiltration system consisting of Phragmites australis and Kyllinga nemoralis: evaluation of efficient removal of metals and pathogenic microorganisms. Desalin Water Treat 1(169):120–32

Olicón-Hernández DR, González-López J, Aranda E (2017) Overview on the biochemical

Placek A, Grobelak A, Kacprzak M (2016) Improving the phytoremediation of heavy metals contaminated soil by use of sewage sludge. International journal of phytoremediation 18(6):605–18. Potential of filamentous fungi to degrade pharmaceutical compounds. Front Microbiol 8:1792. Potentials. Bioresour Bioprocess 6(1):1–19

Prabhakar A, Mishra S, Das AP (2019) Isolation and Identification of Lead (Pb) solubilizing bacteria from automobile waste and its potential for recovery of lead from end of life waste batteries. Geomicrobiological Journal 36(10):894–903. https://doi.org/10.1080/01490451.2019.1654044

Purnomo AS, Mori T, Putra SR, Kondo R (2013) Biotransformation of heptachlor and

Purohit J, Anirudha C, Mohan KB, Singh NK (2018) Mycoremediation of

Raj A, Jhariya MK, Yadav DK, Banerjee A, Meena RS (2019) Soil for sustainable environment and ecosystems management. Sustainable agriculture, forest and environmental management, 189–221

Ramana S, Mounissamy VC, Saha JK (2021) Heavy Metal contaminated soils. Bioremediation Sci Theory to Practice 20:79

Robertson GP, Swinton SM (2005) Reconciling agricultural productivity and environmental integrity: a grand challenge for agriculture. Front Ecol Environ 3(1):38–46

Sarwar N, Imran M, Shaheen MR, Ishaque W, Kamran MA, Matloob A, Rehim A, Hussain S (2017) Phytoremediation strategies for soils contaminated with heavy metals: modifications and future perspectives. Chemosphere 1(171):710–21

Sharma I (2020) Bioremediation techniques for polluted environment: concept, advantages, limitations, and prospects. In: Trace metals in the environment-new approaches and recent advances 2020 Dec 7.

Sharma I (2021) Bioremediation techniques, for polluted environment: concept, advantages, limitations and prospects, trace metals in the environment- recent approaches and advances. IntechOpen. https://doi.org/10.5772/intechopen.90453

Sikhosana ML, Botha A, Mpenyane-Monyatsi L, Coetzee MA (2020) Evaluating the effect of seasonal temperature changes on the efficiency of a rhizofiltration system in nitrogen removal from urban runoff. J Environ Manage 15(274):111192

Singh Y, Saxena MK (2022) Insights into the recent advances in nanoremediation of pesticides from the contaminated soil. Front Microbiol 13:982611

Skenderovic I, Kalac B, Becirovic S (2015) Environmental pollution and waste management. Balkan J Health Sci 3(1):2–10. Soils from industrial by-products for use as growth substrates in mine reclamation. Restor Ecol 25:587–594. https://doi.org/10.1111/rec.12464

Spaczynski M, Seta-Koselska A, Patrzylas P, Betlej AK, Skorzynska-Polit E (2012) Phytodegrdation and biodegradation in rhizosphere as efficient methods of reclamation of soil, contaminated by organic chemicals (a review). Acta Agrophysica 19(1). species. Microbiol Spect 5(4):5–4

Spina F, Tigini V, Romagnolo A, Varese GC (2018) Bioremediation of landfill leachate with fungi: autochthonous vs. allochthonous strains. Life 8(3):27. Superfund remediation and technology innovation. Accessed Feberary 24, 2014

Tan HW, Pang YL, Lim S, Chong WC (2023) A state-of-the-art of phytoremediation approach for sustainable management of heavy metals recovery. Environ Technol Innov 30:103043, ISSN 2352–1864

Tarfeen N, Nisa KU, Hamid B, Bashir Z et al (2022) Microbial remediation: a promising tool for reclamation of contaminated sites with special emphasis on heavy metal and pesticide pollution: a review. Processes 10(7): 358

Tigini V, Varese GC (2018) Biosorption with autochthonous and allochthonous fungal

Tkavc R, Matrosova VY, Grichenko OE, Gostinčar C, Volpe RP, Klimenkova P, Gaidamakova EK, Zhou CE, Stewart BJ, Lyman MG et al (2017) Prospects for fungal bioremediation of acidic radioactive waste sites: characterization and genome sequence of Rhodotorula taiwanensis MD1149. Front Microbiol 8:2528–2528

Ugya AY, Imam TS, Tahir SM (2016) The role of phytoremediation in remediation of industrial waste. World J Pharm Sci 5(12):1403–30

Vishwakarma GS, Bhattacharjee G, Gohil N, Singh V (2020) Current status, challenges and future of bioremediation. Bioremediation Polluta, 403–415

Walker GM, White NA (2017) Introduction to fungal physiology. In: Fungi: Biology and

Watkinson AD, Lock AS, Beckett PJ, Spiers G (2017) Developing manufactured

Waylen KA, Hastings EJ, Banks EA, Holstead KL, Irvine RJ, Blackstock KL (2014) The need to disentangle key concepts from ecosystem-approach jargon. Conserv Biol 28:1215–1224. https://doi.org/10.1111/cobi.12331

Yadav P, Rai SN, Mishra V et al (2021) Mycoremediation of environmental pollutants: a review with special emphasis on mushrooms. Environ Sustain 4:605–618. https://doi.org/10.1007/s42398-021-00197-3

Zhang C, Fang N, Li Y, Wang X, He H, Jiang J, Tang T, Xu Z, Zhao X, Li Y (2022) Uptake, translocation and distribution of cyantraniliprole in rice planting system. J Hazard Mater 15(436):129125

Zhang W, Ricketts TH, Kremen C, Carney K, Swinton SM (2007) Ecosystem services and dis-services to agriculture. Ecol Econ 64(2):253–60

Zhu MJ, Du F, Zhang GQ, Wang HX, Ng TB (2013) Purification a laccase exhibiting dye

Chapter 7
New Insight into Sustainable Crop Production and Soil Contamination: Impact on the Environment and Remediation Strategies

Devi Shidayaichenbi, Vinod Kumar Dubey, Shubham Singh, Arindam Pal, C. Priyadarshini, Priya, Sushmita Thokchom, Sourav Shai, Waqar Majeed, Santosh Kumar Singh, and Mohd. Sayeed Akhtar

Abstract In the agricultural fields, the misutilization of chemical inputs, for maximum crop production from the limited arable land to feed the world's ever-increasing population, exposed the problem of "soil contamination". The dominant soil contaminants include heavy metals such as lead (Pb), cadmium (Cd), zinc (Zn), nickel (Ni), and copper (Cu). They are sunk into the soil through agricultural activities and industrial effluents. Soil contaminants such as heavy metals have received

---

D. Shidayaichenbi (✉) · C. Priyadarshini · S. Shai · S. K. Singh
Department of Soil Science, Post Graduate College of Agriculture, Dr. Rajendra Prasad Central Agricultural University, Pusa (Samastipur), Bihar, India
e-mail: shidayaish@gmail.com

V. K. Dubey
Department of Genetics and Plant Breeding, College of Agriculture, Acharya Narendra Deva University of Agriculture and Technology, Ayodhya, Uttar Pradesh, India

S. Singh
Department of Soil Science and Agricultural Chemistry, College of Post Graduate Studies in Agricultural Sciences, Central Agricultural University (Imphal), Umiam, Meghalaya, India

A. Pal · Priya
Department of Entomology, Post Graduate College of Agriculture, Dr. Rajendra Prasad Central Agricultural University, Pusa (Samastipur), Bihar, India

S. Thokchom
Department of Entomology, Assam Agricultural University, Jorhat, Assam, India

W. Majeed
Department of Zoology, Wildlife and Fisheries, University of Agriculture, Faisalabad, Pakistan

Mohd. S. Akhtar
Department of Botany, Gandhi Faiz-E-Aam College, Shahjahanpur, Uttar Pradesh, India

A. Kuanar et al. (eds.), *Recent Advances in Bioremediation and Phytoremediation*, Environmental Science and Engineering,
https://doi.org/10.1007/978-3-031-77884-1_7

massive attention causing an environmental crisis. The accumulation of soil contaminants has emerged since the Green Revolution (GR). It affects plant growth and development thereby degrading the surrounding environment. This chapter focuses on how agricultural chemical inputs became a source of soil contaminants and the different types of soil contaminants that exist. Furthermore, it focuses on how fertilizers and pesticides produce heavy metals as soil contaminants. Heavy metal bioaccumulation and transformation in the soil, plants, and environment are also explored. Soil contaminants impact sustainable agriculture and human health has been analyzed. Various reclamation measures for heavy metals as soil contaminants in different types of soil have been discussed. We then discussed two biological remedial approaches: phytoremediation and micro-remediation. The importance of biological measures over physical and chemical remedial measures has been explained. In this chapter we also discussed the importance of soil contamination reclamation environmentally, economically, and socially and its role in achieving a sustainable environment. Finally, we concluded the chapter by identifying some research challenges and promising approaches toward sustainable agriculture without compromising crop productivity and environmental friendliness.

Keywords Soil contaminants · Green revolution · Sustainable agriculture · Phytoremediation · Micro · Remediation · Crop productivity

7.1 Introduction

The global ever-increasing population has led the Green Revolution (GR) to fulfill the unlimited food demands from the limited natural resources with the application of chemical inputs such as inorganic fertilizers, pesticides, herbicides, etc. The chemical inputs contain heavy metals, which become contaminants when applied to the soil. The heavy metals are considered "soil contaminants" as they deteriorate the soil health (physical, chemical, and biological properties) thereby plant and threat to animals and human survival. Besides heavy metals, organic pollutants, hydrocarbon, and radioactive metals are also soil contaminants, however, lesser contribution to deteriorating soil health when compared with heavy metals. There are various sources of soil contaminants which may be from agricultural practices, especially conventional farming practices under GR, and the residues (effluent and sludge) of industrial activities.

The heavy metals once enter the soil system, persist for a longer period and become toxic with an increase in concentration; and being high density and high atomic weight, they are harmful to the soil organisms and plants when consume and in contact thereby rendering a threat to the environment. The heavy metals affect the soil properties which negatively impacted the soil fertility thereby affecting plant growth and development. The region where the plant's survival is threatened due to soil contamination is a threat to the sustainable environment as the reduction of trapping of atmospheric Carbon Dioxide (CO_2) by plants leads to global warming and then climate change. Moreover, the Soil Organic Matter (SOM) content also gets reduced when vegetation is affected which deteriorates the soil health and hampers the crop productivity. So, the natural components are interlinked, when a component is disturbed, and the other components malfunction.

The remediation of soil contaminants must be focused on all directions from where it generates to where it sinks. Besides, a complete exclusion of inorganic fertilizers from agricultural practices and the stoppage of industrialization activities is not a possible way to avoid heavy metal contamination in soil. Scientists introduce different soil remediations such as physical, chemical, and biological processes. The effects of the former two remediation processes are short-term whereas biological remediation is considered long-term as it involves the utilization of plants (phytoremediation) and microbes (microremediation). The biological remediation triggers to rejuvenate the contaminated soil into an ecofriendly environment which will support plant growth and development without compromising crop productivity.

7.2 Sources of Soil Contaminants

In nature there are several kinds of sources that can increase soil contaminants, it can be direct or indirect (Table 7.1). Direct sources of soil contaminants involve the immediate introduction of pollutants into the soil, typically through accidental releases, improper practices, or specific industrial, agricultural, or mining activities. It's important to identify and address these direct sources to prevent or mitigate soil contamination and their associated environmental and health risks. Indirect sources of soil contaminants refer to pathways or processes through which pollutants enter the soil. These sources do not directly release contaminants into the soil but rather contribute to soil contamination through various intermediate mechanisms.

Table 7.1 Sources of soil contaminants

Direct sources	Indirect sources
(a) Fertilizers (Pb, and Cu)	(a) Use of phosphate fertilizers
(b) Pesticides (Cu, Hg, Mn, Pb, and Zn compounds)	(b) Through treatments of organisms
(c) Manure (As, Cd, Cu, Hg Pb, Ni, Mo, Zn, and Ti)	(c) Irrigation with untreated wastewater
(d) Biosolids (sewage-sludge)	
(e) Wastewater	
(f) Metal mining, milling processes, and industrial waste	
(g) Air-borne sources (As, Cd, and Pb)	

7.2.1 Direct Inputs of Soil Contaminants

7.2.1.1 Fertilizers

According to Anyakora et al. (2013) the inception of agriculture marked humanity's foremost impact on soil. The nourishment and advancement of plants hinge critically on the absorption of nutrients, encompassing both major nutrients and vital micronutrients. These essential heavy metals, crucial for optimal plant growth, can either be sourced from the soil or administered through foliar spraying (Lasat 2000). For instance, where cereal crops are cultivated in copper deficient soils, occasional soil amendments with copper and manganese might be employed, while manganese may be incorporated into root crop cultivation. To enhance soil fertility in intensive farming practices, fertilizers containing nitrogen, phosphorus, and potassium are commonly employed. However, prolonged fertilizer application can substantially augment the presence of heavy metal impurities in soil, including elements such as lead, and copper.

7.2.1.2 Pesticides

Pesticides that are commonly used in agriculture and horticulture contained substantial metal concentrations. It has been estimated that approximately 10% of the pesticides approved in the UK recently contained copper, mercury, manganese, lead, or zinc compounds. The use of copper-containing fungicidal sprays like bordeaux mixtures, copper chloride, and copper sulfate are examples (Jones and Jarvis 1981) of these pesticides. Australian and New Zealand people use arsenic containing compounds extensively to control cattle ticks and banana pests.

7.2.1.3 Manures

Agricultural manures and compost have unintentionally accumulated heavy metals, including cadmium (Cd), arsenic (As), lead (Pb), chromium (Cr), copper (Cu), zinc (Zn), mercury (Hg), nickel (Ni), molybdenum (Mo),titanium (Ti), and others. Animal wastes like manure from pigs, poultry, and cattle are often applied as solids or slurries to crops, and pastures. Manure from poultry, and pigs can also pollute soil because copper and zinc are added to diets as growth promoters, and arsenic is found in poultry health products (Sumner 2000; Chaney 1996). Using repeated applications over a period can lead to heavy contamination of soil with these metals if animals consuming these diets produce large amounts of manure.

7.2.1.4 Biosolids

USEPA, 1994, states that biosolids (sewage sludge) are solids created during wastewater treatment. In several countries, biosolids are applied to the land, allowing urban populations to reuse these materials (Weggler 2004). As a result of its widespread recognition and regulatory definition, sewage sludge is frequently used in references.

7.2.1.5 Wastewater

The utilization of municipal and wastewater effluents for land irrigation has become a prevalent practice in contemporary times (Reed 1955). It is estimated that approximately 20 million hectares of arable land worldwide are irrigated with wastewater. Particularly in numerous cities across Asia and Africa, wastewater irrigation is believed to play a significant role in supplying vegetables to urban areas (Bjuhr 2007). However, it is important to note that most farmers are primarily focused on enhancing crop yields and profits, often prioritizing these factors over considering potential environmental benefits or hazards.

7.2.1.6 Metal Mining and Milling Processes and Industrial Wastes

Mining operations have caused notable soil metal pollution in numerous countries. Throughout mining activities, dense particles called tailings, which settle at the base of flotation cells, are frequently released directly into natural low areas, including wetlands. This practice results in heightened levels of pollutants (Devolder et al. 2003). Despite various available approaches for rehabilitating these sites, the processes are often prolonged, expensive, and may not completely reinstate soil fertility. The extent of environmental exposure to heavy metals is influenced by how bioavailable these metals are within the soil. There are two routes through which contamination assimilation takes place: the uptake of heavy metals by plants cultivated in polluted soil that subsequently enter the food chain, and direct ingestion of contaminated soil (oral bioavailability).

7.2.1.7 Air-Borne Sources

Airborne sources of metals include stacks and ducts, as well as fugitive sources like dust from storage areas and waste piles. Metals released by airborne sources are usually in the form of particulate matter. High-temperature processing can lead to the volatilization of several metals, including As, Cd, and Pb. These metals must be exposed to a reducing atmosphere to prevent them from converting into oxides and condensing into fine particles (Smith et al. 1995).

7.2.2 *Indirect Inputs of Soil Contaminants*

7.2.2.1 Use Phosphoric Fertilizers

As rocks undergo weathering and atmospheric deposition, the phosphorus content within soils undergoes a gradual replenishment process due to crop uptake and erosion. Consequently, the application of both organic and inorganic forms of phosphorus becomes imperative in crop production (Liu et al. 2021). The primary source of phosphorus for modern agricultural practices is Phosphate Rock (PR), sourced from sedimentary ores (constituting 75–80%), which is a finite resource. Within these sedimentary ore deposits, phosphate rocks are commonly discovered to harbour trace elements and radionuclides (El-Bahi et al. 2017).

7.2.2.2 Through Treatments of Organism

As organic fertilizers sourced from animals or humans are introduced to soils, there is a growing concern regarding the potential emergence of antimicrobial substances and organisms resistant to such agents. The application of antimicrobials serves not only to prevent diseases but also to stimulate growth in animals. It's noteworthy that a significant portion, ranging from 75–90%, of antimicrobials administered to both humans and animals undergo incomplete metabolism, subsequently being excreted and accumulating in animal manure, municipal wastewater, sewage sludge, and biosolids (Bouki et al. 2013; Daghrir and Drogui, 2013).

7.2.2.3 Irrigation with Untreated Waste Water

The utilization of untreated wastewater for irrigation purposes can lead to the buildup of trace elements, the accumulation of organic contaminants, the presence of harmful pathogens, and the contamination of pharmaceuticals and personal care products (Islam et al. 2019). Direct irrigation using non-treated wastewater is a practice observed in only a handful of nations, with the quantities involved exhibiting significant variations. For instance, in Mexico, the annual employment of this approach results in the utilization of up to 4 million cubic meters of wastewater.

7.3 Types of Soil Contaminants

Soil contaminants primarily are either organic or inorganic compounds (Fig. 7.1). They have the potential to cause hazards (Khan et al. 2008). Inorganic soil contaminants that prevail most often are trace elements like Cd, As, Pb, Cr, Cu, Zn, Hg, Ni, Mn, and radionuclides (Mekouar 2020). Depending on the soil pH, inorganic pollutants may be present as dissolved ions in the soil solution (Goldemund 2004). By nature, being persistent, they occur in various forms like oxides, sulphides, salts, or organometallic complexes. Soil pollution can be caused by radionuclides or by trace/heavy metals. The former soil pollution originates either from natural processes like weathering of parent rock and volcanic eruptions or from human activities like radium refinement (for cancer treatment) and uranium, usage of radioactive phosphate, cobalt-ores, as well as nuclear accidents/ nuclear weapon tests (Gad et al. 2019). The major distinction between radionuclides, pollution and pollution from trace elements is that radionuclides can harm live cells by radiation released during radioactive decay, whereas, trace elements' negative impacts are linked to chemical processes occurring within living cells (Monged et al. 2020). Whereas molecules that are carbon based i.e., organic contaminants mostly originated due to anthropogenic activities, and some derived from natural processes like volcanic eruptions or wildfires. Sometimes unintentionally, they might be produced in the form of by-products like industrial emissions, which are mostly obtained from petroleum industries (that release polycyclic aromatic hydrocarbons i.e., PAHs) and mining activities (Valentin et al. 2013). Natural processes like volcanic eruptions and wildfires also contribute to the formation of PAHs (Kim et al. 2011). There are various types of soil contaminants that can negatively impact the quality of soil and harm the environment as well as human health. Some soil contaminants discussed here:

(a) Heavy Metals

These include trace elements such as Cd, Pb, Hg, As, and Cr which can accumulate in the soil through mining, industrial practices, waste disposal, and agriculture. They can persist longer in the environment and have toxic effects on living beings (Alengebawy et al. 2021).

(b) Organic Pollutants

Organic contaminants are derived from petroleum, pesticides, herbicides, solvents, and other chemicals. For instances, Polychlorinated Biphenyls (PCBs), Dioxins Polycyclic Aromatic Hydrocarbons (PAHs), and certain Volatile Organic Compounds (VOCs). These substances can contaminate soil through spills, leaks, improper waste disposal, and agricultural practices (Mekouar 2020).

(c) Pesticides and Herbicides

These are chemical substances used to control pests, insects, weeds, and diseases in agriculture. However, their indiscriminate use can accumulate in the soil, leading to long-term contamination and negative effects on soil ecosystems (Gilden et al. 2010).

(d) Petroleum Hydrocarbons

These contaminants come from petroleum-based goods like diesel, lubricants, industrial oils, and various types of fuel. Spills, storage tank leaks, and incorrect disposal

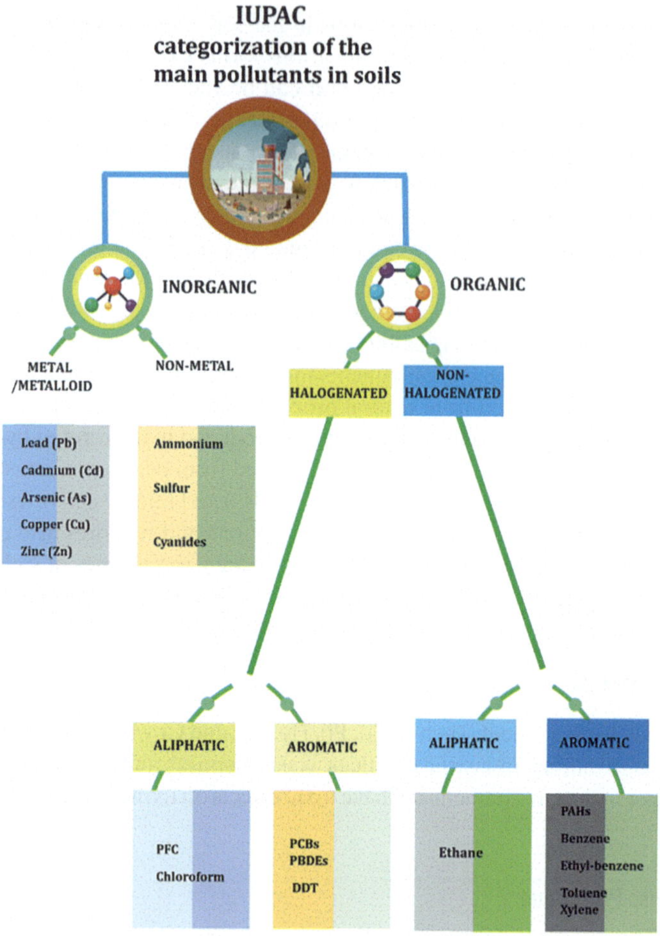

Fig. 7.1 IUPAC categorization of the main pollutants as soil contaminants

of petroleum-based goods can contaminate the soil, which will reduce soil fertility and microbial activity as well (Hussain et al. 2018).

(e) Persistent Organic Pollutants

Persistent organic pollutants (POPs) are a class of hazardous substances that may withstand environmental deterioration and linger in the environment for extended periods of time. Some pesticides, e.g., DDT, industrial chemicals, like PCBs, and unintended industrial byproducts, such as dioxins, are examples. POPs harm the environment and human health because they can build up in the soil and bioaccumulate living things (Namiki et al. 2013).

(f) Radioactive contaminants

Radionuclide contamination of soil may also result from improper nuclear waste disposal. With half-lives of 24, 29, and 30, respectively, strontium-90, plutonium-239, and cesium-137 are the primary radionuclides of concern at polluted sites. Because Radiostrontium and Radiocesium are like calcium and potassium, they are more poisonous and are therefore easily absorbed by plants, animals, and people. Strontium can act as a substitute for calcium, accumulate in the bones, and can result in dystrophic changes to the articular and bone systems.

Some biological contaminants, such as pathogens, bacteria, viruses, and parasites, can contaminate soil through improper disposal of animal waste, sewage, or contaminated water. Humans, animals, and plants can all be harmed by biological pollutants (Bhat et al. 2023). It is significant to highlight that location, industrial activity, agricultural techniques, and previous land use can all affect the presence and amounts of soil pollutants. To lessen the consequences of soil contamination and safeguard the environment and the welfare of people, soil testing and remediation procedures are frequently used (Goldemund 2004).

7.4 Why Heavy Metals Are Considered "Soil Contaminants"

Heavy metals are considered soil contaminants because they have the potential to accumulate in the soil and pose risks to nature and human health (Oves et al. 2012). These are elements that occur naturally with atomic weights and densities being high. They can enter the soil through various sources, including industrial activities, mining, agricultural practices, and atmospheric deposition (Tchounwou et al. 2012). The reasons behind the heavy metals as "soil contaminants" are as follows:

(a) Toxicity

They have toxic effects on living beings like plants, microbes, and animals. When present in high concentrations, they can interfere with soil microbial activity, adversely affect soil fertility, and ultimately hinder plant growth (Singh and Kalamdhad 2011).

(b) Persistence

In the environment, they are generally highly persistent and can prevail there in the soil for a longer period. They are difficult to degrade and can build up over time, leading to increased concentrations that can exceed safe levels (Pingali 2012).

(c) Bioaccumulation

They can be absorbed by plants from the soil. Once in the plants, they can enter the animal or human body via the food chain. As a result of this bioaccumulation, species at higher trophic levels may contain larger quantities of heavy metals, which could be harmful to humans if consumed (Arunakumara and Zhang 2008).

(d) Water Contamination

From contaminated soils, they can leach into groundwater or surface water. This may taint sources of drinking water and harm aquatic habitats, causing extensive environmental harm (Sankhla et al. 2016). Heavy metal exposure can have detrimental impacts on human health. They can enter the human body when in contact with contaminated soil directly, intake of unhygienic drinks or food, or when airborne particles are inhaled. Neurological difficulties, kidney damage, respiratory problems, developmental challenges in children, and in some cases, cancer are some of the possible health repercussions (Kampa and Castanas 2008). For reducing heavy metal contamination and guaranteeing the security of soil resources, proper soil management practices, remediation methods, and regulatory measures are critical (Lin et al. 2022).

7.5 Green Revolution in the Emergence of Heavy Metals as Soil Contaminants

The GR had implications in the emergence of heavy metals as soil contaminants (John and Babu 2021). The GR significantly increased crop yield and food security, but it also brought about several practices that have facilitated the buildup of heavy metals in soils. The main objective of GR was to enhance the production of food grains and ensure that the quantity is sufficient to feed all, however, repercussions on the environment were not considered (Taylor 2020).

One of the key factors is the use of chemical fertilizers and pesticides (Pingali 2012). During the GR, there was widespread adoption of synthetic fertilizers and agrochemicals to enhance crop yields. With discovered levels surpassing the established limits of pesticide presence, it is said that the presence of pesticides in freshwater is a costly concern (Choudhary et al. 2018). These inputs often contain heavy metals like Pb, Hg, and Cd as impurities. Over time, indiscriminate usage of these pesticides and fertilizers resulted in heavy metals' accumulation in the soil (Yargholi and Azarneshan 2014).

In addition to this, the GR also encouraged intensive farming practices, including monocropping and irrigation (Pingali 2012). Monocropping, or the cultivation of a single crop over large areas, can result in the depletion of soil nutrients and increased vulnerability to pests and diseases. Farmers frequently move towards more pesticides and fertilizers to combat these problems, which exacerbates the buildup of heavy metals. Furthermore, the expansion of irrigation systems during the GR also contributed significantly to heavy metal contamination (Singh 2000). Irrigation water may contain dissolved heavy metals, which can infiltrate the soil and accumulate over time. In some cases, the use of poor-quality water sources for irrigation, such as contaminated groundwater, can introduce heavy metals directly into the soil. The GR promoted the use of heavy tillage implements which in the long run led to soil compaction and ultimately helped in the accumulation of heavy metals at the natural sources (Chang et al. 2014).

To mitigate the emergence of heavy metals as soil contaminants, sustainable agricultural practices are crucial (Seleiman and Hafez 2021). This entails using organic farming techniques to lessen the need for synthetic fertilizers and pesticides, applying crop rotation and variety to increase soil fertility, and employing water management techniques to reduce heavy metal contamination through irrigation systems (Lin et al. 2022). Additionally, regular soil testing and monitoring can help identify and manage heavy metal contamination in agricultural soils.

7.6 How Do Heavy Metals Affect Soil Health and Help in Soil Degradation

Heavy metals prevailed in the environment naturally within a non-toxic limit without posing any harmful effect on the environment, rather some of the heavy metals have bio-importance to nature and humans. But anthropogenic initiatives such as chemicals application, mining, industrial establishments, electroplating, smelting operations, and geologic activities resulted in the over-exploitation of the natural environment leading to elevate heavy metal content and its toxic level which causes environmental degradation. Similarly, the soil environment also consists of various heavy metal elements which are indispensable for soil health and crop growth, but their abundance can affect the soil environment and helps in the degradation of physical, chemical, and biological properties of soil.

The elements have more than 20 atomic numbers and show metallic characters despite having high relative density and atomic weight characterized as heavy metals. Heavy metals are the elements that are needed in trace amounts and have important roles in organisms like Zn, Ni, Cu, Fe, Mn, and Co which are also well-known by micronutrients. On the other hand, metallic elements like Pb, Cd, Hg, As etc. are reported to be detrimental to the environment in both animals and plants, considered the "main threats" as higher concentration and accumulation of these metals can induce harmful impacts not only on human health and plant growth but also soil environment.

7.6.1 Heavy Metals in Soil Physical Properties

Heavy metal contamination can affect soil's physical properties. The extent and specific effects can vary depending on the type and concentration of the heavy metals present, the soil's characteristics, and the surrounding environmental conditions. Human activities like mining, industrial establishments, electroplating, and agricultural practices resulted in the release of metals, which can accumulate in soils over time.

The major impact of heavy metal pollution on soil physical properties is through altering the soil structure. Heavy metals can prevent the development of stable soil aggregates, resulting in less porous soil and more compacted soil. As a result, air, water, and nutrient circulation through the soil is reduced which negatively impacts plant growth and overall soil health. Plant development and the condition of the soil may suffer from reduced air, water, and nutrient circulation within the soil.

Heavy metals can also affect the soil's capacity to store and infiltrate water. Because some heavy metals have a strong affinity for binding to soil particles, the soil may retain more water as a result. This may lead to waterlogging and restricted oxygen availability, which might impair plant development and lower soil fertility. However, some heavy metals can also accelerate soil deterioration by raising water infiltration rates and generating excessive nutrient leaching, which can cause nutrient depletion and decreased soil fertility.

7.6.2 Heavy Metals in Soil Chemical Properties

Organic contaminants are generally biodegraded into inorganic form by mineralization, whereas most metals in soils are not degraded by microorganisms or chemicals, unlike organic contaminants. So, these heavy metal elements persist in soil for prolonged periods, which leads to detrimental effects on soil and contributes to soil degradation. The contamination of toxic metal elements can significantly impede the biodegradation of organic pollutants in soil.

When heavy metals like Pb, Cd, Zn, Ni, Cr, and Cu are present in the soil, they may deflect the chemical characteristics of the soil. But Cr (III), Cu (II) and Pb (II), the most well-known hazardous metals, are more likely than Cd, Ni, and Zn to be selectively adsorbed on soil. The soils having comparatively high Cation Exchange Capacity (CEC) are most likely to have adsorbed toxic metals like Cr (III), which are characterized by a high ionic charge. Such absorption of heavy metals to soil particles resulted in a reduction of the availability of essential nutrients such as N, P, and K for plant uptake. This can lead to nutrient deficiencies inplants, affecting their growth and overall health.

In addition, heavy metals can have the potential to alter soil pH. In comparison to Cd^{2+}, Ni^{2+}, Zn^{2+}, and notably Ca^{2+}, the cations like Cr^{3+}, Cu^{2+}, and Pb^{2+} have a stronger proclivity to hydrolyze in solution and these ions also tend to absorb soils. As a result, soil solutions from soils treated with Cr^{3+}, Cu^{2+}, and Pb^{2+} have higher concentrations of hydrogen (H^+) ions than soil solutions from soils modified with the other metals. Basta and Tabatabai (1992) showed that the tendency of reduction of soil pH followed the order-Cu = Pb > Cd = Ni = Zn with a given quantity of metal amended.

Thus, heavy metals significantly affect the chemical characteristics of soil and cause soil deterioration. Their presence can affect the availability of nutrients, change the pH of the soil, damage the soil's structure, and have harmful effects on plants, animals, and people. Implementing appropriate management strategies to lessen their adverse effects and promote healthy and productive soils requires an understanding of how heavy metals affect the soil.

7.6.3 Heavy Metals in Soil Biological Properties

Soil microorganisms, being the living part of the soil, mediate all the biochemical reactions inside the soil and nutrient cycling to maintain the soil health and nutrient availability to plants. The heavy metal contamination in soil results in a heavier impact on soil microorganisms compared to soil plants and animals that thrive on the same soil. The influence of heavy metal pollution on soil biological health is much faster and more sensitive as compared to the chemical and physical properties.

It was at the beginning of this century that the first observations were made regarding the heavy metals' impact on soil microbial processes. However, it wasn't until the 1960s and 1970s that it became clear how seriously elevated metal concentrations could affect soil microorganisms and soil microbial processes, often leading to severe ecosystem disturbance.

Heavy metal pollution (Cr, Zn, and Cd) was reported to have an impact on the metabolism of soil microorganisms in every situation. They have harmful effects on soil microbes, which change their activities, population, and diversity. Generally, a rise in metal content has a negative impact on the microbiological characteristics of the soil, such as soil respiration and enzymatic activity, which appear to be extremely helpful in dicators of soil pollution. However, the degree to which metals have an impact on biological and biochemical features depends significantly on soil characteristics such as soil pH, SOM (soil organic matter), and as well as clay content.

7.7 Role of Agricultural Inputs in Soil Contamination

7.7.1 Heavy Metals from Chemical Fertilizers

Transport of heavy metals to soil is the primary fate of transport from chemical fertilizer. Volatilization, erosion, exporting harvested plants, and leaching are the primary pathways for losing heavy metals from soil. The secondary fate is the chemical conditions inside the soil. Another chemical factor is the soil pH for influencing heavy metals in the environment. When the pH level of the soil is low, metal ionization increases and subsequently increases water mobility and solubility. Secondly, the hydronium ions start displacing other cations and start producing negative charges in surface area. These reasons affect cation exchange and other factors in soil with organic complexation. Heavy metals present in chemical fertilizers are transferred from foliar dust, farm produce, ingestion of fertilizer by animals, fertilizer dust, and sometimes indirectly from the soil by uptake of roots (Sharma et al. 2018). Various sources which facilitate heavy metals are atmospheric deposition, biosolids, and inappropriate fertilizer use in soil systems. Asis a sulfur and oxysalt-containing mineral, not toxic when avail around 0.01–0.10 mg/kg but it is toxic when it is present 40–200 mg/kg (Minnesota Department of Health 2023).

When the pH of the soil is low, Cd increases mobility and most mobile metals inside the soil system. When phosphorous (P) fertilizer is applied, Cd increases in two different ways. One is when both P and Cd are added, plant root uptake starts. Secondly, when the plant vigor and root growth are simulated, Cd uptake increases more and leads to phytotoxicity. In soils that contain high chloride (Cl^-) from the irrigated saline water or salinization of dry land, the Cd uptake starts increasing. Acidic sandy soil having low organic matter has more Cd as compared to clay soil. The concentration of Pb increases during food processing, handling, and storage (Agritopic 2021). Therefore, welded cans should be recommended to reduce lead in the food. Potassium (K), and nitrogen (N) fertilizers are low in heavy metals and do not violate the maximum levels in farm produce. But in P fertilizer, the amounts of heavy metals are high as it is derived from acids and phosphate rock during manufacture. Micronutrients fertilizer such as Zn and Cu may sometimes contain high Pb and Cd (Agritopic 2021). When micronutrients with high content of Pb are applied, it increases the Pb concentration in the soil as compared to root uptake. Thus, the farm produce will have an increased concentration of Pb.

7.7.2 Heavy Metals from Pesticides

Pesticides affect the living biota such as mammals, honeybees, birds, humans, plants, and predators. It creates an imbalance in the environment and affects biodiversity (Paoli et al. 2015). Pesticides that are systemic contaminate the nectar parts through the vascular system of the plant, causing high mortality in nectar-feeding insects and honeybees (El-Wakeil et al. 2013). Pesticides such as endrin, dieldrin, and chlordane remain in the soil and environment for a long time. The pesticides were transferred to the plants by phyto-remediation, phyto-stabilization, and rhizo-filtration. The pesticide residues and their by-product through degradation are usually uptake by the xylem through roots which affect the plant parts causing Root Operating System (ROS) overproduction, Deoxyribonucleic Acid (DNA) damage, oxidative stress, chlorosis, blocking photosynthesis, twisted leaves, and plant death (Sharma et al. 2018). Pesticides can alter biological and physic-chemical properties in the soil. Thiamethoxam decreases the activity of phosphatase by 6.5% (Han et al. 2017). In soil, malathion, chlorpyrifos, endosulfan, and lindane alter the processes of denitrification and nitrification although they are applied in recommended doses. Cypermethrin causes a reduction of 32.8% dehydrogenase enzyme and 74% nitrifying bacteria

(Filimon et al. 2015). Herbicides based on glyphosate affect the microbial fauna and altered the biosynthetic mechanism and physiological system in soil. Insecticides disrupt the microbial protein structure, soil chemistry changes, and enzymatic activity in the soil. In plants, it reduces the grain protein, photosynthetic process, and blocking stomata (Ahemad et al. 2012). In soil, herbicide reduces the soil nutrient, phosphatase suppression, and nitrogenase activity. In plants, it affects the biochemical and physiological efficiency and plants' susceptibility to diseases (Saladin et al. 2005). Fungicides interrupt the phosphatase in soil, dehydrogenase activity, ureases, and nitrifying bacteria. In plants, carotenoid and chlorophyll are reduced, the chloroplast is destroyed, and suppression of electron transfer (Petit et al. 2012).

7.8 Fate of Heavy Metals in Soil

Heavy metals are divided among different environmental compartments such as water, soil, suspended particles, and biota (Gautam et al. 2016). Distribution of heavy metals is controlled by processes viz., dilution, dispersion, sedimentation, adsorption, and desorption (Xu et al. 2023). They reside in dissolved and colloidal ionic particle phases (Evangelou et al. 2007). Warren (1971) revealed a strong affinity for metals in the humic acids, organic clays, and oxides finished using organic compounds. Typically, its dissolved types of heavy metals are ionic as well as unionized organometallic chelates or complications. While pH of groundwater, as well as soil, largely determines how these heavy metals are soluble (McNeil and Waring 1992), quantity made from metals, capability of interchange of cations, content of organic carbon (Elliot et al. 1986), the oxidation state for minerals as well as system of redox potential. The pH level of the soil tends to possess a greater total effect over the dissolution or preservation of heavy metals within soils that any other individual variable (Ghosh and Singh 2005). High soil pH was found to increase the metal cation retention and decrease solubility (Basta et al. 1993). The chemical structure of the soil's mixture along with the level of moisture present in the soil has a substantial impact upon the amounts of heavy metals and their diversity (Fotovat et al. 1997).

7.8.1 Bioaccumulation of Heavy Metals

Both natural and man-made processes introduce pollutants as well as heavy metals into the soil environment. Metal variation in the soil is mostly linked to heavy metal poisoning. The percentage of heavy metals is significantly impacted by soil organic content (Khan et al. 2015). When heavy metals defect the food chain primarily, the soil is a major source of waste management along with rubbish disposal (Zhuang et al. 2009). The soil-crop system in agricultural areas is the main way that heavy metals enter the human body (Liu et al. 2007) and the main sources of contamination are caused by anthropogenic activities. The pH of the soil has a significant influence on heavy metal absorption as well as a hazard for biological organisms (Amini et al. 2005). Soil pH has a considerable impact and a negative correlation with Cu bioaccumulation in plants (Wang et al. 2013). The development, reproduction, occurrence, and survival of several plant species in metal-defected soils are proof these various heavy metals bioaccumulate over time. Various plant species demonstrate varying degrees of toxicology of equivalent contaminants due to the equivalent atmospheric state they have various mechanisms for absorbing components. Heavy metals have a propensity to build up in the tissues of food plants. The introduction of unnecessary substances into our environment could result in plant bioaccumulation (Kieffer et al. 2009). Elements that are traceable possess the ability to bind to roots by absorbing in a contaminated soil atmosphere, increasing the potential for negative impacts on animals and plants alike.

Pollution due to heavy metals is a major problem nowadays and among those Pb accounts for 10% whose accumulation leads to a 42% reduction in the root growth (Collin et al. 2022). Heavy metals have several harmful effects on plants like photosynthetic inhibition, chlorosis, and low biomass accumulation (Balkhair and Ashraf 2016). Heavy metals interfere with the roots and soil. As a result, they accumulate in the root so root vegetables contain the highest percent of toxins. Phytoremediation uses living plants to remove toxins accumulated in the environment (Rigoletto et al. 2020), and rhizofiltration is used to remove heavy metals from soil.

7.8.1.1 Bioaccumulation of Heavy Metals in Fauna and Flora

The impact of heavy metals on biota depends on their lipid content and the composition of biological tissue. Humans are exposed 2to dimethyl mercury through food, and they can easily cross the blood–brain barriers. Hypertension and serious cardiovascular diseases are caused in humans due to As toxicity. They also have a negative impact on spermatogenesis and testosterone release in rats. Selenium (Se) toxicity affects by creating oxidative stress. Some heavy metals support plant growth but in excess concentration, they causes toxicity to the plants.

7.8.1.2 Effect of Heavy Metals on Invertebrates

Many invertebrates are taken as food by birds and mammals. The concentration of heavy metal on isopoda was highest and coleoptera was the lowest. Cu, Pb, Cd concentrations in body increased with soil concentrations. Higher quantity of Cd is found in vegetarian diet like spinach, sunflower seeds etc. (Spungen et al. 2019) as compared to non-vegetarian diet (Li et al. 2019). Cd can lead to breast cancer development due to its ability to bind to oestrogen receptor alpha which leads to breast cancer development (Siewit et al. 2010). As are used in lipsticks and skincare products which may have many effects like cancer, nausea, fatigue, vomiting etc. (Naujokas et al. 2013).

7.8.1.3 Effect of Heavy Metals on Fishes

Since fishes are the topmost consumers in aquatic ecosystems, they are the most affected (Luo et al. 2014). It affects the interaction of fish with the environment by affecting their central nervous system (Baatrup et al. 1991). Fish take up Zn through gills. These heavy metals may transfer to other fish and food chains. Mercury has so much affinity with proteins so it accumulates easily in the human body.

7.8.2 Mode of Action

Regarding the soil's adsorption–desorption or precipitation-dissolution processes, the concentration of heavy metals is dangerous in soil solution. Protons or different metal cations might get released as the outcome of the dissolution of heavy metals in soil. According to Doner et al. (1982), Ni, Cu, and Cd were added, which assisted in releasing Mn through the substrate, and most of this release was probably caused by proton break with Ni and Cu sorption. Through non-specific electrostatic forces, to keep residual metal cations in an approach the bigger silicates in soil offer each consistent values and influenced by pH values. Such minor components contend between Ca as well as Mg among the accessible interchange locations. In lower pH mediums, heavy metal retention is higher, but in higher pH systems, retention is lower. The most frequent and frequent reaction is metal hydrolysis (Brady and Weil 2002). The highest levels of adsorption were found in soils with the highest concentrations of clay content, Cation Exchange Capacity (CEC), organic matter, calcium carbonate ($CaCO_3$), and pH; in contrast, the highest levels of the binding energy constant were found in soils with the lowest concentrations of $CaCO_3$ and total Fe (Rana and Kaushal 1983). Further they noted that the release of adsorbed Cd reduced as pH, organic matter, $CaCO_3$, CEC, and clay concentration rose. Less Cd was liberated from soils with high adsorption maxima and binding energy constants. Revealed that iron hydroxide (Fe-OH) absorbed up to 96% of the Hg from the solution in the pH range of 5–8, showing that the predominant mechanism of Hg adsorption is the sorption of Hg $(OH)_2$ molecules. With increasing alkalinity of solution, the amount of Hg sorption by clay minerals increased, reaching maximum values at a pH > 6 (Obukhovskaya 1982). Compared to Cd and Ni, Zn's adsorption isotherms displayed stronger pH effects, although they were mostly unaffected by an increase in Cl concentration. Additionally, it was discovered that Cd adsorption increased with a rise in Cl-concentration from 151 to 152 Mol Cl-/l and decreased with a fall in pH (Girth et al. 1981).

7.8.3 *Transformation*

Fe (III), Mn (III/IV), and Cr (III/IV) are three hydroxides and oxides of iron, manganese, and chromium respectively that are frequently found in soils and sediments as suspended particles as well as coatings on the surfaces of clay materials. These hydroxides are thermodynamically stable at pH 7 in oxygen-containing environments. In anoxic conditions, reducing agents dissolve oxides and hydroxides through reduction. Mn (IV) and Mn (III) oxides and hydroxides are converted through the most solubility manganous within anoxic as well as acidic situations. Among the initial discernible consequences of decreasing circumstances caused due to waterlogging proved to be a boost in water soluble and disposable Mn. The easily reducible Mn component was found to have decreased by over 80% in less than 40 days following submersion (Patrick and Turner 1968). Nearly all the Mn that was reducible below pH 5 became water soluble and exchangeable. In the soil environment, Cr (IV) is mobile and found as $Cr_2O_7^{2-}$ and CrO_4^{2-} anions (Fendorf and Zasoski 1992). While Cr (III) can be oxidized to Cr (IV) by Mn (III)/(V) oxides, Cr (IV) can also occur as Cr^{3+} and its hydroxide counterparts, CrO^{2-}. According to White and Peterson (1998), iron (II) oxides such magnetite (Fe^{2+}, Fe_2^{3+}) O_4 and ilmenite (Fe^{2+}, Ti) O_3 significantly contribute to the reduction of inorganic pollutants. They claimed that a common Fe^{2+} oxide can reduce Cr (IV) to Cr (III). However, they claim that under field conditions and for longer periods of time, the capability of oxide minerals to reduce matter, the impacts upon surface polishing and opposition, and the toxicity caused by aquatic organisms all affect how effectively contaminants like heavy metals are reduced.

7.9 Effects of Heavy Metals on Different Components

Due to modernization and industrialization, environmental pollution has gained much attention around the world recently. The production of toxic chemicals, pesticides, and petroleum products has increased the number of pollutants in natural resources such as water, soil, and air. As a result, it adversely affects plants, soil, and humans. Several heavy metals are present in the environment in large quantities, such as Cd, Pb, As, Hg, and Ni. Globally, these metals pose a serious environmental threat. Besides being toxic to plants and humans, heavy metals make soil less fertile, slow plant growth, and interfere with photosynthesis, as well as reducing soil nutrient content.

7.9.1 *On Plant*

Plants face various challenges in both biotic and abiotic environments, which have a direct impact on their growth and development. In addition to insects, nematodes, bacteria, and fungi, a wide range of organisms can impose biological stress on plants. Similarly, abiotic stress arises from extreme temperatures, light conditions, droughts, salinity, and, heavy metal pollution. With increasing environmental pollution, heavy metal contamination has emerged as a major concern. Heavy metals, being non-degradable, can exert adverse biological effects (Xu et al. 2023). While plants require certain heavy metals for their growth, the presence of excessive amounts can be detrimental to their survival. High concentrations of metals directly impact plants by inhibiting cytoplasmic enzymes, causing oxidative stress-induced damage to cell structures, disrupting metabolic processes, and hindering the normal functioning of biomolecules. These alterations ultimately affect vital plant processes such as respiration and photosynthesis. Exposure to heavy metals leads to the production of reactive oxygen species (ROS), including superoxide and hydroxyl free radicals, as well as non-free radicals like hydrogen peroxide (H_2O_2), resulting in oxidative stress by disrupting the balance between pro-oxidants and antioxidants in plant cells. This condition leads to protein and lipid oxidation, ion leakage, redox imbalance, denaturation of cell structures, and activation of programmed cell death pathways. While metal ions serve as micronutrients essential for plants at normal concentrations, they become toxic when present in higher amounts. Based on their physiochemical properties, bioactive metals can be classified as redox-active metals and non-redox-active metals.

7.9.2 On Soil

Heavy metal contamination has increased because of the development of economy and society. Because heavy metals are colorless and odorless, they are difficult to detect. When heavy metals contaminate soil, the environment is not explicitly damaged within a short time frame, but if their contamination exceeds environmental tolerance, serious ecological damage can occur. These contaminants are usually chemical time bombs (CTBs). When pollution occurs in the air and water, it can be resolved by dilution and self-purification after the sources are removed. Heavy metal contamination and soil quality can be improved by dilution or self-purification, but they are difficult to use. In some cases, resolving heavy metal-contaminated soil can take a century or more. Heavy metal contamination caused by a variety of heavy metals will always be amplified by separate contamination caused by heavy metals. Several studies have found that soil has been considerably contaminated by metals both deposited on the road surface and within nearby industries as well as metals relocated from the road surface (Harrison et al. 1981). The adverse environmental and ecological effects of heavy metal contamination in soils along major roads have been studied by several researchers (Turer et al. 2003). Among the examples given by Okunola et al. (2007) are plants and crops grown near major roads and in nearby industrial zones that accumulate metals through direct or indirect consumption by people and animals, particularly livestock. Turer et al. (2003) report that within a short period of time it can become lethal. As industrialization and urbanization accelerate in developing countries, heavy metal pollution in agricultural soils is becoming a serious problem. Because heavy metal pollutants don't biodegrade and have long half-lives in soil, they have a high risk of harming soil microbes and other soil life in the long run (Okunola et al. 2007). Several heavy metals have been documented in literature as possible hazards or occurrences in contaminated soils, including Cd, Cr, Pb, Zn, Fe, and Cu (Alloway et al. 1996). According to Long et al. (2002), soil metal concentrations range from less than one milligram per kilogram to 100,000 mg per kilogram. Urban and agricultural soils can become contaminated with heavy metals when synthetic materials are mined, manufactured, or used which includes pesticides, paints, batteries, and industrial wastes. The soils of old landfills may contain contaminants and heavy traffic on roads releases Pb, Cd, Zn, and Ni, which are present in the fuel as anti-knock agents (Atayese et al. 2008). Several studies have proven that heavy metals in soil can be used as powerful tracers for monitoring environmental impacts such as industrial emissions (cement plants, fossil fuel and coal combustion chemical plants), vehicular emissions, and atmospheric deposition. Heavy metal pollution levels are generally determined by collecting the topsoil adjacent to the roads. Some research extended the monitored metals to Cr, and Ni (Chen et al. 2010). Additionally, soil not only acts as a medium for plants to grow and a place for unwanted materials to be dumped, but it also transports pollutants into surface water, groundwater, and the atmosphere.

7.9.3 On Environment

The historical contribution of heavy metals to environmental pollution has been substantial. The escalating traffic linked with rapid industrialization and urban development has led vehicles to discharge heavy metals, exacerbating this issue. These metals permeate the soil, triggering adverse shifts in plant physiology and biochemistry, culminating in diminished growth, performance, and yields. This decline in agricultural productivity, because of reduced plant growth, underlies food insecurity. The interconnectedness between soil-crop systems in farming regions plays a pivotal role in human exposure to heavy metals, which are absorbed by crops. Vigilance in monitoring heavy metal concentrations within the environment is imperative due to their well-documented toxicity, especially for metals like Cd and Pb, along with their enduring and non-biodegradable attributes. The contamination of agricultural land assumes even greater significance as it can have dire health implications when these pollutants infiltrate the food chain.

7.9.4 *On Animal*

Soluble inorganic arsenic can induce immediate toxic effects. In cases of high exposure, alkylating agents can trigger the disintegration of red blood cells, enlargement of the liver, skin discoloration, limb numbness, brain impairment, and tingling sensations. Cd chiefly impacts health by causing harm to the kidneys and bones due to ingestion through diet and inhalation. Inhaling elemental mercury vapor can lead to symptoms like tremors, emotional instability, sleep disturbances, memory decline, neuromuscular alterations, headaches, and dysfunction in the kidneys and thyroid. Methylmercury, a potent neurotoxin, detrimentally influences human brain development. Its easy passage through the placenta and blood–brain barrier renders it especially perilous for fetuses during pregnancy. The International Agency for Research on Cancer categorizes methylmercury as potentially carcinogenic, with a connection to the development of cancer. Immediate excessive intake of copper can result in queasiness, vomiting, and abdominal unease. In higher quantities, copper toxicity can lead to liver harm and even fatality. Elevated copper exposure can also provoke the destruction of red blood cells, leading to anemia. While the body necessitates minimal nickel for proper functioning, excessive amounts can be detrimental. Prolonged exposure to nickel raises the risk of lung, nasal, larynx, and prostate cancers. Consequences of excessive nickel intake encompass asthma, persistent bronchitis, allergic responses (especially to jewelry), respiratory failure, lung embolism, birth anomalies, allergies, and heart ailments. Ni fumes act as respiratory irritants and may induce pneumonitis.

7.10 Soil Contamination and Sustainable Agriculture

Agriculture has undergone significant transformations in recent decades, leading to higher productivity. The improved technologies come along with severe issues like topsoil loss, tainted groundwater, problems of soil contamination, pollution hazards, and even qualitative degradation of the life of farmers including dissolution of rural communities. Sustainable agriculture has emerged as a solution for managing these issues with new and improved strategies. It exclusively focuses on environmental integrity, economic feasibility, and issues related to the equity of society (Brodt et al. 2011). So sustainable agriculture is an all-around solution for better management and protection of resources. It efficiently manages scarce environmental resources especially soil by protecting from excessive chemical use and conserving the integrity of soil health. It is evident that Sustainable agriculture is a frontline defense against soil contamination.

7.10.1 Practices Recommended in Sustainable Agriculture for Mitigation of Soil Contamination

7.10.1.1 Phytoremediation

Sustainable agriculture uses certain plants as a source for bioremediation of soil contaminants. In phytoremediation, in-situ contaminated soil and groundwater treatment is achieved using green plants and their associated microbiota. Five main processes are associated with it, rhizofiltration, phytostabilisation, phytoextraction, phytovolatilization, and phytodegradation. Plants employ three strategies to counter heavy metal contaminated soil, including metal exclusion, metal indication, and metal accumulation (Ghosh and Singh 2005). A total of 87 species from 11 genera belong to the Brassicaceae family, which contains a wide range of hyper accumulative species.

7.10.1.2 Use of Vermicomposting

Vermicompost is prepared by the action of earthworms that degrades the organic matter through bio-oxidation and stabilization. Organic matter can help in immobilization of heavy metals (Basta and McGowen, 2004). It also increases the CEC of soil leading to increasing adsorption of positive ions including heavy metals. It significantly increases the growth of plants that facilitates phytoremediation resulting in the reduction of soil contamination. Vermicompost as a strategy of sustainable agriculture not only decreases the use of synthetic chemical sources but also facilitates several associated mechanisms that ultimately lead to the reduction of heavy metal-related soil contamination (Belliturk et al. 2015).

7.10.1.3 Effective Application of Soil Microorganisms

Plant Growth Promoting Rhizobacteria (PGPR) are inseparable components of sustainable agriculture battling against heavy metal contamination. PGPRs have been shown to aid in increasing plants' tolerance to heavy metals. They can accelerate nutrient cycling, detoxify metals, and improve soil structure by assisting the plant (Karthik et al. 2017). Studies have shown that PGPR increases the process of phytoremediation by producing substances like siderophores and the enzyme ACC deaminase. They can change toxic metals into nontoxic forms, and stimulate the transport of proteins, which improves the plant's ability to withstand heavy metal stress. PGPRs facilitate nutrient uptake, stabilize soil structure, and neutralize heavy metals, benefiting the plant (Kalyanasundaram et al. 2021). Cyanobacteria transform unavailable soil nutrients into plant available forms. They increase the bioavailability of nutrients like P for crop use. Biofertilizers have become an important complement to chemical fertilizers in improving soil fertility and nutrient supply. Cyanobacteria and other microbes based biofertilizers can improve mobilization in soil and yield.

7.10.1.4 Integrated Pest Management Strategies (IPM)

Pesticides are one of the most prolific sources of heavy metal contamination in soil (Alengebawy et al. 2021). IPM strategies focuses on reduction of pesticide usage in agro ecosystems (Ehler 2006). It leads to lesser residues and sharp decline in heavy metal contamination. Nowadays IPM is a leading strategy, accompanied by sustainable agriculture to counter the soil contamination.

7.10.1.5 Cover Crops

Cover crops often have extensive root systems at different depths, rapid growth, and the ability to quickly cover the soil surface compared to other crop types. Certain cover crops can produce chemical compounds that assist crops in absorbing and immobilizing heavy metals into roots and/or shoots (*Ageratum conyzoides* Linn. against cd, *Cynodon dactylon* against Zn, Cu, Cd, Co, Ni, and Pb, *Lantana camara* Linn. against Pb, *Vetiveria zizanoides* against Pb, Zn, Cu and Cd etc.). Cover crops typically have high adaptability to a wide range of environmental and climatic conditions. Perennial cover crops can be more effective for long-term heavy metal remediation since they do not need to be replanted after harvest (Sumiahadiet al. 2019). So, it is obvious that crop covers are not only sustainable, but it also protects against soil contamination.

7.10.1.6 Precision Farming

The implementation of precision agriculture technologies enables farmers to optimize the utilization of crop inputs such as tillage practices, irrigation water and most importantly fertilizers, pesticides that are the most important sources of contamination. Radio Frequency Identification (RFID) sensors combined with biosensors consisting of aptamer and DNA-based characters, can be used for detection of heavy metals at nanoscale (Akhtar et al. 2021). It can be used for screening of the soil contamination and as a base for further remediation.

7.11 Reclamation and Remedial Measures of Soil Contamination in Sustainable Agriculture

Soil contamination has impacted the quality of the soil and degraded the fertility status of the soil. Around 29% of the total geographical area in India comes under degraded soil. Soil contamination accounts for 40% of the losses in the soil fertility status which gradually affects the global growth domestic product. In the past few years, human activities have contributed 15% of the total soil contamination in agricultural land.

A recent study has reported that due to anthropogenic activities, there are around 30 million tons of pesticides and chemicals have contaminated the soil which directly or indirectly affects human health and the environment. So, it's crucial to reclaim the contaminants, organic pollutants, and heavy metals from the affected soil to reduce their concentration to a harmless level and prevent these contaminants from further spreading and make the soil suitable for agriculture and human use in the future.

7.11.1 Physical Remediation

The physical method of soil remediation is the use of physical treatments to eliminate the contaminants and heavy metals from the soil (Fig. 7.2). It depends on the physical properties of the soil and the contaminants. Some of the physical methods for soil reclamation are discussed below:

(a) **Soil Washing**: Soil washing is a liquid-based method that uses water or other aqueous solution with chemical additives to eliminate undesirable contaminants. In this method, soil that needs to be reclaimed is excavated from the polluted area and screened to remove gravel to reduce the damage to equipment. The screened soil is then mixed with washing agents (liquid with additives) in specific washing equipment and stirred. Washed unpolluted soils are placed back in their place after they settle down and the washing agents can be renewed and reprocessed. This is known as the ex-situ soil washing technique. This method is used for reclaiming soils that are contaminated by heavy metals, hydrocarbons, and pesticides.

(b) **Soil Flushing**: This is an in-situ soil washing method. In this method, washing agents are flushed through the contaminated area and extracted through the extraction well, and then deionized water is used for secondary washing to eliminate residual agents. It is suitable when the contaminated area is large. For instance, to reclaim the Cd-contaminated paddy soil, calcium chloride ($CaCl_2$) mixed with water is used in the field to reduce the Cd concentration.

(c) **Soil encapsulation**: It is a stabilization method of soil remediation which separates contaminates by isolation instead of removing or destroying them. In this method of soil reclamation, contaminated soil is covered with a layer of impermeable material to restrict its spreading. It is often used in such areas where reclamation by other methods is difficult. The impermeable barriers used are lime, cement, and concrete which avoids contaminants from touching nearby unpolluted soil. These impervious materials neutralize the contaminants and restrict their movement so that they cannot affect the surrounding soils.

(d) **Thermal desorption**: An in-situ technique that involves the use of high temperatures, even more than the boiling point of the contaminants to separate it from the soil in the form of gas. Thermal desorption can be of two types—low-temperature technique and high-temperature technique. Low-temperature techniques heat the soil in dryers at temperatures ranging from 150 to 550 °C to volatilize soil contaminants. This method is used for the treatment of soil that is contaminated with petroleum hydrocarbons. High-temperature technique uses incineration to destroy the hazardous contaminants by using very high temperatures ranging from 1,400 to 2,200 °F and the vapors are collected and treated to remove the heavy metals.

(e) **Soil Vapor Extraction (SVE)**: SVE is a frequently used and efficient physical remediation method to remove volatile contaminants from unsaturated soils (vadose zone). It is based on the mass transfer of pollutants from the solid and liquid state into the gaseous state, and further gaseous pollutants are collected and extracted to remove the pollutants. This technique uses vacuum blowers and vapor extraction wells to inject gas through the subsurface and collects contaminated soil vapor (Khan et al. 2008). This method of soil reclamation is used for treating soil with volatile organic compounds which have a low boiling point and a high vapor pressure such as mercury, benzene, toluene, ethylbenzene, xylenes, etc.

(f) **Electrokinetic remediation**: The process of applying electric current to facilitate the movement of ions to remove the soil contaminants. This method is used to remove organic, inorganic, and heavy metal particles from the soil by virtue of its electric potential. The two electrodes, a positive charged anode, and a negative charged cathode are inserted between contaminated soil matrix inside a chamber filled with water or electrolyte to facilitate the movement of ions and a low voltage direct current passes through it. The electric field generated thus induces the mobilization and transportation of contaminants through the porous matrix toward the electrodes, where the contaminants are harvested, pumped out, and treated method leads to the transportation of the contaminants through four main mechanisms, Electroosmosis, Electromigration, Electrophoresis, and transportation due to streaming potential.

(g) **Magnetic separation**: Magnetic separation is a remediation technique that uses the magnetic behaviour of the contaminants and eradicates it through a magnetic field. Here the contaminant first adheres to a material that is magnetic in nature (e.g., magnetite), and then, the contaminant-laden carrier is passed through a magnet and separated under a magnetic field. This method is effective for removing heavy metals that are magnetic, such as iron, nickel, and cobalt-containing minerals.

(h) **Solidification and stabilization**: Stabilization/Solidification (S/S) is naturally a process in which contaminated soil is mixed with stabilizers or binding agents such as cement, lime, port land, gypsum, elemental sulfur etc. to avoid leaching of the contaminants in soil by reducing its mobility. Stabilization generally reduces the solubility of the contaminants by changing their physical and chemical properties while solidification specifically treats contaminants to increase their solidity and structural integrity and further convert it into a less toxic level. It can be either in situ, where the binders are injected into the area of contaminated soil, or ex-situ, where the contaminated soil is excavated and mixed with binders or stabilizers in a machine, and then the solidified mass is dumped in a chosen area where contaminants have to be removed.

(i) **Soil aeration**: This method utilizes air to increase the level of oxygen flow in the soil which can improve the activity of microorganisms that degrade contaminants. It is achieved by tilting soil, tillage operations, pumping air into the soil, and adding bulking agents to increase porosity.

(j) **Soil excavation**: In this method, the contaminated soil is excavated from the area and replaced by good soil. The excavated soil is further treated in different ways. It is used when the contaminated area is small or a hotspot for a particular pollutant.

(k) **Soil blending**: Soil blending is a method of physical mixing of contaminated soil with clean soil to dilute the contaminants and decrease their concentration. It is efficient for contaminants that are less hazardous or persistent. In-situ soil blending is also used for the effective distribution of chemical amendments over the contaminated soil to reduce its level. The chemical amendments most often used are oxidants, reductants, biostimulants, or soil stabilizers. This method is more efficient if performed within the top layer of soil above the water table (10–12 feet).

Generally, these methods of physical remediation of contaminated soil are mostly efficient for eliminating heavy metals from the soil rather than for other kinds of contaminants or soil pollution. The prior evaluation of soil and contaminants is of utmost importance before choosing the appropriate method for reclaiming the soil.

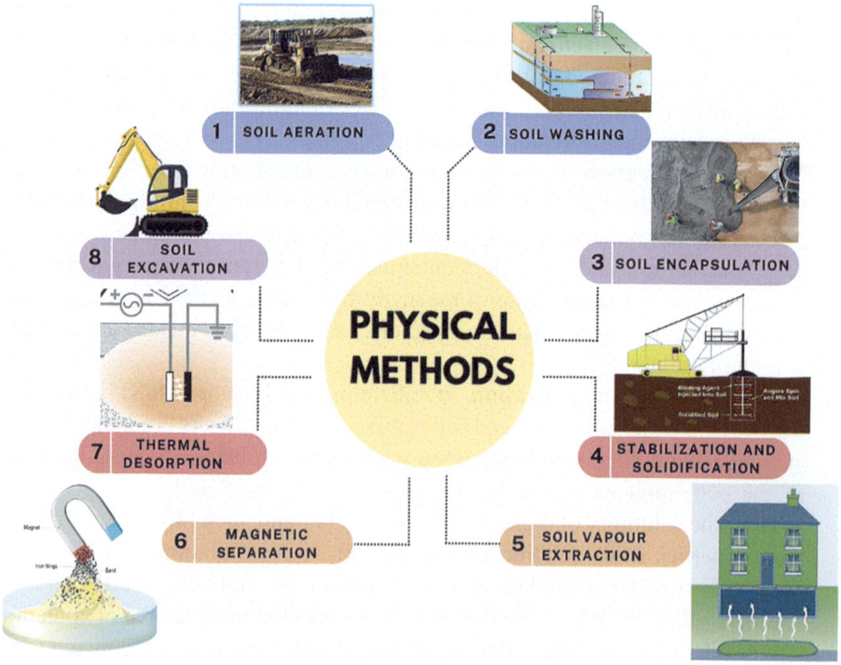

Fig. 7.2 Different physical remediation techniques

7.11.2 Chemical Remediation

The use of different types of chemical agents to alter the properties of the soil and eliminate the concentration of heavy metals is known as chemical method of reclamation of contaminated soil (Fig. 7.3). Its purpose is to add chemicals or solvents into the contaminated soils so that the contaminants stabilize and convert into a less noxious forms that are less harmful to the soil, water bodies, plants, and human beings. The materials such as metallic oxides, clays, or biomaterials are usually used for chemical treatments. Different types of methods are available for soil remediation using chemical agents. These methods are discussed below:

(a) **Chemical precipitation**: In this method, a soluble heavy metal pollutant is converted into insoluble form by reacting with different precipitating agents such as lime, sodium hydroxide, sodium carbonate, sulfides, and phosphates that can be easily removed. The solubility of the precipitants used in this method depends on the pH and concentration.

It is used mainly for the exclusion of metallic ions like fluoride, cyanide, and phosphate; organic molecules like phenols and aromatic amines; and detergents and oily emulsions by $BaCl_2$ (barium chloride). For instance, heavy metals such as zinc, copper, and lead are removed from aqueous solutions of soil by using lime ($Ca(OH)_2$), soda ash (Na_2CO_3), and sodium sulfide (Na_2S). This method

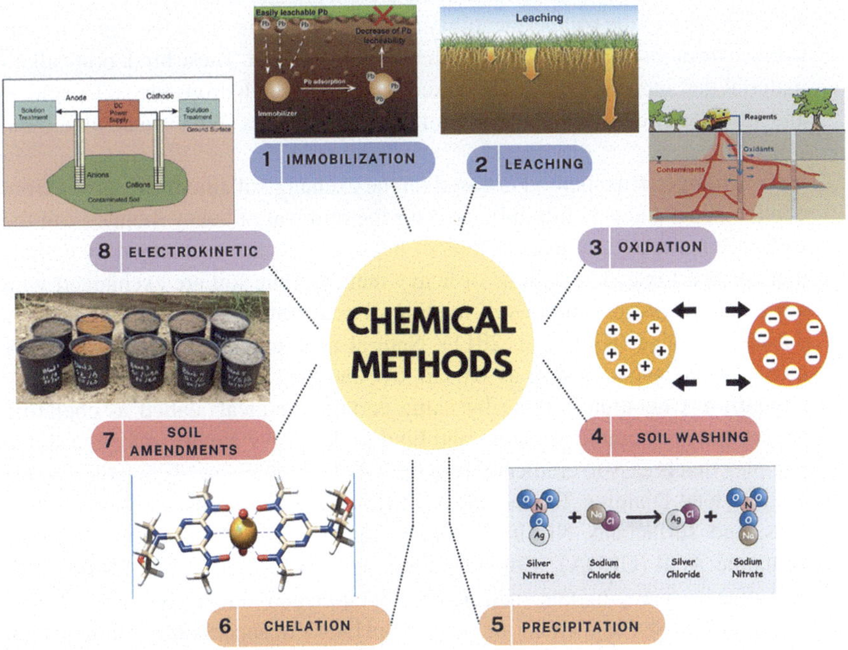

Fig. 7.3 Different chemical remediation techniques

is simple, non-hazardous, cost-efficient, and the most conventional method to reclaim the affected soils.

(b) **Chemical oxidation**: This method uses oxidizing agents to degrade hydrocarbons or heavy metals and convert them into insoluble and non-toxic forms. It is also known as In Situ Chemical Oxidation (ISCO) or "Chemox". Chemox is a system that is extremely operative in eradicating a variety of organic contaminants such as polychlorinated biphenyls, total petroleum hydrocarbons, benzene, toluene, ethylbenzene, and xylenes. The most frequently used oxidizing agents for soil reclamation comprise hydrogen peroxide (H_2O_2), ozone (O_3), potassium permanganate ($KMnO_4$), and sodium persulphate ($Na_2S_2O_8$). The oxidizing reagents can either be used in isolation or in combination with other catalyzing agents to accelerate their activity and optimize their performance.

(c) **Chemical reduction**: This method uses reducing agents to reduce toxicity, solubility, and mobility of heavy metals like chromium (Cr) and reclaim the Cr-contaminated soils. For example, highly toxic soluble hexavalent chromium (Cr^{6+}) is reduced to non-toxic insoluble trivalent chromium (Cr^{3+}) by adding ferrous sulfate or sodium dithionite, which ultimately precipitated into an environment.

(d) **Chemical immobilization**: In this, chemicals such as lime, phosphate, and organic acids are added to the contaminated soil that binds with the contaminants or heavy metals present in that soil and makes them insoluble. This reduces its mobility and restricts them from leaching into the soil or groundwater (Basta et al. 2004).Contaminants are chemically fixed by adding binders which decreases the soil permeability and makes it difficult for the heavy metals to move through the soil and pollute the surrounding soil. These binders are alkali material that increases pH when mixed with soil, which drops the soil's acidity and ultimately reduces the mobility of the contaminants (Wuana and Okieimen, 2011).

(e) **Ion exchange**: This method is based on the exchange of anions or cations from contaminants and is commonly used for the removal of heavy metals from the soil. In this method, unwanted ions of heavy metals are exchanged by cations that are non-toxic. The cations of heavy metals in the soil are exchanged with other cations present in the soil medium thus maintaining the charge balance during exchange (Kim et al. 2011). Natural zeolite is widely used for heavy metal cation exchange such as Zn, Cu, Co, and Mn.

(f) **Chelation**: Chelation is done by using certain chemicals called as chelating agents that are organic in nature and bind to the heavy metals to form a stable complex that is easy to eradicate from the soil. Commonly used chelating agents are Ethylene Diamine Tetraacetic Acid (EDTA) humic substances, cyclodextrins, and surfactants Nitrilo Triacetic Acid (NTA). For example, Diacetic Glutamic Acid (GLDA), a biodegradable chelate is used at pH 4.0 to eradicate diverse heavy metals such as Cu contaminated sludge, Ni, and Cd, and citric acid is used as a chelating agent at pH 4.43 to successfully eliminate Zn from polluted soils. Some other biodegradable chelators such as GLDA (N, N-bis (carboxymethyl)-L-glutamate), IDS (iminodisuccinate), HIDS (3-hydroxy-2,2′-iminodisuccinic acid), EDDS (S,S ethylenediamine-disuccinate), MGDA (methylglycine diacetate), NTA (nitrilotriacetate) can also be used as chelating agents.

(g) **Chemical leaching**: In this method, polluted soil is leached down with chemical reagents, gases, or other liquids to remove contaminates such as Pb, Hg, Zn, Ni, Va, etc. by increasing their mobility. It includes the dissolution of ions from heavy metals into leaching liquid and further collected from the leachate and discarded. Usually, this leaching liquid is acidic by adding inorganic acids such as nitric acid, sulphuric acid, and hydrochloric acid at a pH of 1.5–2.0 to intensify the solubility of metal ions. This technique is preferred when the contaminated area is large and the accumulation of heavy metals is significant in that particular area. For instance, recommended chemical leaching to be an effective way for the removal of Plutonium (Pu) from soils polluted with radioactive waste.

(h) **Soil amendments**: Soil amendments are the techniques used to reclaim contaminated soils with heavy metals and to improve soil quality. This system of soil reclamation is primarily aimed to decline the solubility and mobility of heavy metals by using amendments such as lime, silica, organic matter, zeolites, cement, and fly ash. The principle involved in this method relies upon the adsorption capacity of the soil which is enhanced by adding amendments to the contaminated soils that adsorb heavy metals and restrict their bioavailability. This method is used to remove Cr, As, Pb, Zn, Cd, and Cu from the soil. Soil amendments are also used for increasing soil fertility by using organic materials such as compost, wood chips, biochar, animal manure, straw, husk, geotextile, and sewage manure. Compost and biochar can also be used for reclaiming soils contaminated with heavy metals by reducing their mobility and bioavailability in the soil. Though chemical remediation is a fast and effective method for reclaiming contaminated soils it is costly and not environment-friendly which may pose a serious threat to the environment. Therefore, these methods should be evaluated beforehand and should be used in integration with other remediation methods for sustainable reclamation of affected soils.

7.11.3 Biological Remediation

The practice of using living organisms to remove pollutants from soil and water by the process of degrading, breaking, or transforming them into less toxic forms that can be used by humans for various purposes is known as biological remediation or bioremediation. Bioremediation is based on the natural living process of bacteria, fungi, and plants to modify the toxic substances through various enzymatic processes into less-toxic forms while carrying out their usual life activities. These living organisms use organic pollutants as an energy source for performing their various metabolic

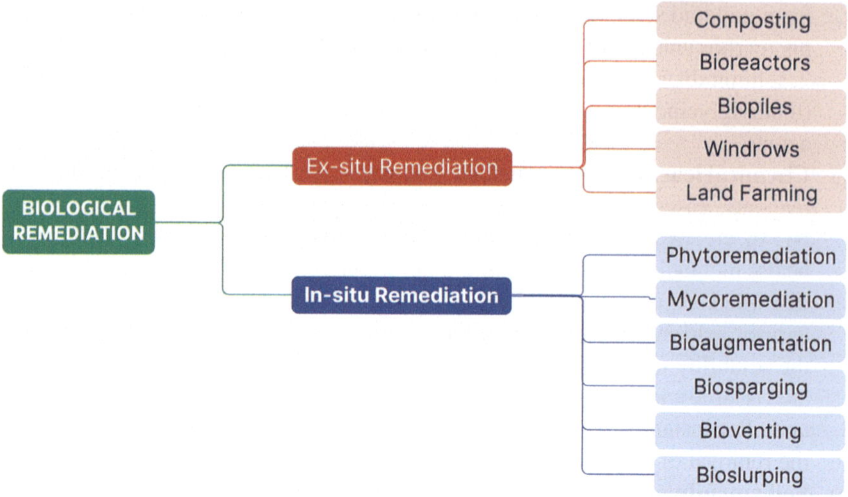

Fig. 7.4 Flowchart showing different types of biological remediation techniques

processes that renders the pollutants into a less-toxic product for the environment. It is a sustainable and ecological method of soil reclamation.

The environment should be favorable to allow the growth of microbes and for the degradation of contaminants. This method is widely used to eradicate various inorganic and organic pollutants such as petroleum, hydrocarbons, halogenated organic solvent, halogenated organic compounds, non-chlorinated pesticides and herbicides, nitrogen compounds, heavy metals such as lead, mercury, chromium, and radionuclides. For example, water hyacinth (*Eichhornia crassipes*) is used for the remediation of crude oil-contaminated soil, and plant growth-promoting halophilic bacteria used for the reclamation of saline, and sodic soils.

Many species of bacteria like *Bacillus, Enterobacter,* and *Pseudomonas* and some fungi like yeast and mould helps in the biological remediation of metal. Microorganisms such as *Rhodobacter sphaeroides, Serratia* sp., *Bacillus* sp., *Pseudomonas putida, Polyporales* sp., *Sporosarcina pasteurii,* and *Cerrena aurantiopora*;and plants such as *Chromolaena odorata, Ocimum gratissimum, Helianthus annuus, Lolium perenne,* and *Hibiscus cannabinus* found prominent to eradicate Zn from zinc polluted soils. Different remediation techniques were mentioned below (Fig. 7.4).

7.11.3.1 Phytoremediation

A biological remediation method that exploits plants and associated microorganisms to eliminate specific contaminants from the soil is known as phytoremediation. It depends upon the basic physiological mechanisms of the plants and related microbes such as absorption from roots, transport and translocation of metabolites in the plant system, transpiration, photosynthesis, and mineralization. Plants that are used to extract noxious substances from the soil absorb toxic organic and inorganic contaminants through roots and stabilize them by binding them with microbes present in the rhizosphere. The absorbed materials are then translocated in the plant system and accumulate in several parts of the plant where they are converted into less-toxic form. The aerial parts of the plants such as leaves take part in a gaseous exchange with the atmosphere and thus release the toxic substances. The leaves can also be collected and further treated to eradicate the accumulated heavy metals to protect the environment. This method is mostly used for reclaiming soils with metals but can also be used for eliminating other organic and inorganic contaminants. There are several strategies for phytoremediation such as phytoextraction, phytostabilisation, phytovolatalisation, phytoexcavation, etc. but the most used strategies are:

(a) **Phytoextraction**: It uses the plant's absorption and translocation mechanism to uptake and translocate metal pollutants from soil to aerial plant parts. Ultimately, the aerial parts of the plants such as leaves are plucked and finally disposed off to remove the metals contaminants (Chaney et al. 1996).
(b) **Phytostabilisation**: In this method, plants accumulate heavy metals in their roots by precipitating them in the root zone. This restricts the movement of heavy metals in the soil and stabilizes them in a particular place that can be treated further by uprooting the plants or by adding amendments in that specific area.

7.11.3.2 Mycoremediation

A bioremediation method that exploits the use of fungi as remediation tool to eradicate toxic contaminants from the soil is known as myco-remediation. It uses the vegetative part of the fungi such as mycelium and cell wall to eliminate the toxic contaminants. Since fungi are called decomposers and thus can help in breaking down large molecules of contaminants into small molecules that can be easily removed from the soil. Fungi release several hydrolytic enzymes that integrate with the contaminants present in the soil and aid in their breakdown by degrading them. It also interferes with the mobility of contaminants in the soil and prevents their spreading.

Several saprotrophic fungi such as *Phanerochaete chrysosporium, Pleurotus ostreatus, Agaricus bisporus,* and *Lentinula edodes*, and biotrophic fungi such as *Amanita* sp., *Boletus* sp., *Gautieria* sp., *Hebeloma* sp., *Lactarius* sp., *Morchella* sp., *Suillus* sp., and *Rhizopogon* sp. degrade insoluble organic and inorganic contaminants and also mobilizes several heavy metals to reduce their concentration in the soil. Several species of *Rhizopus, Paecilomyces, Alternaria, Mucor, Gliocladium, Aspergillus, Fusarium,* and *Penicillium* degrade petroleum and crude oils by using them as a source of carbon in the degradation process. Species of *Aspergillus, Penicillium, Rhizopus, Fusarium, Cladosporium,* and *Trichoderma* can degrade Polycyclic Aromatic Hydrocarbons (PAHs) from the soil.

7.11.3.3 Land Farming

Land farming is a biological remediation technique where polluted soil is excavated from the contaminated area and spread over a surface and nutrients and bulking agents are added to the contaminated soil. This mixture is inverted regularly and tilled periodically for proper aeration to enhance the growth and activity of native microbes present that aerobically degrade the contaminants. It is practiced in upper surface soil up to 15–30 cm where the microbial activity is considerably high. The microorganisms present naturally in the soil degrade the organic pollutants into CO_2, water, and biomass. Land farming is used for reclaiming soil polluted with oily sludges and petroleum products.

7.11.3.4 Composting

It is a bioremediation technique in which contaminated soils are mixed with non-toxic organic materials such as manure and agriculture waste which encourages the growth of microbes. The microbes are mainly mesophiles and thermophiles that can tolerate high temperature and moisture level and degrades toxic hazardous contaminants in the soil.

7.11.3.5 Vermiremediation

The practice of using earthworms for eradicating toxic substances from the soil is known as vermiremediation. Earthworms move through the soil and change the physical, chemical, and biological conditions of soil by mixing the soil in their gut, as a result, it increases the activity of soil microbes and their interaction with the toxic substances and thus facilitates their biodegradation. This biological process is commonly used to eradicate organic contaminants such as petroleum, hydrocarbons, and herbicides. An earthworm, *Eisenia fetida* along with bacteria *Pseudomonas, Azotobacter,* and *Clostridium* and fungi like *Aspergillus* and *Penicillium* is used to eradicate petroleum and diesel from contaminated soil.

7.12 Importance of Biological Remediation Over Physicochemical Remediation in Sustainable Agriculture

Biological remediation and physicochemical remediation are two different types of approaches used in ecological remediation, as well as sustainable agriculture. While both approaches have their importance and uses, biological remediation offers numerous benefits over physicochemical remediation in the context of sustainable agriculture. Some of the advantages of biological remediation are discussed below:

(a) **Environmental Friendliness**: Bioremediation depends on the exploitation of living organisms and their natural processes for breaking, degrading, and removing contaminants from the soil and environment. It generally uses microorganisms, plants, or the enzymes secreted by them to destroy the contaminants. On the other hand, physicochemical methods use synthetic, hazardous chemicals that cause damage to the environment. Biological remediation is usually regarded as more environmental-friendly than physicochemical methods.

(b) **Cost-effectiveness**: In biological remediation plants or microbes used on the reclamation site to degrade the pollutants stays for a longer period without any extra care and attention, thus reducing the cost. Whereas, physicochemical methods are highly costly which uses expensive chemicals regularly and requires heavy equipment for reclamation purposes. Hence, biological remediation can be proven to be a cost-effective solution where reclamation has to be done on a large scale.

(c) **Sustainability**: By encouraging natural processes and reducing the use of synthetic chemicals, biological remediation aligns with the principles of sustainable agriculture. It can lessen the effect on the environment by preserving biodiversity and the health of the soil. In some instances, biological remediation can help restore damaged lands, increasing their sustainability and long-term production. On the other hand, physical and chemical remediation do not rely on the principles of sustainable agriculture, and using chemicals and other non-biological substances poses harmful effects on the environment. The physicochemical method also damages the soil's physical and chemical structure and makes it unfit for agricultural purposes, and hinders the growth and mobility of micro-organisms present in the soil, ultimately reducing soil fertility.

(d) **Versatility**: Biological remediation may be helpful in treating huge variety of contaminants and agricultural pollutants that includes natural compounds, pesticides, heavy metals, and nutrients whereas physical and chemical remediation methods are confined to treat a limited number of contaminants only.

(e) **Long-term effectiveness**: Biological remediation frequently produces lasting and durable effects. This can offer long-term advantages and aid in avoiding environmental problems in the future. Whereas physicochemical remediation is a short term reclamation method of contaminated soil where the physical and chemical agents are used regularly as their effect is for a short period of time.

(f) **Ecosystem benefits**: Plants used in phyto-remediation for reclaiming contaminated soil takes part in a gaseous exchange with the atmosphere along with reclaiming the soil. Also, they support ecological biodiversity by providing food sources and shelter for several beneficial insects, birds, and other wildlife. Microorganisms that are used in reclamation procedures improve the overall health of the soil by improving nutrient cycling and soil fertility by decomposing the organic matter present in the soil. In this way, biological remediation has a positive impact on the surrounding ecosystem.

While biological remediation offers numerous advantages, it is essential to consider site-specific factors, contaminant characteristics, and the overall goals of remediation when selecting the most suitable approach. In some cases, a combination of biological and physicochemical methods may be necessary for optimal results.

7.13 Current Challenges and Future Perspectives

Determining how plants respond to heavy metal toxicity advancement, tolerance, along with hyperaccumulation on the molecular and physiological levels serves as essential towards the successful outcome of phytoremediation. Since the vegetation utilized during phytoremediation must endure tension to their initial phase of expansion, most of the research conducted till now had concentrated on examining the impact of heavy metal effect during the mature development of plants phase. However, it is extremely important to learn about the plant's reaction to heavy metal stress starting during the embryonic phase of development (Thakur and Sharma 2016; Dubey et al. 2022) such are the key periods in the plant development process which are especially responsive to outside stress. The best times of biomonitoring occur during the period of seed germination along with within the initial weeks of seedling development. It is also essential that plants growing from polluted soil must endure many heavy metal stresses instead of just one is another crucial consideration. It seems obvious while metals feature within a combination, they behave differentially. Various heavy metal ions are competing at their respective binding positions about soil particles or plant metal transporters (Clemens et al. 1998). For the successful implementation of phytoremediation during agricultural circumstances, it is, therefore, crucial to comprehend the reaction method of the target plant during various metal stresses.

A relatively novel method is to make utilization of rhizobacteria that encourage plant development in polluted soils to increase the susceptibility of plants to phytoremediation. For instance (Babu et al. 2013) demonstrated that inoculating *Alnus firma* using endophytic, *Bacillus thuringiensis* GDB-1 enhanced its plant's capability to eliminate heavy metals from the environment. By altering the bioavailability of metal ions over being absorbed, promoting root growth, solubilizing phosphate, producing siderophores, fixing nitrogen, and producing 1-aminocyclopropane-1-carboxylate deaminase (ACC), for example, the application of metal-resistant rhizobacteria can speed up the method of phytoremediation. The phytoremediation capability of plants can be increased by rhizobacteria by reducing the development of root restriction resulting from the exposure of heavy metal ions. The application of plant-growth-promoting bacteria (PGPR) for phytoremediation must be studied more thoroughly because this is an emerging approach. This includes studying how plants along with microorganisms behave in the natural environment, employing modified genetic bacteria, etc. (Arshad et al. 2007).

7.14 Conclusion

Soil contamination outbreaks the present crisis of declining the sustainable environment. The remediation of heavy metals content from the soil is the only target despite other soil contaminants. The biological remediation became a popular method as it heals the soil contamination as a long term effect. It includes the growing of vegetation for phytoremediation which further supports soil microbial population for microremediation promoting bioaccumulation of heavy metals into unavailable forms. The introduction of a sustainable environment involves good soil health with the soil contaminants below the critical limits for a sustainable crop production. The reduced utilization of inorganic fertilizers containing heavy metals in agricultural practices and safe disposal of industrial waste are the promising approach for a sustainable environment rendering a good soil health for a sustainable agriculture without sacrificing crop productivity.

References

Agritopic (2021) Heavy metals in fertilizers and agriculture. Incitec pivot fertilisers. https://www.incitecpivotfertilisers.com.au/~/media/Files/IPF/Documents/Agritopics/61%20Heavy%20Metals%20Agritopic.pdf. Accessed on 02 Jan 2024

Ahemad M, Khan MS (2012) Evaluation of plant-growth-promoting activities of rhizobacterium *Pseudomonas putida* under herbicide stress. Ann Microbiol 62(4):1531–1540. https://doi.org/10.1007/s13213-011-0407-2

Akhtar MN, Shaikh AJ, Khan A, Awais H, Bakar EA, Othman AR (2021) Smart sensing with edge computing in precision agriculture for soil assessment and heavy metal monitoring: a review. Agricul 11(6):475. https://doi.org/10.3390/agriculture11060475

Alengebawy A, Abdelkhalek ST, Qureshi SR, Wang MQ (2021) Heavy metals and pesticides toxicity in agricultural soil and plants: ecological risks and human health implications. Toxic 9(3):42. https://doi.org/10.3390/toxics9030042

Alloway BJ (1996) Soil pollution and land contamination. Pollution, causes, effects and control, pp 352–368

Amini M, Khademi H, Afyuni M, Abbaspour KC (2005) Variability of available cadmium in relation to soil properties and landuse in an arid region in central Iran. Water, Air, Soil Pollut 162:205–218. https://doi.org/10.1007/s11270-005-6273-4

Anyakora C, Ehianeta T, Umukoro O (2013) Heavy metal levels in soil samples from highly industrialized Lagos environment. Afr J Environ Sci Technol 7(9):917–924

Arshad M,Saleem M, Hussain S (2007) Perspectives of bacterial ACC deaminase in phytoremediation. TRENDS Biotechnol 25(8):356–362. https://doi.org/10.1016/j.tibtech.2007.05.005

Arunakumara KKIU, Zhang X (2008) Heavy metal bioaccumulation and toxicity with special reference to microalgae. J Ocean Univ China 7:60–64. https://doi.org/10.1007/s11802-008-0060-y

Atayese MO, Eigbadon AI, Adesodun JK (2008) Heavy metal contamination of *Amaranthus* grown along major highways in Lagos, Nigeria. Afr Crop Sci J 16(4):225–235. https://doi.org/10.4314/acsj.v16i4.54390

Baatrup E (1991) Structural and functional effects of heavy metals on the nervous system, including sense organs, of fish. Compar Biochem Physiol Part C: Compar Pharmacol 100(1–2):253–257. https://doi.org/10.1016/0742-8413(91)90163-N

Babu AG, Kim JD, Oh BT (2013) Enhancement of heavy metal phytoremediation by Alnus firma with endophytic *Bacillus thuringiensis* GDB-1. J Hazard Mater 250:477–483. https://doi.org/10.1016/j.jhazmat.2013.02.014

Balkhair KS, Ashraf MA (2016) Field accumulation risks of heavy metals in soil and vegetable crop irrigated with sewage water in western region of Saudi Arabia. Saudi J Biol Sci 23(1):S32–S44. https://doi.org/10.1016/j.sjbs.2015.09.023

Basta NT, McGowen SL (2004) Evaluation of chemical immobilization treatments for reducing heavy metal transport in a smelter-contaminated soil. Environ Pollut127(1):73–82. https://doi.org/10.1016/S0269-7491(03)00250-1

Basta NT, Tabatabai MA (1992) Effect of cropping systems on adsorption of metals by soils: II. Effect of pH. Soil Sci 153(3):195–204

Basta NT, Pantone DJ, Tabatabai MA (1993) Path analysis of heavy metal adsorption by soil. Agron J85(5):1054–1057. https://doi.org/10.2134/agronj1993.00021962008500050018x

Belliturk K, Shrestha P, Görres JH (2015) The importance of phytoremediation of heavy metal contaminated soil using vermicompost for sustainable agriculture. J Rice Res3(2):6–e114. https://doi.org/10.4172/2375-4338.1000e114

Bhat SA, Cui G, Li W, Ameen F, Yaseera N, Wei Y, Li F (2023) Fate of bio-contaminants in soil systems and available remediation methods. In: Fate of biological contaminants during recycling of organic wastes. Elsevier, pp 213–227. https://doi.org/10.1016/B978-0-323-95998-8.00014-5

Bjuhr J (2007) Trace metals in soils irrigated with waste water in a peri-urban area downstream Hanoi City, Vietnam, pp 1–38

Bouki C, Venieri D, Diamadopoulos E (2013) Detection and fate of antibiotic resistant bacteria in wastewater treatment plants: a review. Ecotoxicol Environ Saf 91:1–9. https://doi.org/10.1016/j.ecoenv.2013.01.016

Brady NC, Weil RR (2002) The nature and properties of soils, 13th edn. Hall Upper Saddle River, New Jersey, pp 55–71

Brodt S, Six J, Feenstra G, Ingels C, Campbell D (2011) Sustainable Agriculture. Nat Educ Knowl 3(1):1–41

Chaney RL, Oliver DP (1996) Sources, potential adverse effects and remediation of agricultural soil contaminants. In: Contaminants and the soil environment in the Australasia-Pacific region: proceedings of the first Australasia-Pacific conference on contaminants and soil environment in the Australasia-Pacific Region, held in Adelaide, Australia, 18–23 February 1996. Springer Netherlands, Dordrecht, pp 323–359

Chang TJ, Cui XQ, Ruan Z, Zhao XL (2014) Long-term effects of tillage methods on heavy metal accumulation and availability in purple paddy soil. Huan Jing ke Xue= Huanjing Kexue 35(6):2381–2391

Chen X, Xia X, Zhao Y, Zhang P (2010) Heavy metal concentrations in roadside soils and correlation with urban traffic in Beijing, China. J Hazard Mater 181(1–3):640–646. https://doi.org/10.1016/j.jhazmat.2010.05.060

Choudhary S, Yamini NR, Yadav SK, Kamboj M, Sharma A (2018) A review: pesticide residue: cause of many animal health problems. J Entomol Zool Stud 6(3):330–333

Clemens S, Antosiewicz DM, Ward JM, Schachtman DP, Schroeder JI (1998) The plant cDNA LCT1 mediates the uptake of calcium and cadmium in yeast. Proced Nation Acad Sci 95(20):12043–12048. https://doi.org/10.1073/pnas.95.20.12043

Collin S, Baskar A, Geevarghese DM, AliMNVS, Bahubali P, Choudhary R, Swamiappan, S (2022) Bioaccumulation of lead (Pb) and its effects in plants: a Rev J Hazard Mater Lett 2:100064. https://doi.org/10.1016/j.hazl.2022.100064

Daghrir R, Drogui P (2013) Tetracycline antibiotics in the environment: a review. Environ Chem Lett 11:209–227. https://doi.org/10.1007/s10311-013-0404-8

DeVolder PS, Brown SL, Hesterberg D, Pandya K (2003) Metal bioavailability and speciation in a wetland tailings repository amended with biosolids compost, wood ash, and sulfate. J Environ Qual 32(3):851–864. https://doi.org/10.2134/jeq2003.8510

Doner HE, Traina SJ, Pukite A (1982) Interaction of nickel, copper, and cadmium with soil manganese in saline solutions. Soil Sci 133(6):369–377

Dubey VK, Sahoo SK, Sujatha B, Das A (2022) Impact of heavy metals on honey bees. Vig var 3(1):101–103

Ehler LE (2006) Integrated pest management (IPM): definition, historical development and implementation, and the other IPM. Pest Manag Sci 62(9):787–789. https://doi.org/10.1002/ps.1247

El-Bahi SM, Sroor A, Mohamed GY, El-Gendy NS (2017) Radiological impact of natural radioactivity in Egyptian phosphate rocks, phosphogypsum and phosphate fertilizers. Appl Radia Isotop 123:121–127. https://doi.org/10.1016/j.apradiso.2017.02.031

Elliott HA, Liberati MR, Huang CP (1986) Competitive adsorption of heavy metals by soils. Am Soc Agron Crop Sci Soc Am Soil Sci Soc Am 15(3):214–219. https://doi.org/10.2134/jeq1986.00472425001500030002x

El-Wakeil N, Gaafar N, Sallam A, Volkmar C (2013) Side effects of insecticides on natural enemies and possibility of their integration in plant protection strategies. Agricultural and biological sciences "insecticides-development of safer and more effective technologies". Intech, Rijeka, Croatia, pp 1–54

Evangelou MW, Ebel M, Schaeffer A (2007) Chelate assisted phytoextraction of heavy metals from soil. Effect, mechanism, toxicity, and fate of chelating agents. Chemosph 68(6):989–1003. https://doi.org/10.1016/j.chemosphere.2007.01.062

Fendorf SE, Zasoski RJ (1992) Chromium (III) oxidation by delta—manganese oxide (MnO_2). 1. Characterization. Environ Sci Technol 26(1):79–85. https://doi.org/10.1021/es00025a006

Filimon MN, Voia SO, Popescu ROXANA, Dumitrescu GABI, Ciochina LP, Mituletu M, Vlad DC (2015) The effect of some insecticides on soil microorganisms based on enzymatic and bacteriological analyses. Roman Biotechnol Lett 20(3):10439–10447

Fotovat A, Naidu R, Sumner ME (1997) Water: soil ratio influences aqueous phase chemistry of indigenous copper and zinc in soils. Soil Res 35(4):687–710. https://doi.org/10.1071/S96086

Gad A, Saleh A, Khalifa M (2019) Assessment of natural radionuclides and related occupational risk in agricultural soil, southeastern Nile Delta. Egypt. Arab J Geosci 12:1–15. https://doi.org/10.1007/s12517-019-4356-6

Gautam PK, Gautam RK, Banerjee S, Chattopadhyaya MC, Pandey JD (2016) Heavy metals in the environment: fate, transport, toxicity and remediation technologies. Nova Sci Publ 60:101–130

Ghosh M, Singh SP (2005) A review on phytoremediation of heavy metals and utilization of it's by products. Asian J Energy Environ 6(4):18

Gilden RC, Huffling K, Sattler B (2010) Pesticides and health risks. J Obstet Gynecol Neonatal Nur 39(1):103–110. https://doi.org/10.1111/j.1552-6909.2009.01092.x

Girth J, Schimming CG, Brummer G (1981) Effect of formation of chloro-complex on the solubility and adsorption of nickel, zinc and cadmium.Mitteilungen der Deutschen. Roden Kundlic Gesellschaff 30:19–29

Goldemund H (2004) Soil development and properties. Soil Contamination and Amelioration. Encyclop Forest Scipp 1241

Han Y, Mo R, Yuan X, Zhong D, Tang F, Ye C, Liu Y (2017) Pesticide residues in nut-planted soils of China and their relationship between nut/soil. Chemosphr 180:42–47. https://doi.org/10.1016/j.chemosphere.2017.03.138

Harrison RM, Laxen DP, Wilson SJ (1981) Chemical associations of lead, cadmium, copper, and zinc in street dusts and roadside soils. Environ Sci Technol 15(11):1378–1383. https://doi.org/10.1021/es00093a013

Hussain I, Puschenreiter M, Gerhard S, Schöftner P, Yousaf S, Wang A, Reichenauer TG (2018) Rhizoremediation of petroleum hydrocarbon-contaminated soils: improvement opportunities and field applications. Environ Exp Bot 147:202–219. https://doi.org/10.1016/j.envexpbot.2017.12.016

Islam MMM, Shafi S, Bandh SA, Shameem N (2019) Impact of environmental changes and human activities on bacterial diversity of lakes. In: Freshwater microbiology. Academic Press, pp 105–136. https://doi.org/10.1016/B978-0-12-817495-1.00003-7

John DA, Babu GR (2021) Lessons from the aftermaths of green revolution on food system and health. Front Sus Food Sys 5:644559. https://doi.org/10.3389/fsufs.2021.644559

Jones LHP, Jarvis SC (1981) The fate of heavy metals. The chemistry of soil processes, pp 593–620

Kalyanasundaram GT, Syed N, Subburamu K (2021) Recent developments in plant growth-promoting rhizobacteria (PGPR) for sustainable agriculture. Rec Dev Appl Microbiol Biochempp 181–192.https://doi.org/10.1016/B978-0-12-821406-0.00017-5

Kampa M, Castanas E (2008) Human health effects of air pollution. Environ Pollut 151(2):362–367. https://doi.org/10.1016/j.envpol.2007.06.012

Karthik C, Elangovan N, Kumar TS, Govindharaju S, Barathi S, Oves M, Arulselvi PI (2017) Characterization of multifarious plant growth promoting traits of rhizobacterial strain AR6 under Chromium (VI) stress. Microbiol Res 204:65–71. https://doi.org/10.1016/j.micres.2017.07.008

Khan S, Cao Q, Zheng YM, Huang YZ, Zhu YG (2008) Health risks of heavy metals in contaminated soils and food crops irrigated with wastewater in Beijing, China. Environ Pollut 152(3):686–692. https://doi.org/10.1016/j.envpol.2007.06.056

Khan S, Waqas M, Ding F, Shamshad I, Arp HPH, Li G (2015) The influence of various biochars on the bioaccessibility and bioaccumulation of PAHs and potentially toxic elements to turnips (*Brassica rapa* L.). J Hazard Mater 300:243–253. https://doi.org/10.1016/j.jhazmat.2015.06.050

Kieffer P, Schröder P, Dommes J, Hoffmann L, Renaut J, Hausman JF (2009) Proteomic and enzymatic response of poplar to cadmium stress. J Proteom 72(3):379–396. https://doi.org/10.1016/j.jprot.2009.01.014

Kim EJ, Choi SD, Chang YS (2011) Levels and patterns of polycyclic aromatic hydrocarbons (PAHs) in soils after forest fires in South Korea. Environ Sci Pollut Res 18:1508–1517. https://doi.org/10.1007/s11356-011-0515-3

Lasat MM (2000) Phytoextraction of metals from contaminated soil: a review of plant/soil/metal interaction and assessment of pertinent agronomic issues. J Hazard Subst Res 2:1–25

Li H, Fagerberg B, Sallsten G, Borné Y, Hedblad B, Engström G, Andersson EM (2019) Smoking-induced risk of future cardiovascular disease is partly mediated by cadmium in tobacco: Malmö diet and cancer cohort study. Environ Health 18(1):1–10. https://doi.org/10.1186/s12940-019-0495-1

Lin H, Wang Z, Liu C, Dong Y (2022) Technologies for removing heavy metal from contaminated soils on farmland: a review. Chemosph 305:135457. https://doi.org/10.1016/j.chemosphere.2022.135457

Liu WX, Shen LF, Liu JW, Wang YW, Li SR (2007) Uptake of toxic heavy metals by rice (*Oryza sativa* L.) cultivated in the agricultural soil near Zhengzhou City, People's Republic of China. Bull Environ Contamin Toxicol 79:209–213. https://doi.org/10.1007/s00128-007-9164-0

Liu X, Chen S, Yan X, Liang T, Yang X, El-Naggar A, Chen H (2021) Evaluation of potential ecological risks in potential toxic elements contaminated agricultural soils: correlations between soil contamination and polymetallic mining activity. J Environ Manag 300:113679. https://doi.org/10.1016/j.jenvman.2021.113679

Long XX, Yang XE, Ni WZ (2002) Current status and perspective on phytoremediation of heavy metal polluted soils. J Appl Ecol 13(6):757–762

Luo J, Ye Y, Gao Z, Wang W (2014) Essential and nonessential elements in the red-crowned crane *Grus japonensis* of Zhalong Wetland, northeastern China. Toxicol Environ Chem 96(7):1096–1105. https://doi.org/10.1080/02772248.2015.1007989

McNeill KR, Waring S (1992) Vitrification of contaminated soils. In: International conference on contaminated land treatment technologies, pp 143–159

Mekouar MA (2020) Food and agriculture organization of the United Nations (FAO). Yearb Int Environ Law 31(1):326–340. https://doi.org/10.1093/yiel/yvab061

Minnesota Department of Health (2023). https://www.health.state.mn.us/communities/environment/risk/studies/metals.html. Accessed on 02 Jan 2024

Monged MH, Hassan HB, El-Sayed SA (2020) Spatial distribution and ecological risk assessment of natural radionuclides and trace elements in agricultural soil of northeastern Nile Valley, Egypt. Water, Air, Soil Pollut 231(7):338. https://doi.org/10.1007/s11270-020-04678-9

Moore PA Jr, Patrick WH Jr (1989) Manganese availability and uptake by rice in acid sulfate soils. Soil Sci Soc Am J 53(1):104–109. https://doi.org/10.2136/sssaj1989.03615995005300010019x

Namiki S, Otani T, Seike N (2013) Fate and plant uptake of persistent organic pollutants in soil. Soil Sci Plant Nut 59(4):669–679. https://doi.org/10.1080/00380768.2013.813833

Naujokas MF, Anderson B, Ahsan H, Aposhian HV, Graziano JH, Thompson C, Suk WA (2013) The broad scope of health effects from chronic arsenic exposure: update on a worldwide public health problem. Environ Health Perspect 121(3):295–302. https://doi.org/10.1289/ehp.1205875

Obukhovskaya JD (1982) Mercury sorption by soil minerals Soviet. Soil Sci 14(3):49–58

Okunola OJ, Uzairu A, Ndukwe G (2007) Levels of trace metals in soil and vegetation along major and minor roads in metropolitan city of Kaduna, Nigeria. Afr J Biotechnol 6(14):1703–1709

Oves M, Khan MS, Zaidi A, Ahmad E (2012) Soil contamination, nutritive value, and human health risk assessment of heavy metals: an overview. Springer Vienna, pp 1–27. https://doi.org/10.1007/978-3-7091-0730-0_1

Paoli D, Giannandrea F, Gallo M, Turci R, Cattaruzza MS, Lombardo F, Gandini L (2015) Exposure to polychlorinated biphenyls and hexachlorobenzene, semen quality and testicular cancer risk. J Endocrinolog Invest 38:745–752. https://doi.org/10.1007/s40618-015-0251-5

Patrick WH, Turner FT (1968) Effect of redox potential on manganese transformation in waterlogged soil. Nat 220(5166):476–478. https://doi.org/10.1038/220476a0

Petit AN, Fontaine F, Vatsa P, Clément C, Vaillant-Gaveau N (2012) Fungicide impacts on photosynthesis in crop plants. Photosynth Res 111:315–326. https://doi.org/10.1007/s11120-012-9719-8

Pingali PL (2012) Green revolution: impacts, limits, and the path ahead. Proceed Nation Acad Sci 109(31):12302–12308. https://doi.org/10.1073/pnas.0912953109

Rana RP, Kansal BD (1983) Sorption and release of cadmium in some sewage irrigated soils. J Ind Soc Soil Sci 31(3):439–443

Reed SC, Crites RW, Middlebrooks EJ (1995) Natural systems for waste management and treatment (No. Ed. 2). McGraw-Hill, Inc, pp 1–433

Rigoletto M, Calza P, Gaggero E, Malandrino M, Fabbri D (2020) Bioremediation methods for the recovery of lead-contaminated soils: a review. Appl Sci 10(10):3528. https://doi.org/10.3390/app10103528

Saladin G, Clément C (2005) Physiological side effects of pesticides on non-target plants. In: Livingston JV (ed) Agriculture and soil pollution: new research. Nova Science Publishers, Inc., Hauppauge, NY, USA, pp 53–86

Sankhla MS, Kumari M, Nandan M, Kumar R, Agrawal P (2016) Heavy metals contamination in water and their hazardous effect on human health—a review. Int J Curr Microbiol App Sci 5(10):759–766. https://doi.org/10.20546/ijcmas.2016.510.082

Seleiman MF, Hafez EM (2021) Optimizing inputs management for sustainable agricultural development. Mitigating environmental stresses for agricultural sustainability in Egypt, pp 487–507. https://doi.org/10.1007/978-3-030-64323-2_18

Sharma A, Kumar V, Kumar R, Shahzad B, Thukral AK, Bhardwaj R (2018) Brassinosteroid-mediated pesticide detoxification in plants: a mini-review. Cogent Food Agricul 4(1):1436212. https://doi.org/10.1080/23311932.2018.1436212

Siewit CL, Gengler B, Vegas E, Puckett R, Louie MC (2010) Cadmium promotes breast cancer cell proliferation by potentiating the interaction between ERα and c-Jun. Mol Endocrinol 24(5):981–992. https://doi.org/10.1210/me.2009-0410

Singh J, Kalamdhad AS (2011) Effects of heavy metals on soil, plants, human health and aquatic life. Int J Res Chem Environ 1(2):15–21

Singh RB (2000) Environmental consequences of agricultural development: a case study from the green revolution state of Haryana, India. Agricul Ecosys Environ 82(1–3):97–103. https://doi.org/10.1016/S0167-8809(00)00219-X

Smith LA, Means JL, Chen A (1995) Remedial options for metals-contaminated sites. Lewis Publishers, Boca Raton, Fla, USA, p 1130282272814553900

Spungen JH (2019) Children's exposures to lead and cadmium: FDA total diet study 2014–16. Food Addit Contamin Part A 36(6):893–903. https://doi.org/10.1080/19440049.2019.1595170

Sumiahadi A, Acar R, Direk M (2019) The potential use of cover crops for phytoremediation process of heavy metals contaminated soils, pp 1–14

Sumner ME (2000) Beneficial use of effluents, wastes, and biosolids. Commun Soil Sci Plant Analys 31(11–14):1701–1715. https://doi.org/10.1080/00103620009370532

Taylor M (2020) Hybrid realities: making a new green revolution for rice in south India. The J Peasant Stud 47(3):483–502. https://doi.org/10.1080/03066150.2019.1568246

Tchounwou PB, Yedjou CG, Patlolla AK, Sutton DJ (2012) Heavy metal toxicity and the environment. Environ Toxicol 3:133–164. https://doi.org/10.1007/978-3-7643-8340-4_6

Thakur S, Sharma SS (2016) Characterization of seed germination, seedling growth, and associated metabolic responses of *Brassica juncea* L. cultivars to elevated nickel concentrations. Protoplasm 253:571–580. https://doi.org/10.1007/s00709-015-0835-0

Turer DG, Maynard BJ (2003) Heavy metal contamination in highway soils. Comparison of corpus Christi, Texas and Cincinnati, Ohio shows organic matter is key to mobility. Clean Technol Environ Policy4(4):235–245. https://doi.org/10.1007/s10098-002-0159-6

Valentin L, Nousiainen A, Mikkonen A (2013) Introduction to organic contaminants in soil: concepts and risks. Emerging organic contaminants in sludges: analysis, fate and biological treatment, pp 1–29.https://doi.org/10.1007/698_2012_208

Wang C, Yang Z, Yuan X, Browne P, Chen L, Ji J (2013) The influences of soil properties on Cu and Zn availability in soil and their transfer to wheat (*Triticum aestivum* L.) in the Yangtze River delta region, China. Geoderma 193:131–139. https://doi.org/10.1016/j.geoderma.2012.10.004

Warren C (1971) Biology and water pollution control W.B. Sanders Company, Philadelphia, 434 pp.

Weggler K, McLaughlin MJ, Graham RD (2004) Effect of chloride in soil solution on the plant availability of biosolid-borne cadmium. J Environ Qual 33(2):496–504.https://doi.org/10.2134/jeq2004.4960.

White B, Peterson ML (1988) The reduction of aqueous metal species on the surface of iron (11) contaminants. J Environ Qual 27:181–183

Wuana RA, Okieimen FE (2011) Heavy metals in contaminated soils: a review of sources, chemistry, risks and best available strategies for remediation. Int Schol Res Not 2011:1–20. https://doi.org/10.5402/2011/402647

Xu J, Li Y, Wang S, Long S, Wu Y, Chen Z (2023) Sources, transfers and the fate of heavy metals in soil-wheat systems: the case of lead (Pb)/zinc (Zn) smelting region. J Hazard Mater 441:129863. https://doi.org/10.1016/j.jhazmat.2022.129863

Yargholi B, Azarneshan S (2014) Long-term effects of pesticides and chemical fertilizers usage on some soil properties and accumulation of heavy metals in the soil (case study of Moghan plain's (Iran) irrigation and drainage network). Int J Agricul Crop Sci 7(8):518

Yekeen TA, Onifade TO (2012) Evaluation of some heavy metals in soils along a major road in Ogbomoso. South West Niger Evaluat 2(8):1–9

Zhuang P, McBride MB, Xia H, Li N, Li Z (2009) Health risk from heavy metals via consumption of food crops in the vicinity of Dabaoshan mine South China. Sci Total Environ 407(5):1551–1561. https://doi.org/10.1016/j.scitotenv.2008.10.061

Chapter 8
Paper and Paperboard from Agricultural Waste—A Sustainable Waste Management Approach

M. Vishnuvarthanan

Abstract The proper disposal of agricultural waste is essential for both environmental sustainability and human well-being. It plays a crucial role in preventing soil and water pollution and preserving the health of ecosystems. Effective waste management also contributes to reducing greenhouse gas emissions, which, in turn, helps to maintain climate stability. Moreover, it enhances agricultural productivity by recycling organic matter, ensuring fertile and sustainable land for future generations. There is a growing interest in utilizing agricultural waste for paper and paperboard production due to its environmental and economic benefits. This chapter discusses the methods for converting agricultural residues such as straw, bagasse and husks into high-quality paper products. The advanced techniques such as pulping, bleaching and pressing these agricultural wastes can be turned into sustainable and biodegradable paper materials. This research underscores the potential of this eco-friendly approach for minimizing environmental pollution, conserving forest resources, and fostering a circular economy.

8.1 Introduction

In recent days, sustainable practices have become increasingly vital and the management of electronic waste, microplastics, and heavy metals has emerged as a major concern (Das et al. 2011; Mishra et al. 2019). The propagation of the global population has driven agricultural operations, leading to a notable surge in agricultural waste generation. Therefore, it has become imperative to explore innovative and eco-friendly approaches to address this issue. The use of agricultural waste for paper and paperboard production is one such method that is gaining momentum, offering a

M. Vishnuvarthanan (✉)
Department of Mechanical Engineering, Kalasalingam Academy of Research and Education, Krishnankovil, Srivilliputtur, Tamil Nadu, India
e-mail: vishnuvarthanan.india@gmail.com

A. Kuanar et al. (eds.), *Recent Advances in Bioremediation and Phytoremediation*, Environmental Science and Engineering,
https://doi.org/10.1007/978-3-031-77884-1_8

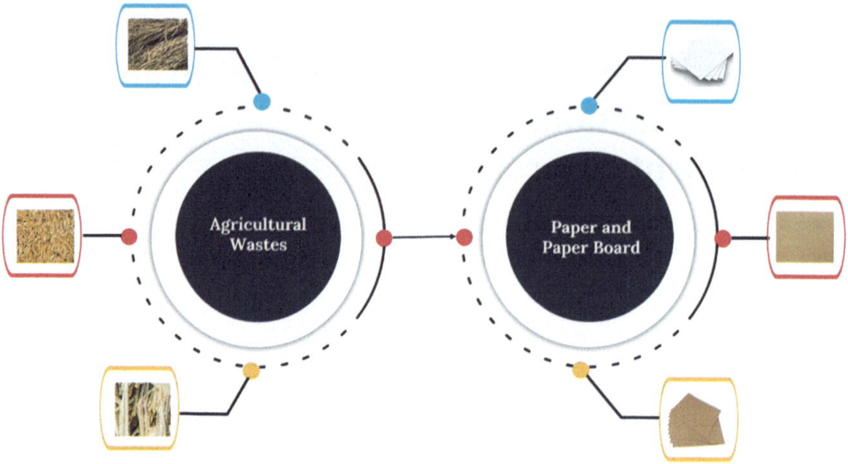

Fig. 8.1 Agricultural wastes to paper and paper board

promising avenue for effective waste management and sustainable resource utilization (Gupta et al. 2023). Agricultural waste encompasses a wide array of byproducts generated from various farming activities, including crop residues, husks, straw, bagasse and other nonedible parts of plants (Koul et al. 2022). Generally, agricultural waste has been incinerated, left to decompose or disposed of in landfills, contributing to environmental pollution and posing health hazards (Mishra et al. 2023). However, with increasing environmental awareness and the pressing need to reduce carbon footprints, there is a growing interest in exploring alternative methods to manage this abundant resource more sustainably (Zhou et al. 2023) (Fig. 8.1).

The paper and paperboard industry are a significant consumer of wood pulp and has long complained about its high environmental impact, including deforestation, water consumption, and chemical pollution (Talattov et al. 2021; Bhattacharjee et al. 2021). By addressing these concerns, the researchers and industry experts have sought alternative sources of raw materials that are renewable, abundant and environmentally friendly. Agricultural waste is a viable candidate, offering a sustainable solution to mitigate the ecological footprint of paper production. The use of agricultural waste for paper and paperboard production has several distinct advantages. It provides a renewable and abundant source of raw material, reducing the dependence on finite resources such as wood pulp (Blasi et al. 2023). This helps to relieve pressure on natural forests and ecosystems, promotes biodiversity conservation and mitigates deforestation. The repurposing of agricultural waste for paper production offers a more cost-effective alternative than traditional methods, thereby enhancing economic viability for both producers and consumers (Freitas et al. 2021). Moreover, it contributes to the development of rural economies by creating new avenues for income generation and employment opportunities in agricultural regions.

The conversion of agricultural waste into paper and paperboard offers significant environmental benefits. Diverting organic waste from landfills and incineration helps to reduce greenhouse gas emissions and mitigate climate change. The use of agricultural residues for paper production can minimize the need for chemical additives and bleaching agents, leading to a reduction in water pollution and energy consumption throughout the manufacturing process (Worku et al. 2023). Thus, the adoption of this sustainable waste management approach not only addresses the pressing issue of agricultural waste management but also promotes a circular economy model wherein waste is repurposed into valuable resources. The various aspects of producing paper and paperboard from agricultural waste, exploring the technological processes involved, the environmental implications, the economic feasibility and the policy frameworks supporting this sustainable initiative are discussed.

8.2 Overview of Agricultural Waste

Agricultural waste has become a significant issue due to its large volume, varied composition and environmental impact. This postharvest phase generates various waste materials such as crop residues, spoiled produce and packaging materials. These materials present challenges for farmers but also create opportunities for adopting environmentally friendly waste management practices (Ayilara et al. 2020). Residual crop materials such as stalks, husks, leaves and stems are common agricultural waste remaining after harvesting. These materials remain in fields after the main crop is collected and play a crucial role in maintaining soil health and improving fertility. Some residues breakdown naturally, adding to soil organic carbon levels, others can persist for a long time (Dynarski et al. 2020). It is important to develop effective management strategies to prevent soil erosion and nutrient depletion.

The common practice for managing crop residues is tillage, where farmers plough or cultivate the soil to incorporate residues into the soil surface (Wang et al. 2020). Tillage can help accelerate decomposition and improve soil structure and may also contribute to soil erosion, nutrient leaching and loss of soil organic carbon. As a result, the conservation tillage practices such as no-till or reduced tillage, have gained popularity among farmers seeking to minimize soil disturbance and preserve crop residues on the soil surface. Another method for managing crop residues is mulching, which involves spreading residues on the soil surface to provide ground cover and mitigate soil erosion (Prem et al. 2020). Mulching aids in conserving soil moisture, inhibiting weed growth and enhancing soil health by increasing the organic matter content and promoting microbial activity (Mechergui et al. 2021). Moreover, the crop residues can be repurposed for livestock feed, bedding material or the production of bioenergy, which offers supplementary economic and environmental advantages (Fig. 8.2).

In addition to crop residues, postharvest losses significantly contribute to agricultural waste after the harvest of crops (Das et al. 2020). These losses transpire during the processes of handling, storage and transportation, restricting from various factors including pests, diseases, mechanical damage and inadequate storage conditions. Although certain postharvest losses are unavoidable, the mitigation of such

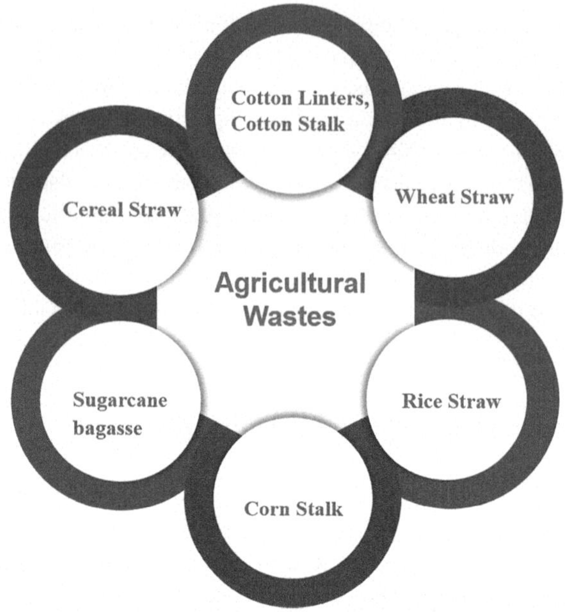

Fig. 8.2 Overview of agricultural wastes

losses through enhanced harvesting practices, appropriate storage infrastructure, and effective transportation systems can play a pivotal role in waste reduction and the enhancement of food security (Dora et al. 2021). Packaging materials also contribute to agricultural waste after harvesting crops particularly in the case of fresh produce destined for the market (Birania et al. 2022). Packaging materials such as cardboard boxes, plastic bags and foam trays are used to protect and transport agricultural products from farm to table. However, excessive or nonrecyclable packaging can contribute to environmental pollution and waste generation prompting efforts to reduce packaging waste through sustainable packaging alternatives and recycling initiatives (Coelho et al. 2020).

The agricultural waste after harvesting crops presents both challenges and opportunities for sustainable management practices (Adejumo and Adebiyi 2020). Crop residues, postharvest losses and packaging materials all contribute to the overall volume of agricultural waste and require attention from farmers, policymakers and consumers. The implementation of practices such as conservation tillage, mulching and waste reduction strategies, farmers can minimize waste generation, conserve resources, and promote environmental sustainability in the postharvest phase of agricultural production (Patel et al. 2020).

8.3 Importance of Sustainable Waste Management and Circular Economy

Sustainable waste management refers to practices that minimize waste generation, maximize resource recovery, and reduce environmental impacts (Mostaghimi and Behnamian 2023; Pradhan et al. 2023). This approach is essential for several reasons ranging from environmental conservation to public health and economic sustainability (Fig. 8.3).

Sustainable waste management plays a critical role in environmental conservation. The improper disposal of waste such as open dumping or burning can lead to soil, water and air pollution and harm ecosystems and biodiversity (Siddiqua et al. 2022; Paul et al. 2023). Plastics and other nonbiodegradable materials can persist in the environment for hundreds of years posing a threat to wildlife and marine ecosystems (Al-Thawadi 2020). Pollution can reduce and minimize our ecological footprint by following sustainable waste management practices such as recycling, composting, and waste-to-energy technologies (Sharma et al. 2020). Sustainable waste management is essential for mitigating climate change. Organic wastes such as food scraps and yard trimmings emit potent greenhouse methane gas when decomposing in landfills (Yasmin et al. 2022). Methane has a much greater global warming potential than carbon dioxide, contributing to climate change and its associated impacts, such as rising temperatures, extreme weather events, and sea-level rise. Through the diversion of organic waste from landfills through composting or anaerobic digestion, methane emissions can be captured, and renewable energy can be generated, thereby reducing greenhouse gas emissions and combating climate change (Saravanakumar

Fig. 8.3 Circular economy

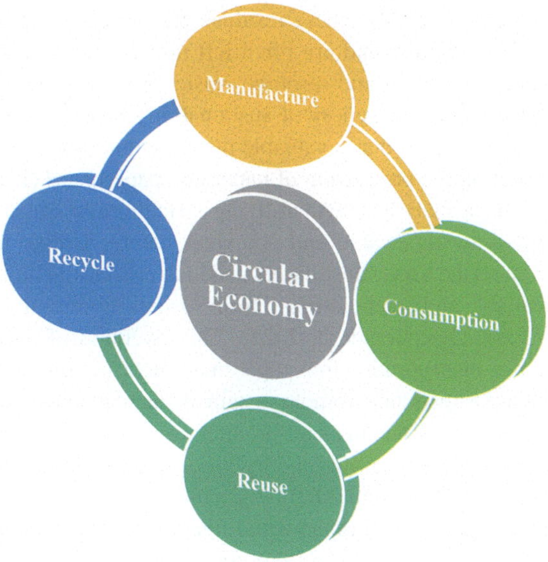

et al. 2023). The sustainable waste management promotes public health and safety (Singh et al. 2020). Improperly managed waste such as open dumps or uncontrolled landfills can attract pests, vermin, and disease vectors. It increases the risk of infectious diseases and health hazards. Inadequate waste management infrastructure can also lead to contaminated water sources and the spread of waterborne diseases. The implementation of proper waste collection, treatment and disposal systems can protect public health and create a safer and healthier environment.

The sustainable waste management contributes to economic sustainability and resource efficiency (Fatimah et al. 2020). Waste represents a significant loss of valuable resources including energy, water and raw materials. Recycling and resource recovery initiatives not only conserve natural resources but also create economic opportunities (Xavier et al. 2021). The transition towards a circular economy model where waste is viewed as a resource rather than a liability. The importance of sustainable waste management cannot be overstated. The adoption of practices that prioritize waste reduction, resource recovery and environmental protection. It can address pressing global challenges such as climate change, pollution and resource depletion.

8.4 Utilization of Agricultural Waste in Paper and Paperboard Production

Agricultural waste presents a significant challenge and also holds immense potential as a valuable resource for various industries including the paper and paperboard manufacturing sectors. Agricultural waste encompasses a wide range of materials generated during farming activities such as crop residues, straw, bagasse and husks (Maji et al. 2020). These residues are typically considered byproducts of agricultural production and are often left unused or discarded. It contributes to environmental pollution and resource waste. However, with advancements in technology and growing awareness of sustainability issues, agricultural waste is increasingly being recognized as a valuable resource that can be harnessed for various purposes, including the production of paper and paperboard (Kumar and Verma 2024).

The primary benefit of utilizing agricultural waste in paper and paperboard production is its renewable and abundant nature. Unlike traditional wood pulp, which requires the harvesting of trees and long processing times, agricultural waste is readily available and can be sourced from various agricultural activities throughout the year (Aliaño-González et al. 2022). The diversion of agricultural waste from landfills and its incorporation into paper manufacturing processes can reduce the pressure on natural forests and promote sustainable land management practices. The utilization of agricultural waste in paper production offers environmental advantages by reducing the carbon footprint and energy consumption associated with traditional papermaking methods (Man et al. 2020). Agricultural residues such as straw and bagasse have a lower lignin content than wood fibers, making them easier to process and requiring less energy-intensive pulping techniques. The incorporation of agricultural waste into paper and paperboard production can result in products with improved strength, durability and recyclability contributing to circular economy principles and reducing the generation of waste (Savio et al. 2022).

Several agricultural residues have been successfully utilized in paper and paperboard production, each offering unique advantages and characteristics. Rice straw is a byproduct of rice cultivation and is abundant in regions with significant rice production. Rice straw fibers are relatively long and possess good tensile strength, making them suitable for papermaking applications (Jayanti et al. 2023). Similarly, sugarcane bagasse (Singh et al. 2021), a residue from sugar production, contains short fibers and high cellulose content, making it ideal for producing pulp and paper products. In addition to crop residues, other agricultural waste materials, such as banana stems (Ferdous et al. 2021), corn husks (Ratna et al. 2022) and wheat straw (Malik et al. 2020) have also been explored as potential raw materials for paper and paperboard production. However, the utilization of agricultural waste in paper manufacturing is not without challenges. One of the main challenges is the variability in the composition and quality of agricultural residues, which can affect the properties and performance of paper products.

The logistics of collecting, transporting and processing agricultural waste can be complex and costly particularly in rural areas with limited infrastructure. To overcome these challenges, collaborative efforts between farmers, researchers, industry stakeholders and policymakers are essential for developing innovative solutions and creating value chains that maximize the utilization of agricultural waste in paper and paperboard production. Despite these challenges, the utilization of agricultural waste in paper manufacturing holds significant promise for promoting sustainable development and addressing environmental concerns. The harnessing of renewable resources available in agricultural residues can reduce the reliance on finite natural resources, minimize waste generation and mitigate the environmental impacts associated with traditional papermaking processes. The utilization of agricultural waste in paper production can create economic opportunities for rural communities, providing additional income streams for farmers and supporting local industries. The utilization of agricultural waste in paper and paperboard production offers numerous benefits including environmental sustainability, resource efficiency and economic development. The harnessing of renewable resources available in agricultural residues can reduce the environmental footprint, promote circular economy principles and create a more sustainable future (Barros et al. 2020).

8.5 Advantages of Using Agricultural Waste for Paper and Paperboard

The primary advantages of using agricultural waste for paper and paperboard production are its abundance and renewability. Agricultural wastes such as crop residues, straw, bagasse and husks are generated in large quantities worldwide as a byproduct of farming activities (Dey et al. 2021). Traditional wood pulp which requires the harvesting of trees can contribute to deforestation and habitat loss. Agricultural waste is readily available and does not require additional land use or resource extraction. The utilization of agricultural waste as a raw material for paper production can reduce the reliance on finite natural resources and promote sustainable land management practices (Antar et al. 2021). In addition to environmental advantages, the use of agricultural waste for paper production can also lead to economic benefits for farmers, industries and communities. Farmers can generate additional income streams by selling their agricultural residues to paper mills or participating in cooperative ventures for waste collection and processing. Moreover, the utilization of agricultural waste as a raw material for paper production can create employment opportunities in rural areas where agricultural activities are dominant.

The other advantage of using agricultural waste for paper and paperboard production is its versatility and adaptability. Agricultural residues come in various forms and compositions with unique characteristics and properties. Rice straw, sugarcane bagasse, wheat straw and other agricultural residues can be processed into pulp and paper products with different qualities and applications (Thongsomboon et al. 2023). Rice straw fibers are relatively long and possess good tensile strength, making them suitable for producing high-quality printing paper (Saini et al. 2022). The sugarcane bagasse fibers are shorter and can be used to manufacture packaging materials and board products (Liu et al. 2020). The utilization of a diverse range of agricultural residues can meet the diverse needs of consumers and industries while minimizing waste and maximizing resource efficiency. The advantages of using agricultural waste for paper and paperboard production are diverse, encompassing environmental, economic and social dimensions. The abundance and renewability of agricultural residues can reduce our environmental footprint, promote sustainable land management practices and create economic opportunities. Moreover, the use of agricultural waste as a raw material for paper production offers versatility and adaptability.

8.6 Challenges and Limitations

One of the primary challenges of converting agricultural waste into paper and paperboard is the variability in the composition and quality of agricultural residues. Wood fibers have consistent properties and characteristics but agricultural residues such as straw, bagasse and husks can vary significantly depending on factors such as crop type, harvesting method and environmental conditions (Neitzel et al. 2022). This variability can affect the processing and performance of paper products leading to inconsistencies in quality and durability. Therefore, there is a need to develop standardized methods for collecting, processing and treating agricultural waste to ensure consistent and reliable paper production. Moreover, agricultural residues often contain higher levels of impurities such as dirt, sand, and silica than traditional wood pulp (Haile et al. 2021). These impurities can cause equipment wear and tear, increase processing costs and degrade the quality of paper products. The removal of impurities from agricultural residues requires additional processing steps such as cleaning, screening and refining which can be time-consuming and resource-intensive (Phiri et al. 2023). Therefore, it is necessary to find efficient and cost-effective methods for removing impurities from agricultural waste to optimize the process of paper production.

The availability and accessibility of raw materials is another challenge for converting agricultural waste into paper and paperboard. Agricultural residues are typically generated seasonally and are often located in rural or remote areas, making collection and transportation logistics challenging. Moreover, the competition for agricultural residues from other sectors, such as bioenergy production and animal feed can increase prices and limit the availability of raw materials for paper production. Therefore, it is essential to establish efficient supply chains and collaboration between farmers, waste collectors and paper mills to ensure a steady and reliable supply of agricultural waste for paper production. The conversion of agricultural waste into paper and paperboard requires specialized equipment and technologies that may not be readily available or affordable for small-scale producers or developing countries. Traditional papermaking processes such as pulping and refining may need to be adapted or optimized to accommodate the unique properties of agricultural residues (Gupta and Shukla 2020). The conversion of agricultural waste into paper and paperboard offers numerous benefits, including environmental sustainability, resource efficiency and economic viability (Angulo-Mosquera et al. 2021). The composition and quality, impurities, availability and accessibility of raw materials, technological limitations and regulatory barriers are among the key challenges that need to be overcome.

8.7 Techniques for Processing Agricultural Waste into Paper and Paperboard

The most common technique for processing agricultural waste into paper and paperboard is pulping which involves breaking down raw materials into fibers that can be used to form paper sheets (Mboowa 2021). Agricultural residues such as straw, bagasse and husks contain cellulose fibers that are suitable for papermaking but specialized equipment and extraction processes are needed. Mechanical pulping methods such as grinding and refining are often used to separate fibers from agricultural residues resulting in a pulp mixture that can be further processed into paper products. Mechanical pulping is relatively simple and cost-effective and may produce fibers of lower quality and strength than other pulping methods (Jahan et al. 2021).

Chemical pulping is another technique commonly used to process agricultural waste into paper and paperboard (Sežun et al. 2023). This method involves treating raw materials with chemicals such as sodium hydroxide (caustic soda) and sodium sulfide to dissolve the lignin and extract cellulose fibers. Chemical pulping processes such as the kraft and sulfite processes are highly efficient at removing lignin and producing high-quality pulp suitable for a wide range of paper products (Yadav et al. 2022) (Fig. 8.4).

An alternative to traditional pulping methods is the use of enzymatic hydrolysis to break down agricultural residues into pulp. Enzymes such as cellulases and hemicellulases can selectively degrade lignin and hemicellulose components in raw materials leaving behind cellulose fibers that can be used for papermaking (Michelin et al. 2020). Enzymatic hydrolysis offers several advantages over conventional pulping methods, including lower energy consumption, reduced chemical usage and improved fiber quality. However, the enzymatic processes are still relatively expensive and require further optimization to be economically viable on a large scale (Fig. 8.5).

In addition to pulping techniques, various mechanical and chemical treatments can be used to enhance the properties and performance of agricultural waste-based paper and paperboard. Mechanical refining processes such as beating and refining

Fig. 8.4 Types of pulping

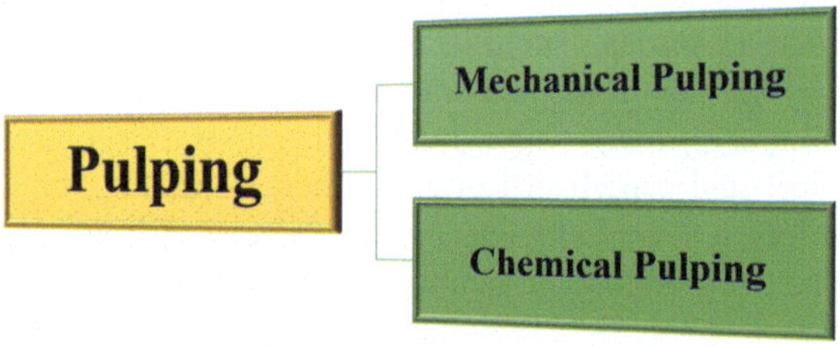

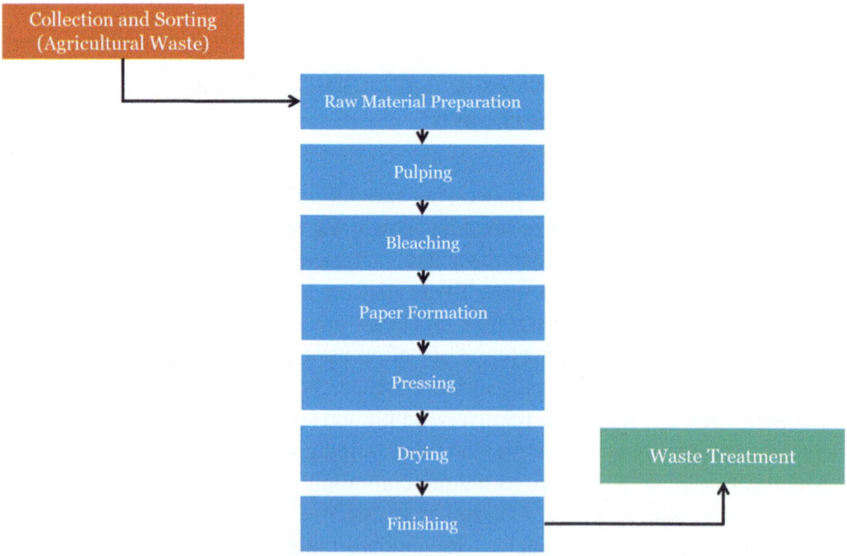

Fig. 8.5 Steps for making paper from agricultural waste

can improve fiber bonding and strength resulting in paper products with enhanced mechanical properties. Chemical treatments such as sizing and surface coating can improve the water resistance, printability, and other functional characteristics of paper and paperboard. Bio pulping involves the use of fungi or bacteria to degrade lignin in raw materials to reduce the need for harsh chemicals and energy-intensive processes (Oforibika and Alalibo 2020). Nanocellulose derived from agricultural residues through mechanical or enzymatic methods can be used to produce lightweight and high-strength paper and paperboard products with unique properties such as transparency, barrier properties and conductivity. The techniques for processing agricultural waste into paper and paperboard offer promising opportunities for promoting sustainability, resource efficiency and innovation in the paper industry.

8.8 Environmental Impact Assessment

The primary environmental impact of converting agricultural waste into paper and paperboard is the potential for air pollution. The processing of agricultural residues such as straw and husks into pulp and paper products may release particulate matter, volatile organic compounds (VOCs) and greenhouse gases into the atmosphere (Gaur et al. 2020). These emissions can contribute to air pollution, smog formation and respiratory health problems. The conversion of agricultural waste into paper and paperboard may have implications for water quality and aquatic ecosystems. The pulping and papermaking processes require significant amounts of water for washing, bleaching and processing the raw materials. The discharges from paper mills including wastewater and chemical effluents can contain many pollutants such as suspended solids, organic compounds and toxic chemicals which may contaminate water bodies and harm aquatic organisms (Shakil and Mostafa 2021). Therefore, it is necessary to implement wastewater treatment systems and water recycling and to adopt closed-loop water management practices to mitigate water pollution and minimize the environmental impact on aquatic ecosystems.

The utilization of agricultural waste for paper production may affect soil health and land. Agricultural residues from fields can disturb the soil, reduce the organic matter content and contribute to soil erosion and degradation. The conversion of agricultural waste into paper and paperboard may have indirect environmental impacts through energy consumption and resource utilization. Agricultural residues are renewable resources, and the processing and transportation of these materials require energy inputs such as electricity, fuel and water. The use of chemicals such as bleaching agents and adhesives in the papermaking process may have environmental consequences such as energy consumption, emissions and waste generation (Wang et al. 2021). Therefore, optimizing energy efficiency, reducing resource consumption and promoting the use of eco-friendly chemicals are essential strategies for minimizing the environmental footprint of converting agricultural waste into paper and paperboard.

8.9 Economic Viability and Market Potential

The economic viability of converting agricultural waste into paper and paperboard is evident through several factors. The utilization of agricultural residues as raw materials for paper production can significantly reduce production costs compared to traditional wood pulp. Agricultural waste is often available at low or even no cost to paper manufacturers because it is considered a byproduct of farming activities (Donner et al. 2021). The local sourcing of raw materials avoids the transportation costs associated with wood pulp.

The conversion of agricultural waste into paper and paperboard offers opportunities for revenue diversification and value addition along the supply chain. Farmers and waste collectors can generate additional income streams by selling agricultural residues to paper mills. The market potential for paper and paperboard products made from agricultural waste is significant and continues to grow as consumers and businesses increasingly prioritize sustainability and environmental responsibility (Cristofoli et al. 2023). With increasing awareness of the environmental impacts of traditional paper production methods, there is a growing demand for alternatives that utilize renewable resources and minimize waste generation.

8.10 Future Directions and Research Opportunities

One future direction is the development of advanced technologies and processes for efficiently converting agricultural waste into high-quality paper and paperboard products. Research efforts should focus on optimizing pulping methods, refining techniques and chemical treatments to enhance the strength, durability and recyclability of paper products made from agricultural residues. It is necessary to explore novel approaches such as enzymatic hydrolysis, bio pulping and nanocellulose extraction to provide new possibilities for improving process efficiency and product performance (Pirich et al. 2020). There is a need for research on the environmental and socioeconomic impacts of converting agricultural waste into paper and paperboard. Life cycle assessments (LCAs) can provide valuable insights into the environmental footprint of different paper production processes by comparing the resource consumption, emissions and waste generation associated with traditional wood pulp and agricultural waste-based paper production (Moretti et al. 2021). Another important research opportunity lies in exploring the potential of integrating agricultural waste utilization with other industries and sectors to create synergies and value-added products.

8.11 Conclusion

In conclusion, the proper disposal of agricultural waste is a critical imperative for safeguarding both environmental sustainability and human well-being. Effective waste management practices can mitigate soil and water pollution and thereby preserve the health and integrity of ecosystems. These measures serve to reduce greenhouse gas emissions and contribute to climate stability and the mitigation of global warming effects. Efficient agricultural waste management has the added benefit of enhancing agricultural productivity through the recycling of organic matter, ensuring the fertility and sustainability of our land for future generations. The utilization of agricultural waste in the production of paper and paperboard is an important solution that offers significant environmental and economic advantages. Through the use of innovative processing techniques such as pulping, bleaching and pressing, agricultural residues such as straw, bagasse and husks can be transformed into high-quality paper products. This eco-friendly approach not only reduces environmental pollution but also conserves forest resources, thereby promoting a circular economy and sustainable resource management.

References

Adejumo IO, Adebiyi OA (2020) Agricultural solid wastes: causes, effects, and effective management. In: Strategies of sustainable solid waste management, vol 8

Aliaño-González MJ, Gabaston J, Ortiz-Somovilla V, Cantos-Villar E (2022) Wood waste from fruit trees: biomolecules and their applications in agri-food industry. Biomolecules 12(2):238

Al-Thawadi S (2020) Microplastics and nanoplastics in aquatic environments: challenges and threats to aquatic organisms. Arab J Sci Eng 45(6):4419–4440

Angulo-Mosquera LS, Alvarado-Alvarado AA, Rivas-Arrieta MJ, Cattaneo CR, Rene ER, García-Depraect O (2021) Production of solid biofuels from organic waste in developing countries: a review from sustainability and economic feasibility perspectives. Sci Total Environ 795:148816

Antar M, Lyu D, Nazari M, Shah A, Zhou X, Smith DL (2021) Biomass for a sustainable bioeconomy: An overview of world biomass production and utilization. Renew Sustain Energy Rev 139:110691

Ayilara MS, Olanrewaju OS, Babalola OO, Odeyemi O (2020) Waste management through composting: challenges and potentials. Sustainability 12(11):4456

Barros MV, Salvador R, de Francisco AC, Piekarski CM (2020) Mapping of research lines on circular economy practices in agriculture: from waste to energy. Renew Sustain Energy Rev 131:109958

Bhattacharjee J, Mishra S, Das AP (2021) Recent advances in sensor-based detection of toxic dyes for bioremediation application: a review. Appl Biochem Biotechnol 2021:1–20. https://doi.org/10.1007/S12010-021-03767-7

Birania S, Kumar S, Kumar N, Attkan AK, Panghal A, Rohilla P, Kumar R (2022) Advances in development of biodegradable food packaging material from agricultural and agro-industry waste. J Food Process Eng 45(1):e13930

Blasi A, Verardi A, Lopresto CG, Siciliano S, Sangiorgio P (2023) Lignocellulosic agricultural waste valorization to obtain valuable products: an overview. Recycling 8(4):61

Coelho PM, Corona B, ten Klooster R, Worrell E (2020) Sustainability of reusable packaging—current situation and trends. Resour, Conserv Recycl: X 6:100037

Cristofoli NL, Lima AR, Tchonkouang RD, Quintino AC, Vieira MC (2023) Advances in the food packaging production from agri-food waste and by-products: market trends for a sustainable development. Sustainability 15(7):6153

Das AP, Sukla LB, Pradhan N, Nayak S (2011) Manganese biomining: a review. Biores Technol 102(16):7381–7387. https://doi.org/10.1016/J.BIORTECH.2011.05.018

Das PK, Das BP, Dash P (2020) Potentials of postharvest rice crop residues as a source of biofuel. In: Refining biomass residues for sustainable energy and bioproducts. Academic Press, pp 275–301

Dey T, Bhattacharjee T, Nag P, Ghati A, Kuila A (2021) Valorization of agro-waste into value added products for sustainable development. Bioresour Technol Rep 16:100834

Donner M, Verniquet A, Broeze J, Kayser K, De Vries H (2021) Critical success and risk factors for circular business models valorising agricultural waste and by-products. Resour Conserv Recycl 165:105236

Dora M, Biswas S, Choudhary S, Nayak R, Irani Z (2021) A system-wide interdisciplinary conceptual framework for food loss and waste mitigation strategies in the supply chain. Ind Mark Manag 93:492–508

Dynarski KA, Bossio DA, Scow KM (2020) Dynamic stability of soil carbon: reassessing the "permanence" of soil carbon sequestration. Front Environ Sci 8:514701

Fatimah YA, Govindan K, Murniningsih R, Setiawan A (2020) Industry 4.0 based sustainable circular economy approach for smart waste management system to achieve sustainable development goals: a case study of Indonesia. J Clean Prod 269:122263

Ferdous T, Quaiyyum MA, Jin Y, Bashar MS, Yasin Arafat KM, Jahan MS (2021) Pulping and bleaching potential of banana pseudo stem, banana leaf and banana peduncle. Biomass Convers Biorefinery 1–12

Freitas LC, Barbosa JR, da Costa ALC, Bezerra FWF, Pinto RHH, de Carvalho Junior RN (2021) From waste to sustainable industry: how can agro-industrial wastes help in the development of new products? Resour Conserv Recycl 169:105466

Gaur VK, Sharma P, Sirohi R, Awasthi MK, Dussap CG, Pandey A (2020) Assessing the impact of industrial waste on environment and mitigation strategies: a comprehensive review. J Hazard Mater 398:123019

Gupta AP, Upadhyay P, Sen T, Dutta J (2023) Agricultural waste as a resource: the lesser travelled road to sustainability. In: Agriculture waste management and bioresource: the circular economy perspective, pp 1–20

Gupta GK, Shukla P (2020) Insights into the resources generation from pulp and paper industry wastes: challenges, perspectives and innovations. Biores Technol 297:122496

Haile A, Gelebo GG, Tesfaye T, Mengie W, Mebrate MA, Abuhay A, Limeneh DY (2021) Pulp and paper mill wastes: utilizations and prospects for high value-added biomaterials. Bioresour Bioprocess 8:1–22

Jahan MS, Rahman MM, Ni Y (2021) Alternative initiatives for non-wood chemical pulping and integration with the biorefinery concept: a review. Biofuels, Bioprod Biorefin 15(1):100–118

Jayanti MA, Ardhana A, Husna N, Karimah A, Rahmi Dn M, Ariyanta HA, Eko BS, Fatriasari W (2023) Biocomposites of rice straw paper with chitosan: hydrophobicity and mechanical properties. Biomass Convers Biorefinery 1–14

Koul B, Yakoob M, Shah MP (2022) Agricultural waste management strategies for environmental sustainability. Environ Res 206:112285

Kumar V, Verma P (2024) Pulp-paper industry sludge waste biorefinery for sustainable energy and value-added products development: a systematic valorization towards waste management. J Environ Manag 352:120052

Liu C, Luan P, Li Q, Cheng Z, Sun X, Cao D, Zhu H (2020) Biodegradable, hygienic, and compostable tableware from hybrid sugarcane and bamboo fibers as plastic alternative. Matter 3(6):2066–2079

Maji S, Dwivedi DH, Singh N, Kishor S, Gond M (2020) Agricultural waste: its impact on environment and management approaches. In: Emerging eco-friendly green technologies for wastewater treatment, pp 329–351

Malik S, Rana V, Joshi G, Gupta PK, Sharma A (2020) Valorization of wheat straw for the paper industry: pre-extraction of reducing sugars and its effect on pulping and papermaking properties. ACS Omega 5(47):30704–30715

Man Y, Li J, Hong M, Han Y (2020) Energy transition for the low-carbon pulp and paper industry in China. Renew Sustain Energy Rev 131:109998

Mboowa D (2021) A review of the traditional pulping methods and the recent improvements in the pulping processes. Biomass Convers Biorefinery 1–12

Mechergui T, Pardos M, Jhariya MK, Banerjee A (2021) Mulching and weed management towards sustainability. In: Ecological intensification of natural resources for sustainable agriculture, pp 255–287

Michelin M, Gomes DG, Romaní A, Polizeli MDLT, Teixeira JA (2020) Nanocellulose production: exploring the enzymatic route and residues of pulp and paper industry. Molecules 25(15):3411

Mishra S, Ghosh S, van Hullebusch ED, Singh S, Das AP (2023) A critical review on the recovery of base and critical elements from electronic waste-contaminated streams using microbial biotechnology. Appl Biochem Biotechnol 1–30. https://doi.org/10.1007/S12010-023-04440-X/METRICS

Mishra S, charan Rath C, Das AP (2019) Marine microfiber pollution: a review on present status and future challenges. Mar Pollut Bull 140:188–197. https://doi.org/10.1016/J.MARPOLBUL.2019.01.039

Moretti C, Corona B, Hoefnagels R, Vural-Gürsel I, Gosselink R, Junginger M (2021) Review of life cycle assessments of lignin and derived products: lessons learned. Sci Total Environ 770:144656

Mostaghimi K, Behnamian J (2023) Waste minimization towards waste management and cleaner production strategies: a literature review. Environ Dev Sustain 25(11):12119–12166

Neitzel N, Hosseinpourpia R, Walther T, Adamopoulos S (2022) Alternative materials from agro-industry for wood panel manufacturing—a review. Materials 15(13):4542

Oforibika GA, Alalibo IK (2020) Biopulping—an ecofriendly technology. J Sci Ind Stud 15(1)

Patel SK, Sharma A, Singh GS (2020) Traditional agricultural practices in India: an approach for environmental sustainability and food security. Energy, Ecol Environ 5:253–271

Paul A, Dey S, Ram DK, Das AP (2023) Hexavalent chromium pollution and its sustainable management through bioremediation. Geomicrobiol J 1–11. https://doi.org/10.1080/01490451.2023.2218377

Phiri R, Rangappa SM, Siengchin S, Marinkovic D (2023) Agro-waste natural fiber sample preparation techniques for bio-composites development: methodological insights. Facta Univ, Ser: Mech Eng 21(4):631–656

Pirich CL, Picheth GF, Fontes AM, Delgado-Aguilar M, Ramos LP (2020) Disruptive enzyme-based strategies to isolate nanocelluloses: a review. Cellulose 27:5457–5475

Pradhan G, Tripathy B, Ram DK, Digal AK, Das AP (2023) Bauxite mining waste pollution and its sustainable management through bioremediation. Geomicrobiol J 1–10. https://doi.org/10.1080/01490451.2023.2235353

Prem M, Ranjan P, Seth N, Patle GT (2020) Mulching techniques to conserve the soil water and advance the crop production—a review. Curr World Environ 15:10–30

Ratna AS, Ghosh A, Mukhopadhyay S (2022) Advances and prospects of corn husk as a sustainable material in composites and other technical applications. J Clean Prod 133563

Saini S, Kardam SK, Kadam AA, Kumar V, Gaikwad KK (2022) Green and energy-efficient extraction of cellulose nano-fibrils from rice straw and its coating to improve functional properties of rice straw paperboard made via refiner mechanical pulping. Prog Org Coat 165:106747

Saravanakumar A, Sudha MR, Chen WH, Pradeshwaran V, Ashokkumar V, Selvarajoo A (2023) Biomethane production as a green energy source from anaerobic digestion of municipal solid waste: a state-of-the-art review. Biocatal Agric Biotechnol 102866

Savio L, Pennacchio R, Patrucco A, Manni V, Bosia D (2022) Natural fibre insulation materials: use of textile and agri-food waste in a circular economy perspective. Mater Circ Econ 4(1):6

Sežun M, Karlovits I, Kavčič U (2023) Chemical and enzymatic deinking efficiency of agricultural and industrial waste fiber-based paper packaging. J Sci Food Agric 103(3):1069–1076

Shakil MSZ, Mostafa MG (2021) Water quality assessment of paper mills effluent discharge areas. Al-Nahrain J Sci 24(3):63–72

Sharma S, Basu S, Shetti NP, Kamali M, Walvekar P, Aminabhavi TM (2020) Waste-to-energy nexus: a sustainable development. Environ Pollut 267:115501

Siddiqua A, Hahladakis JN, Al-Attiya WAK (2022) An overview of the environmental pollution and health effects associated with waste landfilling and open dumping. Environ Sci Pollut Res 29(39):58514–58536

Singh N, Tang Y, Ogunseitan OA (2020) Environmentally sustainable management of used personal protective equipment. Environ Sci Technol 54(14):8500–8502

Singh SP, Jawaid M, Chandrasekar M, Senthilkumar K, Yadav B, Saba N, Siengchin S (2021) Sugarcane wastes into commercial products: processing methods, production optimization and challenges. J Clean Prod 328:129453

Talattov AP, Wibowo MR, Kallman M (2021) Corporate transformation towards sustainability: the case of the Indonesian pulp and papers industry. Sustain Sci Resour 1:107–134

Thongsomboon W, Baimark Y, Srihanam P (2023) Valorization of cellulose-based materials from agricultural waste: comparison between sugarcane bagasse and rice straw. Polymers 15(15):3190

Wang H, Wang S, Yu Q, Zhang Y, Wang R, Li J, Wang X (2020) No tillage increases soil organic carbon storage and decreases carbon dioxide emission in the crop residue-returned farming system. J Environ Manag 261:110261

Wang T, Chang D, Huang D, Liu Z, Wu Y, Liu H et al (2021) Application of surfactants in paper-making industry and future development trend of green surfactants. Appl Microbiol Biotechnol 1–16

Worku LA, Bachheti A, Bachheti RK, Rodrigues Reis CE, Chandel AK (2023) Agricultural residues as raw materials for pulp and paper production: overview and applications on membrane fabrication. Membranes 13(2):228

Xavier LH, Giese EC, Ribeiro-Duthie AC, Lins FAF (2021) Sustainability and the circular economy: a theoretical approach focused on e-waste urban mining. Resour Policy 74:101467

Yadav V, Kumar A, Bilal M, Nguyen TA, Iqbal HM (2022) Lignin removal from pulp and paper industry waste streams and its application. In: Nanotechnology in paper and wood engineering. Elsevier, pp 265–283

Yasmin N, Jamuda M, Panda AK, Samal K, Nayak JK (2022) Emission of greenhouse gases (GHGs) during composting and vermicomposting: measurement, mitigation, and perspectives. Energy Nexus 7:100092

Zhou H, Awosusi AA, Dagar V, Zhu G, Abbas S (2023) Unleashing the asymmetric effect of natural resources abundance on carbon emissions in regional comprehensive economic partnership: what role do economic globalization and disaggregating energy play? Resour Policy 85:103914

Chapter 9
Phytoremediation of Environmental Emerging Pollutants

Haet Desai, Shreya Ojha, Anargha Nambiar, and Uma Hapani

Abstract Emerging pollutants (or contaminants) (EPs) are a diverse range of organic and inorganic chemicals that have entered into the environment as a result of human activity. Their sources include everyday domestic activities, human and animal health protection, and industrial and agricultural output. The majority of these chemicals have been present in the environment for decades, but their detection has only recently been possible due to the development of highly sensitive analytical techniques. The group of emerging pollutants includes; human and animal pharmaceuticals, personal care products, surfactants, and flame retardants, in addition to compounds used in the production of plastics, hormones and steroids, pesticides and fertilizers, nanoparticles, microplastics, and trace metals. Despite relatively low quantities in the environment, these pollutants can cause endocrine disruption and other health effects in humans, as well as oxidative stress in organisms. Phytoremediation, an environmentally sound and sustainable strategy, is based on utilizing the power of plants to mitigate the negative effects of contaminants on our ecosystems. This novel strategy takes advantage of plants' natural ability to absorb and detoxify contaminants and accumulate them, efficiently detoxifying soil, water, and air. This chapter emphasized on the sources of emerging pollutants, the phytoremediation process, the status of phytoremediation in the global context, and current research and prospects.

Keywords Accumulation · Emerging contaminants · Phytoremediation · Pollutants · Plant uptake · Biotransformation · Ecotoxicity

H. Desai · S. Ojha · U. Hapani (✉)
Department of Environmental Studies, The Maharaja Sayajirao University of Baroda, Vadodara, Gujarat, India
e-mail: umahapani18@gmail.com

A. Nambiar (✉)
Department of Biotechnology and Bioengineering, Institute of Advanced Research, Koba Institutional Area, Gandhinagar, Gujarat, India
e-mail: nambiaranargha@gmail.com

A. Kuanar et al. (eds.), *Recent Advances in Bioremediation and Phytoremediation*, Environmental Science and Engineering,
https://doi.org/10.1007/978-3-031-77884-1_9

9.1 Introduction

Abnormal expansion in environmental deterioration is one of the major threats impacting the existence of our planet and its various life forms. A number of anthropogenic activities, including waste creation, urbanization, industrialization and agricultural activities have significantly contributed to the soil, air, and water contamination, respectively, using harmful chemicals such as heavy metals. Apart from disturbing the ecosystem, these contaminants enter the food chain and cause health risks and also decline the biodiversity. The need to clean up the environment and get rid of harmful contaminants in a sustainable way has become essential. A variety of methods have been used to free contaminants from the environment. Amongst different approaches, phytoremediation is one of the potential techniques that uses plants to remove toxins from the environment (Khan et al. 2024). According to Pilon-Smits, phytoremediation is regarded as a highly acceptable, economical, and environmentally beneficial alternative to traditional physical and chemical remediation methods. In order to fix the damaged soil, control soil health and productivity, plants are extended to its roots in order to create a rhizosphere environment. This will make the soil capable to absorb heavy metals, break down organic pollutants and modify their bioavailability.

Emerging pollutants (EPs) are chemicals that have been shown to have significant effects on terrestrial, aquatic, and human health due to their chronic and acute toxicity (Mishra et al. 2021) Although there are many studies found in literature on EPs behaviour in the aquatic environment, however, there is an urgent need to conduct toxicological investigations, develop and execute the efficient ecological strategies for their removal. This paper therefore discusses some pertinent issues concerning EP's contamination in water environment, their risks associated to aquatic life and humans, and lastly, opportunities to mitigate the impacts of pollution through EP removal. The compounds of concern, their sources, ill effects and the fate has been classified in several categories. Moreover, extensive research on EPs category, their occurrence, fate and elimination in the environment have become a necessity. Additionally, EPs released into aquatic environment directly from wastewater treatment plants are thoroughly investigated. The prevalence of EP-related environmental and human health hazards is also considered. The removal of EPs from the environment as a risk mitigation strategy is discussed, with an emphasis on many non-traditional procedures involving biological EP removal. This chapter takes a critical look at existing issues faced by the existence of emerging pollutants in the aquatic environment. Certainly, including critical remarks and recommendations for future research to eradicate the impact of EPs on water and human health, as developed ways for their removal (Vasilachi et al. 2021).

Phytoremediation consumes less materials, energy, and equipment than physical and chemical remediations, that is seen in developed natural attenuations. The success of natural processes like river self-purification and natural attenuations as particular examples of phytoremediation has been demonstrated beyond the scope of our research. An effective method for treating various contaminants on-site or in-situ is generally phytoremediation. Emerging pollutants, also known as emerging contaminants, are defined as artificial or naturally occurring pollutants that have the potential to negatively impact the environment and/or human health and are not currently regulated in the environment or included in regular environmental monitoring programs (Unesco.org; NORMAN network 2019). More than 1036 such compounds are listed as EPs by the NORMAN network (NORMAN network 2019).

According to Sousa et al. EPs are (a) substances that are recently added to the environment; (b) compounds that have been there for a while but have not yet been proven to be dangerous to humans or ecosystem; and (c) compounds that have been released into the environment for a long time but have not been discovered until the development of new detection techniques (Sousa et al. 2018; Geissen et al. 2015; Belver et al. 2017). Many organic pollutants, also known as emerging pollutants (EPs), exists widely in various environmental matrices, including water resources. These pollutants comprise active residues from hormones, insecticides, medicines, surfactants, and personal care items. The persistent EPs are termed as micropollutants (MPs) because they pose harmful effects on the environment and human health even at extremely low concentrations of micrograms per litre or even less. Although, effective reduction of EPs and MPs in the environment is often difficult and the present methods do not have enough potential to treat them completely. Various degradation techniques and treatment processes, including advanced oxidation processes (AOPs), ultrafiltration, nanofiltration, and enzyme-based treatments in conjunction with membrane filtrations, have been introduced and applied to different EPs and/or MPs (Morsi et al. 2020).

9.2 Sources of Emerging Pollutants

The awareness of "emergence" for emerging contaminants should probably be attributed to Rachel Carson for her 1962 book "Silent Spring". The author clearly articulated the consequences of widespread DDT use to eliminate mosquitoes and other pests that led to deaths and disappearance of many birds. It is to be known that DDT was first synthesized about a hundred years before Carson's book and began to be spread generously during the Second World War Although, Carson was heavily criticized at her time for challenging the societal benefits of pesticides in general and more specifically DDT, however, later in the history, ill effects of such compounds were proved. In this context, DDT was banned and this incident became an example of how an environmentalist rang the alarm bell that was continued by academic researchers for collecting factual data to uncover the truth and risks associated with DDT (Carson R: Silent Spring—40thAnniversary Edition. Boston: Mariner; 2002). Figure 9.1 illustrating four sources of emerging pollutants and Table 9.1 listed different sources of emerging pollutants.

The laboratory and research operations, medical excretion - other factors, hospital wastewaters containing a wide range of harmful or persistent compounds, including radionuclides, solvents, disinfectants, and pharmaceuticals for medical purposes are

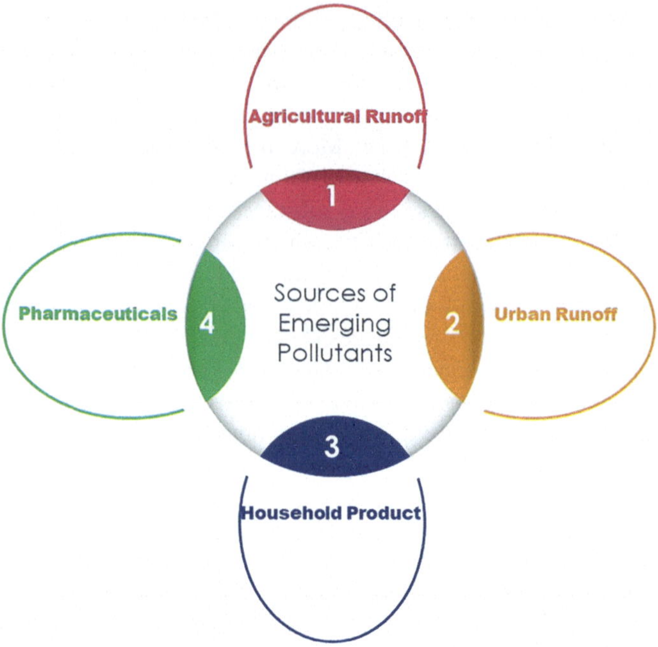

Fig. 9.1 Sources of emerging pollutants

Table 9.1 Different sources of emerging pollutants

AIR (VOC's)	WATER	SOIL
Acrylonitrile	Trichloropropane (TCP)	Per- and polyfluoroalkyl substances (PFAS)
1,1,2,2-tetrachloroethane	Dioxane, trinitrotoluene (TNT)	Micro-plastic
1 – 3-butadiene	Polybrominated diphenyl ethers (PBDEs)	Nano-plastic
Dichloromethane	N-nitroso-dimethylamine (NDMA)	Biphenyl
Naphthalene	Tungsten	Pthalates

more common. The majority of these substances are classified as "emerging contaminants," which are typically unregulated pollutants that could potentially be regulated in the future by studying their possible health impacts and occurrence tracking. These studies suggest that due to high rates of transformation and removal, such substances can be easily introduced into the environment, negating the need for them to remain there and do harm. Additionally, urban wastewaters may also contain some of these substances, however, majority of them are from pharmaceuticals and personal care items. The range of their concentrations in the effluents is ng L^{-1} to μg L^{-1} (Verlicchi et al. 2010).

More than 700 new contaminants, metabolites and transformed products, these are found to be present in the European aquatic environment (www.norman-network.net). Emerging pollutants (EPs) are chemicals that are not widely monitored in the environment but have the potential to infiltrate the ecosystem and produce known or suspected detrimental environmental and (or) human health consequences. In few cases, developing contaminants though released into the environment were not detected due to the unavailability of detection tools which, now have been improved in recent times with newer techniques in the market. Possibly, new sources of emerging pollutants might be generated with the synthesis of new compounds or changes in the usage and disposal of existing chemicals (www.norman-network.net) (Geissen et al. 2015).

Landfills have historically remained the most common methods of organized waste disposal and still remain top in many regions of the world. Nevertheless, they may contain wastes resulting from several decades of disposal and decomposition with subsequent release of organic compounds that may have environmental, wildlife and human health consequences. Products containing different types of additives with unique benefits with improved properties are in daily use. However, when these products are decomposed the additives are released into the environment in its disintegrated forms. Unfortunately, some of them shown negative environmental impacts, resulting in the ban or at least restricted application of some chemicals. It is to note that new and emerging compounds are continuously discovered in the environment (Eggen et al. 2010).

These substances, which include goods used daily in households, industry, and other anthropogenic activities (surfactants and degradation products, pharmaceuticals and personal care products, gasoline additives and plasticizers, and so on), are classified as developing contaminants. According to literature studies, current techniques for gathering precise information on the dynamics of emerging pollutants in water need to be updated and modified in order to avoid dangers to river basin water quality, flora, and fauna. Extensive research highlights the emergence of "new" xenobiotic substances over the past decade, which now is widespread in aquatic environments. These substances collectively termed as "emerging pollutants," encompasses a diverse range of compounds, including metabolites and transformation products known as degradates. Advanced analytical techniques can detect these substances at remarkably low levels, often in parts per trillion (ppt), even within complex environmental matrices like sludge.

Given the multitude of potential compounds, numerous studies have investigated various classes of emerging pollutants, guided by priority lists considering consumption, predicted environmental concentrations, as well as ecotoxicological, pharmacological, and physicochemical data. Despite this, the environmental impact associated with the occurrence of emerging pollutants, including their ecotoxicological effects and potential health risks for humans, remains largely unknown. Although reported concentrations are generally low, concerns have been raised about the potential of long-term impacts on human and animal health following continuous exposure.

Furthermore, the effectiveness of existing water and wastewater treatment infrastructures in adequately removing these compounds remains largely uncertain, contributing to a serious environmental challenge. This inefficiency underscores the urgency of addressing the topic, which has become increasingly significant from both research and regulatory perspectives. The pervasive presence of emerging pollutants in aquatic environments is a consequence of various anthropogenic activities, reflecting the complexity of contemporary waste production, urbanization, industrial processes, and agricultural practices. These substances, previously unrecognized or understudied, have entered the environment and now pose potential risks to ecological systems and human health.

The term "emerging pollutants" encompasses a broad spectrum of chemical entities, including not only the original compounds but also their metabolites and degradation products. The advancement in analytical methodologies has enabled the detection of these substances at remarkably low concentrations, even in challenging environmental matrices like sludge. The diversity and abundance of these compounds have prompted focused research on different classes of emerging pollutants, guided by priority lists that consider factors such as consumption patterns, predicted environmental concentrations, and various toxicological and physicochemical parameters. Despite significant strides in identifying and quantifying emerging pollutants, the true extent of their environmental impact remains elusive. While concentrations in the environment are generally reported at low levels, the cumulative effects of long-term exposure are not well understood, raising concerns about potential consequences for both human and animal health. This uncertainty underscores the importance of thorough investigation and risk assessment.

Moreover, the efficiency of current water and wastewater treatment facilities in removing these emerging pollutants is a critical concern. The inadequate removal of these substances poses a considerable environmental challenge, as conventional treatment methods may not be designed to address the unique characteristics of emerging pollutants. This knowledge gap has spurred growing interest from both the research community and regulatory bodies, emphasizing the need for comprehensive studies and the development of effective strategies to mitigate the impact of emerging pollutants (Thomaidis et al. 2012).

9.3 The Processes of Phytoremediation

See Fig. 9.2.

9.3.1 Phytoextraction (Use of Plants to Reduce the Bioavailability of Pollutants in Soil)

This, also known as phytoaccumulation, refers to the uptake of metal pollutants in the soil by plant roots into the aboveground sections of the plants. Phytoextraction is

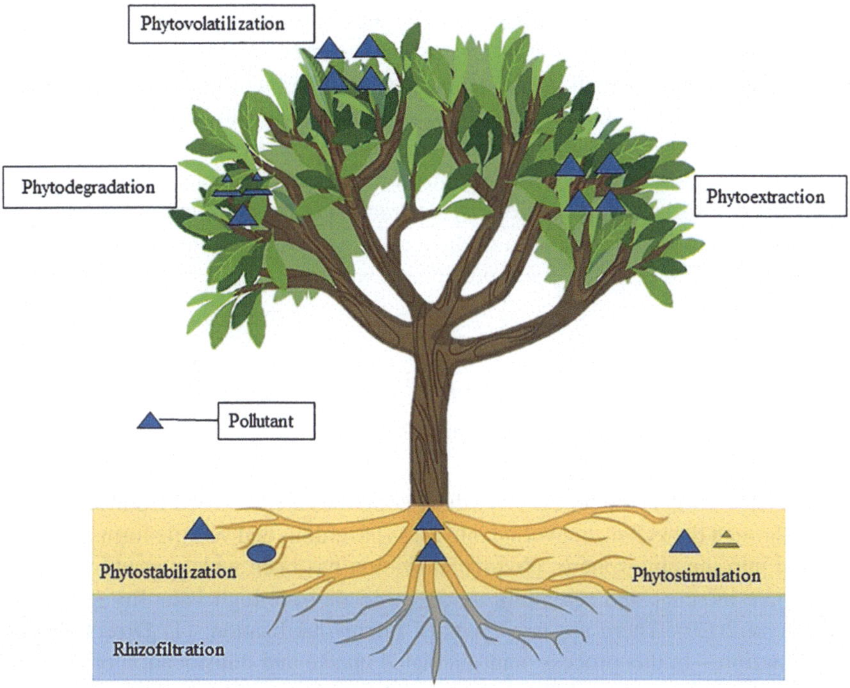

Fig. 9.2 The processes of phytoremediation

generally used to treat polluted soils. Plants are used in this method to absorb, concentrate, and precipitate hazardous metals from polluted soils into aboveground biomass (shoots, leaves, and so on) (Sharma and Pandey 2014). Discovery of metal hyperaccumulator species demonstrates that plants have the potential to remove metals from contaminated soils (Wuana et al. 2010). More than 500 plant species have shown to absorb metals or metalloids from soils and concentrating them at exceptionally high levels in leaf tissues (Goolsby and Mason 2015). Hyperaccumulators are conventionally defined as species capable of accumulating metals at levels 100-fold greater than those typically measured in common non-accumulator plants. Metals such as nickel, zinc, and copper are the ideal candidates for the removal of phytoextractants. Since, it demonstrates its likeliness by majority of plants (about 400) which assimilate and absorb substantial amounts of metals (Sharma and Pandey 2014). There are mainly two factors affecting the extent of metal phytoextraction. **Edaphic Factors**: Various aspects of this kind such as sorptive capacity of soil, heavy metal content, cation exchange capacity, soil pH and organic matter content affect metal hyperaccumulation in plants. **Environmental Factors:** The climate at a particular site poses the greatest and most obvious limitation to the applicability of phytoextraction. Temperature influences transpiration, water chemistry, growth and metabolism of plants which therefore defines both uptake and elimination of pollutants. The removal rates of metals by plants are known to increase linearly with increase in temperature.

Environmental pollution may also exert complex influence on mineral accumulation. This is because the plant roots absorb heavy metals from the soils and aerosols from the atmosphere which penetrates to the plants through the surface of leaves. Minute metal particles stick to the leaves but only some are absorbed. For example, Pb remains as surface precipitant, while Cu, Cd and Zn can partially penetrate into leaves (Bhargava et al. 2012). **Bioavailability:** Contaminants must be bioavailable (ready to be absorbed by roots) for phytoextraction to occur. Some metals, such as Zn and Cd, are primarily found in exchangeable and bioavailable forms. However, other elements like Pb exist as soil precipitate, which is in far less accessible form.

9.3.2 Phytovolatilization (Use of Plants to Absorb Pollutants and Release Into Atmosphere)

Phytovolatilization is the process by which pollutants are absorbed from soil, transported through the xylem, converted into a less hazardous and volatile form, and later released into the atmosphere. Phytovolatilization has been widely used to remove metals like mercury and selenium, as these metals have high volatility (Awa and Hadibarata 2020). There are two types of Phytovolatilization: (1) **Direct Phytovolatilization**—In this process plant-mediated uptake and translocation of contaminants to the shoot portion diffuses across hydrophobic barriers such as cutin in the epidermis or suberin in woody dermal tissues (Arya et al. 2017) and (2) **Indirect phytovolatilization**—It is the increase in volatile contaminant flux from the subsurface resulting from plant root activities (Limmer and Burken, 2016). Hofmeister proposed in 1894 that selenium in animals is detoxified by emitting volatile dimethyl selenide from the lungs, based on the detection of dimethyl telluride odour in the breath of dogs injected with sodium tellurite. According to the same rationale, the garlicky odour of plants that accumulate selenium may signal the release of volatile selenium molecules. This is the most contentious phytoremediation technology. Hg and Se are poisonous, and it is debatable whether their volatilization into the atmosphere is desirable or safe (Padmavathiamma and Li 2007).

9.3.3 Phytodegradation (Use of Plants for the Degradation of Pollutants)

Phytodegradation is the process by which organic pollutants are degraded by compounds produced by plants during metabolic processes. Soil, clay, sediment, and subsurface waters can all benefit from vegetal degradation. The method's most advantageous feature is that reduction and degradation occur within the plant as a physiological process and do not rely on microorganisms. Whereas, the emergence of toxic intermediate and end-use products is difficult to detect making it a major drawback (Adiloğlu 2018). The phytodegradation uses the hydrolysis in which hydroxyl functional group is attached to the pollutant's molecule or it is oxidised. It also removes toxic chemicals through the use of detoxification enzymes. By the process of phytodegradation the contaminants like pesticides, PCPs, atrazine, and also some petrochemicals are treated (Anerao et al. 2022a, b), (Lahiri et al. 2021).

9.3.4 *Rhizofiltration*

Rhizofiltration involves the use of plant roots to adsorb pollutants from the surrounding root zone (rhizosphere), concentrating and precipitating them on or inside the root. Root exudates (secondary metabolites) released by plant roots will be subjected to biogeochemical processes that will precipitate pollutants on the roots or in the water body. Thereafter, depending on the kind of plant, contaminant species and concentration, pollutants will adsorb on the roots, within the roots, or translocate to the phyllosphere (plant organs above ground) (Kritanti et al. 2021). During root breakdown, exudates such as simple phenolics and other organic acids can be discharged from living cells or from the complete cell contents. Roots or full plants are collected for disposal as they get saturated with metal pollutants. Essentially, particular genes in the plant body make it a good metal accumulator. The metal tolerance in plants is influenced by glutathione and organic acid metabolism. Also, the plants grown hydroponically are transplanted into metal-polluted water, where they absorb and concentrate the metals in their roots and shoots. The root system has a massive surface area that absorbs and accumulates the water and nutrients required for growth, as well as other non-essential pollutants.

9.3.5 *Phytostabilization (Migration of Contaminants is Prevented by Control of Hydraulic Gradient)*

The use of metal-tolerant plants to prevent bulk erosion and limit airborne movement and leaching of pollutants is known as phytostabilization. Revegetation stabilizes the polluted site by creating a cover crop, which prevents metal-contaminated soil particles from being dispersed by water or wind erosion. This process limits the metal mobility and availability through rhizosphere-induced adsorption and precipitation processes. Most importantly, site revegetation improves contaminated soil's physicochemical and biological qualities by boosting organic matter content, nutrient levels, cation exchange capacity, and biological activity (Alkorta et al. 2010).

When different ameliorants boosting metal sorption and deposition are applied first in chemo-phytostabilization, the rate of binding metals rises. Thus, the goal of chemo-phytostabilization is to inactivate the metals-pollutants rather than to purify contaminated soil.

9.3.6 *Rhizodegradation*

Plant rhizospheres are known for their incredible complexity; they are made up of various living and non-living parts that interact in sophisticated ways, making the complex component like soils even more difficult when subjected to analysis. Despite the intricacy of rhizospheres, humans have historically attempted to influence them in order to achieve desired results like increased agricultural output. Phytostimulation is another term for Rhizodegradation, which is the breakdown of pollutants within the plant root zone, often known as the rhizosphere. Bacteria or other microbes are thought to be responsible for such processes. It is reported that in rhizosphere soil about 100 times as many microorganisms are observed than that compared to non-rhizosphere soil. Microorganisms may be so abundant in the rhizosphere as plants produce sugars, amino acids, enzymes, and other substances that promotes bacterial development. The roots also give more surface area for microorganisms to grow on and a channel an oxygen transfer route from the surrounding environment (Sharma and Pandey 2014).

9.4 Status of Phytoremediation in India and World Scenario

9.4.1 *Phytoremediation in Africa*

Africa has an abundance of natural resources. Various materials are available in varied degrees of abundance. Mineral resource extraction, exploration, and trading are key sources of revenue for the government. According to statistics, the natural resource mining sector has received the greatest direct foreign investment into Africa during the last five decades. The exploitation of these minerals, however, has raised socio-environmental and public health concerns about the overall ecosystem. Phytoremediation could be one of the most suitable remediation approaches for developing economies given its net socio-economic and environmental benefits (Bardos et al. 2016; Zabbey et al. 2017). According to Watanabe (1997) phytoremediation is appropriate for environmental rehabilitation in Africa where over 40% of her population live below poverty line (Table 9.2).

In Africa, pilot phytoremediation projects, such as those in Kenya, have been initiated to restore degraded sites. Three species, namely, Casuarina equisetifolia Forst, Conocarpus lancifolius Engl., and Cocosnucifera L., have shown resistance to extreme environmental conditions.

Table 9.2 Types of plants used for removal of contaminants and application

Application	Contaminants	Types of plants	References
Phytoextraction	Zn and Cd	*T. caerulescens, Willows, R. crispus* etc.	Sakakibara et al. (2010) and Zhuang et al. (2007)
Phytovolatilization	As, Se, Hg	*Pteris vittata*	Sakakibara et al. (2010)
Phytodegradation	Munitions and Atrazine	*Erythrinacrista-galli* L	Susarla et al. (2002)
Rhizofiltration	Lead and Uranium	*Carex pendula, Helianthus anuus and Phaseolus vulgaris*	Yadav et al. (2011) and Lee and Yang (2010)
Phytostabilization	Zn, Cd, Pb	*Miscanthus sinensis, Miscanthus giganteus, Cynara cardunculus and Brassica juncea*	Pavel et al. (2014) and Lacalle et al. (2023)
Rhizodegradation	Total petroleum hydrocarbons	Seepweed, Sealavander, Central Asia Saltbush and Reed	(Wang et al. 2011)

In South Africa, phytoremediation projects have been utilized to clean up places polluted by mineral extraction (gold and uranium tailings) and to naturally recolonize surrounding damaged soils with plant species. In 1977, this phytotechnology was employed to kickstart the natural growth of an indigenous dune forest in Zululand. A variety of fast-germinating species (e.g., Helianthus annuus L., Sorghum spp., Pennisetum americanum L., Crotalaria juncea L.) were planted in polluted abandoned sites, resulting in the reclamation of over 400 hectares since 1978. Research was conducted in Nigeria on the use of phytotechnology to remediate petroleum hydrocarbon polluted (PHC) areas (Anyasi and Atagana 2018). In their investigation, 28 native plant species from various families were grown in and surrounding hydrocarbon-contaminated soil. Interestingly, Odoh et al. in his work found that C. odorata, A. africana, and U. chamae had the highest bioaccumulation coefficients (BACs) for maximum efficiency (Odoh et al. 2019).

Human-caused soil and groundwater contamination in Tanzania is a major concern due to improper treatment of Municipal Solid Waste, increased pesticide use in agriculture, and acid-mine drainage. To address these issues, Tanzania urgently needs a cost-effective and ecologically friendly clean-up procedures, such as phytoremediation, which uses plants to degrade harmful substances in the environment (Mwegoha 2008).

9.4.2 Phytoremediation in Australia and New Zealand

Phytoremediation in Australia has centred on the utilization of plants as bio pumps. Plants use the sun's energy to dewater contaminated areas, regulate leaching, and increase organic matter and microbial activity in the rhizosphere. As a result, these root-zone mechanisms accelerate pollutant degradation while reducing heavy metal mobility. As a result, the term phytoremediation refers to a variety of plant-based environmental applications, including mine-site revegetation, riparian management, and phytoextraction. In Australia, phytoremediation serves mostly to minimize contaminant mobility and decompose organic contaminants, rather than to recover heavy metals. (Robinson et al. 2003).

New Zealand's economy is characterized by a "clean-green" image, with agricultural exports and tourism contributing to 16 and 9% of its GDP, respectively. Australia, with mining as its largest exporter, lacks overarching environmental regulation. Contaminated sites in New Zealand are mainly related to agricultural and silvicultural production. Phytoremediation is best suited for long-term clean up of low-value land, making it suitable for contaminated sites in Australia's large production networks. However, low population densities in New Zealand and Australia reduce land prices and remediation needs (Robinson and Anderson 2007).

The main issue for small and growing Pacific Island states can be attributed to the persistent toxicants such as heavy metals (HMs) in the environment. The environmental proliferation of trace elements, notably HMs, in terrestrial and marine ecosystems has shown to cause a variety of health disorders and hamper overall ecosystem function (Rai et al. 2019). In total, fifty-four sites expanded over entire South Pacific region have been identified and designated as "contaminated," with an estimated cost of US$110 million for clean-up and remediation. These massive amounts impose a challenge for small island states largely by addressing and controlling hazardous waste pollution (Mohanty et al. 2024). As a result, these encounters have highlighted the importance of comprehensive, practical, cost-effective, long-term and sustainableroutesof dealing with hazardous contaminants. Phytoremediation has been the focus of considerable research in the last two decades and is considered one of the safest, cleanest, cost-effective, and least disruptive methods for treating sites contaminated with refractory pollutants like HMs.

Phytoremediation throughs light on basic understanding of biochemical processes and selecting the appropriate species and agronomic approaches to promote plant growth in polluted sites. Efficient implementation requires accurate assessment of soil and plant materials before and after treatment as well as long-term monitoring. The traditional detection methods include ICP-MS/AES, AAS, ASV, and LIBS, each with varying costs and applicability for specific elements. More importantly, long-term monitoring is crucial for tracking progress (Diarra et al. 2022a, b).

9.4.3 Phytoremediation in Europe

Compared to the United States and Canada, Europe has a restricted use of phytotechnologies. In some countries, private enterprises have been established to use plant resources to mitigate pollution. Phytoremediation is still used with caution in Europe, according to norms and public opinion. The following are the demonstrated setbacks:

- Limited understanding and inadequate communication have led to public uncertainties and limited adoption.
- Current regulations do not explicitly include phytoremediation within applicable methods.
- Competition with traditional clean-up approaches that have a proven track record of success.
- Insufficient funding to promote private initiative.
- Proprietary rights may prevent the implementation of successful ideas in the USA.

One of the most significant impediments, however, is a dearth of fundamental and applied research. In fact, more study on the core mechanisms of phytotechnologies, as well as case studies, are required to employ these technologies coherently (Sharma and Pandey 2014). Goat willow (Salix caprea) is a common pioneer plant found at former ore mining sites in Eastern and Middle Europe. Its high tolerance and capacity to absorb and accumulate hazardous components suggest that this native species could be used to remediate polluted environments. According to Panday et al. (2015), the use of naturally colonizing, economically valuable, and perennial plants that are unpalatable to livestock (e.g., Jatropha curcas, Miscanthus giganteus, and Ricinus communis) is the best strategy for long-term phytoremediation of contaminated soils.

9.4.4 Phytoremediation in Asia

Very few reports are available on phytoremediation under Indian field conditions and the existing studies are mostly confined to water bodies. In India, phytoremediation technology has been used only on certain areas i.e., revegetation of mine spoils. There is also lack of scientific studies on phytoremediation of heavy metal contaminated soils in India. Notably, the application of phytoremediation has gained attention in recent years as a sustainable approach to address various environmental issues in India. Several research studies and initiatives have been conducted to explore the potential of plants in mitigating pollution. India faces challenges related to heavy metal contamination in soil and water due to industrial activities, mining, and improper waste disposal (Mishra et al. 2022). Phytoremediation has been studied as a potential solution to address heavy metal pollution. Research studies have focused on identifying hyper accumulator plant species that can efficiently accumulate heavy metals from contaminated sites (Mandal et al. 2014). Phytoremediation is also being explored to treat organic pollutants, including pesticides and industrial chemicals. Studies have investigated the ability of certain plant species to break down or absorb organic pollutants, contributing to the overall remediation process (Anerao et al. 2022a, b). Various research institutions, universities, and environmental organizations in India have been actively involved in studying and promoting phytoremediation techniques. Collaborative projects between government agencies, academic institutions, and industry stakeholders have been initiated to assess the feasibility and effectiveness of phytoremediation in different regions of the country (Suresh and Ravishankar 2004). Indian government has shown interest in promoting sustainable environmental practices. Thus, policies and regulations related to environmental conservation indirectly supports the adoption of phytoremediation techniques. Although, phytoremediation holds promise, but the challenges such as plant selection, site-specific conditions, and long-term effectiveness needs to be focussed. Public awareness and education regarding the benefits and limitations of phytoremediation are essential for its widespread adoption.

In Pakistan, the potential of phytoremediation is not explored much irrespective of its rich diversity of plants species. Most studies conducted here are limited in scope and the reports obtained for the studies contradict the results. Furthermore, a few field studies were conducted to clean up ecosystems and most studies were carried out in lab scale in the country. Although, in recent years, the research on phytoremediation holds a good place, however, there is still a dire need for exploring the opportunities for practical implications of phytoremediation of heavy metals, especially on field scale (Akhtar et al. 2023). Concepts and categorization of PHYTOPET (phytoremediation for petroleum hydrocarbons) and PHYTOREM (phytoremediation of metals and metalloids) are highly recommended in the country by considering climatic, biotic, and edaphic factors. Research on phytoremediation also highlighted disadvantages and challenges that phytoremediation faced in the country and addresses glaring difficulties in managing biomass after successful implementation of phytoremediation on heavy metals. Different parts of plant have different concentrations of metals accumulated (Jamali et al. 2007). The disposal of plants post phytoremediation is problematic due to their high toxic metal concentrations, making them unsuitable for human consumption or animal feed. Strict guidelines are to be followed while disposing harvested plants. Pakistani scientists estimate that phytoremediation may take longer period to clean up contaminated sites. Integrating solar-driven techniques with engineering-based systems will probably fasten the decontamination process. Despite its cost-effectiveness, phytoremediation remains a safe and environmentally friendly method for removing contaminants from the environment (Jabeen et al. 2009).

The study investigated the phytoaccumulation of arsenic by Brassica juncea (L.) in Bangladeshi soil at varying concentrations. Also, B. juncea was grown in hydroponic media under greenhouse conditions with different concentrations of sodium arsenite. In Bangladesh, B. juncea can be employed for arsenic phytoaccumulation, but at higher concentrations, it can be considered as good accumulator that uptakes upto 1% of total plant's biomass. Coagulation-flocculation and Fenton process in conjunction with phytoremediation by water hyacinth (Eichhornia crassipes) were applied to treat the most frequently occurred contaminants in textile wastewater. Textile effluents treatment and management is considered as one of the most significant issues in Bangladesh, therefore combination of chemical and phytoremediation technologies could be a promising sustainable low-cost alternative for Bangladesh's textile industrial sector (Tasneem et al. 2021).

9.4.5 Phytoremediation in America

In the US, phytoremediation is widely accepted as a practical and affordable approach to managing environmental contamination, especially contaminated soil and water. The versatility of the technique is validated by the employment of phytoremediation technologies in a range of contexts, such as cities, agricultural areas, industrial sites, and Superfund sites (Doty et al. 2017). The Environmental Protection Agency (EPA) and other US government agencies fund the development and application of phytoremediation. According to Juberg et al. (1997), there are regulatory frameworks in place to direct the use of phytoremediation for environmental cleanup. At Superfund sites around the nation, phytoremediation is actively employed to address complex environmental pollution issues brought on by hazardous materials (Pivetz 2001). Cooperation between governmental organizations, commercial businesses, and academic institutions fosters creativity and expedites the creation and application of phytoremediation techniques. Improved plant tolerance to pollutants, improved understanding of plant–microbe interactions, and optimal plant species selection for particular contaminants are the goals of ongoing scientific research. The use of native plant species in phytoremediation projects is being prioritized in an effort to increase ecological sustainability and plant adaptability to local environmental conditions (Stephenson and Black 2014).

More efficient phytoremediation solutions are produced by genetically modifying plants to boost their capacity to absorb pollutants and to withstand it. Our knowledge of how plants react to contaminants is being improved by the integration of omics technologies, such as metabolomics and genomics, which enables us to choose and maximize certain plant species for phytoremediation (Nayeri et al. 2023). In order to create nanoparticles that enhance plant pollutant uptake, transport, and detoxification and raise overall efficiency, nanotechnology is also employed in phytoremediation (Kumar and Das 2017). In order to optimize the root-soil interface and promote favorable plant–microbe interactions that enhance nutrient uptake and pollution remediation, the study focuses on rhizosphere engineering (Nwadinigwe and Ugwu 2018). Advances in sensor technology, such as real-time monitoring and remote sensing, have been documented in the literature as helping to improve the management of phytoremediation projects by monitoring plant health, pollutant levels, and the surrounding environmental conditions. Additionally, a number of studies concentrate on creating plant varieties that can withstand climate change, and research and educational initiatives generate a trained labour force. Plant selection, monitoring, and implementation tactics can all be adjusted with the help of adaptive management solutions and ongoing assessment (USEPA report).

9.5 Conclusion and Future Prospects

Phytoremediation is a sustainable and environmentally friendly strategy to improve the damaged settings that use plants to reduce the impact of contaminants. Emerging pollutants, also known as contaminants of emerging concern (CECs), include a wide range of substances, including pharmaceuticals, personal care items, endocrine disrupting chemicals, and other novel contaminants that are not often monitored in the environment.

Conclusion:

There is a growing global concern regarding the escalating presence of emerging contaminants in various locations worldwide. Numerous studies have highlighted substantial evidence of contamination in soils and surface water that are attributed to discharged treated wastewater. However, there remains a significant gap in our comprehension towards the fate of these contaminants in treated wastewater when utilized for irrigation, encompassing aspects such as crop absorption, bioaccumulation, integration into the food chain, and potential health risks to both humans and other organisms. There is a lack of clarity pertaining to the efficacy of different wastewater treatment methods in removing emerging contaminants and their metabolites from influent water. Consequently, there is an urgent need to establish standards and policy guidelines delineating acceptable concentrations of these emerging contaminants in soil and water. Phytoremediation has showed promise in the removal and degradation of a variety of developing contaminants. The process entails the uptake, accumulation, and transformation of toxins by plants, which can be especially effective for specific types of developing pollutants. The selection of plant species is critical for successful phytoremediation. Some plants have natural potential to accumulate specific pollutants, whilst others may increase microbial activity in the rhizosphere, aiding in the breakdown of toxins. Plant-microorganism synergy in the rhizosphere is critical for phytoremediation of developing contaminants. This interaction improves the degradation, transformation, or immobilization of pollutants, hence increasing the process efficiency. Despite its potential, phytoremediation has several problems, including sluggish remediation rates, limited plant species appropriateness, and the requirement for long-term monitoring. Furthermore, the efficacy of phytoremediation varies according to site-specific variables. While laboratory studies have shown that phytoremediation is beneficial for emerging contaminants, more field-scale applications are required to evaluate its feasibility, scalability, and economic viability in real-world circumstances.

Future Prospects:

This would ensure efficient effluent management in accordance with the standards governing the discharge of treated effluents. In this context, efficient methods for advanced treatment of EP-polluted effluents (sand and media filtration; chlorination; advanced oxidation processes, AOPs; adsorption using granular activated carbon, zeolite, or other clay materials; hydrolysis processes; constructed wetland

(CW); membrane bioreactors, phytoremediation, biosorption) are being developed. Numerous studies have been conducted in recent decades to develop innovative, sustainable technology for the removal of EPs from wastewater, which is the most significant source of EP pollution in the aquatic environment. Biological processes have proved to be inefficient towards some of the EPs degradation, while many extremely refractory contaminants remain difficult to (bio)degrade, if not impossible to remove. However, effective new methods can be used in hybrid systems to overcome the shortcomings of current biological technology (Vasilachi et al. 2021). Advances in genetic engineering may result in the generation of hyperaccumulating plants with improved uptake and tolerance to specific emerging contaminants. This could considerably increase the efficacy of phytoremediation techniques. Combining phytoremediation with other remediation technologies, such as microbial bioremediation or electrokinetics, may improve overall efficiency and cater the limitations accompanying the individual approaches. Creating and executing policies and regulations that promote the use of phytoremediation for emerging pollutants can help to gain its real power on bigger scale. Incentives for research and development in this area may also encourage innovation. Educating communities for the benefits of phytoremediation and involving them in decision-making processes can help to increase the acceptability and successful implementation of this ecologically friendly remediation technology. Additional research is required to improve the monitoring and evaluation processes for phytoremediation projects, ensuring that the method's long-term efficacy and environmental safety are adequately studied.

To summarize, phytoremediation shows promise for addressing new toxins, continuous study, technical developments, and collaborative efforts are required to overcome difficulties and understand its full potential as a long-term environmental remediation solution.

References

Adiloğlu S (2018) Heavy metal removal with phytoremediation. In: Advances in bioremediation and phytoremediation. InTech. https://doi.org/10.5772/intechopen.70330

Akhtar MS, Hameed A, Aslam S, Ullah R, Kashif A (2023) Phytoremediation of metal-contaminated soils and water in Pakistan: a review. Water Air Soil Pollut. https://doi.org/10.1007/s11270-022-06023-8

Alkorta I, Becerril JM, Garbisu C (2010) Phytostabilization of metal contaminated soils. Rev Environ Health

Anerao P, Kaware R, Khedikar AK, Kumar M, Singh L (2022a) Phytoremediation of persistent organic pollutants: concept challenges and perspectives. In: Phytoremediation technology for the removal of heavy metals and other contaminants from soil and water. Elsevier, pp 375–404. https://doi.org/10.1016/B978-0-323-85763-5.00018-0

Anerao P, Kaware R, Khedikar AK, Kumar M, Singh L (2022b) Phytoremediation of persistent organic pollutants: concept challenges and perspectives. In: Phytoremediation technology for the removal of heavy metals and other contaminants from soil and water. Elsevier, pp 375–404. https://doi.org/10.1016/B978-0-323-85763-5.00018-0

Arya SS, Devi S, Angrish R, Singal I, Rani K (2017) Soil reclamation through phytoextraction and phytovolatilization. In: Volatiles and food security: role of volatiles in agro-ecosystems. Springer, Singapore, pp 25–43. https://doi.org/10.1007/978-981-10-5553-9_3

Awa SH, Hadibarata T (2020) Removal of heavy metals in contaminated soil by phytoremediation mechanism: a review. Water Air Soil Pollut. https://doi.org/10.1007/s11270-020-4426-0

Bardos RP, Bone BD, Boyle R, Evans F, Harries ND, Howard T, Smith JWN (2016) The rationale for simple approaches for sustainability assessment and management in contaminated land practice. Sci Total Environ 563–564:755–768. https://doi.org/10.1016/j.scitotenv.2015.12.001

Belver C, Han C, Rodriguez JJ, Dionysiou DD (2017) Innovative W-doped titanium dioxide anchored on clay for photocatalytic removal of atrazine. Catal Today 280:21–28

Bhargava A, Carmona FF, Bhargava M, Srivastava S (2012) Approaches for enhanced phytoextraction of heavy metals. J Environ Manage. https://doi.org/10.1016/j.jenvman.2012.04.002

Diarra I, Kotra KK, Prasad S (2022a) Application of phytoremediation for heavy metal contaminated sites in the south pacific: strategies, current challenges and future prospects. Appl Spectrosc Rev. https://doi.org/10.1080/05704928.2021.1904410

Diarra I, Kotra KK, Prasad S (2022b) Application of phytoremediation for heavy metal contaminated sites in the south pacific: strategies, current challenges and future prospects. Appl Spectrosc Rev. https://doi.org/10.1080/05704928.2021.1904410

Doty SL, Freeman JL, Cohu CM, Burken JG, Firrincieli A, Simon A, Khan Z, Isebrands JG, Lukas J, Blaylock MJ (2017) Enhanced degradation of TCE on a superfund site using endophyte-assisted poplar tree phytoremediation. Environ Sci Technol 51(17):10050–10058

Eggen T, Moeder M, Arukwe A (2010) Municipal landfill leachates: a significant source for new and emerging pollutants. Sci Total Environ 408(21):5147–5157

Geissen V, Mol H, Klumpp E, Umlauf G, Nadal M, van der Ploeg M, van de Zee SE, Ritsema CJ (2015) Emerging pollutants in the environment: a challenge for water resource management. Int Soil Water Conserv Res 3(1):57–65

Goolsby EW, Mason CM (2015) Toward a more physiologically and evolutionarily relevant definition of metal hyperaccumulation in plants. Front Plant Sci. https://doi.org/10.3389/fpls.2015.00033

Jabeen R, Ahmad A, Iqbal M (2009) Phytoremediation of heavy metals: physiological and molecular mechanisms. Bot Rev 75:339–364. https://doi.org/10.1007/s12229-009-9036-x

Jamali MK, Kazi TG, Arain MB, Afridi HI, Jalbani N, Memon AR, Shah A (2007) Heavy metals from soil and domestic sewage sludge and their transfer to sorghum plants. Environ Chem Lett 5:209–218. https://doi.org/10.1007/s10311-007-0101-6

Khan Z, Deka BK, Chanu TT, Chakraborty MB, Upadhyaya H (2024) Phytoremediation of heavy metals: a review. In: Phytoremediation potential of medicinal and aromatic plants, pp 1–13

Kristanti RA, Ngu WJ, Yuniarto A, Hadibarata T (2021) Rhizofiltration for removal of inorganic and organic pollutants in groundwater: a review. Biointerface Res Appl Chem. https://doi.org/10.33263/BRIAC114.1232612347

Kumar MS, Das AP (2017) Emerging nanotechnology based strategies for diagnosis and therapeutics of urinary tract infections: a review. Adv Colloid Interface Sci 249:53–65. https://doi.org/10.1016/J.CIS.2017.06.010

Lahiri D, Nag M, Dey A, Sarkar T, Joshi S, Pandit S, Das AP, Pati S, Pattanaik S, Tilak VK, Ray RR (2021) Biofilm mediated degradation of petroleum products. Geomicrobiol J 1–10. https://doi.org/10.1080/01490451.2021.1968979

Lee M, Yang M (2010) Rhizofiltration using sunflower (Helianthus annuus L.) and bean (Phaseolus vulgaris L. var. vulgaris) to remediate uranium contaminated groundwater. J Hazard Mater 173(1–3):589–596

Limmer M, Burken J (2016) Phytovolatilization of organic contaminants. Environ Sci Technol. https://doi.org/10.1021/acs.est.5b04113

Mandal A, Purakayastha TJ, Ramana S, Neenu S, Bhaduri D, Chakraborty K, Manna MC, Rao AS (2014) Status on phytoremediation of heavy metals in India—a review. Int J Bio-Resour Stress Manag 5:553. https://doi.org/10.5958/0976-4038.2014.00609.5

Mishra S, Dash D, Al-Tawaha ARMS, Das AP (2022) A review on heavy metal ion adsorption on synthetic microfiber surface in aquatic environments. Appl Biochem Biotechnol 194:4639–4654. https://doi.org/10.1007/S12010-022-04029-W/METRICS

Mishra S, Rout PK, Das AP (2021) Emerging microfiber pollution and its remediation. In: Prasad R (ed) Environmental and microbial biotechnology. Springer, Singapore, pp 247–266. https://doi.org/10.1007/978-981-15-5499-5_9

Mohanty M, Mohanty J, Dey S, Dutta K, Shah MP, Das AP (2024) The face mask: a tale from protection to pollution and demanding sustainable solution. Emerg Contam 10:100298. https://doi.org/10.1016/J.EMCON.2023.100298

Morsi R, Bilal M, Iqbal HM, Ashraf SS (2020) Laccases and peroxidases: the smart, greener and futuristic biocatalytic tools to mitigate recalcitrant emerging pollutants. Sci Total Environ 714:136572

Mwegoha WJS (2008) The use of phytoremediation technology for abatement soil and groundwater pollution in Tanzania: opportunities and challenges. J Sustain Dev Africa 10

Nayeri S, Dehghanian Z, Lajayer BA, Thomson A, Astatkie T, Price GW (2023) CRISPR/Cas9-mediated genetically edited ornamental and aromatic plants: a promising technology in phytoremediation of heavy metals. J Clean Prod 428:139512

NORMAN network (2019) NORMAN network-glossary of terms. http://www.normannetwork.net/?q=node/9. Accessed 13 Nov 2019

Nwadinigwe AO, Ugwu EC (2018) Overview of nano-phytoremediation applications. Phytorem Manag Environ Contam 6:377–382

Odoh CK, Zabbey N, Sam K, Eze CN (2019) Status, progress and challenges of phytoremediation—an African scenario. J Environ Manage. https://doi.org/10.1016/j.jenvman.2019.02.090

Padmavathiamma PK, Li LY (2007) Phytoremediation technology: hyper-accumulation metals in plants. Water Air Soil Pollut. https://doi.org/10.1007/s11270-007-9401-5

Pavel PB, Puschenreiter M, Wenzel WW, Diacu E, Barbu CH (2014) Aided phytostabilization using miscanthus sinensis × giganteus on heavy metal-contaminated soils. Sci Total Environ 479:125–131

Pivetz BE (2001) Phytoremediation of contaminated soil and ground water at hazardous waste sites. US Environmental Protection Agency, Office of Research and Development, Office of Solid Waste and Emergency Response

Rai PK, Lee SS, Zhang M, Tsang YF, Kim KH (2019) Heavy metals in food crops: health risks, fate, mechanisms, and management. Environ Int. https://doi.org/10.1016/j.envint.2019.01.067

Robinson B, Green S, Mills T, Clothier B, Van Der Velde M, Laplane R, Fung L, Deurer M, Hurst S, Thayalakumaran T, Van Den Dijssel C (2003) Phytoremediation: using plants as biopumps to improve degraded environments. Aust J Soil Res:599–611. https://doi.org/10.1071/SR02131

Sakakibara M, Watanabe A, Inoue M, Sano S, Kaise T (2010) Phytoextraction and phytovolatilization of arsenic from as-contaminated soils by pteris vittata. In: Sediments, water and energy

Sharma P, Pandey S (2014) Status of phytoremediation in world scenario. Int J Environ Bioremediat Biodegrad 2:178–191. https://doi.org/10.12691/ijebb-2-4-5

Sousa JC, Ribeiro AR, Barbosa MO, Pereira MFR, Silva AM (2018) A review on environmental monitoring of water organic pollutants identified by EU guidelines. J Hazard Mater 344:146–162

Stephenson C, Black CR (2014) One step forward, two steps back: the evolution of phytoremediation into commercial technologies. Biosci Horizons 7:hzu009

Suresh B, Ravishankar GA (2004) Phytoremediation—a novel and promising approach for environmental clean-up. Crit Rev Biotechnol. https://doi.org/10.1080/07388550490493627

Susarla S, Medina VF, Mccutcheon SC (2002) Phytoremediation: an ecological solution to organic chemical contamination, Ecol Eng

Tasneem A, Sarker P, Akter S, Mouna SSP, Rahaman MS, Mohinuzzaman M, Uddin MK, Kabir MM (2021) Textile wastewater treatment by combination of chemical and phytoremediation processes. Pollution 7:43–54. https://doi.org/10.22059/poll.2020.304569.835

Thomaidis NS, Asimakopoulos AG, Bletsou AA (2012) Emerging contaminants: a tutorial mini-review. Global Nest J 14(1):72–79

Vasilachi IC, Asiminicesei DM, Fertu DI, Gavrilescu M (2021) Occurrence and fate of emerging pollutants in water environment and options for their removal. Water 13(2):181

Verlicchi P, Galletti A, Petrovic M, Barceló D (2010) Hospital effluents as a source of emerging pollutants: an overview of micropollutants and sustainable treatment options. J Hydrol 389(3–4):416–428

Wang Z, Xu Y, Zhao J, Li F, Gao D, Xing B (2011) Remediation of petroleum contaminated soils through composting and rhizosphere degradation. J Hazard Mater 190(1–3):677–685

Wuana RA, Okieimen FE, Imborvungu JA (2010) Removal of heavy metals from a contaminated soil using organic chelating acids. Int J Environ Sci Tech 7:485–496

Yadav BK, Siebel MA, van Bruggen JJ (2011) Rhizofiltration of a heavy metal (lead) containing wastewater using the wetland plant Carex pendula. Clean–Soil Air Water 39(5):467–474

Zabbey N, Sam K, Onyebuchi AT (2017) Remediation of contaminated lands in the Niger Delta, Nigeria: prospects and challenges. Sci Total Environ. https://doi.org/10.1016/j.scitotenv.2017.02.075

Zhuang P, Yang QW, Wang HB, Shu WS (2007) Phytoextraction of heavy metals by eight plant species in the field. Water Air Soil Pollut 184:235–242. https://doi.org/10.1007/s11270-007-9412-2

Chapter 10
Role of Enzymes in Phyto-Transformation for Degradation of Organic and Inorganic Pollutants

Anamika Basu

Abstract Phytodegradation, also known as phytotransformation, is the process by which organic pollutants found in the soil or inside plant bodies are broken down by plants or microorganisms. The term "soil enzyme" refers to both extracellular and intracellular enzymes that are produced in the soil by plant roots, other biological cells, and soil microorganisms. These soil enzymes are important in the breakdown of organic matter and the recycling of nutrients. Enzymes can be either extracellular or intracellular, depending on where they are located. Moreover, that, rhizosphere is a unique soil microenvironment in which the characteristics and activities of microorganisms, plant roots, and soil interact in a coordinated way. The root-soil interface inside rhizospher is significantly influenced by rhizosphere soil, and the conversion of soil C, N, P, S, and other nutrient elements is intimately correlated with soil enzyme activity. Enzyme activities isolated from roots, other soil organisms, and the rhizosphere-soil are sensitive markers of heavy metal contamination of the soil and nutrient cycling. Because heavy metals pose serious health risks to humans once they enter the food chain, reclamation of soils tainted by heavy metals has become an environmental concern of great urgency. Many remediation strategies have been tested with differing degrees of success in dealing with this environmental concern. To reduce the amount of pollutants to acceptable levels, biological methods are considered as more effective and biocompatible. Enzyme activities isolated from roots, other microorganisms in the soil, and the rhizosphere-soil are sensitive markers of heavy metal contamination of the soil and nutrient cycling. Lead, chromium, mercury, different polymers, azo dyes, PAHs, and other inorganic and organic pollutants can all be remedied by enzymes. Different enzymes like mono- or dioxygenases, halogenases, peroxidases, phosphotriesterases, hydrolases, laccase, transferases, and oxidoreductases from various species of bacteria, fungi, algae, and plants have been used for the bioremediation of pollutants. A family of enzymes known as ligninolytic enzymes has several uses in bioremediation. The main reason for lead's toxicity is the ROS it produces. The main enzymatic antioxidants are ascorbate–glutathione

A. Basu (✉)
Department of Biochemistry, Gurudas College, Affiliated to University of Calcutta, Kolkata, India
e-mail: anamika.biochem@gurudas.education

A. Kuanar et al. (eds.), *Recent Advances in Bioremediation and Phytoremediation*, Environmental Science and Engineering,
https://doi.org/10.1007/978-3-031-77884-1_10

cycle enzymes, peroxidase, superoxide dismutase, and catalase (Prabhakar et al., Geomicrobiol J 36:894–903, 2019). In the transcriptome experiment the profile gene expression has been studied in above ground parts and in the roots of arabidopsis plants in response to toxic ions of lead. The enzymes, whose gene expression are increased many folds with peroxidase activities, in presence of heavy metal Pb, are summarized here. These antioxidant enzymes can scavenge reactive oxygen species (ROS), when their relative gene expression level is increased as a result of lead metal toxicity (Raj and Das, Environ Chem Ecotoxicol 5:79–85, 2023). During remodeling of plant root cell wall, the uronic acid, production is increased under Pb^{+2} stress condtion. The carboxyl group fractions of pectins and hemicelluloses are also increased in stressed conditions such as heavy metal toxicity. During lead toxicity condition, these increased number of carboxyl groups, provides more Pb^{+2} metal ion binding sites in plant cell wall. Thus, metal ion chelation via COO- groups in demethylesterified pectin is made possible by the process of demethylesterification of pectin. Similarly, pectin methylesterase activity is suppressed during heavy metal contamination, but this enzyme is released when metal ions bind to pectin. One of the primary mechanisms by which plants regulate harmful substances, such as As and Cd, is vacuolar sequestration. Several transporter channel proteins are involved in the uptake of Pb^{+2} ions from soil through plasma membrane of root cell of arabidopsis along with Ca+2 ions. This non-selective cation transporter protein can function as passive transporter of cations between the vacuole and cytosol and plays an important role in vacuolar sequestration of heavy metal lead in arabidopsis.

Graphical Abstract

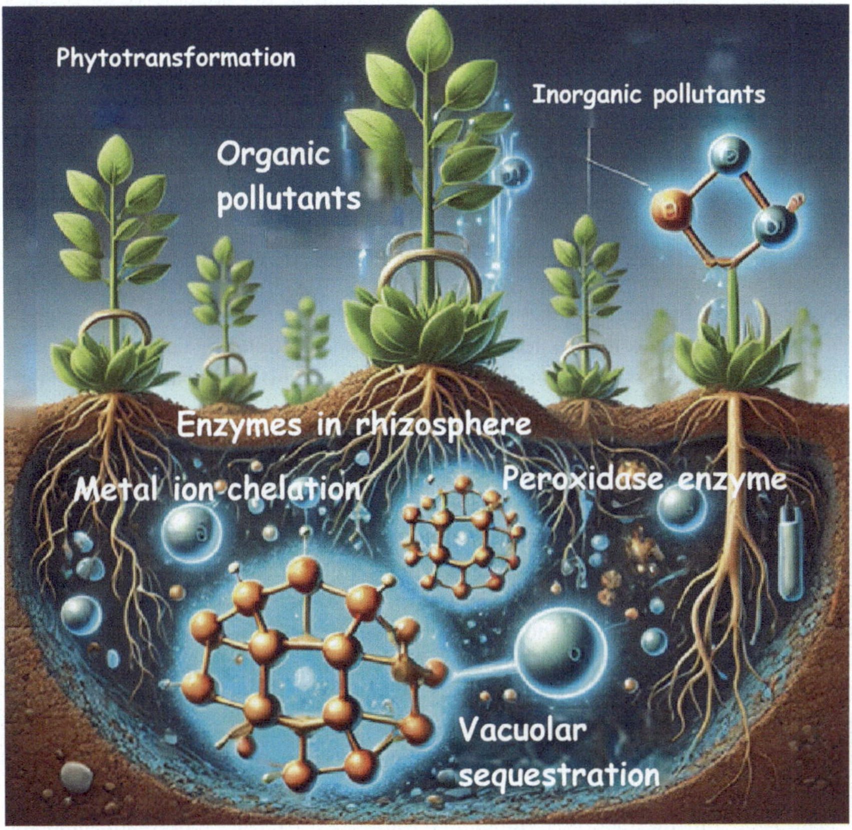

10.1 Introduction

Phytodegradation, also known as phytotransformation, is the process by which organic pollutants found in the soil or inside plant bodies are broken down by plants or microorganisms (Vroblesky 2005). Enzymes secreted by plant roots break down organic compounds (Dietz and Schnoor 2001). The plant then absorbs these molecules and releases them through transpiration. Herbicides and other organic pollutants are the best candidates for this process.

The term "soil enzyme" refers to both extracellular and intracellular enzymes that are produced in the soil by plant roots, other biological cells, and soil microorganisms. The majority of soil microorganisms and plants, along with a minor quantity from earthworms, ants, and other soil animals, are the primary sources of soil enzymes (Liu et al. 2015).Through by the method of mineralization (by amidase, protease, urease, phosphatase and sulfatase) and by oxidation (using hydrolase, dehydrogenase, and glucosidase), soil enzymes are important in the breakdown of organic matter and the recycling of nutrients. As a result, these enzymes can be referred to as "rhizozymes" as a group (rhizo = root, and zymes = enzymes) (Inamdar et al. 2022). The following soil parameters affect rhizozymes: pH,temperature, agricultural practices, availability of water, organic matter content,soil depth/horizons, cropping history, and last but not the least, presence of heavy metals.

The biochemical process of recycling organic matter in the soil system is primarily carried out by soil enzymes (Rao et al. 2014), whose activities are intimately linked to soil organic matter (SOM), soil physical properties, microbial activity, and/or biomass. Enzymes can be either extracellular or intracellular, depending on where they are located. Intracellular enzymes are present in the cytoplasm of living, metabolically active cells as well as in dead, viable, but non-proliferating cells (like spores), and their cell walls. Extracellular enzymes are released into the soil and become "permanently" immobilised on humic and clay colloids through a variety of mechanisms, including entrapment, covalent bonds, hydrogen bonding, and ionic interactions.

The rhizosphere is a unique soil microenvironment in which the characteristics and activities of microorganisms, plant roots, and soil interact in a coordinated way. Through altering the biogeochemical parameters of soil (such as pollutants, nutrient concentrations, pH and redox potential, complexing or chelating compound concentrations, partial pressures of O_2 and CO_2, etc.), all biological functions of plant roots, including uptake, respiration, and exudation, have an impact in the rhizosphere (Gianfreda 2015). Every physiological activity of an organism in the rhizosphere will alter the parameters of the soil and the processes that control the availability of nutrients. In addition, depending on a plant's capacity to modify rhizosphere processes and characteristics, nutrient bioavailability differs between species and even between genotypes within a single species.

The root-soil interface is significantly influenced by rhizosphere soil, and the conversion of soil C, N, P, S, and other nutrient elements is intimately correlated with soil enzyme activity. Enzyme activities isolated from roots, other soil organisms, and the rhizosphere-soil are sensitive markers of heavy metal contamination of the soil and nutrient cycling. As a result, they can be employed to track the ecological well-being of the soil system in the presence of heavy metal pollution (Liang et al. 2014; Mishra et al. 2022). Particularly, rhizosphere enzyme activities show how plants and microorganisms interact and can provide insight into the makeup and role of the microbiome (Li et al. 2017; Razavi et al. 2016).

Furthermore, the activity of soil enzymes in the rhizosphere is highly responsive to drought and is extremely sensitive to it. It is a sensitive early warning system for changes in the soil ecosystem. Among these, two types of enzymes that are more studied in soil are oxidoreductases (Cárdenas-Morenoet al. 2023) (such as polyphenol oxidase, dehydrogenase, catalase, sulphate reductase, nitrate reductase) and hydrolases (Speir and Ross 2002) (such as amylase, sucrase, urease, phosphatase, protease, β-glucosidase, cellulase, etc.).

Large volumes of waste are disposed of in conjunction with human activities like mining (Edraki et al. 2014), and these wastes frequently contain high concentrations of heavy metals (Chilesha et al. 2020). Furthermore, depending on the method used to process the ore, mine waste lands can have high levels of soil compaction, high acidity, and low nutrient content, all of which inhibit the growth of vegetation in post-mining landscapes (Chilesha et al. 2020; Festin et al. 2019). Because heavy metals pose serious health risks to humans once they enter the food chain, reclamation of soils tainted by heavy metals has become an environmental concern of great urgency. Many remediation strategies have been tested with differing degrees of success in dealing with this environmental concern (Festin et al. 2019).

10.2 Activity of Soil Enzymes Combating Environmental Pollution

Environmental pollution is becoming a major concern for the health of humans and other living things. To reduce the amount of pollutants to acceptable levels, conventional methods of eliminating pollutants such as membrane filtration as well as ion exchange are insufficiently effective. Because biological methods are more effective and biocompatible, so, they are the preferred method for pollutant remediation. Using microorganisms and plants, phytoremediation lessens the harmful effects of heavy metals on the soil (Mendez and Maier 2008). The capacity of the plant to produce a high biomass and tolerate potentially toxic concentrations of heavy metals, while simultaneously accumulating and translocating those metals, is critical to the efficacy of phytoremediation (Anning and Akoto 2018).

Enzyme activities isolated from roots, other microorganisms in the soil, and the rhizosphere-soil are sensitive markers of heavy metal contamination of the soil and nutrient cycling. As a result, they can be employed to track the ecological well-being of the soil system in the presence of heavy metal pollution (Liang et al. 2014). Particularly, rhizosphere enzyme activities show how plants and microorganisms interact and can provide insight into the makeup and role of the microbes (Li et al. 2017; Razavi et al. 2016). According to earlier research, the activities of the enzymes catalase, urease, dehydrogenase, and phosphatase that are extracted from soil (and possibly from the rhizosphere) are affected by contamination with heavy metals (Marzadori et al. 1996; Pattanaik and Equeenuddin 2016; Daun et al. 2018). The alterations in these enzyme functions in response to certain heavy metals, however, are not consistent.

10.3 Bioremediation with Enzymes

Bioremediation is the term for these economical and environmentally acceptable methods of lowering pollution. In bioremediation techniques, enzymes are the most important component (Sharma et al. 2018). Lead, chromium, mercury, different polymers, azo dyes, PAHs, and other inorganic and organic pollutants can all be remedied by enzymes. Every year, a significant amount of organic pollutants can contaminate the air, soil, and water. These pollutants include plastics (Sahoo et al. 2022), herbicides, pesticides, dyes (Lellis et al. 2019; Shindhal et al. 2021), various types of medicationssuch as antibiotics (Kumar et al. 2019), steroids, polymers (Bhatt et al. 2021), aromatic molecules, chlorinated hydrocarbons, polycyclic aromatic hydrocarbons (PAHs), enzymes are the most efficient bioremediation tools and can control all chemical changes on pollutants (Saravanan et al. 2021). Enzymes' specificity is usually broad enough to act on different molecules with similar structures. Using enzymes in bioremediation could be either individually that the isolated enzyme used and added to the contaminated area or as a whole cell, e.g., bacteria, fungi, or algae. In a second way, continuous aeration, inoculation, and nutrition are necessary. Different enzymes (Mousavi et al. 2021) like mono- or dioxygenases, halogenases, peroxidases, phosphotriesterases, hydrolases, laccase (Bhatt et al. 2023; Dong et al. 2023), transferases, and oxidoreductases from various species of bacteria, fungi (Temporiti et al. 2022), algae, and plants have been used for the bioremediation of pollutants (Yaashikaa et al. 2022).

Hydrolase enzymes, including lipase, cutinase, nitrilases, aminohydrolases, and organophosphorus hydrolase, are employed in the bioremediation of various chemicals, including polymers, herbicides, pesticides, and organophosphorus compounds (Dash and Osborne 2023). Enzymes that can be used for the bioremediation of organophosphorus compounds include phosphotriesterase (Thakur et al. 2019), also referred to as organophosphorus hydrolase. Although its fungal version is expressed in *Aspergillus niger* and *Penicilliumlilacinum*, it is primarily isolated from *Pseudomonas diminuta.* It can affect bonds that are P–S, P–O, and P–F. Zn2+ is a cofactor in this enzyme's native form, but experiments revealed that substituting Co2+ has the greatest anti-paraoxon activity (Thakur et al. 2019).

A family of enzymes known as ligninolytic enzymes has several uses in bioremediation (Novotný et al. 2004). This class of enzymes, known as "ligninolytic," is generated by white-rot fungi (WRF) under nutrient limitation (Winquist 2014). For instance, when acetosyringone is present as a mediator, laccase (Zainith et al. 2020) can decolorize azo, indigo, and as well asanthraquinone dyes by 80% in less than one hour.

A family of heme-containing enzymes known as cytochrome p450 monooxygenases (CYP) catalyses (Syed and Yadav 2012) a variety of reactions, including N-hydroxylation, oxidative dehalogenation, O-dealkylation, and hydroxylation reaction of C–H bonds. Pollutants that CYPs can degrade and bioremediate include PAHs, dioxins, aliphatic hydrocarbons, PCBs (polychlorinated biphenyls), PCDFs (polychlorinated dibenzofurans), PCDDs (polychlorinated dibenzo-p-dioxins), and even Cr (VI) compounds (Coelho-Moreira et al. 2013).

Most microorganisms produce metal-binding peptides, such as phytochelatins and metallothioneins (Sharma et al. 2016), in the presence of toxic heavy metals (Goldbrough 2020), such as arsenic, lead, mercury, chromium, etc. These peptides reduce the toxicity of the heavy metals by sequestering them.

One of the most tough sustainable development goals to date is environmental pollution. Some of the pollutants that pose the greatest threat to human health are toxic metals, such as lead (Pb) (Usman et al. 2020). This is particularly true in nations that are rapidly industrializing such as India. Pbis ranked as the top heavy metal on the Earth in the year 2015. At soil concentrations greater than 30 mg/kg, it is usually toxic to the majority of plants.

Natural processes like geochemical weathering, sea spray emissions, volcanic activity, and the remobilization of silt, soil, and water from mining (Das and Ghosh 2022). areas have all been linked to lead contamination in air, soil, and water resources (Singh and Steinnes 2020). Different countries and regions of the world have different sources of lead contamination, which can be found in agricultural soils, crops, and water. The primary sources of lead contamination in the environment are thought to be human-made products (Triantafyllidou and Edwards 2012) and processes (e.g. oil-processing industries, paint, mining (Punia and Singh 2021; Obasi and Akudinobi 2020), cosmetics, smelting, refining, peeling of window and door frames, jewellery, ceramics, toys, pottery, plumbing materials, alloys, water from old pipes, stained glass vinyl mini-blinds, lead-glazed dishes, batteries, lead-filled firearms, radiators for cars, andcolours of ink).

10.4 Role of Heavy Metal Lead in Soil Pollution

Lead tends to be stored in soil where it forms complexes that are stable with various compounds (Meena et al. 2020; Laidlow et al. 2017). Lead's form, solubility, mobility, and bioavailability all influence its fate and behaviour in soil, which is governed by a variety of biogeochemical factors including pH, redox conditions, cation-exchange capacity, and soil mineralogy, the presence and kind of organic and inorganic ligands, the biological and microbiological circumstances, and the competing cations. Pb present in the soil through any sources and finds its way into the roots of plants (Zeng et al. 2007), where it can either build up or move to parts of the plant that are above ground, known as aerial plant part (APP) (Dotaniya et al. 2020). Pb primarily accumulates (approximately 95%) in plant species' roots, with very little of it moving to APP. Lead primarily enters plants through the apoplast pathway as well as calcium ion channels in the roots. Lead can also, in trace amounts, enter plants through their leaves. Lead tends to store in root cells once it enters the roots. Typically, plants use their roots to absorb metal ions from the soil. In many plant species, Pb from the soil is adsorbed (in uneven ways) through the roots (Rizwan et al. 2021) and bound to the uronic acid/polysaccharide of the rhizoderm.

The main reason for lead's toxicity (Zulfiqar et al. 2019) is the ROS it produces. These free radicals cause irreversible metabolic malfunction and cell death by upsetting the redox balance of cells, oxidative stress, and DNA damage through oxidation. Reactive oxygen species (ROS) are produced when lead (Pb) interacts with plants (Sharma and Dubey 2005). Pb and ROS have the ability to harm biomolecules, cellular components, and tissue ultrastructure, which can result in phytotoxicity (Kumar and Prasad 2018). These detrimental effects might potentially lead to the inhibition of regular physiological responses, cellular activity, and overall plant performance (Verma and Dubey 2003). In the plant defence system, ROS serve a dual purpose and function as a signalling molecule. Both enzymatic and non-enzymatic antioxidative pathways are included in this system. The main enzymatic antioxidants are ascorbate–glutathione cycle enzymes, peroxidase, superoxide dismutase, and catalase. Non-enzymatic antioxidants areflavonoids, phenols, ascorbic acid, and glutathione. Utilizing plants to remove lead from contaminated areas depends on the plants' ability to accumulate lead, their ability to combat phytotoxicity caused by lead, and the detoxification and tolerance strategies they have developed (Pourrut et al. 2011).

10.5 A Transcriptome Study with the Differentially Expressed Genes in Presence of Lead

In the transcriptome experiment the profile gene expression has been studied [55] in above ground parts and in the roots of arabidopsis plants in response to toxic ions of lead. Utilizing transcriptomic, proteomic, and metabolomics analysis to examine the expression profiles of genes or proteins induced by heavy metals, studies on plants for phytoremediation have discovered that the plants have their own homeostatic mechanisms in their tissues as well as organelles to reduce the damage caused by various toxic metals.

Profile expression of genes involved in plant reaction to lead in GEO dataset (GSE6157) (Liu et al. 2009) from NCBI has been analyzed for top differentially expressed genes with GEO2R analyser (Smyth 2005). In response to lead, following enzymes are expressed in increased order from root, phosphatidatecytidylyltransferase (AT4G26770), pyruvate decarboxylase 1(AT4G33070), nudix hydrolase 10 (NUDT10), oxidoreductase, 2OG-Fe(II) oxygenase family protein (AT4G35820), sulfotransferase 3 (AT4G26280), acetyl-CoA carboxylase 2 (ACC2), xanthine dehydrogenase 2 (XDH2), cytokinin oxidase 2 (CKX2), cysteine-rich receptor-like protein kinase 37 (CRK37), basic chitinase B (HCHIB), GDSL esterase/lipase (AT1G54020), 1-aminocyclopropane-1-carboxylate synthase (ACS1), purple acid phosphatase 11 (PAP11), peroxidase 13 (AT1G77100).

Top differentially expressed genes (7943) from GSE6157 dataset are analyzed with DAVID (Sherman et al. 2022), a web server for gene functional enrichment analysis. Functional annotation of these genes according to their enzymological functions is classified in Table 10.1.

The increased gene expression of AT1G77100, having peroxidase activity and GSTU3 gene having glutathione transferase activity in root part of the plant species in presence of lead, compared to root control from gene expression profile of GSE6157 dataset, is shown in Fig. 10.1.

Table 10.1 Functional annotation of top differentially expressed genes represented in chart

Category of enzymes	Number of genes
Oxidoreductase	567
Peroxidase	88
Serine/threonine-protein kinase	343
Lyase	179
Helicase	94
Glycosidase	157
Multifunctional enzyme	36
Kinase	517
Monooxygenase	117
Tyrosine-protein kinase	14
Glycosyltransferase	203
Dioxygenase	50
Decarboxylase	20

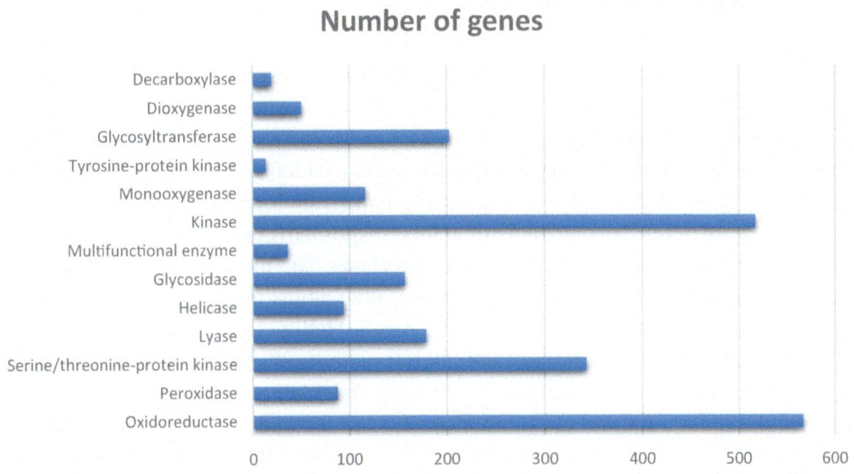

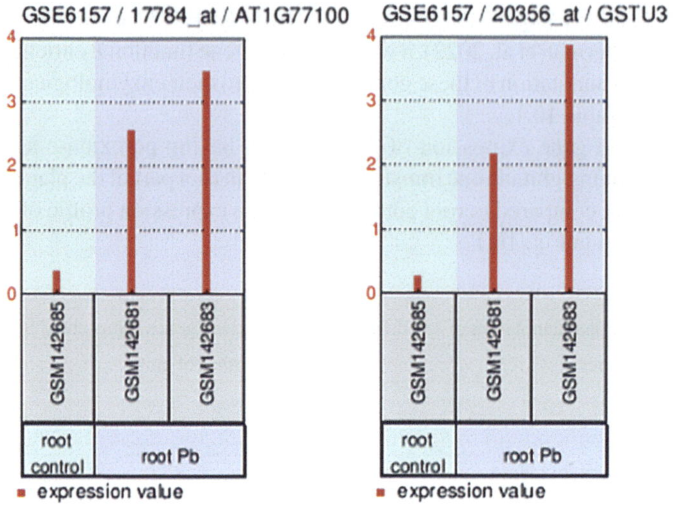

Fig. 10.1 Gene expression values for peroxidase and glutathione transferase genes in presence of lead in soil

The enzymes, whose gene expression are increased many folds (indicating by the values of t in Table 10.2) with peroxidase activities, in presence of heavy metal Pb, are summarized in Table 10.2.

10.6 Significance of Increased Gene Expression of Antioxidant Enzymes

The role of these enzymes in phyto-transformation in response to lead is analyzed in this study. These antioxidant enzymes are activated indirectly due to exposure to different heavy metals such as lead. The most typical reaction to lead toxicity in plants is the generation of reactive oxygen species (ROS), which causes oxidative stress in the plants and interferes with normal metabolism (Kumar and Prasad 2018). Plants have a variety of defensive mechanisms at their disposal to control the stress that these harmful substances cause and preserve homeostasis. The most widely used defence strategy against ROS produced by lead metals is to adjust the levels of antioxidants and as well as antioxidative enzymes in the Asada-Halliwell pathway (Kanwar et al. 2015). The relative gene expression and catalytic activity of antioxidant enzymes such as glutathione peroxidase, ascorbate peroxidase and other peroxidases are increased significantly in the root part of arabidopsis plant under lead exposure. These antioxidant enzymes can scavenge reactive oxygen species (ROS), when their relative gene expression level is increased as a result of lead metal toxicity. Among them glutathione peroxidase (GSTF9, GPX6) and ascorbate peroxidase (SAPX, APX1,

Table 10.2 List of differentially expressed peroxidase enzymes from root in presence of lead metal

Sl no.	Id	adj. P. val	P.value	t	logFC	Gene.symbol	GO.Function
1	17784_at	0.854	0.00948	−6.81	−2.91	AT1G77100	Peroxidase activity
2	20356_at	0.854	0.01436	−5.77	−3.29	GSTU3	Glutathione transferase activity
3	20375_at	0.854	0.04096	−3.74	−1.45	PER1	Thioredoxin peroxidase activity
4	17339_at	0.854	0.06376	−3.08	−1.33	AT2G41480	Peroxidase activity
5	16430_at	0.854	0.07406	−2.87	−1.06	GSTF9	Glutathione peroxidase activity glutathione transferase activity
6	12883_at	0.883	0.10944	−2.37	−9.22E-01	SAPX	L-ascorbate peroxidase activity
7	12333_at	0.883	0.12723	−2.2	−1.93	AT4G36430	Peroxidase activity
8	17806_at	0.883	0.12865	−2.18	−9.46E-01	AT4G33870	Peroxidase activity
9	20366_at	0.883	0.12892	−2.18	−1.08	RHS18	Peroxidase activity
10	12355_at	0.883	0.13637	−2.12	−1.6	AT2G35380	Peroxidase activity
11	14638_at	0.883	0.13812	−2.1	−4.55	PRXCB	Peroxidase activity
12	16462_s_at	0.883	0.13926	−2.09	−1.02	AT2G38380	Peroxidase activity
13	16429_at	0.883	0.1395	−2.09	−8.39E-01	APX1	L-ascorbate peroxidase activity
14	17411_s_at	0.893	0.16503	−1.9	−7.08E-01	TAPX	L-ascorbate peroxidase activity
15	15787_s_at	0.896	0.17812	−1.82	−8.10E-01	RBOHB	NAD(P)H oxidase activity peroxidase activity
16	15660_s_at	0.924	0.23576	−1.53	−6.12E-01	GPX6	Glutathione peroxidase activity

TAPX) enzymes are involved in Halliwell- Asada enzyme pathway or ascorbate–glutathione cycle. Superoxide dismutase (SOD) converts superoxide radical (O_2) into H_2O_2, acting as a first line of defense in scavenging reactive oxygen species (ROS). This produced H_2O_2 is further scavenged by membrane-bounded peroxidase (POD) and catalase (CAT) in peroxisomes. Through the ascorbate–glutathione cycle, glutathione reductase (GR) converts dehydro-ascorbate to ascorbate, controlling the pool of glutathione in the reduced state. They thus play a crucial role in plants that are under a variety of stresses including heavy metal toxicity such as lead (Navabpour et al. 2020).

The Ascorbate–Glutathione (AsA-GSH) pathway, also referred to as the Asada–Halliwell pathway, is made up of four enzymes that are essential for detoxifying reactive oxygen species (ROS): ascorbate peroxidase, dehydroascorbate reductase, monodehydroascorbate reductase, and glutathione reductase (Malecka et al. 2009). In addition to detoxifying ROS, they also work with other defence mechanisms in plants to shield them from harm caused by a variety of abiotic stresses such as heavy metal toxicity. According to a number of plant studies, plants that had their AsA and GSH levels increased and their AsA-GSH pathway enzymes overexpressed or upregulated were better able to withstand abiotic stresses by lowering reactive oxygen species (ROS).

Similarly, AT1G77100 (PER13orperoxidase 13) can function as remover of H_2O_2, oxidation of toxic reductants, biosynthesis and degradation of lignin, response to environmental stresses such as wounding, pathogen attack and oxidative stress. Moreover, that it has a metal ion binding activity for calcium ion with the binding site in positions 65, 68, 70, 72 and 74, according to Uniprot database. Lead ion binding in this specific enzyme cleft can be proposed during lead toxicity in soil.

10.7 A Detoxification Mechanism for Lead Through Remodeling of Plant Root Cell Wall

Plant tall fescue (FestucaarundinaceaSchreb) exhibits Pb^{+2} resistance by using a special detoxification mechanism through remodeling of plant root cell wall (Zhang et al. 2020). Here, the root cell wall remodeling in this plant species is regulated by ROS signalling pathway. Thus, the plant with high resistance for Pb-induced toxicity, can accumulate a huge amount of Pb in their roots. They can be predicted as an important tool of 'phytoremediation' due to their activity as hyper accumulators. This type of green technology by using phytoremediation, can be extended for other heavy metals such as Cu, Cd and Zn also. Under Pb^{+2} stress condition H_2O_2 production is also increased. Zhang and coworker in 2020, has been revealed that during remodeling of plant root cell wall, the uronicacid, production is increased under Pb^{+2} stress condtion. This uronic acid is present in plant cell wall carbohydrate molecules, such as pectin, hemicellulose.

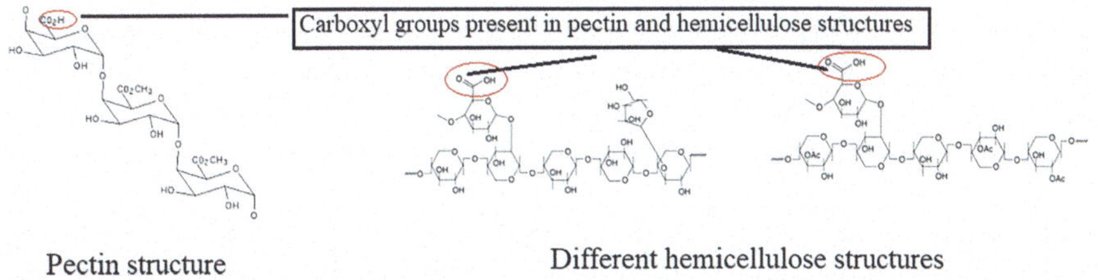

Fig. 10.2 (Figure size would be changed) Presence of carboxyl groups in different carbohydrates present in plant cell wall structure

The complex polysaccharide pectins are made up of 1,4-linked α-D-galactosyluronic residues. From the primary cell walls of plants, three pectic polysaccharides have been isolated: homogalacturonan, rhamnogalacturonan-I, and substituted galacturonans. A linear chain of 1,4-linked α-D-galactosyluronic residues with some carboxyl groups methyl esterified is called homogalacturonan (HG). Additionally, they could have O-acetylated groups at C-2 and C-3. Although homogalactanans have been isolated from apple pectin and sunflower heads, it's possible that they were liberated from a heterogeneous pectic polysaccharide during the extraction processes that were used to obtain them.

Nevertheless, pectin is a heteropolysaccharide because it contains 1–4% substitution of rhamnopyranosyl residues in the galactosyluronicbackbone (Zdunek et al. 2021). Large substituted side chains are the other key characteristic of these rhamnogalacturonan-I (RG-I) chains. Depending on the plant source and isolation technique, 20–80% of the rhamnopyranosyl residues have neutral and acidic oligosaccharide side chains substituted at C-4. Large linear and branching L-arabinofuranosyl and/or D-galactopyranosyl residues make up the predominant side chains; the relative proportion and length of these residues vary depending on the source plant. There are also other, rarer side chains that are typically shorter. These are demonstrated in Fig. 10.2.

The minor component of the backbone is rhamnogalacturonan-II (RG-II). This is not structurally related to RG-I since its backbone is composed of 1,4-linked D-galactosyluronic residues like HG. At approximately 30 glycosyl residues long it has a non-saccharide and an octasaccharide side chain attached to C-2 of some of the backbone residues and two structurally different disaccharides attached to C-3 of the backbone. RG-II is of interest as it occurs in relatively high amounts in wine and other fruit juices and it has been demonstrated that it binds heavy metals.

Side chains that are affixed to the primary backbone of hemicelluloses may contain uronicacids (Rao et al. 2023). For instance, 4-O-Methyl-D-Glucuronic Acid partially substitutes the xylan main chain in the arabinoxylans of softwoods, and the ratio of xylose to uronic acids varies from 4:1 to 9:1. In addition to galactuornic acid, other softwood hemicelluloses may also contain glucuronic acid. 4-O-Methyl-D-Glucuronic Acid can replace the xylan backbone of the glucuronoxylans in hardwoods. The ratio of xylose to uronic acid can range from 3:1 to 20:1, with a typical ratio between 7:1 and 12:1. Hemicelluloses in grasses also contain uronic acids, which are usually more prevalent in the early stages of growth. In addition, uronic acid concentrations are generally highest in nodes, lowest in internodes, and intermediate in leaves.

Structural polysaccharides known as pectins are mostly found in the primary cell wallstructures of land plants. They are the most highly branched polysaccharides and are polyuronides mainly composed of galacturonic acid units. A noteworthy component is also rhamnose (6-deoxymannose). Despite being acidic polysaccharides, pectins lack the hemicellulose-only uronic acids such as glucuronic acid and 4-O-Methyl-D-Glucuronic acid.

Uronic acids can make up a significant amount of the mass of certain seaweeds, despite being generally insignificant components of lignocellulosic biomass. Moreover, mannuronic acid and guluronic acid are the most significant uronic acids in seaweed, whereas glucuronic acid, 4-O-Methyl-D-Glucuronic Acid, and galacturonic Acid are by far the most abundant uronic acids in the hemicelluloses of terrestrial plants. These are present in the seaweeds' alignate or alginic acid polysaccharides.

Both guluronic and mannuronic acids are present in these polymers in linear chains, with the ratios of mannuronic to guluronic acid ranging from 1.2 to 2.1 or higher. These hexuronic acids typically occur in C5 epimer blocks of one or the other within this linear chain. Seasonal variations have been noted in the abundance and composition of uronic acids found in seaweeds.

These carboxyl group fractions of pectins and hemicelluloses are also increased in stressed conditions such as heavy metal toxicity. During lead toxicity condition, these increased number of carboxyl groups, provides more Pb^{+2} metal ion binding sites in plant cell wall, resulting the hyper-accumulation of heavy metal ions (e.g. Pb^{+2}) without affecting metabolic activity of plants (Zhang et al. 2020).

The enzyme, pectin methylesterase plays an important role in regulating the binding of heavy metals in plant cell wall remodeling. The increased gene expression of PME38, AT2G19150, AT3G47380, AT4G15980, ATPMEPCRC, AT2G47280, AT4G33230, AT3G47380, ATPMEPCRB, AT2G1610, AT3G47380, AT4G03930, AT4G02300, AT4G15750, PME44, PE11, PME1, AT5G62350, Sks16 and AT2G36700, belonging to the aspartyl esterase and pectinesterase activity, is observed in root part of arabidopsis plant due to exposure to lead.

The components of pectin include galactans, arabinans, homogalacturonan (HGA), rhamnogalacturonans I and II, and other polysaccharides. Typically, HGA's carboxyl groups are methylesterified. Pectin's HGA's negatively charged carboxylic groups have a strong affinity for metal ions (Zhu et al. 2015). On the other hand, pectin's degree of de-methylesterification, which is regulated by the activity of pectin methylesterases (PMEs) (Qi et al. 2022), determines the quantity of negatively charged carboxyl groups in the substance. PMEs are widely distributed enzymes that fall under the carbohydrate esterase family's class 8. PMEs generate a carboxylate group (–COO–), release protons and methanol, and partially demethylate HGA. Zhu et al. (2015) also suggested that metal ion chelation via COO- groups in demethylesterified pectin is made possible by the process of demethylesterification of pectin.

PMEs release both protons and methanol, converting the methoxyl groups on the polygalacturonic acid chain into negatively charged carboxyl groups. These newly form carboxyl groups participate in chelation reaction with metal ions such as Pb^{+2} (Qi et al. 2022). Therefore, during lead metal ion exposure, increased level of PME1, PME38 gene expression, facilitates phytoremediation in plants. It was shown by Schmohl and Horst (2000) that cations can boost PME activity. Metal ions have been shown by Moustacas et al. (1991) to affect PME activity when combined with pectin. They proposed that the "blocks" of carboxylic groups on pectin may trap PME, preventing PME and pectin from interacting. PME activity is therefore suppressed, but PME is released when metal ions bind to pectin. As Schmohl and Horst (2000) have shown, the binding of cations to the carboxylic groups of pectin not only modifies the porosity of the cell wall but also prevents it from binding Ca2+ normally.

While inorganic pollutants do not break down and instead build up in the environment, the majority of organic pollutants are gradually broken down, primarily by bacteria. Heavy metals (such as cadmium (Cd), lead (Pb), chromium (Cr), mercury (Hg), and nickel (Ni)) and metalloids (such as arsenic (As)) are examples of inorganic pollutants that cause environmental pollution and severely impede plant growth and development, thereby lowering crop plant productivity Moustacas et al. (1991).

Heavy metal and metalloid contamination of soil, including cadmium (Cd) and arsenic (As), is a serious risk to human health and the environment. One of the primary mechanisms by which plants regulate harmful substances, such as As and Cd, is vacuolar sequestration (Clemens 2006; Zhang et al. 2018). Comprehending the mechanisms underlying heavy metal accumulation and tolerance can prove beneficial for the development of safe crops as well as phytoremediation.

10.8 Vacuolar Sequestration of Heavy Metals in Plant

Within the mature plant cell, the vacuole is the largest organelle. It serves as a significant ion and metabolite reservoir and is essential for both healthy cell development and the detoxification process (Martinoia et al. 2012). The vacuole is primarily used to store toxic substances like Cd and As in order to detoxify the cytosol.

Vacuolar sequestration of heavy metals has been linked to a number of transporter families, including ATP-binding cassette subfamily C proteins (ABCCs), natural resistance-associated macrophage proteins (NRAMPs), Ca2+ exchangers (CAXs), and heavy metal ATPases (HMAs). P-type heavy metal ATPase HMA3 mediates the vacuolar sequestration of multiple heavy metals in different plant species, such as cobalt (Co), zinc (Zn), cadmium (Cd), and lead (Pb). HMA3's precise roles, however, differ amongst plants in terms of substrate selectivity, expression pattern, and degree of expression. With its wide substrate selectivity, Arabidopsis HMA3 is able to transfer Zn, Pb, Cd, and Co into the vacuole (Morel et al. 2009).

The mechanisms underlying Pb uptake remain poorly understood (Zhao et al. 2022). According to some research, Ca-permeable channels like cyclic nucleotide-gated channels might play a role in how Pb is absorbed by Arabidopsis roots (Moon et al. 2019). Cyclic nucleotide-gated channels (CNGCs) in plants are a kind of Ca2+ permeable channel that may be involved in the uptake of toxic and necessary cations. In order to identify the potential candidates in charge of heavy metal ion transport, the reaction of several Arabidopsis CNGC mutants to Pb^{+2} and Cd2+ ions have been studied. Based on the examination of primary root length, the results showed that AtCNGC1, AtCNGC10, AtCNGC13, and AtCNGC19 are involved in Pb^{+2} toxicity in Arabidopsis, whereas AtCNGC11, AtCNGC13, AtCNGC16, and AtCNGC20 are involved in Cd2+ toxicity. Ion content analysis confirmed that while mutations of AtCNGC11, AtCNGC15, and AtCNGC19 resulted in less Pb^{+2} and Cd2+ accumulation in plants, mutations of AtCNGC1 and AtCNGC13 resulted in reduced Pb^{+2} accumulation.

AtCNGC11, AtCNGC13, AtCNGC15 transporter channel proteins are involved in the uptake of Pb^{+2} ions from soil through plasma membrane of root cell of arabidopsis along with Ca+2 ions. But the involvement of transporter protein AtCNGC19 has been recorded for the transport of Pb^{+2} ion in presence of Ca+2 ion, towards vacuole. Therefore, this non-selective cation transporter protein canfunction as passive transporter of cations between the vacuole and cytosol and plays an important role invacuolar sequestration of heavy metal lead in arabidopsis.

10.9 Discussion

Waste products may contain pollutants like phenolic agents, synthetic dyes, volatile organic compounds, polycyclic aromatic hydrocarbons, and other compounds that are particularly detrimental to human health and the environment, such as pesticides, drug molecules, and polychlorinated biphenyls. As members of the class oxidoreductases, peroxidases (Sellami et al. 2022) can carry out a broad range of oxidation reactions, including those triggered by radical mechanisms, oxidation of two-electron substrate and oxygen insertion into C–H bonds.

Plants use a variety of avoidance or detoxification strategies to protect themselves from lead toxicity. Plants bind lead to their cell walls or other ligands to prevent lead from entering their cells through exclusion. Plants resist elevated lead levels by turning on a variety of antioxidant enzymes, the generation of ROS (Gawronski 2007). Plants with effective detoxification systems are typically classified as hyperaccumulators. These plants can be used to bioremediate soil for a variety of metals.

References

Anning AK, Akoto R (2018) Assisted phytoremediation of heavy metal contaminated soil from a mined site with typhalatifolia and chrysopogonzizanioides. Ecotoxicol Environ Saf 148:97–104

Bhatt P, Bhatt K, Chen WJ, Huang Y, Xiao Y, Wu S, Chen S et al (2023) Bioremediation potential of laccase for catalysis of glyphosate, isoproturon, lignin, and parathion: molecular docking, dynamics, and simulation. J Hazard Mater 443:130319

Bhatt P, Gangola S, Bhandari G, Zhang W, Maithani D, Mishra S, Chen S (2021) New insights into the degradation of synthetic pollutants in contaminated environments. Chemosphere 268:128827

Cárdenas-Moreno Y, González-Bacerio J, García Arellano H, del Monte-Martínez A (2023) Oxidoreductase enzymes: characteristics, applications, and challenges as a biocatalyst. Biotechnol Appl Biochem 70(6):2108–2135

Chileshe MN, Syampungani S, Festin ES, Tigabu M, Daneshvar A, Odén PC (2020) Physico-chemical characteristics and heavy metal concentrations of copper mine wastes in Zambia: implications for pollution risk and restoration. J for Res 31:1283–1293

Clemens S (2006) Toxic metal accumulation, responses to exposure and mechanisms of tolerance in plants. Biochimie 88(11):1707–1719

Coelho-Moreira JDS, Bracht A, Souza ACDSD, Oliveira RF, Sá-Nakanishi ABD, Souza CGMD, Peralta RM (2013) Degradation of diuron by Phanerochaetechrysosporium: role of ligninolytic enzymes and cytochrome P450. BioMed Res Int

Das A, Ghosh S (2022) Role of microorganisms in extenuation of mining and industrial wastes. Geomicrobiol J 39:173–175. https://doi.org/10.1080/01490451.2022.2038953

Dash DM, Osborne WJ (2023) A systematic review on the implementation of advanced and evolutionary biotechnological tools for efficient bioremediation of organophosphorus pesticides. Chemosphere 313:137506

Dietz AC, Schnoor JL (2001) Advances in phytoremediation. Environ Health Perspect 109(suppl 1):163–168

Dong CD, Tiwari A, Anisha GS, Chen CW, Singh A, Haldar D, Singhania RR et al (2023) Laccase: a potential biocatalyst for pollutant degradation. Environ Pollut:120999

Dotaniya ML, Dotaniya CK, Solanki P, Meena VD, Doutaniya RK (2020) Lead contamination and its dynamics in soil–plant system. In: Lead in plants and the environment, pp 83–98

Duan C, Fang L, Yang C, Chen W, Cui Y, Li S (2018) Reveal the response of enzyme activities to heavy metals through in situ zymography. Ecotoxicol Environ Saf 156:106–115

Edraki M, Baumgartl T, Manlapig E, Bradshaw D, Franks DM, Moran CJ (2014) Designing mine tailings for better environmental, social and economic outcomes: a review of alternative approaches. J Clean Prod 84:411–420

Festin ES, Tigabu M, Chileshe MN, Syampungani S, Odén PC (2019) Progresses in restoration of post-mining landscape in Africa. J for Res 30(2):381–396

Gawronski S (2007) Profile expression of genes involved in plant reaction to lead: uptake tolerance and defense

Gianfreda L (2015) Enzymes of importance to rhizosphere processes. J Soil Sci Plant Nutr 15(2):283–306

Goldsbrough P (2020) Metal tolerance in plants: the role of phytochelatins and metallothioneins. In: Phytoremediation of contaminated soil and water. Crc Press, pp 221–233

Inamdar A, Sangawe V, Adhapure N (2022) Enzymes in rhizosphere engineering. In: Rhizosphere engineering. Academic Press, pp 259–272

Kanwar MK, Poonam Pal S, Bhardwaj R (2015) Involvement of Asada-Halliwell pathway during phytoremediation of chromium (VI) in *Brassica juncea L.* plants. Int J Phytorem 17(12):1237–1243

Kumar A, Prasad MNV (2018) Plant-lead interactions: transport, toxicity, tolerance, and detoxification mechanisms. Ecotoxicol Environ Saf 166:401–418

Kumar M, Jaiswal S, Sodhi KK, Shree P, Singh DK, Agrawal PK, Shukla P (2019) Antibiotics bioremediation: perspectives on its ecotoxicity and resistance. Environ Int 124:448–461

Laidlaw MA, Filippelli GM, Brown S, Paz-Ferreiro J, Reichman SM, Netherway P, Mielke HW et al (2017) Case studies and evidence-based approaches to addressing urban soil lead contamination. Appl Geochem 83:14–30

Lellis B, Fávaro-Polonio CZ, Pamphile JA, Polonio JC (2019) Effects of textile dyes on health and the environment and bioremediation potential of living organisms. Biotechnol Res Innov 3(2):275–290

Li X, Meng D, Li J, Yin H, Liu H, Liu X, Yan M et al (2017) Response of soil microbial communities and microbial interactions to long-term heavy metal contamination. Environ Pollut 231:908–917

Liang Q, Gao R, Xi B, Zhang Y, Zhang H (2014) Long-term effects of irrigation using water from the river receiving treated industrial wastewater on soil organic carbon fractions and enzyme activities. Agric Water Manag 135:100–108

Liu R, Dai Y, Sun L (2015) Effect of rhizosphere enzymes on phytoremediation in PAH-contaminated soil using five plant species. PLoS ONE 10(3):e0120369

Liu T, Liu S, Guan H, Ma L, Chen Z, Gu H, Qu LJ (2009) Transcriptional profiling of Arabidopsis seedlings in response to heavy metal lead (Pb). Environ Exp Bot 67(2):377–386

Małecka A, Derba-Maceluch M, Kaczorowska K, Piechalak A, Tomaszewska B (2009) Reactive oxygen species production and antioxidative defense system in pea root tissues treated with lead ions: mitochondrial and peroxisomal level. Actaphysiologiaeplantarum 31:1065–1075

Martinoia E, Meyer S, De Angeli A, Nagy R (2012) Vacuolar transporters in their physiological context. Annu Rev Plant Biol 63:183–213

Marzadori C, Ciavatta C, Montecchio D, Gessa C (1996) Effects of lead pollution on different soil enzyme activities. Biol Fertil Soils 22:53–58

Meena V, Dotaniya ML, Saha JK, Das H, Patra AK (2020) Impact of lead contamination on agroecosystem and human health. In: Lead in plants and the environment, pp 67–82

Mendez MO, Maier RM (2008) Phytoremediation of mine tailings in temperate and arid environments. Rev Environ Sci Bio/technol 7:47–59

Mishra S, Dash D, Al-Tawaha ARMS, Das AP (2022) A Review on heavy metal ion adsorption on synthetic microfiber surface in aquatic environments. Appl Biochem Biotechnol 194(10):4639–4654. https://doi.org/10.1007/S12010-022-04029-W/METRICS

Moon JY, Belloeil C, Ianna ML, Shin R (2019) Arabidopsis CNGC family members contribute to heavy metal ion uptake in plants. Int J Mol Sci 20(2):413

Morel M, Crouzet J, Gravot A, Auroy P, Leonhardt N, Vavasseur A, Richaud P (2009) AtHMA3, a P1B-ATPase allowing Cd/Zn/co/Pb vacuolar storage in arabidopsis. Plant Physiol 149(2):894–904

Mousavi SM, Hashemi SA, ImanMoezzi SM, Ravan N, Gholami A, Lai CW, Behbudi G et al (2021) Recent advances in enzymes for the bioremediation of pollutants. Biochem Res Int

Moustacas AM, Nari J, Borel M, Noat G, Ricard J (1991) Pectin methylesterase, metal ions and plant cell-wall extension: the role of metal ions in plant cell-wall extension. Biochem J 279(2):351–354

Navabpour S, Yamchi A, Bagherikia S, Kafi H (2020) Lead-induced oxidative stress and role of antioxidant defense in wheat (*Triticumaestivum* L.). Physiol Mol Biol Plants 26:793–802

Novotný Č, Svobodová K, Erbanová P, Cajthaml T, Kasinath A, Lang E, Šašek V (2004) Ligninolytic fungi in bioremediation: extracellular enzyme production and degradation rate. Soil Biol Biochem 36(10):1545–1551

Obasi PN, Akudinobi BB (2020) Potential health risk and levels of heavy metals in water resources of lead–zinc mining communities of Abakaliki, southeast Nigeria. Appl Water Sci 10(7):1–23

Pattnaik BK, Equeenuddin SM (2016) Potentially toxic metal contamination and enzyme activities in soil around chromite mines at Sukinda ultramafic complex, India. J Geochem Explor 168:127–136

Pourrut B, Shahid M, Dumat C, Winterton P, Pinelli E (2011) Lead uptake, toxicity, and detoxification in plants. Rev Environ Contam Toxicol 213:113–136

Prabhakar A, Mishra S, Das AP (2019) Isolation and identification of lead (Pb) solubilizing bacteria from automobile waste and its potential for recovery of lead from end-of-life waste batteries. Geomicrobiol J 36(10):894–903. https://doi.org/10.1080/01490451.2019.1654044

Punia A, Singh SK (2021) Contamination of water resources in the mining region. In: Contamination of water. Academic Press, pp 3–17

Qi W, Ye T, Xiaolong Z, Xiaoying D, Jixing X, Renfang S, Xiaofang Z (2022) Pectin methylesterases enhance root cell wall phosphorus remobilization in rice. Rice Sci 29(2):179–188

Raj K, Das AP (2023) Lead pollution: Impact on environment and human health and approach for a sustainable solution. Environ Chem Ecotoxicol 5:79–85. https://doi.org/10.1016/J.ENCECO.2023.02.001

Rao J, Lv Z, Chen G, Peng F (2023) Hemicellulose: structure, chemical modification, and application. Prog Polym Sci:101675

Rao MA, Scelza R, Gianfreda L (2014) Soil enzymes. In: Enzymes in agricultural sciences. OMICS Group eBooks, Foster City, pp 10–43

Razavi BS, Zarebanadkouki M, Blagodatskaya E, Kuzyakov Y (2016) Rhizosphere shape of lentil and maize: spatial distribution of enzyme activities. Soil Biol Biochem 96:229–237

Rizwan MS, Imtiaz M, Zhu J, Yousaf B, Hussain M, Ali L, Hu H et al (2021) Immobilization of Pb and Cu by organic and inorganic amendments in contaminated soil. Geoderma 385:114803

Sahoo PP, Singh S, Rout PK, Mishra S, Das AP (2022) Microbial remediation of plastic pollutants generated from discarded and abandoned marine fishing nets. Biotechnol Genet Eng Rev. https://doi.org/10.1080/02648725.2022.2152629

Saravanan A, Kumar PS, Vo DVN, Jeevanantham S, Karishma S, Yaashikaa PR (2021) A review on catalytic-enzyme degradation of toxic environmental pollutants: microbial enzymes. J Hazard Mater 419:126451

Schmohl N, Horst WJ (2000) Cell wall pectin content modulates aluminium sensitivity of Zea mays (L.) cells grown in suspension culture. Plant Cell Environ 23(7):735–742

Sellami K, Couvert A, Nasrallah N, Maachi R, Abouseoud M, Amrane A (2022) Peroxidase enzymes as green catalysts for bioremediation and biotechnological applications: a review. Sci Total Environ 806:150500

Sharma B, Dangi AK, Shukla P (2018) Contemporary enzyme based technologies for bioremediation: a review. J Environ Manage 210:10–22

Sharma P, Dubey RS (2005) Lead toxicity in plants. Braz J Plant Physiol 17:35–52

Sharma R, Bhardwaj R, Handa N, Gautam V, Kohli SK, Bali S, Vig AP et al (2016) Responses of phytochelatins and metallothioneins in alleviation of heavy metal stress in plants: an overview. Plant Metal Interact:263–283

Sherman BT, Hao M, Qiu J, Jiao X, Baseler MW, Lane HC, Chang W et al (2022) DAVID: a web server for functional enrichment analysis and functional annotation of gene lists (2021 update). Nucleic Acids Res 50(W1):W216–W221

Shindhal T, Rakholiya P, Varjani S, Pandey A, Ngo HH, Guo W, Taherzadeh MJ et al (2021) A critical review on advances in the practices and perspectives for the treatment of dye industry wastewater. Bioengineered 12(1):70–87

Singh BR, Steinnes E (2020) Soil and water contamination by heavy metals. In: Soil processes and water quality. CRC Press, pp 233–271

Smyth GK (2005) Limma: linear models for microarray data. In: Gentleman R, Carey V, Dudoit S, Irizarry R, Huber W (eds) Bioinformatics and computational biology solutions using R and bioconductor

Speir TW, Ross DJ (2002) Hydrolytic enzyme activities to assess soil degradation and recovery. In: Burns RG (eds) Enzymes in the environments: activity, ecology and applications. RP Dick, pp 407–431

Syed K, Yadav JS (2012) P450 monooxygenases (P450ome) of the model white rot fungus *phanerochaetechrysosporium*. Crit Rev Microbiol 38(4):339–363

Temporiti MEE, Nicola L, Nielsen E, Tosi S (2022) Fungal enzymes involved in plastics biodegradation. Microorganisms 10(6):1180

Thakur M, Medintz IL, Walper SA (2019) Enzymatic bioremediation of organophosphate compounds—progress and remaining challenges. Front Bioeng Biotechnol 7:289

Triantafyllidou S, Edwards M (2012) Lead (Pb) in tap water and in blood: implications for lead exposure in the United States. Crit Rev Environ Sci Technol 42(13):1297–1352
Usman K, Abu-Dieyeh MH, Zouari N, Al-Ghouti MA (2020) Lead (Pb) bioaccumulation and antioxidative responses in Tetraenaqataranse. Sci Rep 10(1):17070
Verma S, Dubey RS (2003) Lead toxicity induces lipid peroxidation and alters the activities of antioxidant enzymes in growing rice plants. Plant Sci 164(4):645–655
Vroblesky DA (2005) Phytoremediation: transformation and control of contaminants. Ground Water 43(1):6–8
Winquist E (2014) The potential of ligninolytic fungi in bioremediation of contaminated soils
Yaashikaa PR, Devi MK, Kumar PS (2022) Advances in the application of immobilized enzyme for the remediation of hazardous pollutant: a review. Chemosphere 299:134390
Zainith S, Chowdhary P, Mani S, Mishra S (2020) Microbial ligninolytic enzymes and their role in bioremediation. In: Microorganisms for sustainable environment and health. Elsevier, pp 179–203
Zdunek A, Pieczywek PM, Cybulska J (2021) The primary, secondary, and structures of higher levels of pectin polysaccharides. Compr Rev Food Sci Food Saf 20(1):1101–1117
Zeng LS, Liao M, Chen CL, Huang CY (2007) Effects of lead contamination on soil enzymatic activities, microbial biomass, and rice physiological indices in soil–lead–rice (Oryza sativa L.) system. Ecotoxicol Environ Saf 67(1):67–74
Zhang J, Martinoia E, Lee Y (2018) Vacuolar transporters for cadmium and arsenic in plants and their applications in phytoremediation and crop development. Plant Cell Physiol 59(7):1317–1325
Zhang J, Qian Y, Chen Z, Amee M, Niu H, Du D, Sun J et al (2020) Lead-induced oxidative stress triggers root cell wall remodeling and increases lead absorption through esterification of cell wall polysaccharide. J Hazard Mater 385:121524
Zhao FJ, Tang Z, Song JJ, Huang XY, Wang P (2022) Toxic metals and metalloids: uptake, transport, detoxification, phytoremediation, and crop improvement for safer food. Mol Plant 15(1):27–44
Zhu XF, Wang ZW, Wan JX, Sun Y, Wu YR, Li GX, Zheng SJ et al (2015) Pectin enhances rice (*Oryza sativa*) root phosphorus remobilization. J Exp Bot 66(3):1017–1024
Zulfiqar U, Farooq M, Hussain S, Maqsood M, Hussain M, Ishfaq M, Anjum MZ et al (2019) Lead toxicity in plants: impacts and remediation. J Environ Manage 250:109557

Chapter 11
Green Technologies for Sustainable Agricultural Management

Debapam Saha, Florence Akangle Panme, and Abhishek Pradhan

Abstract Green technologies in sustainable agricultural waste management address a multifaceted exploration of environmentally sound methods, practices and innovations employed to reduce, recycle and manage agricultural wastes to minimize their adverse impact on human health and ecosystem. The chapter delves into distinct areas of agricultural waste management, offering a rich tapestry of green solutions. Green technologies in wastewater management, especially biological treatments involving different microorganisms and microflora sets real-world examples of bio-remediation. The innovative potential of phytoremediation in wastewater treatment is also commendable as substantiated through compelling case studies in the text. Moreover, the staggering scale of the global food waste problem have also been discussed, suggesting anaerobic digestion, composting, biomethane production, and the valorisation of industrial-grade food waste; as possible green solutions. Lastly some lights are also thrown on the emerging green technologies like 3D printing and its potential for agricultural waste reduction, design sustainability and optimize production. Thus, the chapter underpins the pressing need to embrace green technologies, that aim to achieve a balance between agricultural productivity and environmental stewardship along with a roadmap for the challenges that lie ahead.

D. Saha (✉) · A. Pradhan
Department of Agricultural and Food Engineering, IIT Kharagpur, Kharagpur, West Bengal 721302, India
e-mail: sdebapam97@kgpian.iitkgp.ac.in

F. A. Panme
Department of Agricultural Engineering, Assam University Silchar, Silchar, Assam 788011, India

A. Kuanar et al. (eds.), *Recent Advances in Bioremediation and Phytoremediation*, Environmental Science and Engineering,
https://doi.org/10.1007/978-3-031-77884-1_11

Graphical Abstract

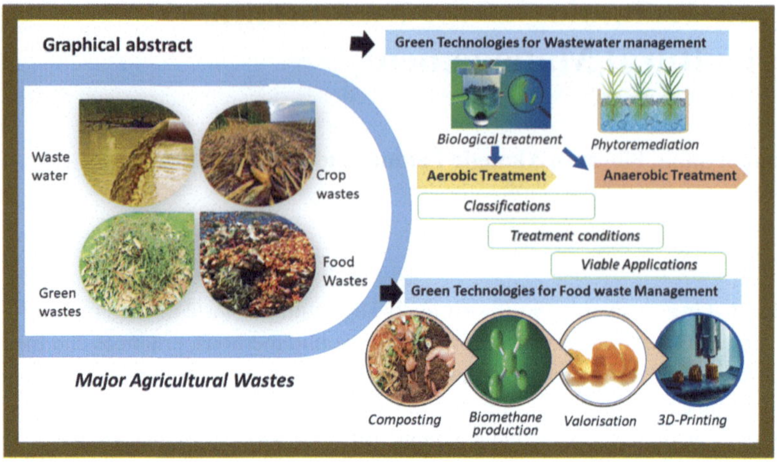

11.1 Introduction

Intensification of agricultural practices has led to a substantial increase in crop yields, meeting the rising demand for food among growing global populations. However, alongside this boost in agricultural production comes the generation of large quantities of waste known as "**agricultural wastes**" (Matharu et al. 2016). So, it is very important to note that the continued and sustained advancement of humanity relies on global cooperation, symbiosis, and circulation. To address these needs, United Nations Member States formulated the Sustainable Development Goals (SDGs) in 2015 (Kobayashi and Nakajima 2021). The SDGs serve as a blueprint for achieving a more sustainable future for all, encapsulated in "The 2030 Agenda for Sustainable Development." These goals, numbering 17 in total, aim to eradicate poverty and deprivation of natural resources worldwide. Interestingly, it has been observed that "Green technologies" for agri-waste management directly contribute to achieving the SDGs by promoting resource efficiency, environmental protection, poverty alleviation, and climate action. We define "**Green tech–or green technology**"–as an umbrella term that refers to the application of scientific knowledge and technological innovations aimed at mitigating human effects on the natural environment. It spans various fields of research such as energy, agriculture, hydrology, atmospheric science and material science.

Green technology is essential to wastewater management because it provides long-term fixes that reduce pollution to the environment, preserve resources, and enhance ecosystem health (Neoh et al. 2016). Through techniques like water recycling, nutrient recovery, and biogas generation, green technologies concentrate on recovering useful resources from wastewater, such as water, nutrients, and energy.

Green solutions eliminate the need for chemical fertilizers and energy-intensive treatment processes, limit nutrient runoff, and assist protect freshwater resources by recycling and reusing treated wastewater. Adoption of biological measures to treat wastewater will provide low-carbon alternatives that help reduce climate change by reducing energy consumption and greenhouse gas emissions. Examples of these technologies include solar-powered treatment systems, anaerobic digestion with biogas recovery, and wetland-based treatment systems. In the current scenario, industries and municipalities are under pressure to develop sustainable and environmentally friendly wastewater management techniques due to regulatory scrutiny and environmental rules governing wastewater discharge and treatment standards. Bioremediation provide sustainable, compliant solutions that improve corporate social responsibility, meet legal requirements, and help achieve long-term environmental sustainability objectives (Khalidi-Idrissi et al. 2023). All things considered, green technology is essential to making wastewater treatment a sustainable and eco-friendly procedure that safeguards ecosystems, saves resources, slows down global warming, improves public health, and enhances community well-being (Singh and Kumar 2017). We can create resilient and regenerative wastewater infrastructure that serves the requirements of both the current and future generations while preserving the health of the environment by embracing green technologies and practices.

Also, green technologies play a pivotal role in addressing the pressing issue of food waste management, offering innovative solutions globally. These technologies encompass a diverse range of approaches, including composting, anaerobic digestion, hydrothermal carbonization and biomethane production from food waste, all aimed at minimizing waste generation and promoting sustainable practices across the food supply chain (FSC). Additionally, emerging technologies like 3D printing are increasingly being explored as a means to repurpose food waste into valuable products, further advancing the circular economy model and contributing to a more sustainable future.

11.2 Green Technologies for Sustainable Wastewater Treatment

11.2.1 Biological Wastewater Treatment

In the quest for sustainable environmental management, the treatment of wastewater stands out as a critical challenge. As populations grow and industrial activities expand, the generation of wastewater increases, presenting a significant threat to water resources and ecosystems if left untreated. Fortunately, advancements in biological wastewater treatment (WWT) offer promising solutions that not only effectively remove contaminants but also promote environmental sustainability. Biological processes stand out as the predominant conventional techniques for wastewater treatment, offering unparalleled advantages that remain unchallenged. Renowned for their cost-effectiveness and extensive research-backed knowledge, these methods can be readily tailored to address specific local requirements. Biological treatment processes are often more cost-effective compared to chemical or physical methods (Mishra et al. 2022; Mishra and Das 2021, 2023a). Once established, they require minimal energy input and rely on naturally occurring microorganisms, reducing operational expenses over time. Unlike chemical treatments, which can introduce harmful substances into the environment, biological processes utilize natural microbial communities to degrade pollutants, resulting in environmentally friendly treatment outcomes. Biological treatment systems can be tailored to various types of pollutants, including organic matter, nutrients (such as phosphorus and nitrogen), and even some toxic compounds. This versatility makes them suitable for diverse wastewater streams, from municipal sewage to industrial effluents. Nevertheless, biological processes exhibit significant limitations in degrading hazardous and/or refractory organic contaminants, particularly as wastewater discharge regulations become increasingly stringent. Consequently, there is an exigent requirement for comprehensive and efficacious treatment approaches to meet evolving regulatory standards (Babuponnusami et al. 2023). Biological treatment employs microorganisms to biodegrade organic matter. The principle behind biological waste water treatment lies in harnessing the power of microorganisms such as bacteria and algae, to break down organic matter in the water into harmless byproducts such as carbon dioxide and water.

11.2.1.1 Importance of Microorganism in WWT

Among the various components and processes within WWTPs, microorganisms stand out as indispensable agents driving the treatment efficiency and effectiveness (Gonzalez-Martinez et al. 2018). These microscopic organisms, including bacteria, fungi, protozoa, and viruses, play multifaceted roles in the complex biological processes involved in wastewater treatment. Understanding their importance sheds light on the functioning and optimization of wastewater treatment systems. Organic pollutants, such as proteins, carbohydrates, and fats, serve as substrates for microbial metabolism (Mishra and Das 2023a). Bacteria, in particular, are adept at utilizing these organic compounds as energy sources through various biochemical pathways, including aerobic and anaerobic respiration, fermentation, and denitrification. By breaking down complex organic molecules into simpler, more stable compounds such as carbon dioxide, water, and microbial biomass, microorganisms facilitate the removal of organic pollutants from wastewater, reducing its biochemical oxygen demand (BOD) and preventing oxygen depletion in receiving water bodies. Nitrogen compounds, such as ammonia and nitrate, are converted by nitrifying bacteria through the processes of nitrification, which transforms ammonia into nitrate, and denitrification, which reduces nitrate to nitrogen gas. Similarly, phosphorus is precipitated and immobilized by phosphate-accumulating organisms (PAOs) under anaerobic conditions, leading to its removal from the wastewater stream. These microbial-mediated processes are essential for preventing eutrophication and maintaining water quality in receiving environments. Beyond organic and nutrient removal, microorganisms also contribute to the removal of pathogens and the disinfection of wastewater. While some pathogenic microorganisms are removed through physical processes such as sedimentation and filtration, others are inactivated or destroyed through biological mechanisms. Predatory microorganisms, including certain protozoa and bacteriophages, can feed on pathogenic bacteria, reducing their concentration in the treated effluent. Moreover, disinfection methods such as chlorination or ultraviolet (UV) irradiation target microbial pathogens, ensuring that the treated wastewater meets regulatory standards for safe discharge or reuse. Microbial communities adapt to changes in environmental conditions, such as fluctuations in temperature, pH, and substrate availability, through a process known as acclimation. This adaptive capacity enables WWTPs to maintain consistent treatment performance despite variations in influent characteristics. Moreover, microbial diversity within treatment systems contributes to redundancy and functional stability, reducing the risk of treatment failure in response to disturbances or shocks (Fig. 11.1).

Therefore, microorganisms are indispensable allies in the quest for effective and sustainable wastewater treatment. Their metabolic activities drive the degradation of organic matter, removal of nutrients, inactivation of pathogens, and maintenance of treatment system stability. Harnessing the potential of microbial communities through proper process design, operation, and management is essential for optimizing the performance of wastewater treatment plants and safeguarding human health and environmental quality.

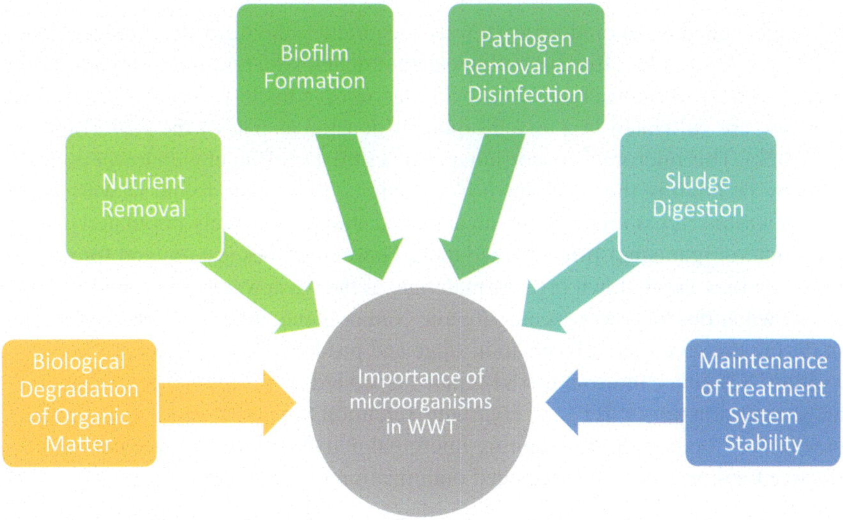

Fig. 11.1 Role of microorganisms in WWT

11.2.1.2 Methods of Biological Waste Water Treatment

Instead of using chemicals to clean wastewater, modern biological wastewater treatment uses microorganisms. Biological wastewater treatment (WWT) can stop the negative consequences of chemical wastewater treatment, like chemical buildup in water bodies and algae blooms. Numerous aerobic and anaerobic microbes can be employed in biological wastewater treatment to decrease the variety of contaminants present in water as well as the water's biochemical oxygen demand (Mishra and Das 2022). A myriad of microorganisms, encompassing yeasts, fungi, bacteria and algae, demonstrate the capacity to act as biologically active methylators, effectively modifying toxic species. Biological WWT can be performed either in the presence of oxygen (aerobically) or in the absence of oxygen (anaerobically).

11.2.1.3 Aerobic Waste Water Treatment

The process of treating wastewater via aerobic biological treatment uses aerobic microorganisms' activity to decompose organic contaminants in the presence of oxygen. Because of how well this technology works to remove organic debris and improve the quality of the water, it is frequently employed in both municipal and industrial wastewater treatment plants. The different types of biological WWT that uses aerobic principles are discussed briefly.

Activated Sludge Process (ASP)

One of the mainstays of modern wastewater treatment is the Activated Sludge Process (ASP), which provides effective biological treatment for the removal of organic pollutants. A variety of microorganisms, including aerobic bacteria, protozoa, and other microbes, are cultivated in a mixed liquid suspension as part of the ASP (Márquez et al. 2022). This microbial consortium produces treated effluent with lower pollutant levels by facilitating the breakdown of organic contaminants found in wastewater. In ASP, wastewater is injected into an activated sludge (a densely populated population of microorganisms)-filled aerated tank. These microbes use oxygen for respiration while they break down organic materials in the wastewater. Microbial flocs are created when microorganisms and organic contaminants interact. These flocs settle out during the secondary clarification stage and produce cleared effluent. Dissolved oxygen (DO) concentration, mixed liquid suspended solids (MLSS) concentration, sludge retention time (SRT) and Hydraulic retention time (HRT) are important variables in ASP operation. Maintaining microbial activity in the system and ensuring effective treatment performance need optimization of these parameters.

Aerobic Membrane Bioreactors

Membrane filtration and biological treatment methods are combined in an aerobic membrane bioreactor (AMBR) to efficiently treat wastewater (Rahman et al. 2023a, b). Wastewater is first put into a bioreactor vessel to start the procedure. Bacteria and other aerobic microorganisms disintegrate the organic contaminants in the wastewater inside this vessel. For these microbes to break down organic matter and create simpler, less toxic compounds, they need oxygen to do so. The wastewater in the bioreactor is continuously aerated to maintain the aerobic conditions required for microbial activity. The process of aerating wastewater usually involves the use of mechanical agitators or diffusers to add air. By promoting the development and metabolism of aerobic microorganisms, the oxygen added during aeration aids in the breakdown of organic contaminants. The mixture comprising suspended biomass and treated wastewater is directed towards the membrane filtration unit as the biological treatment advances (Holloway et al. 2015). The membrane functions as a physical barrier that lets clean water molecules pass through while trapping suspended solids, microbes, and colloidal particles (Yılmaz et al. 2020). It can be constructed of materials such polymeric hollow fibrers or flat sheets. Wastewater is forced to permeate through the membrane's pores as it travels through the membrane under pressure. A layer of cake is created on the membrane surface by the accumulation of bacteria and residual materials. Over time, membrane fouling brought on by this layer may lower filtration efficiency. To keep the membrane functioning and get rid of accumulated fouling elements, it must be cleaned or backwashed on a regular basis (Kumar et al. 2023). The clean water that passes through the membrane is gathered as treated wastewater. Usually, this effluent has low concentrations of pathogens, suspended particles, and organic pollutants, and meets high water quality criteria. The treated effluent may require extra disinfection or polishing procedures before to disposal or reuse, depending on the particular requirements. AMBR creates concentrated sludge with biomass and collected solids in addition to treated effluent. This sludge can be treated and dewatered further for disposal or reuse, or it can be recycled back into the bioreactor to sustain microbial activity.

Trickling Filter

A trickling filter is a reasonably easy and efficient biological wastewater treatment method that removes organic contaminants from wastewater by using microbial biofilms adhered to support media (Shareef 2022). In the trickling filter system, wastewater is dispersed equally across a bed of support media made of pebbles or plastics. Microorganisms adhere to the support media's surface and create a biofilm when the wastewater percolates down the filter bed. With the help of oxygen from the surrounding air or water, microorganisms in the biofilm metabolize and decompose organic contaminants found in wastewater through aerobic processes. At the bottom of the trickling filter bed is treated effluent, which has lower concentrations of pollutants and organic debris. Before being released into receiving water bodies or used for other purposes, the treated effluent may go through additional treatment or polishing.

Constructed Wetlands

In order to clean rainwater and wastewater, artificial wetlands are designed to resemble natural wetlands. Wetland plants, pebbles, and soil are found in shallow ponds or channels where wastewater or stormwater is introduced. Natural processes like sedimentation, filtration, adsorption, and biological degradation take place inside the artificial wetland (Parde et al. 2021). Through mechanisms like absorption, adsorption, and transformation, microbial communities in the soil and the root systems of wetland plants assist in the removal of pollutants. As the wastewater passes through the wetland, suspended solids settle out, organic matter breaks down, and nutrients are taken up by microorganisms and plants and then converted into new forms. After leaving the constructed wetland, treated wastewater is gathered and either discharged into receiving water bodies or used again. Though they don't require much upkeep, constructed wetlands may occasionally need to be monitored and managed, such as by managing the vegetation and removing silt. By combining the actions of plants, soil, and microbial populations to improve water quality and remove pollutants, constructed wetlands offer a sustainable and natural method of treating stormwater and wastewater (Al Hadidi 2021).

11.2.1.4 Anaerobic Treatment of Wastewater

Anaerobic Membrane Bioreactors

Renowned for its benefits like energy recovery and decreased sludge production, anaerobic digestion is a well-established method for treating organic wastewaters. Anaerobic membrane bioreactors (AnMBRs) are the result of the increased interest in integrating membranes within anaerobic digestion systems in recent years (Al-Asheh et al. 2021). Anaerobic membrane bioreactors (AnMBRs) offer enhanced effluent quality and may produce more biogas than traditional anaerobic systems because they integrate the advantages of membrane filtration and anaerobic digestion. Newer studies have concentrated on creating membranes with enhanced permeability, durability, and fouling resistance for AnMBR applications. Ceramic membranes, hydrophilic polymers, and mixed matrix membranes have been researched as potential novel membrane materials to address fouling and improve long-term performance. The logical order of the steps in a typical anaerobic treatment, which is a key step in AnMBR, is shown in Figure. The generation of volatile fatty acids (VFAs), which are short-chain components, is the outcome of the acidogenic phase and hydrolysis process (An et al. 2023). Complex chemical substances and molecules are broken down into smaller, water-soluble ones during this process. These complex molecules are made up of proteins, lipids, and carbohydrates. The process of facultative and/or obligatory anaerobic hydrolytic bacteria disintegrates these molecules intoamino acids, fatty acids, and monosaccharides. Acidogenesis, which is the second stage of a normal anaerobic treatment, is produced from the fatty acid, amino acid, and monosaccharide products of the first stage. These compounds undergo conversion into short-chain compounds such as alcohols, VFAs, aldehydes, carbon dioxide and hydrogen, as illustrated in Fig. 11.2. Acetogenesis is the third phase of the anaerobic process. The process of converting alcohol, hydrogen, and VFAs into acetate is carried out by the action of acetogenic bacteria. The acetogenesis process is primarily driven by two types of bacteria: those that produce hydrogen gas (H_2)

oxidize propionate, long-chain fatty acids and butyrate, to form acetate, while other bacteria, known as homoacetogens, produce acetate by consuming CO_2 and H_2. Homoacetogens can also use alcohol and sugar to produce acetate and H_2. This phase involves three species of bacterium consortia: Clostridium, Sporomusa, and Acetobacterium. Methanogenesis, the process that produces methane gas, is the final stage of anaerobic treatment. There are two distinct ways to create methane. Methanogens like Methanosaeta and Methanosarcina can produce acetate in the first instance. Aceticlastic methanogenesis produces about 70% of methane gas. However, hydrogenotrophic methanogens, which are produced when carbon dioxide and hydrogen react to make methane, can also produce methane. The availability of methane formers and substrate determines the rate at which ethane is produced.

Anaerobic Digesters for Wastewater Treatment

Anaerobic digestion is a potent, environmentally friendly wastewater treatment method that breaks down organic contaminants by using microorganisms' transformational properties. An essential part of contemporary wastewater treatment facilities, anaerobic digesters provide resource recovery, energy production, and pollution control (Bandgar et al. 2021). The organic matter in wastewater is broken down by anaerobic microorganisms, which are able to survive in conditions lacking in oxygen. Anaerobic digestion has enormous potential to spearhead the shift toward wastewater treatment methods that are more resilient, effective, and ecologically friendly as technology develops. Large tanks or digesters are usually used for anaerobic digestion, which takes place in a closed, oxygen-free environment. Organic-containing

Fig. 11.2 Fundamental process of anaerobic degradation

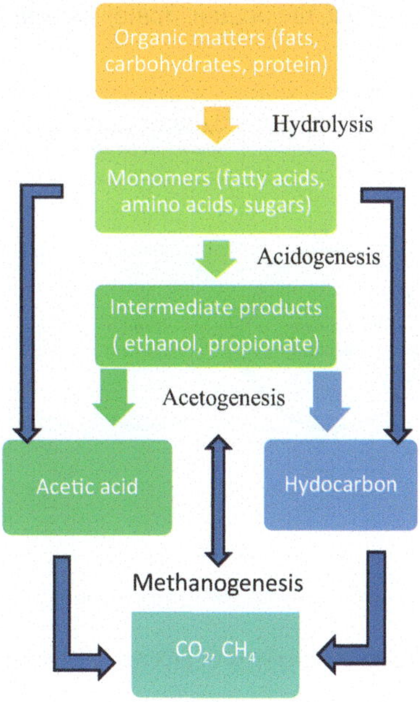

wastewater is gathered and sent to the anaerobic digester. The wastewater may go through pre-treatment procedures to get rid of big particles and change the pH to make the environment more conducive to anaerobic bacteria before it enters the digester. Anaerobic bacteria in the digester use a sequence of metabolic events to break down organic molecules in wastewater. Biogas, mostly methane (CH_4) and carbon dioxide (CO_2), as well as some digestate—a residue solid—are produced as a result of these interactions (Mir et al. 2016). Anaerobic digestion produces biogas, which is collected from the digester's top. It can be used as a renewable energy source for several purposes, such as heating, power generation, and transportation fuel. After additional processing, the digestate is cleaned to get rid of any persisting moisture and germs. In order to close nutrient loops and advance the ideas of the circular economy, efforts are being made to optimize the use of digestate, the left-over material from anaerobic digestion, as a nutrient-rich fertilizer or soil conditioner (Fig. 11.3).

Anaerobic sequencing Batch Reactors (Chaturvedi et al. 2022)

The use of anaerobic sequencing batch reactors (ASBRs) for wastewater treatment has grown in popularity because of their adaptability, effectiveness, and capacity to handle a wide range of influent characteristics. The typical steps of how anaerobic sequencing batch reactors (ASBRs) operate is given below (Dutta and Sarkar 2015):

1. Fill Phase: This stage involves adding influent wastewater to the ASBR. The influent is introduced into the reactor until a particular level is reached.
2. Anaerobic Reaction Phase: The influent is held in the reactor for a predetermined amount of time once it is full, which is referred to as the reaction phase. In this stage, the organic matter present in the wastewater undergoes decomposition into

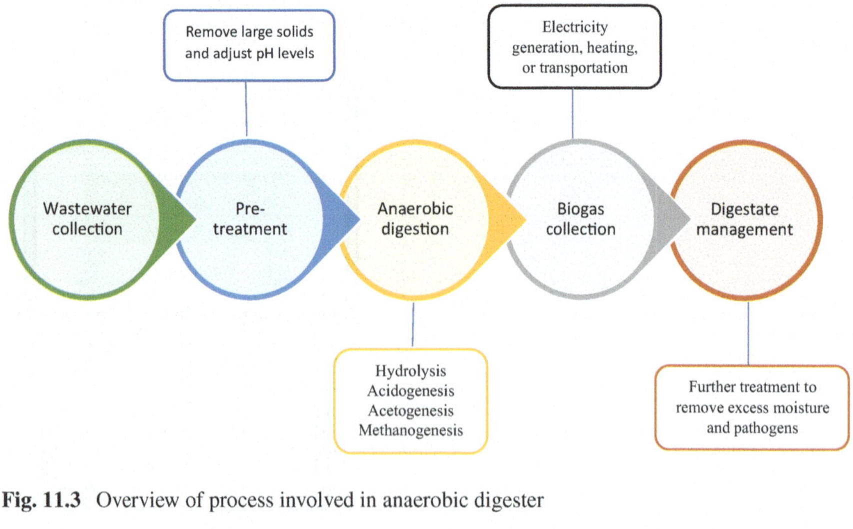

Fig. 11.3 Overview of process involved in anaerobic digester

simpler constituents, like carbon dioxide and methane by anaerobic microbes in the absence of oxygen.

3. Phase of Settling: Following the reaction stage, the reactor is given time to settle. During this stage, the biomass and suspended solids form a sludge blanket as they fall to the reactor's bottom due to gravity.
4. Decant Phase: The treated effluent is either decanted or removed from the reactor after the settling phase is completed. Having undergone extensive treatment, this effluent is comparatively clear and contains lower levels of organic contaminants.
5. Waste Sludge Removal: To preserve the intended sludge retention duration and avoid excessive buildup, any excess sludge or biomass that has accumulated at the reactor's bottom may be periodically removed.
6. Idle Phase: The reactor may go into an idle phase after the decant phase and, if necessary, the removal of sludge, during which it is either left empty or partially filled until the beginning of the following cycle. This stage guarantees the reactor is operating properly and enables preparation for the subsequent batch cycle.
7. Repeat: Next, the reactor is filled with influent wastewater for the subsequent batch cycle, and the entire cycle is repeated.

Up-Flow Anaerobic Sludge Blanket (UASB) Reactors

Wastewater passes through a floating bed of anaerobic sludge in UASB reactors using the anaerobic digestion principle. Organic contaminants in wastewater are broken down by microorganisms in the sludge blanket, producing biomass and biogas in the process. To aid with sludge retention and settling, UASB reactors normally consist of a rectangular or cylindrical tank with an inverted cone-shaped base. Within the UASB reactor, gas–liquid-solid separation is accomplished using a combination of settling and gas capture technologies (Kong and Shi 2022). Anaerobic digestion-produced gas bubbles rise through the sludge blanket and are caught by gas collection hoods or domes that are positioned above the reactor. The top of the reactor releases treated effluent, which is comparatively clear and free of organic contaminants, for additional processing or release. In order to guarantee consistent flow distribution and improve treatment efficiency, advanced reactor designs integrate influent distribution systems at the reactor bottom. Under Indian conditions, the UASB reactors are a reliable and effective technique for sewage pre-treatment (removing 51% of COD, 56% of BOD, and 54% of TSS). They may also produce biogas (yielding 0.20–0.40 m^3 kg^{-1} COD removed), which is a renewable energy source (Gaur et al. 2017). UASB reactors are useful for treating a variety of industrial and municipal wastewaters because of their many benefits, which include high organic matter removal efficiency, biogas production, and a small footprint (Mainardis et al. 2020). They have proven effective in a variety of sectors, including decentralized wastewater treatment systems, food processing, pharmaceuticals, paper and pulp, and chemical manufacture.

11.2.2 Phytoremediation in Wastewater Treatment

An additional biological technique for treating wastewater is phytoremediation. The term phytoremediation originates from the combining of two Latin words: cure and plant. The pollution from soil or water is taken up by the plant, bacteria originating from the plant, or the related microbiota. Either retention, removal, or deterioration in a manner similar to how it occurs naturally in an ecosystem are the methods used to accomplish remediation. A less expensive, environmentally responsible, and practically sustainable approach to pollutant removal is phytoremediation. In addition, the technique has an aesthetic requirement and little nutritional expense. But the phytoremediation approach is likewise not without its limits. The slow pace of the ecological cleanup process—which could take more than ten years—is the main drawback of the phytoremediation approach. Additionally, there can be a significant drop in phytoremediation, which typically occurs in the winter when plant development is slower. Vegetation damage could also result from weather-related issues, pest infestations, or plant diseases. Through their roots, wetlands plants like reeds, cattails, and water hyacinths take up nutrients and water from the wastewater. Additionally, these plants have the capacity to absorb and retain contaminants in the form of volatile compounds produced by transpiration, as well as in their leaves, stems and roots (Khan et al. 2022). Microbialcommunities interact with plant roots in the rhizosphere, the soil area affected by plant roots, and are essential to the breakdown and transformation of pollutants. Plant roots are home to microorganisms that can degrade organic pollutants and change them into less dangerous forms. Physical activities like filtration, sorption, and sedimentation are beneficial to phytoremediation systems in addition to microbial activity and plant uptake. Certain contaminants may stick to the surfaces of plants or soil particles, while particulate matter may drop out of the water column. Pollutants are eliminated or changed as the wastewater moves through the phytoremediation system, producing treated effluent with lower contaminant levels (Farraji 2014). For phytoremediation systems to function at their best, continuous monitoring and maintenance are necessary. To prevent the re-release of stored pollutants, this may involve harvesting and disposing of biomass on a regular basis, controlling vegetation growth, and keeping an eye on plant health. Using the inherent powers of plants and related microbial populations, phytoremediation treats wastewater and eliminates impurities, providing a sustainable and green approach to water purification (Gonzalez-Martinez et al. 2018).

11.3 Green Technologies for Sustainable Food Waste Management

In scientific literature, the terms "food loss (FL)" and "food waste (FW)" are commonly employed todescribe the disposal, degradation, or contamination of materials originally intended for human consumption. According to the Food and Agriculture Organization of the United Nations (FAO), FW or FL is defined as any alteration in the edibility, wholesomeness quality, or availability of edible substances that renders them unfit for human consumption (Girotto et al. 2015). The disposal of food stuff takes place throughout the entirety of the Food Supply Chain (FSC), encompassing all stages of waste management from collection to disposal. A thorough examination of an FSC system reveals the impact of waste material generation on every sector engaged in food production, distribution, and consumption. Thus, a large margin of agricultural produce gets wasted in the different stages of the FSC as depicted in the Fig. 11.4.

Interestingly, a staggering 1.3 billion tonnes of food getslost or wasted worldwideper year. Developed nations, with a population of 1.4 billion, discard approximately 670 million tonnes of consumable foods, while developing countries, with a population of 6.2 billion, waste around 630 million tonnes. Alarmingly, despite this abundance, between 690 to 829 million people suffer from hunger globally, while 3 billion individuals are unable to afford a nutritious diet (Roy et al. 2023). Globally, onanaveragevegetables and fruitscontributeda major proportion of 39%tothetotalFW (Fig. 11.4). Surprisingly, these inevitable food waste, particularly comprising peels, skins, and fruit twigs, contains a higher concentration of bioactive compounds compared to the generally consumed parts. Moreover, food processing waste is reported to be abundant in protein, lipids, and carbohydrates, showcasing potential for valorisation—transforming FW into value-added products such as animal feed, cosmetics, prebiotics and chemicals. This utilization of food waste can be possible only with the deployment of green technologies. These technologies will serve as a crucial measure in mitigating issues linked with food waste without any adverse environmental impact.

11.3.1 Sustainable Food Waste Reduction Strategies

Green technologies are poised to be deployed in the sustainable transformation of underutilized biomass, agricultural residues, FW or byproducts into valuable products suitable for diverse applications. A spectrum of conversion techniques is employed to generate value-added products catering to various needs. Figure 11.5 delineates a selection of commonly utilized green technologies, products, and potential applications in valorisation processes. Typically, food waste tends to have a

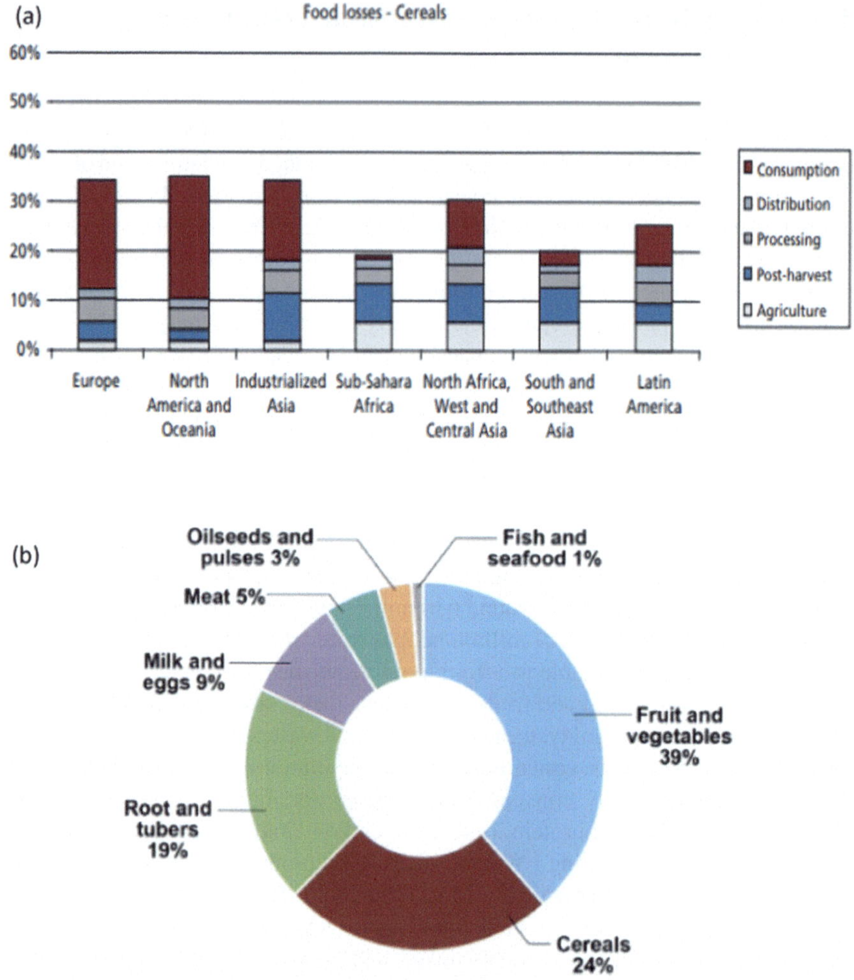

Fig. 11.4 **a** Bar graph showing trends in cereal losses (staple food loss) in different continents; **b** Share of various global food wastes (Roy et al. 2023)

high moisture content, which can offer advantageous effects in certain conversion processes like anaerobic digestion (AD), pyrolysis, hydrothermal carbonization (HTC), hydrothermal liquefaction and fermentation (solid-state/enzymatic) (Carmona-Cabello et al. 2018). However, in some cases, it may require comminution or pretreatment methods such as pulsed electric field, microwave, or ultrasoundtreatment to enhance the conversion process and extract greater yields of value-added products (Arshad et al. 2021). Thus, the product yields have been found to rely on the valorisation techniques (stage 1 and/or stage 2), types of FW and pretreatments,

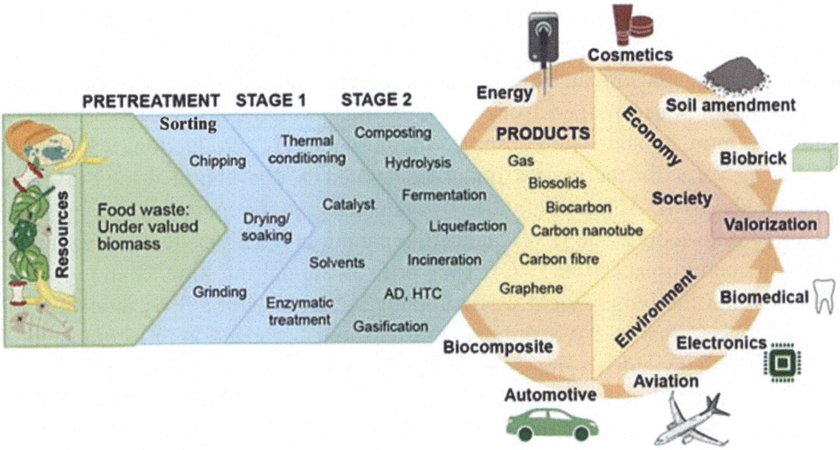

Fig. 11.5 Schematic of the various green conversion technologies for food waste management (Roy et al. 2023) *(*HTC: hydrothermal carbonization; AD: anaerobic digestion)*

as shown in Fig. 11.5. The following subsection addresses some of the widely applicable sustainable green technologies for the management of agricultural and food wastes and byproducts.

11.3.1.1 Green Technologies for Biochemical Conversion

Landfilling

FW constitutes a significant portion of Municipal Solid Waste (MSW) and is commonly directed to landfills. Depending on the jurisdiction, MSW can contain anywhere from 25 to 65% food waste (Roy et al. 2023). However, food waste containing both lignin and lignocellulosic materials, are resistant to breakdown in absence of oxygen (Kibler et al. 2018). This resistance hinders the bio-accessibility of these materials for biogas production, resulting in lower methane yields without any pretreatments. Therefore, landfilling food waste is not at all the optimal method for sustainable management.

Anaerobic Digestion (AD)

Anaerobic digestion is a biological process in which microorganisms break down organic matter in the absence of oxygen. This process occurs naturally in environments such as wetlands, marshes, and the digestive systems of animals, and it is also harnessed in various engineered systems for waste treatment and biogas production. Anaerobic digestion (AD) of organic waste stands out as an attractive and effective green technology for valorisation, enabling both energy and resource recovery. As noted by Banks et al. (2018), food waste (FW) is composed of proteins, fats, carbohydrates, and essential nutrients, showcasing considerable potential for biogas production. Resultantly, AD or industrial utilization within a biorefinery presents the foremost opportunities for energy and resource recovery, making it the primary choice for food waste management. Through thermal treatment, the CH_4 yield from mildly treated kitchen waste ranged from 342–398 mL/g volatile solid (VS). However, significant improvements were observed in hydrogen (H2) production, reaching up to 113 mL/g VS, as the thermal treatment accelerated the conversion of soluble sugars (Banks et al. 2018). Abbas et al. (2020) conducted a study on anaerobic digestion, examining both food waste (FW) and cattle manure in a carbon steel reactor under mesophilic conditions (30 ± 1 °C). They utilized animal intestine as an inoculum and found that the biogas yield was optimally achieved with a 30% inoculum for both materials. Notably, FW exhibited a faster biogas production rate compared to cattle manure and yielded a greater quantity of biogas overall. Moreover, the addition of trace metals and biochar enhanced methane and biogas yield from FW digestion. For instance, with a biochar dose of 2.0 and 5.0 g/L, biogas yield increased from 11.2 ± 6.5% to 27.3 ± 9.5% and methane yield from 8.3 ± 6.8% to 33.2 ± 2.8%, respectively (Akturk and Demirer 2020). Furthermore, co-digestion has also been proved to be very effective in reducing toxicity potential, stabilizing the digestate, and enhancing the methane yield through synergistic effects of different substrates. The nutrient-rich digestate in particular optimal proportions resulting from this process holds promise for soil amendment purposes.

Fermentation

Fermentation of FW is a bio-process that involves the use of microorganisms to biologically transform the carbohydrates and proteins present in them into simpler compounds through an anerobic metabolism. This process is characterized by the creation of organic acids, gases, flavor compounds, nutritional components (such as gamma-aminobutyric acid and bioactive peptides), and the decrease of antinutritional components (phytates or saponins) (Castro-Alba et al. 2019; Rollán et al. 2019). While fermentation stands as a promising technology for the valorisation of FW, the high cost associated with the enzymes essential for FW fermentation poses a primary barrier to its widespread adoption. However, combining onsite enzyme production with fermentation plants could serve as a viable solution to mitigate this challenge (Roy et al. 2023). The yield of products in fermentation of FW also relies on the methods of inoculation. A two-step approach has been implemented by Carmona-Cabello et al. (2021) for the production of biodiesel using potato peels. Initially, a fungal strain (*Aspergillus awamori*) was utilized in the solid-state fermentation of potato peels to generate protease and glucoamylase. The crude hydrolysates obtained from restaurant food waste were subjected to fermentation in a shake flask using yeast (*Rhodosporidiumtoruloides*) to produce microbial oil, subsequently undergoing transesterification to yield biodiesel. Additionally, hydrolysate derived from solid-state fermentation of FW canbe employed in biohydrogen production. While utilizing solid-state fermentation to produce flavoring compounds from food waste (FW) can help reduce FW itself, the associated processing costs are significant. Moreover, scaling up this process incurs a high capital expenditure.

Composting

Composting is a natural process that involves the decomposition of organic materials into a nutrient-rich soil amendment known as compost. During composting, microorganisms such as bacteria, fungi, and insects break down organic matter like food scraps, yard waste, leaves, and other biodegradable materials. As the organic materials decompose, heat is generated, which helps to accelerate the process and kill off pathogens and weed seeds (Bong et al. 2017). Thisdark, crumbly, earthy-smelling substanceimproves soil structure, fertility, and moisture retention in gardens, landscaping, agriculture, and horticulture. It helps to replenish soil nutrients, promote healthy plant growth, and reduce the need for chemical fertilizers and pesticides. Composts derived from agricultural practices commonly utilize animal excreta, while urban composting processes predominantly involve the inclusion of yard waste and food scraps. Interestingly, it has been found that compost derived from food waste (FW) is richer in nutrients compared to vermicompost, making it a promising option for soil amendment (Kelley et al. 2020). Green technologies like composting divert organic waste from landfills, reduce greenhouse gas emissions, and contribute to environmental conservation.

11.3.1.2 Green Technologies for Thermochemical Conversion

Pyrolysis/Co-Pyrolysis

Pyrolysis is a thermal decomposition process that occurs in the absence of oxygen. It involves heating organic materials, such as biomass or waste, to high temperatures (typically between 300–800 °C) in a controlled environment. During pyrolysis, the organic material undergoes chemical and physical changes, breaking down into three main products: biochar (a carbon-rich solid), bio-oil (a liquid mixture of organic compounds), and syngas (a gaseous mixture primarily composed of hydrogen, carbon monoxide, and methane) (Kim et al. 2020). Pyrolysis represents an emerging technology utilized to transform food waste (FW) into innovative products, whether with or without the aid of catalysts. Low-temperature pyrolysis of FW compost demonstrates superior carbon fixation compared to pyrolysis at higher temperatures (400–500 °C) (Roy et al. 2023). Additionally, catalytic pyrolysis of FW demonstrates improvements in both product yield and quality.

Co-pyrolysis, on the other hand, is a process in which two or more different types of organic materials are pyrolyzed simultaneously. This can include various combinations of biomass, waste materials, or even plastics. By combining different feedstocks, co-pyrolysis aims to optimize product yields, enhance the quality of end products, or tailor specific properties to meet desired specifications (Lee et al. 2021). It has been seen thatFW can also co-pyrolyzedwith herbal medicine byproducts, wood bark etc. led to a greater yield of hydrogen (H2) on account of synergistic interactions (Lee et al. 2021; Park et al. 2021).

Hydrothermal Carbonization (HTC) and Liquefaction (HTL)

HTC is a promising green technology that is being widely utilized now-a-days for FW (that are classified to be high moisture stuffs) rather than simply incinerate them conventionally. This is because, incineration of FW is characterized by high energy consumption and the emission of hazardous gases such as heavy metals and dioxins, particularly for combustion below 1000 °C (Roy et al. 2023). Hydrothermal carbonization (HTC) is a thermochemical conversion process that transforms biomass or organic waste materials into a carbon-rich solid called hydrochar or hydrothermal carbon. This process occurs under temperatures (typically between 180 to 250 degrees Celsius) and elevated pressure in the presence of water (Yao et al. 2020). Hydrothermal liquefaction (HTL) is another green technology recognized as a potential path for the recovery of nutrient and energy from FW. Liquefaction of food waste (FW) under elevatedpressure (35.3 MPa) and temperature conditions (600 °C) resulted in the transfer of 68% and 50% of P and N, respectively, to the process water as reported by Motavaf and Savage (2021). This process also generates biocrude with a high energy content (30 wt%, 36.5 MJ/kg); however, the efficiency of the process varies depending on the specific process parameters.

11.3.2 Case Studies on Successful Food Waste Management

Further exploration is required to unlock the chemical and material potential inherent in food waste, paving the way for their conversion into novel and innovative products. FW streams are rich in functionalized molecules, including biopolymers and essential nutrients, which can be recovered, concentrated, and converted into valuable precursors (Arancon et al. 2013). These precursors have applications across various industries, including chemical, cosmetic, pharmaceutical, and food. Efforts have been made to valorise these waste streams by transforming them into new bio-based products such as enzymes, oligosaccharides, pectin, biosurfactants, as well as high-value chemical intermediates like lactic acid and furfural (Ong et al. 2018). Some of these successful efforts (case-studies) across different countries globally have been enumerated in the Table 11.1.

11.4 3D Printing: An Emerging Green Technology for Sustainable Waste Management

In today's world, where sustainability is a growing concern, innovative technologies are being explored to address environmental challenges. One such technology that has gained significant attention is 3D printing, also known as additive manufacturing. This technique constructs three-dimensional items by stacking material in a precise and controlled way, providing extensive design flexibility and customisation options

Table 11.1 Case Studies of different countries on successful adoption of green technologies for food waste management

Case Studies	Problems/challenges	Product and yield	Adopted Green Technologies (treatment conditions)	Reference(s)
Case Study-1 **(Implementing co-composting of GW and FW: Colombia)**	Cellulose and lignin prolong the processing time for composting of GW	Compost	• UF + PF (co-substrates for GW) + PR (for quality improvement) • TSCto accelerate the degradation (composition of mixture taken: 48% GW + 21% UF + 18% PF + 13% SW) 	Hernández-Gómez et al. (2021)
Case study-II **(Valorisation of fruit residue: India)**	Among agro-produce, fruit and vegetable peels and residuesare the most common waste streams in India	Orange peel oil *(0.68–2.22% yield),* biosurfactant: 1.8 g/L yield)	• Supercritical CO_2 extraction of high-quality orange-peel oil with improved shelf life • Solid-state fermentation of orange peel residue with *Bacillus licheniformi* resulted in efficient biosurfactant  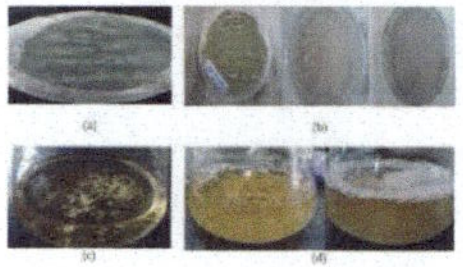	Nautiyal and Tiwari (2011) Kumar et al. (2016)
Case study-III **(Bioconversion of rice wastes: Thailand)**	Traditional rice harvesting practicesrenders huge amount of rice straw on the field (40.8 million tonnes yearly)	Power generation (electricity) (*20,000 kW/ year of electricity produced*)	The "A.T. Biopower" project, overseen by the EPPO of the Ministry of Energy in Thailand, (agricultural waste into energy). Specifically, it harnesses rice husk-pelletized by briquetting technology 	Ong et al. (2018)

(continued)

Table 11.1 (continued)

Case Studies	Problems/challenges	Product and yield	Adopted Green Technologies (treatment conditions)		Reference(s)
Case Study-IV **(Palm oil waste to energy production: Indonesia)**	Most food wastes in Indonesia are subjected to either landfill disposal or combustion—not sustainable	Biogas *(56%-83%. Energy yield)*	Torrefaction (a mild pyrolysis process under an anaerobic inert atmosphere @ 200–300 °C) of palm waste oil to generate biogas		Fauzianto (2014)
Case study-V **(Bioconversion of pineapple waste: Malaysia)**	Mostly incineration and landfills are carried out as waste management techniques that quite cost-intensive in terms of ROI and sustainability	Vanillic acid and vanillin (flavouring agent): *5 g ofpineapple cannery waste yields 0.163 g/L vanillic acid and 0.141 g/L vanillin* Bioethanol *(5.4 g/L of liquid pineapple waste)*	Pineapple waste (Ferulic acid) ↓ *(Aspergillus niger)* ↓ Vanillic acid *(P. cinnabarinus)* vanillin • Bioethanol production from liquid pineapple waste by immobilized *Saccharomyces cerevisiae*	 	Lun et al. (2014) Zain et al. (2012)

* *GW: Green Waste; FW: Food Waste; UF: Unprocessed food; PF: Processed food; PR: Phosphate rock; TSC: Two-stage composting; SW: Sawdust; EPPO:Energy Policy and Planning Office; FSC: Food supply chain; F&V: Fruits and vegetables; ROI: Return on Investment*

while reducing material waste common in traditional subtractive manufacturing methods (Mansi et al. 2023). Unlike traditional manufacturing methods, which often contribute to waste generation and resource depletion, 3D printing offers a promising solution for sustainable waste management (Olawumi et al. 2023). By fundamentally changing the way products are designed, manufactured, and distributed, 3D printing has the potential to minimize material waste, utilize recycled materials, and reduce carbon emissions associated with transportation.

3D printing has various environmental impacts, mainly by reducing material use through additive methods and enabling the use of recycled materials in production, which helps decrease dependence on new resources and reduce waste. Furthermore, the localised and on-demand capabilities of 3D printing show potential for improving energy efficiency by avoiding the logistical inefficiencies linked to centralised production and distribution networks, therefore reducing related carbon emissions (Gebler et al. 2014). Additionally, the inherent adaptability of 3D printing allows for design optimisation, resulting in the production of lightweight, resource-efficient components that help reduce environmental impacts across product life cycles. Comprehensive evaluation requires taking into account additional factors such as energy consumption of printing technologies, sustainability metrics of materials used, and strategies for managing end-of-life and material recycling (Peng et al. 2018). This highlights the intricate complexities related to its environmental impact. Advancements in technology and materials are driving the development of 3D printing, making it a key tool for sustainable waste management and environmentally friendly industrial practices.

11.4.1 Sustainable 3D Printing Materials

Sustainable 3D printing materials prioritise environmental responsibility while maintaining performance and innovation in additive manufacturing. The materials are obtained from renewable or recycled sources, decreasing the dependence on limited resources and lowering the carbon footprint linked to manufacturing (Zheng and Suh 2019). These materials consist of several substrates including bioplastics, recovered plastics, bio-based resins, and sustainable metals, providing a range of characteristics designed to fulfil particular application needs (Patel et al. 2023). Their formulation often highlights biodegradability and recyclability, to ensure that end-of-life factors are in line with circularity principles and waste minimization. Sustainable 3D printing materials promote environmentally conscious production methods and encourage innovation and technical progress in the additive manufacturing sector.

11.4.1.1 Recycled Plastics and Bio-Based Materials

Sustainable 3D printing utilises recycled plastics and bio-based materials, each offering environmental benefits and specific purposes. Post-consumer or post-industrial plastics are transformed into filament or resin for 3D printing after undergoing processing (Hassan et al. 2023). Redirecting plastic waste away from landfills decreases pollution and the production of new plastic. Various recycled polymers such as PET, ABS, PLA, and nylon are used in 3D printing (Mikula et al. 2021). Conversely, bio-based products come from renewable plant-based polymers, starches, or cellulose. Maize, sugarcane, and algae-derived polymers serve as sustainable substitutes for petroleum-based polymers. Bio-based goods decompose or compost, resulting in environmentally beneficial end-of-life disposal. PLA, a biopolymer derived from maize starch or sugarcane, is widely used in 3D printing due to its biodegradability, flexibility, and ease of use (Bhagia et al. 2021). Recycled plastics and bio-based materials lessen the environmental footprint of 3D printing by promoting resource preservation, waste reduction, and sustainable material sourcing. The widespread use of additive manufacturing demonstrates the commitment of the industry to sustainable and circular, eco-friendly production.

11.4.1.2 Application and Feasibility

Recycled plastics are used in the automobile industry to help achieve lightweighting and fuel economy objectives, while bio-based materials are used for interior components and trim (Rahman et al. 2023a, b). In medicine, bio-compatible materials sourced sustainably are used to create customised implants and prostheses, in line with personalised treatment and environmental responsibility (Ngo et al. 2018). Sustainable 3D printing materials in consumer goods and packaging provide prospects for personalised, environmentally friendly products and just-in-time production, meeting changing customer expectations for sustainability (Haleem et al. 2023). Examples include 3D-printed shoes made from recycled materials, eco-friendly packaging solutions, and sustainable architectural models built using bio-based filaments. The wide-ranging usefulness and increasing practicality of sustainable 3D printing materials highlight their capacity to stimulate innovation, productivity, and environmental responsibility in several economic sectors.

11.4.2 Waste Reduction in 3D Printing

11.4.2.1 Material Efficiency: Zero Waste Processing

Waste reduction in 3D printing, particularly within the framework of zero waste processing, is facilitated by several key strategies. Firstly, the additive manufacturing principle inherently minimizes waste by depositing material only where necessary, unlike traditional subtractive methods that generate significant scrap (Jayawardane et al. 2023). Advanced slicing software and optimization algorithms further optimize material usage by generating minimal support structures, reducing post-processing waste. Closed-loop systems enable the recycling and reuse of excess or failed prints, promoting a circular approach to material usage. For example, filament makers like the Filabot can recycle plastic waste into usable filament for future prints (Woern et al. 2018). Design for additive manufacturing (DFAM) principles also play a crucial role by eliminating unnecessary material and reducing waste generation from the outset (Egan 2023). Additionally, real-time monitoring and control systems identify and address issues during printing, minimizing errors and reducing the likelihood of print failures. Finally, initiatives for material recycling and repurposing, such as the use of biodegradable materials like PLA derived from renewable sources like cornstarch or sugarcane, further contribute to waste reduction and environmental sustainability (Reichert et al. 2020). Overall, these strategies collectively drive 3D printing towards achieving zero waste processing, fostering a more sustainable manufacturing ecosystem.

11.4.2.2 Design for Sustainability (Intricate Designs with Optimal Material Feed)

3D printing as design for sustainability embodies a transformative approach to manufacturing, emphasizing the creation of intricate designs with optimal material feed to minimize waste and environmental impact (Hegab et al. 2023). One notable example is the use of generative design software in aerospace engineering. Engineers utilize these tools to create lightweight and structurally efficient components, such as aircraft brackets or engine parts, by iteratively generating and evaluating designs based on specified performance criteria (Klahn et al. 2014). By leveraging 3D printing technology, these intricate designs can be translated into physical objects with precise material deposition, ensuring that only the necessary amount of material is used to achieve structural integrity. This results in lightweight components that contribute to fuel efficiency and emissions reduction in aircraft, while also minimizing material waste compared to traditional manufacturing methods. Additionally, the ability to incorporate recycled or bio-based materials further enhances the sustainability profile of these 3D printed parts. Overall, by optimizing material feed and embracing intricate designs, 3D printing enables a more sustainable approach to manufacturing across various industries, including aerospace, automotive, and consumer goods (Fidan et al. 2023).

11.5 Conclusion

In conclusion, the utilization of green technologies holds immense promise for addressing the challenges of agro-food waste management in a sustainable manner. By harnessing innovative approaches such as biological wastewater treatment and phytoremediation, we can effectively mitigate water pollution while promoting environmental health. Similarly, sustainable strategies for food waste management, including composting, anaerobic digestion, and biomethane production, offer viable solutions for reducing waste and promoting resource recovery. Moreover, the emerging technology of 3D printing presents exciting opportunities for waste reduction and material efficiency. By prioritizing sustainability in our waste management practices, we can contribute significantly to achieving the Sustainable Development Goals (SDGs) and creating a more environmentally conscious future.

References

Abbas Y, Jamil F, Rafiq S, Ghauri M, Khurram MS, Aslam M, Bokhari A, Faisal A, Rashid U, Yun S (2020) Valorization of solid waste biomass by inoculation for the enhanced yield of biogas. Clean Technol Environ Policy 22:513–522

Akturk AS, Demirer GN (2020) Improved food waste stabilization and valorization by anaerobic digestion through supplementation of conductive materials and trace elements. Sustainability 12(12):5222

Al-Asheh S, Bagheri M, Aidan A (2021) Membrane bioreactor for wastewater treatment: a review. Case Studies Chem Environ Eng 4:100109

Al Hadidi L (2021) Constructed wetlands a comprehensive review

An Z, Zhu J, Zhang M, Zhou Y, Su X, Lin H, Sun F (2023) Anaerobic membrane bioreactor for the treatment of high-strength waste/wastewater: a critical review and update. Chem Eng J, 144322

Arancon RAD, Lin CSK, Chan KM, Kwan TH, Luque R (2013) Advances on waste valorization: new horizons for a more sustainable society. Energy Sci Eng 1(2):53–71

Arshad RN, Abdul-Malek Z, Roobab U, Qureshi MI, Khan N, Ahmad MH, Liu ZW, Aadil RM (2021) Effective valorization of food wastes and by-products through pulsed electric field: a systematic review. J Food Process Eng 44(3):e13629

Babuponnusami A, Sinha S, Ashokan H, Paul MV, Hariharan SP, Arun J, Gopinath K, Le QH, Pugazhendhi A (2023) Advanced oxidation process (AOP) combined biological process for wastewater treatment: a review on advancements, feasibility and practicability of combined techniques. Environ Res, 116944

Bandgar P, Jain S, Panwar N (2021) Current status, opportunities and challenges in anaerobic digestion in Indian context: an overview. Bioresource Technol Reports 16:100830

Banks C, Heaven S, Zhang Y, Baier U (2018) Food waste digestion: anaerobic digestion of food waste for a circular economy

Bhagia S, Bornani K, Agrawal R, Satlewal A, Ďurkovič J, Lagaňa R, Bhagia M, Yoo CG, Zhao X, Kunc V (2021) Critical review of FDM 3D printing of PLA biocomposites filled with biomass resources, characterization, biodegradability, upcycling and opportunities for biorefineries. Appl Mater Today 24:101078

Bong CP-C, Goh RKY, Lim J-S, Ho WS, Lee C-T, Hashim H, Mansor NNA, Ho CS, Ramli AR, Takeshi F (2017) Towards low carbon society in Iskandar Malaysia: implementation and feasibility of community organic waste composting. J Environ Manage 203:679–687

Carmona-Cabello M, García I, Papadaki A, Tsouko E, Koutinas A, Dorado M (2021) Biodiesel production using microbial lipids derived from food waste discarded by catering services. Biores Technol 323:124597

Carmona-Cabello M, Garcia IL, Leiva-Candia D, Dorado MP (2018) Valorization of food waste based on its composition through the concept of biorefinery. Current Opinion Green Sustain Chem 14:67–79

Castro-Alba V, Lazarte CE, Perez-Rea D, Carlsson NG, Almgren A, Bergenståhl B, Granfeldt Y (2019) Fermentation of pseudocereals quinoa, canihua, and amaranth to improve mineral accessibility through degradation of phytate. J Sci Food Agric 99(11):5239–5248

Chaturvedi A, Singh RS, Jaiswal RP (2022) Hybrid bioreactor in combination with ozone-based technologies for industrial wastewater treatment. In: Development in wastewater treatment research and processes, pp 629–650. Elsevier

Dutta A, Sarkar S (2015) Sequencing batch reactor for wastewater treatment: recent advances. Curr Pollut Reports 1:177–190

Egan PF (2023) Design for additive manufacturing: recent innovations and future directions. Designs 7(4):83

Farraji H (2014) Wastewater treatment by phytoremediation methods. Wastewater engineering: advanced wastewater treatment systems, 194

Fauzianto R (2014) Implementation of bioenergy from palm oil waste in Indonesia. J Sustai Dev Studies 5(1)

Fidan I, Huseynov O, Ali MA, Alkunte S, Rajeshirke M, Gupta A, Hasanov S, Tantawi K, Yasa E, Yilmaz O (2023) Recent inventions in additive manufacturing: Holistic review. Inventions 8(4):103

Gaur RZ, Khan AA, Lew B, Diamantis V, Kazmi AA (2017) Performance of full-scale UASB reactors treating low or medium strength municipal wastewater. Environ Process 4:137–146

Gebler M, Uiterkamp AJS, Visser C (2014) A global sustainability perspective on 3D printing technologies. Energy Policy 74:158–167

Girotto F, Alibardi L, Cossu R (2015) Food waste generation and industrial uses: a review. Waste Manage 45:32–41

Gonzalez-Martinez A, Sihvonen M, Muñoz-Palazon B, Rodriguez-Sanchez A, Mikola A, Vahala R (2018) Microbial ecology of full-scale wastewater treatment systems in the polar arctic circle: Archaea. Bacteria and Fungi. Scientific Reports 8(1):2208

Haleem A, Javaid M, Rab S, Singh RP, Suman R, Kumar L (2023) Significant potential and materials used in additive manufacturing technologies towards sustainability. Sustain Oper Computers 4:172–182

Hassan M, Mohanty AK, Misra M (2023) 3D printing in upcycling plastic and biomass waste to sustainable polymer blends and composites: a review. Mater Design, 112558

Hegab H, Khanna N, Monib N, Salem A (2023) Design for sustainable additive manufacturing: a review. Sustain Mater Technol, e00576

Hernández-Gómez A, Calderón A, Medina C, Sanchez-Torres V, Oviedo-Ocaña ER (2021) Implementation of strategies to optimize the co-composting of green waste and food waste in developing countries. a case study: Colombia. Environ Sci Pollut Res 28:24321–24327

Holloway RW, Achilli A, Cath TY (2015) The osmotic membrane bioreactor: a critical review. Environ Sci: Water Res Technol 1(5):581–605

Jayawardane H, Davies IJ, Gamage J, John M, Biswas WK (2023) Sustainability perspectives–a review of additive and subtractive manufacturing. Sustain Manuf Serv Econ, 100015

Kelley A, Wilkie AC, Maltais-Landry G (2020) Food-based composts provide more soil fertility benefits than cow manure-based composts in sandy soils. Agriculture 10(3):69

Khalidi-Idrissi A, Madinzi A, Anouzla A, Pala A, Mouhir L, Kadmi Y, Souabi S (2023) Recent advances in the biological treatment of wastewater rich in emerging pollutants produced by pharmaceutical industrial discharges. Int J Environ Sci Technol, 1–22

Khan AU, Khan AN, Waris A, Ilyas M, Zamel D (2022) Phytoremediation of pollutants from wastewater: a concise review. Open Life Sci 17(1):488–496

Kibler KM, Reinhart D, Hawkins C, Motlagh AM, Wright J (2018) Food waste and the food-energy-water nexus: a review of food waste management alternatives. Waste Manage 74:52–62

Kim S, Lee Y, Lin K-YA, Hong E, Kwon EE, Lee J (2020) The valorization of food waste via pyrolysis. J Clean Prod 259:120816

Klahn C, Leutenecker B, Meboldt M (2014) Design for additive manufacturing–supporting the substitution of components in series products. Procedia Cirp 21:138–143

Kobayashi T, Nakajima L (2021) Sustainable development goals for advanced materials provided by industrial wastes and biomass sources. Curr Opin Green Sustain Chem 28:100439

Kong L, Shi X (2022) Effect of antibiotic mixtures on the characteristics of soluble microbial products and microbial communities in upflow anaerobic sludge blanket. Chemosphere 292:133531

Kumar AP, Janardhan A, Viswanath B, Monika K, Jung J-Y, Narasimha G (2016) Evaluation of orange peel for biosurfactant production by Bacillus licheniformis and their ability to degrade naphthalene and crude oil. 3 Biotech 6:1–10

Lee Y, Kim S, Lee J (2021) Co-pyrolysis for the valorization of food waste and oriental herbal medicine byproduct. J Anal Appl Pyrol 154:105016

Lun OK, Wai TB, Ling LS (2014) Pineapple cannery waste as a potential substrate for microbial biotranformation to produce vanillic acid and vanillin. Int Food Res J 21(3):953

Mainardis M, Buttazzoni M, Goi D (2020) Up-flow anaerobic sludge blanket (UASB) technology for energy recovery: a review on state-of-the-art and recent technological advances. Bioengineering 7(2):43

Mansi, Kumar H, Singholi A, Moona G (2023) Additive manufacturing: a brief introduction. In: Handbook of metrology and applications, pp 1–23. Springer

Márquez P, Gutiérrez M, Toledo M, Alhama J, Michán C, Martín M (2022) Activated sludge process versus rotating biological contactors in WWTPs: evaluating the influence of operation and sludge bacterial content on their odor impact. Process Saf Environ Prot 160:775–785

Matharu AS, de Melo EM, Houghton JA (2016) Opportunity for high value-added chemicals from food supply chain wastes. Biores Technol 215:123–130

Mikula K, Skrzypczak D, Izydorczyk G, Warchoł J, Moustakas K, Chojnacka K, Witek-Krowiak A (2021) 3D printing filament as a second life of waste plastics—a review. Environ Sci Pollut Res 28:12321–12333

Mir MA, Hussain A, Verma C (2016) Design considerations and operational performance of anaerobic digester: a review. Cogent Eng 3(1):1181696

Mishra S, Das AP (2021) Current treatment technologies for removal of microplastic and microfiber pollutants from wastewater. Wastewater TREATMENT: CUTTING-EDGE MOLECULAR TOOLS, TECHNIQUES AND APPLIED ASPECTS 237–251. https://doi.org/10.1016/B978-0-12-821881-5.00011-8

Mishra S, Das AP (2022) Treatment of the wastewater polluted with synthetic microfiber released from washing machine. In: Das BB, Hettiarachchi H, Sahu PK, Nanda S (Eds), Lecture notes in civil engineering (Vol 207, pp 109–117). Springer Science and Business Media Deutschland GmbH. https://doi.org/10.1007/978-981-16-7509-6_9/COVER

Mishra S, Das AP (2023) Microbial remediation of synthetic microfiber contaminated wastewater. Microbial technologies in industrial wastewater treatment 337–350. https://doi.org/10.1007/978-981-99-2435-6_17/COVER

Mishra S, Das AP (2023b) Microbial remediation of synthetic microfiber contaminated wastewater. In: Shah MP (Ed), Microbial technologies in industrial wastewater treatment, pp 337–350. Springer, Singapore. https://doi.org/10.1007/978-981-99-2435-6_17

Mishra S, Singh RP, Rout PK, Das AP (2022) Membrane bioreactor (MBR) as an advanced wastewater treatment technology for removal of synthetic microplastics. Development in wastewater treatment research and processes 45–60. https://doi.org/10.1016/B978-0-323-85583-9.00022-3

Motavaf B, Savage PE (2021) Effect of process variables on food waste valorization via hydrothermal liquefaction. ACS ES&T Eng 1(3):363–374

Nautiyal OH, Tiwari KK (2011) Supercritical carbon dioxide extraction of Indian orange peel oil and hydro distillation comparison on their compositions. Sci Technol 1:29–33

Neoh CH, Noor ZZ, Mutamim NSA, Lim CK (2016) Green technology in wastewater treatment technologies: Integration of membrane bioreactor with various wastewater treatment systems. Chem Eng J 283:582–594

Ngo TD, Kashani A, Imbalzano G, Nguyen KT, Hui D (2018) Additive manufacturing (3D printing): a review of materials, methods, applications and challenges. Compos B Eng 143:172–196

Olawumi MA, Oladapo BI, Ikumapayi OM, Akinyoola JO (2023) Waste to wonder to explore possibilities with recycled materials in 3D printing. Sci Total Environ 905:167109

Ong KL, Kaur G, Pensupa N, Uisan K, Lin CSK (2018) Trends in food waste valorization for the production of chemicals, materials and fuels: case study South and Southeast Asia. Biores Technol 248:100–112

Parde D, Patwa A, Shukla A, Vijay R, Killedar DJ, Kumar R (2021) A review of constructed wetland on type, treatment and technology of wastewater. Environ Technol Innov 21:101261

Park C, Lee N, Kim J, Lee J (2021) Co-pyrolysis of food waste and wood bark to produce hydrogen with minimizing pollutant emissions. Environ Pollut 270:116045

Patel KS, Shah DB, Joshi SJ, Patel KM (2023) Developments in 3D printing of carbon fiber reinforced polymer containing recycled plastic waste: a review. Cleaner Mater, 100207

Peng T, Kellens K, Tang R, Chen C, Chen G (2018) Sustainability of additive manufacturing: an overview on its energy demand and environmental impact. Addit Manuf 21:694–704

Rahman MZ, Rahman M, Mahbub T, Ashiquzzaman M, Sagadevan S, Hoque ME (2023) Advanced biopolymers for automobile and aviation engineering applications. J Polym Res 30(3):106

Rahman TU, Roy H, Islam MR, Tahmid M, Fariha A, Mazumder A, Tasnim N, Pervez MN, Cai Y, Naddeo V (2023) The advancement in membrane bioreactor (MBR) technology toward sustainable industrial wastewater management. Membranes 13(2):181

Reichert CL, Bugnicourt E, Coltelli M-B, Cinelli P, Lazzeri A, Canesi I, Braca F, Martínez BM, Alonso R, Agostinis L (2020) Bio-based packaging: materials, modifications, industrial applications and sustainability. Polymers 12(7):1558

Rollán GC, Gerez CL, LeBlanc JG (2019) Lactic fermentation as a strategy to improve the nutritional and functional values of pseudocereals. Front Nutr 6:98

Roy P, Mohanty AK, Dick P, Misra M (2023) A review on the challenges and choices for food waste valorization: Environmental and economic impacts. ACS Environ Au 3(2):58–75

Shareef N (2022) Wastewater treatment applications by using trickling filter as a low energy consumption solution: case studies worldwide. In: Cost-efficient wastewater treatment technologies: engineered systems, pp 83–99. Springer

Singh R, Kumar S (2017) Green technologies and environmental sustainability. Springer

Woern AL, McCaslin JR, Pringle AM, Pearce JM (2018) RepRapable recyclebot: open source 3-D printable extruder for converting plastic to 3-D printing filament. HardwareX 4:e00026

Yao G, Guo Y, Le Y, Jin B, He R, Zhong H, Jin F (2020) Energy valorization of food waste: rapid conversion of typical polysaccharide components to formate. Ind Eng Chem Res 59(39):17069–17075

Yılmaz F, Otuzaltı M, Perendeci N, Karatay M, Ünşar EK, Ateş M, Akin R, Yıldız O, Orhon D (2020) Potential of aerobic membrane bioreactor for recycling and reuse of domestic wastewater for irrigation. Environ Earth Sci 79:1–11

Zain NM, Shen N, Suhaimi M, Hasan N, Aziman S (2012. Potential use of liquid pinapple waste for bioethanol production by immobilized bakers 'yeast. J Teknol 59(1)

Zheng J, Suh S (2019) Strategies to reduce the global carbon footprint of plastics. Nat Clim Chang 9(5):374–378

Chapter 12
An Overview of Phytoremediation for the Treatment of Wastewater

Sweta Sinha

Abstract Phytoremediation has emerged as a cost-effective and eco-friendly approach for treating wastewater using plants. This review article provides an overview of the mechanisms, applications, and advantages of phytoremediation for wastewater treatment. The key processes involved in phytoremediation, including phytoextraction, rhizofiltration, phytostabilization, phytodegradation, and phytovolatilization, are discussed. The ability of plants to remove or degrade organic pollutants, heavy metals, and excess nutrients from wastewater through their roots, leaves, and associated microorganisms is highlighted. Key plants used in phytoremediation, like typha, eichhornia, pistia, lemna, azolla, and wolffia, are evaluated. The applications of phytoremediation for the treatment of sewage, industrial effluents, landfill leachate, and agricultural runoff are reviewed. The advantages of phytoremediation in terms of cost, aesthetics, sustainability, and environmental friendliness compared to conventional remediation methods are analyzed. Finally, the challenges and future directions for wider adoption of phytoremediation are explored. The review suggests that phytoremediation provides an effective green solution for wastewater treatment with tremendous potential for future applications.

Keywords Phytoremediation · Wastewater treatment · Physico-chemical methods · Water Pollutants · Heavy metals · Sustainability · Green technology

12.1 Introduction

The release of wastewater from domestic, industrial, and agricultural sources containing harmful contaminants is a major cause of water pollution worldwide (Sinha et al. 2024; Sinha 2024a). Removing toxic pollutants and excess nutrients before discharging wastewater into natural water bodies is crucial to mitigating environmental damage. Conventional wastewater treatment methods involving physical,

S. Sinha (✉)
Department of Chemistry, Amity University Jharkhand, Ranchi, Jharkhand, India
e-mail: sweta.sinha2203@gmail.com

A. Kuanar et al. (eds.), *Recent Advances in Bioremediation and Phytoremediation*, Environmental Science and Engineering,
https://doi.org/10.1007/978-3-031-77884-1_12

chemical, and biological processes can effectively remediate contaminants but have high operational costs and intensive infrastructure requirements (Deep et al. 2025; Sinha 2024b; Tilley et al. 2014). Phytoremediation has emerged as a cost-effective and eco-friendly alternative that uses plants and their associated microbiomes to remove, degrade, or stabilize pollutants from contaminated environments (Ali et al. 2013). This review article provides a comprehensive overview of the mechanisms, applications, and advantages of phytoremediation for the treatment of different types of wastewater.

A wide spectrum of organic and inorganic pollutants can be extracted, transformed, or immobilized from contaminated matrices using phytoremediation, which makes use of physiological, biochemical, and microbiological processes, that occur naturally or are produced in plants (Megharaj et al. 2011). The key mechanisms involved include phytoextraction, rhizofiltration, phytostabilization, phytodegradation, and phytovolatilization. Plants can uptake, concentrate, and detoxify pollutants through their roots, stems, leaves, or aquatic surface features (Ghosh and Singh 2005). The plant rhizosphere also harbors symbiotic microorganisms that aid in the breakdown of complex contaminants (Kuiper et al. 2004). A range of aquatic plants, terrestrial grasses, trees, and algae have shown potential for phytoremediation applications (Prasad et al. 2010).

There is growing research on the use of phytotechnologies to treat sewage, landfill leachate, industrial effluents containing heavy metals, dyes, petroleum hydrocarbons, and agricultural runoff rich in nutrients like nitrogen and phosphorus (Vymazal 2011). Field studies have shown that phytoremediation for ecological wastewater treatment can be successfully implemented on a pilot scale in wetlands, ponds, and irrigation channels (Ibrahim et al. 2013).

12.2 Challenges of Conventional Physico-Chemical Approaches and Advantages of Phytoremediation for Wastewater Treatment

Traditional wastewater treatment, though successful in removing pollutants, presents several challenges. The methods heavily rely on physico-chemical processes, like coagulation, Flocculation (Sinha 2018, 2022; Sinha et al. 2018) and membrane filtration, which have a significant energy footprint, contributing to concerns about their environmental impact and long-term sustainability (Sinha 2023). Additionally, these methods often involve the use of chemicals, raising concerns about potential harm to the environment and the associated costs of acquiring, utilizing, and safely disposing of them. Furthermore, while these approaches effectively address various contaminants, they may not be universally applicable to all pollutant types. This can necessitate additional or combined treatment steps, increasing the complexity and potentially impacting the cost-effectiveness of the overall process.

Phytoremediation offers a compelling alternative to traditional wastewater treatment, merging economic and environmental benefits (Dushenkov et al. 1995; Shilev et al. 2007). Compared to conventional methods, phytoremediation systems reduce costs by relying on natural processes, which minimizes the need for expensive chemicals and energy-intensive machinery. This leads to lower long-term financial burdens. Phytoremediation systems also promote sustainability by harnessing the power of nature. They utilize solar energy for plant growth and avoid the use of harmful chemicals, thus minimizing environmental impact and contributing to a cleaner ecological footprint (Pazos et al. 2011). Additionally, phytoremediation systems enhance biodiversity by creating green spaces and fostering beneficial microorganisms (Mishra and Das 2023).

Phytoremediation systems can treat diverse pollutants because plants employ various mechanisms to absorb, degrade, or immobilize a wide range of contaminants, including heavy metals, organic matter, and nutrients. This versatility makes phytoremediation a robust solution for treating wastewater from diverse sources. Finally, phytoremediation systems enhance aesthetics by being designed as attractive green spaces that blend seamlessly with the landscape (Kumar et al. 2023). This visual advantage makes them more socially acceptable, particularly in urban areas where aesthetics play a crucial role.

By combining cost-effectiveness, environmental sustainability, broad-spectrum treatment capabilities, and visual appeal, phytoremediation emerges as a promising and advantageous solution for future wastewater management.

12.3 Mechanisms of Phytoremediation

Phytoremediation utilizes a range of complex natural and induced mechanisms in plants to extract, degrade, metabolize, or immobilize contaminants from polluted environments. The key phytoremediation processes include (Table 12.1).

12.3.1 Phytoextraction

Phytoextraction, also known as phytoaccumulation, involves the uptake and translocation of pollutants like metals or organics from soil or water into the plant roots and above-ground shoots and leaves (Salt et al. 1995). Pollutants get concentrated in the biomass, which can be harvested and disposed of. This process is suitable for inorganic contaminants and some organic compounds.

Table 12.1 Various techniques for phytoremediation

Technique	Application	Contaminant	Mechanism	Description	Accumulation	References
Phytoextraction	Soil, water	Metals, radionuclides, inorganics	Uptake and translocation to shoots	Uptake from soil or water into roots and accumulation in above ground biomass	Shoots, leaves	(Salt et al. 1995)
Rhizofiltration	Water	Metals, radionuclides	Adsorption on roots	Adsorption, precipitation and concentration of contaminants from water on root surface	Roots	(Dushenkov et al. 1995; Shilev et al. 2007)
Phytostabilization ǀ	Soil	Metals, organics	Immobilization in soil	Reducing bioavailability and mobility using plant roots	Roots, Roots,	(Mench et al. 2010)
Phytodegradation	Soil, water	Organics, xenobiotics	Metabolic degradation	Breakdown of complex organics inside plant or via phytostimulated microbial degradation	Leaves, roots, rhizosphere	(Pazos et al. 2011; Gerhardt et al. 2009)
Phytovolatilization	Soil	Volatile metals, organics	Release into atmosphere	Uptake, translocation and volatilization of contaminants into atmosphere via leaves	Leaves, atmosphere	(Meagher 2000; Van Aken 2009)

12.3.2 *Rhizofiltration*

Rhizofiltration uses plant roots to absorb, concentrate, and precipitate toxic metals and radionuclides from polluted aqueous sources (Dushenkov et al. 1995). Root biosorption, bioaccumulation, and precipitation mechanisms are involved. It is an efficient approach to treating contaminated groundwater, surface water, and industrial effluents.

12.3.3 *Phytostabilization*

Phytostabilization refers to the immobilization of pollutants like metals or radionuclides in the soil using plant roots, thereby reducing their mobility and bioavailability (Mench et al. 2010). Metal-tolerant plant species can sequester metals in their roots through adsorption, accumulation, or precipitation.

12.3.4 *Phytodegradation*

Phytodegradation is the process by which complex organic pollutants, such as chlorinated solvents, organic pesticides, petroleum hydrocarbons, and polyaromatic hydrocarbons (PAHs), break down inside plants or by microbial action connected to the plant (Schnoor et al. 1995). Plant-derived microbial activity and plant-enzyme interactions promote the degradation process.

12.3.5 *Phytovolatilization*

Phytovolatilization entails the uptake, translocation, and release of volatile contaminants like mercury or selenium by plants into the atmosphere, thereby removing them from the environment (Meagher 2000). Genetically engineered plants can enhance the volatilization and removal of pollutants.

Plants provide a large surface area and ecological niche to support the phytoaccumulation and degradation of a wide range of contaminants using these mechanisms. However, success depends on selecting suitable plant species and optimizing plant–microbe interactions.

12.4 Applications of Phytoremediation

Phytoremediation has been studied for the treatment of various types of wastewater, including.

12.4.1 Municipal and Domestic Sewage

Numerous studies have shown that sewage from homes and municipalities can be polished by using artificial wetlands that are planted with aquatic plants such as lemna, typha, cyperus, eichhornia, and pistia (Vymazal 2005). These systems can remove significant amounts of organic matter, suspended solids, pathogens, nutrients, and heavy metals from sewage through phytoremediation mechanisms (Mishra et al. 2022a, b).

12.4.2 Industrial Wastewater

Utilizing plants like water hyacinth, duckweed, and azolla, phytoremediation has been used to treat effluents from dyeing, tanning, textile, paper, dairy, distillery, and various other industries that contain heavy metals, dyes, oils, phenols, and chlorinated compounds, among other pollutants (Kanwal et al. 2015). Plant-based bioreactors and constructed wetlands are being created.

12.4.3 Landfill Leachate

Studies have assessed the phytoremediation capacity of trees like willow and poplar, as well as aquatic plants like lemna, azolla, and algae, for landfill leachate treatment, with modest results for organics, ammonia, and heavy metal removal (Juwarkar et al. 2010). There are integrated biosystems.

12.4.4 Agricultural Wastewater

Phytoremediation of agricultural runoff containing excess nitrates, phosphates, pesticides, and antibiotics using floating aquatic plants in drainage canals and ponds to prevent eutrophication of water bodies has gained interest (Sharma et al. 2019). Plant-microbe systems have shown good nutrient removal efficiency.

Phytoremediation systems have been successfully implemented on a pilot scale to treat different wastewaters. Research is ongoing to develop and optimize integrated plant-microbial phytotechnologies for field-level remediation of diverse wastewater contaminants.

12.5 Pollutants Removal from Wastewater by Plant Species Used in Phytoremediation

A variety of aquatic and terrestrial plant species have shown potential for the removal of different pollutants from wastewaters through phytoremediation mechanisms. Water hyacinth (*Eichhornia crassipes*), duckweed (*Lemna minor*), and cattail (*Typha latifolia*) are examples of floating and emergent aquatic plants that can absorb nutrients, metals, and organic contaminants from wastewater via their roots and shoots. Submerged aquatic plants like elodea, hydrilla, and valisneria take up dissolved pollutants like nitrogen, phosphorus, and heavy metals efficiently (Das et al. 2011). Trees like poplar, willow, and aquatic legumes like azolla and wolffia exhibit good pollutant removal capabilities and can be used in constructed wetlands. Grasses, like vetiver, have fibrous root systems that immobilize metals and degrade organic xenobiotics (Das et al. 2011; Khatun et al. 2023). Algae also have excellent biosorption capacity for dyes, antibiotics, hydrocarbons, and radionuclides. Pollutant removal efficiency is dependent on several factors, like plant biomass, growth rate, tolerance mechanisms, and bioconcentration factors, and needs to be optimized through phytotechnology designs. Table 12.2 summarizes the role of plants in removing specific pollutants like dyes, inorganic elements, heavy metals, and pesticides from wastewater through phytoremediation.

12.6 Overcoming Limitations and Future Outlook

Phytoremediation provides an environmentally sustainable wastewater treatment solution but needs systematic research to address limitations for successful field-level implementation. Microbiome engineering and transgenic plants with enhanced remediation capabilities can optimize plant-microbe interactions, speeding up and improving phytoremediation. Developing integrated ecological models that incorporate phytotechnologies along with physical, chemical, or biological treatment methods could help counteract phytoprocesses' slowness (Vymazal 2011). Compact floating treatment wetlands, in-situ rhizofiltration, and phytoreactor networks can be designed to minimize space requirements and allow industrial and urban settlement applications (Shardendu et al. 2019). Greenhouse culturing of plant inoculum in conjunction with climate-resilient green infrastructure can reduce performance variability (Brisson and Chazarenc 2009). The field-testing of diverse aquatic plants,

Table 12.2 The role of different plants in removing specific pollutants from wastewater through phytoremediation

Pollutant	Plant species	Mechanism	References
Heavy metals	Water lettuce, elodea	Phytoextraction, rhizofiltration, chelation	(Mishra and Tripathi 2008; Dixit and Puranik 2018)
Dyes	Water hyacinth, aquatic moss	Phytoextraction, enzymatic degradation	(Davies et al. 2005; Jadhav et al. 2013)
Pharmaceuticals	Water spinach, water celery	Uptake, accumulation, degradation	(Chen et al. 2019; Yuan et al. 2019)
Inorganic nutrients (N, P)	Duckweed, azolla, water fern	Assimilation and accumulation in biomass	(Cheng et al. 2002; Choudhury and Kennedy 2005)
Pathogens	Canna, chrysanthemum	Inactivation, antagonism	(Rezania et al. 2015; Chen et al. 2019)
Pesticides	Parrot feather, bulrush	Metabolic degradation, dehalogenation, nitroreduction	(Harms 1975; Mirgain et al. 1993)
Petroleum hydrocarbons	Iris, rushes	Rhizodegradation, phytovolatilization, accumulation	(Huang et al. 2004; Aprill and Sims 1990)

terrestrial grasses, and trees in constructed wetlands and bioreactors using different configurations needs to be expanded across wider geographies (Uggetti et al. 2012). This will help build a portfolio of phytotechnological solutions tailored to region-specific wastewater pollution profiles. Policy incentives, public engagement initiatives, and decentralized community-scale systems can promote adoption by municipalities and industries for sustainable wastewater management (Pandey et al. 2019). Advances across these interdisciplinary domains can enable phytoremediation to become a mainstream technology to treat sewage, industrial effluents, landfill leachate, and agricultural runoff in an eco-friendly and economically viable manner (Mishra et al. 2022a, b).

12.7 Conclusions

Phytoremediation has emerged as a promising green technology for the treatment of various wastewaters using natural physiological and biological processes in plants and their associated microbiomes. This review article provided a comprehensive overview of the mechanisms, applications, advantages, limitations, and future prospects of phytoremediation for sustainable wastewater treatment. Phytotechnologies such as phytoextraction, rhizofiltration, phytostabilization, phytodegradation, and phytovolatilization use complex processes in aquatic plants, terrestrial grasses, and trees to absorb, degrade, metabolize, or immobilize a wide range of organic and inorganic pollutants. Constructed wetlands, ponds, and bioreactors based on phytoremediation have been successfully used to polish sewage, treat industrial effluents, landfill leachate, and treat agricultural runoff. Compared to conventional remediation methods, phytoremediation provides tremendous cost, aesthetics, sustainability, and environmental benefits. However, relatively slow remediation rates, large space requirements, and variations in plant growth cycles pose challenges to field-level implementation. Further advances are required in plant selection, microbiome engineering, integrated system designs, field testing, and policy incentives to facilitate widespread adoption of phytotechnologies. With systematic interdisciplinary research efforts, phytoremediation can become an integral component of ecological wastewater treatment frameworks. This can provide a decentralized, low-cost, and environmentally friendly solution for managing growing wastewater pollution worldwide. The integration of phytoremediation with existing physical, chemical, and biological treatment infrastructures can usher in a sustainable transformation in wastewater treatment practices in both urban and rural areas.

References

Ali H, Khan E, Sajad MA (2013) Phytoremediation of heavy metals—concepts and applications. Chemosphere 91(7):869–881

Aprill W, Sims RC (1990) Evaluation of the use of prairie grasses for stimulating polycyclic aromatic hydrocarbon treatment in soil. Chemosphere 20(1–2):253–265

Brisson J, Chazarenc F (2009) Maximizing pollutant removal in constructed wetlands: Should we pay more attention to Macrophyte species selection? Sci Total Environ 407(13):3923–3930

Chen Z, Chen B, Song J (2019) Aquatic Macrophytes in phytoremediation of pollutants in surface water and wastewater. In Emerging contaminants in the environment, Elsevier, pp 677–699

Cheng J, Landesman L, Bergmann BA, Classen JJ, Howard JW, Yamamoto YT (2002) Nutrient removal from swine lagoon liquid by Lemna minor. Trans ASAE 45(4):1003

Choudhury ATMA, Kennedy IR (2005) Prospects and potentials for systems of biological nitrogen removal from wastewaters using water fern Azolla. Environ Technol 26(5):545–556

Das AP, Sukla LB, Pradhan N, Nayak S (2011) Manganese biomining: A review. Biores Technol 102(16):7381–7387. https://doi.org/10.1016/J.BIORTECH.2011.05.018

Davies LC, Carias CC, Novais JM, Martins-Dias S (2005) Phytoremediation of textile effluents containing azo dye by using Phragmites australis in a vertical flow intermittent feeding constructed wetland. Ecol Eng 25(5):594–605

Deep A, Bhardwaj A, Sinha S, Singh R, Srivastava M (2025) Endophytic-induced nanotechnology: An eco-friendly and potential approach toward sustainable future. In Nanoparticles Synthesis by Soil Microbes, Academic Press, pp 119–139

Dixit R, Puranik S (2018) Bioremediation of Nickel by Elodea canadensis. Int J Curr Microbiol App Sci 7(5):3261–3269

Dushenkov V, Kumar PBAN, Motto H, Raskin I (1995) Rhizofiltration: the use of plants to remove heavy metals from aqueous streams. Environ Sci Technol 29(5):1239–1245

Erakhrumen AA (2007) Phytoremediation: an environmentally sound technology for pollution prevention, control and remediation in developing countries. Educ Res Rev 2(7):151–156

Gerhardt KE, Huang XD, Glick BR, Greenberg BM (2009) Phytoremediation and Rhizoremediation of organic soil contaminants: Potential and challenges. Plant Sci 176(1):20–30

Ghosh M, Singh SP (2005) A review on phytoremediation of heavy metals and utilization of its byproducts. Appl Ecol Environ Res 3(1):1–18

Harms H (1975) Degradation of 1, 2-dibromoethane (ethylene dibromide) by soil bacteria. Appl Microbiol 29(6):803–808

Huang XD, El-Alawi Y, Penrose DM, Glick BR, Greenberg BM (2004) A multi-process phytoremediation system for removal of polycyclic aromatic hydrocarbons from contaminated soils. Environ Pollut 130(3):465–476

Ibrahim S, Jamil NN, Hainida KI (2013) Application of constructed wetland for wastewater treatment: A review. In: Proceedings of the 4th International Conference on Engineering Technology and Technoprenership (ICE2T). IEEE, pp 316–322

Jadhav JP, Hasan S (2013) Bioremediation potential of cyanobacterial mats for decolorization of textile dyes. Bioremediat J 17(1):43–51

Joshua WD, Akpor OB, Afolabi OO (2020) Phytoremediation: A nexus of plant-soil-microbes for sustainable environmental clean-up. Science of the Total Environment 138956

Juwarkar AA, Singh SK, Mudhoo A (2010) A comprehensive overview of elements in bioremediation. Rev Environ Sci Bio/Technol 9(3):215–288

Kanwal P, Kiran B, Pal A, Arora S (2015) Phytoremediation: A sustainable tool for environmental cleaning. Phytoremediation: Management of Environmental Contaminants, Vol 2, pp 1–27

Khatun R, Singh S, Dubey NK, Prasad Das A (2023) A review on marine-based phytochemicals and their application in biomedical research. Recent Frontiers of Phytochemicals: Applications in Food, Pharmacy, Cosmetics, and Biotechnology, pp 383–395. https://doi.org/10.1016/B978-0-443-19143-5.00013-X

Kuiper I, Lagendijk EL, Bloemberg GV, Lugtenberg BJ (2004) Rhizoremediation: A beneficial plant-microbe interaction. Mol Plant Microbe Interact 17(1):6–15

Kumar M, Pandit S, Patel V, Khanna N, Nag M, Lahiri D, Ray RR, Das A, Das D (2023) Maximization of energy recovery from starch processing wastewater by thermophilic dark fermentation coupled with microbial fuel cell technology. Geomicrobiol J. https://doi.org/10.1080/01490451.2023.2209555

Meagher RB (2000) Phytoremediation of toxic elemental and organic pollutants. Curr Opin Plant Biol 3(2):153–162

Meers E (2020) The importance of plant-microbe interactions in phytoremediation of heavy metal-contaminated soils. In Impacts of Pollution on Human Health and Environment. Springer, Singapore, pp 37–53

Megharaj M, Ramakrishnan B, Venkateswarlu K, Sethunathan N, Naidu R (2011) Bioremediation approaches for organic pollutants: A critical perspective. Environ Int 37(8):1362–1375

Memon AR, Schröder P (eds) (2009) Impacts of climate change, eutrophication and chemical pollution on water quality. Nova Publishers

Mench M, Lepp N, Bert V, Schwitzguébel JP, Gawronski SW, Schröder P, Vangronsveld J (2010) Successes and limitations of phytotechnologies at field scale: Outcomes, assessment and outlook from COST Action 859. J Soils Sediments 10(6):1039–1070

Mirgain I, Green GA, Monteil H (1993) Degradation of 2, 4-dichlorophenoxyacetic acid by hairy root cultures of Daucus carota L. Plant Soil 151(1):43–50

Mishra S, Das AP (2023) Microbial remediation of synthetic microfiber contaminated wastewater. Microb Technol Ind Wastewater Treat 337–350. https://doi.org/10.1007/978-981-99-2435-6_17

Mishra S, Dash D, Das AP (2022) Detection, characterization and possible biofragmentation of synthetic microfibers released from domestic laundering wastewater as an emerging source of marine pollution. Mar Pollut Bull 185:114254. https://doi.org/10.1016/J.MARPOLBUL.2022.114254

Mishra S, Singh RP, Rout PK, Das AP (2022) Membrane bioreactor (MBR) as an advanced wastewater treatment technology for removal of synthetic microplastics. Development in Wastewater Treatment Research and Processes, 45–60. https://doi.org/10.1016/B978-0-323-85583-9.00022-3

Mishra VK, Tripathi BD (2008) Concurrent removal and accumulation of heavy metals by the three aquatic macrophytes. Biores Technol 99(15):7091–7097

Pandey VC, Bajpai O, Singh N (2019) Energy crops in sustainable phytoremediation. Renew Sustain Energy Rev 72:62–73

Pazos M, Fink G, Mestrot A, Meckenstock RU, Sanchez J (2011) Identification of chlorinated solvents degradation zones in clay till by high resolution chemical, microbial and compound specific isotope analysis. J Contam Hydrol 126(3–4):235–243

Prasad MNV, Freitas H, Fraenzle S, Wuenschmann S, Markert B (2010) Knowledge explosion in phytotechnologies for environmental solutions. Environ Pollut 158(1):18–23

Rezania S, Park J, Din MFM, Taib SM, Farrokhian Farsangi, A., Kamyab, H., & Jongboum, O. (2015). Removal of organic compounds and nutrients from tannery wastewater by using constructed wetlands at pilot scale. Water Sci Technol 71(5):690–698

Salt DE, Blaylock M, Kumar NPBA, Dushenkov V, Ensley BD, Chet I, Raskin I (1995) Phytoremediation: A novel strategy for the removal of toxic metals from the environment using plants. Nat Biotechnol 13(5):468–474

Schnoor JL, Licht LA, McCutcheon SC, Wolfe NL, Carreia LH (1995) Phytoremediation of organic and nutrient contaminants. Environ Sci Technol 29(7):318A-323A

Shardendu S, S., & Prasad, S. M. (2019) Constructed wetlands for mitigation of endocrine disruptors from paper and pulp mill effluent. J Environ Manage 250:109457

Sharma S, Juwarkar AA, Pandey H, Prasad M (2019) Phytoremediation technology: A sustainable green clean-up approach for removal of pesticides and personal care products from aquatic systems. In Ecotoxicology and risk assessment of nanopesticides in the aquatic environment, Elsevier, pp 407–421

Shilev S, Russev R, Trichkova N, Mekenyan O (2007) Phytoextraction of Cu, Zn, Pb and Cd from contaminated soils using poplar. Wood Production, Wood Technology, and Biotechnological Impacts. Universitätsverlag Göttingen, 197–204

Sinha S (2018) Derivatization of milk protein using poly (Acrylamide): Its characterization and application. In Macromolecular Symposia, Vol. 381, No. 1, p. 1800120

Sinha S, Sen G, Mishra S (2018) Chemical modification of major milk protein (casein): Synthesis, characterization, and application. Functionalized Engineering Materials and Their Applications, 205

Sinha S (2022) Textile wastewater clarification using milk based formulations as an effective dye absorbent and flocculant. Materials Today: Proceedings 67:1304–1309

Sinha S (2023) Synthesis of biopolymer based superabsorbent: An eco-friendly approach towards future sustainability. In Bio-based superabsorbents: recent trends, types, applications and recycling, Springer Nature Singapore, Singapore, pp 29–49

Sinha S (2024a) Food security challenges and microplastics: A Comprehensive review. Microplastics in African and Asian Environments: The Influencers, Challenges, and Solutions, pp 361–371

Sinha S (2024b) Advancements in Biopolymer-Derived Graft Copolymers: A Comprehensive Review on Their Application in Sustainable Wastewater Treatment. Environ Qual Manag 34(2):e22320

Sinha S, Singh S, Dey KP (2024) Microplastics: A significant threat to freshwater ecosystems. In Sustainable Microbial Technology for Synthetic and Cellulosic Microfiber Bioremediation. Springer Nature Switzerland, Cham, pp 91–105

Suresh B, Ravishankar GA (2004) Phytoremediation—a novel and promising approach for environmental clean-up. Crit Rev Biotechnol 24(2–3):97–124

Tilley E, Ulrich L, Lüthi C, Reymond P, Zurbrügg C (2014) Compendium of sanitation systems and technologies. Swiss Federal Institute of Aquatic Science and Technology (Eawag).

Uggetti E, Hughes-Riley T, Morris RH, Newton MI, Trabi CL, Hawes P, Puigagut J (2012) Intermittent application of wastewater effluent improves efficiency of pollutant removal in treatment wetlands. Science of the Total Environment, pp 233–242

USEPA (2000) Introduction to phytoremediation. EPA/600/R-99/107. Cincinnati, OH: National Risk Management Research Laboratory. Office of Research and Development. U.S. Environmental Protection Agency

Van Aken B (2009) Transgenic plants for enhanced phytoremediation of toxic explosives. Curr Opin Biotechnol 20(2):231–236

Vymazal J (2005) Horizontal sub-surface flow and hybrid constructed wetlands systems for wastewater treatment. Ecol Eng 25(5):478–490

Vymazal J (2011) Plants used in constructed wetlands with horizontal subsurface flow: A review. Hydrobiologia 674(1):133–156

Weis JS, Weis P (2004) Metal uptake, transport and release by wetland plants: implications for phytoremediation and restoration. Environ Int 30(5):685–700

Weyens N, Thijs S, Popek R, Witters N, Przybysz A, Espenshade J, Vangronsveld J (2015) The role of phytohormones in plant-microbe symbioses in the rhizosphere. In Beneficial plant-microbial interactions, CRC Press, pp 159–196

Yuan F, Wu X, Li H, Chen S, Zhu D, Hu X (2019) Occurrence, fate and ecological risk of five typical pharmaceuticals in a constructed wetland. Sci Total Environ 647:206–214

Chapter 13
Microbial Bioremediation of Pesticide Pollution in Agricultural Soils: A Review

Somanath Behera and Maminee Panda

Abstract The post green revolution era has witnessed over-use of soil and rapid depletion nutrients in it. Being an Agrarian country, India ranks second worldwide in farm outputs. As per the Indian economic survey 2020–21, agriculture employed more than 50% of the Indian workforce and contributed 20.2% to the country's GDP. To ensure the mandate of green revolution i.e. increase food production, alleviate extreme poverty and malnourishment in the country, the productivity has to increase without being damaged from pests. Side by side an increasing number of pesticides have been used in agriculture for protecting the crops from pests, weeds, and diseases but as much as 80–90% of applied pesticides hit non-target vegetation and stay as pesticide residue in the soil which is potentially a grave risk to the agricultural ecosystem. This review gives an overview of the pollution in agricultural soils by pesticides, and the remediation techniques for pesticide-contaminated soils. Currently, the remediation techniques involve physical, chemical, and biological remediation as well as combined ways for the removal of contaminants. But a long-term monitoring and evaluation of in situ remediation techniques are the need of the time for long-term sustainability and practical applications in the field. Microbial bioremediation, the treatment that uses microorganisms to transform hazardous substances into lesser or non-toxic compounds, is an eye catching, eco-friendly, economical and efficient way to clean soil polluted by chemical pesticides. This review also gives a broad idea on types, mechanism and factors affecting microbial bioremediation.

Keywords Agricultural ecosystem · Pesticide-contaminated soils · In situ remediation techniques · Microbial bioremediation

---

S. Behera (✉)
School of Life Sciences, Sambalpur University, Burla, India
e-mail: som7071@gmail.com

M. Panda
P.G. Department of Zoology, Samanta Chandra Sekhar Autonomous College, Puri, India

A. Kuanar et al. (eds.), *Recent Advances in Bioremediation and Phytoremediation*, Environmental Science and Engineering,
https://doi.org/10.1007/978-3-031-77884-1_13

13.1 Introduction

Environmental pollution is the introduction of hazardous substances into the environment and the substances are regarded as pollutants. Pollutants may natural (fire, volcanic ash) or may be human made (debris from factories). Notably, anthropogenic activities foster abrupt harmful changes and agitations in the environment causing pollution (air pollution, water pollution, and soil pollution), by which the living organisms suffers adversely in the biosphere (Pathak and Mandalia 2011) (Fig. 13.1).

13.1.1 Soil Pollution

Focusing on pollution of soil, it can be defined as the presence of any harmful substance at higher concentration than normal that may have irresistible negative impacts on any non-targeted organism (FAO and ITPS 2015). Profound usage of pesticide and fertilizers in agriculture, inefficient disposal of industrial effluents make soil contaminated which requires removal of toxic elements from by remediation process. The resources for remediation are determined as per the level of contamination.

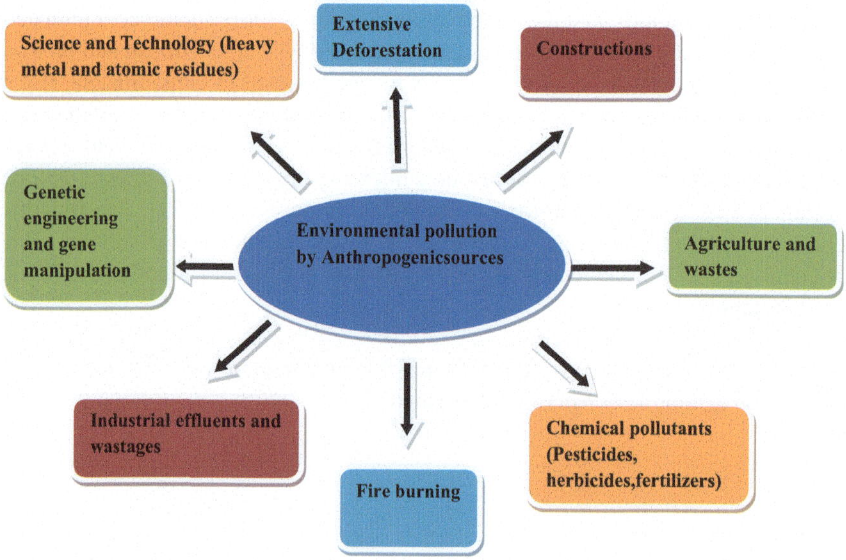

Fig. 13.1 Depicting environmental pollution by anthropogenic activities

13.1.2 Main Reasons Behind Soil Pollution

At broad spectrum, a number of pollutants penetrate the soils ecosystem hampering their physico-chemical and functional activity. Pesticides, heavy metals, fertilizers, and industrial chemicals act as soil pollutants which play unavoidable role in Environmental toxicology. However, it is explained in both soil and aquatic ecosystems by visualizing the populations of the inhabitants. However, natural turmoil like fire, earthquakes, temperatures and flooding affect the organisms, but environmental pollutants like acid deposition and agrochemicals bring about long-term chronic effects on the flora and fauna of soil ecosystems (Fig. 13.2).

Soil gets polluted by chemical toxicants by direct application, leaching, drifting, atmospheric fall-out, disposal of waste and discharge from industry. Out of them, some are short-lived and fugitive having low toxicity and some are persistent and hence pass to the living organisms through food chains and bio-concentrate inside them. These wastes polluting soil are of 3 types.

13.1.2.1 Agrochemical Waste

Soil is a dynamic resource plants as well as animals and also witnessed for sustainable survival of human beings. The quality of crops depends upon the physio-chemical and also biological properties of soil. Being the largest ecosystems, soil can be polluted by both organic as well as inorganic pollutants (Shahid et al. 2015). Toxic pollutants such as aluminum, copper, zinc, nickel, lead and arsenic produced by inoculation of

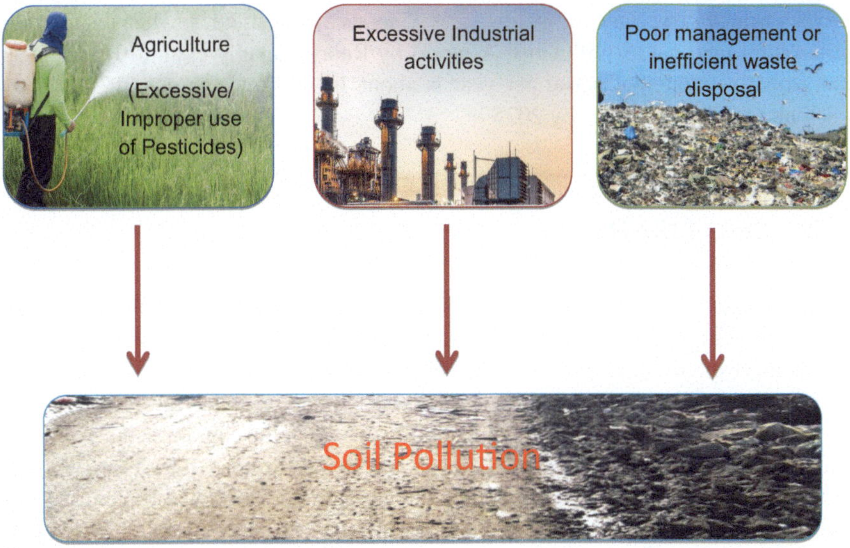

Fig. 13.2 Depicting the main reasons of soil pollution

pesticides, herbicides and various forms of fertilizers into agricultural field polluting soil environment (Prabagar et al. 2021). Similarly, contaminate water is drifted away from agro-food industries into various water bodies can have detrimental toxicity on their biodiversity (AL-Huqail et al. 2022). Ecologically, these pollutants affect agricultural systems at various level: (1) infra-individual level (gene function, enzymatic reactions); (2) individual level (natality, numbers, growth, mortality, density, fecundity and behavior); (3) community level (population size, species diversity, species dominance); (4) ecosystem level (primary productivity and secondary productivity and their nutrient cycling); (5) landscape level (spatial heterogeneity of plants and soil organisms, hydrologic mobility of nutrients).

13.1.2.2 Industrial Chemicals

Many industrial effluents, gaseous emissions, phthalate like plasticizers effect nearby ecosystems and their biodiversity. Heavy metals from automobile exhausts and smelter plants, nitrous oxide (NO_2), sulfur dioxide (SO_2), volatile organic compounds (VOCs), polychlorinated biphenyls, and oils (fats and petroleum) cause acidic depositions and bioaccumulation severely affecting soil ecosystems. The human created sources of heavy metals are industrial and mine residues, leaded oil, gasoline and paints, coal combustion remnants, spillage of petrochemicals from many sources (Alloway 2012).

13.1.2.3 Urban Wastes

Urban wastes like sewage bio-solids, heavy metals and pesticides, when disposed to agricultural land act as pollutants. A report by Rajaganapathy et al. (2011) depicts that globally municipal waste per year is about 1.3 billion tons in 2012 and may reach up to 2.2 billion tons by 2025.

13.2 Aim of the Review

The review has been started with an aim to highlight pesticide pollution due to agriculture. Further, the study focuses on the techniques of reclamation of farmland soil from pesticide, and evaluates the mechanisms of the remediation. Currently, the remediation of pesticide contaminated soil is done by involving techniques of physical, chemical, and biological remediation to extract the contaminants from soil. But evaluation of appropriate remediation techniques is the need of the time for long-term sustainability and practical applications in the field. Microbial bioremediation that uses microorganisms for conversion of highly toxic substances into lesser or non-toxic compounds, is an eye catching, eco-friendly, economical and efficient way to clean soil polluted by pesticides. This review also gives a broad idea on types, mechanism and factors affecting microbial bioremediation.

13.3 Pesticide as Agricultural Soil Pollutant

In the 1960s the so called "Green Revolution" came to rise in developing countries (India, China, Brazil, etc.) by the development of hybrid plants and seeds and the high usage of fertilizers, pesticides and mechanized agriculture, which obviously helped in the reduction of food crisis in India and other countries but also caused unavoidable major environmental pollution, wildlife biodiversity loss (Conway 2019). Some pesticides such as nematicides remain as such in soil for a few days; likewise some organo-chlorine insecticides may abide to the soil for decades also (Edwards 2002). Konradsen et al. (2003) found that eutrophication doubles at every ten years since 1960s and every year it causes desertification of 75.000 square miles of arable land due to vast use of pesticides and fertilizers.

Soil pollution is being promoted by widespread use of pesticides in agricultural field, and therefore, Narayanan et al. (2016) mentioned India stood as a leading nation in terms of production of pesticides. Although pesticides acts as a boosting agent for economy (Gollin et al. 2018), but a large amount of them are used unnecessarily in several industrialized and developing countries. For example, it is reported that the detected concentration of pesticides within freshwater is much more than the set limits of pesticide (Choudhary et al. 2018). Excessive amount pesticides stay as residues causing soil contamination and drastic water pollution. Besides this, due to overuse of pesticide, the population statistics of prey-predator pests becomes disturbed, and hence there is an overpopulation of a particular pest that would attack certain crops, and crop farming is hampered. These crops would need more strong pesticides or of new variety to tackle the pests. Ultimately, the flow of pesticides would damage the food chain (Narayanan et al. 2016).

13.3.1 Pesticide Classification

Pesticide is mostly synthetic chemicals consisting organic and inorganic compound used in agriculture to control pests and its classification based on its origin, chemical composition and target.

As per origin, pesticides can be of two types—chemical or bio-pesticides. Chemical pesticides are usually organic compounds that may be of natural sources or produced synthetically. On the other hand bio-pesticides are natural substances act by non-toxic mechanisms. Depending upon the target organisms; pesticides may be of different types such as insecticide (acts against insect pest), nematicides (acts against nematodes), fungicides (acts against fungi), weedicides (acts against weed), bactericides (acts against bacteria), acaricides (acts against arachnid), rodenticides

Table 13.1 Showing the chemical classification of pesticides and their uses

Chemical type	Functional group	Examples	Used as
Organo-phosphates	Phosphoric acid esters or thio-phosphoric acid esters	Hexaethyl tetraphosphate (HETP), chlorpyrifos, malathion, dioxathion	Insecticide
Organo-chlorines	Hydrocarbon chains with covalently bonded chlorine	Dichlorodiphenyltrichloroethane (DDT), DDD, aldrin, endosulfan, BHC	Insecticide
Carbamates	Esters of carbamic acids or thiocarbamic acids	Aldicarb, benomyl, carbofuran, fluometuron, isoproturon, chlorbromuron	Biocide
Pyrethroids	Extracted from the chrysanthemum flower containing pyrethrins compounds (esters of chrysanthemic acid and pyrethric acid with alcohols)	Cyfluthrin, bifenthrin, allethrin	Insecticide

(acts against rodents) etc. Globally, it was estimated that out of total used pesticides, 3 categories of the pesticides are mostly used. They are herbicides, insecticides and fungicides with a percentage use of 47.5, 29.5 and 17.5% while the rest pesticides accounts for 5.5% (Sharma et al. 2019). Moreover, the chemical composition based classification of pesticides include "organophosphates, organochlorines, carbamates, and pyrethroids" (Yadav and Devi 2017), illustrated in Table 13.1.

13.3.2 Mechanism of Action of Pesticide

The main function of pesticides is to control pests by either killing them or by reducing their metabolic activities. New generation pesticides are having features like low solubility, low volatility, low surface tension, highly lipophilic and an average octanol–water partition coefficient (Wang et al. 2019).

The persistence, behaviour and transport of pesticide calculate the mechanisms of sorption–desorption, volatilization, chemical and biological degradation, and percolation in soil (Arias-Estévez et al. 2008) as shown in Fig. 13.3.

About 1000 pesticides are used worldwide, out of which 292 are registered in India as per the Pesticide Management Bill, 2020. In India, pesticides like methiocarb, benomyl, malathion, phorate, DDT, 2,4-D, diazinon, dichlorvos, endosulfan, cartap, carbaryl, chlorpyrifos etc. are commonly used. Excessive and frequent use of the pesticides interferes with the nature creating major influences on its biotic and abiotic

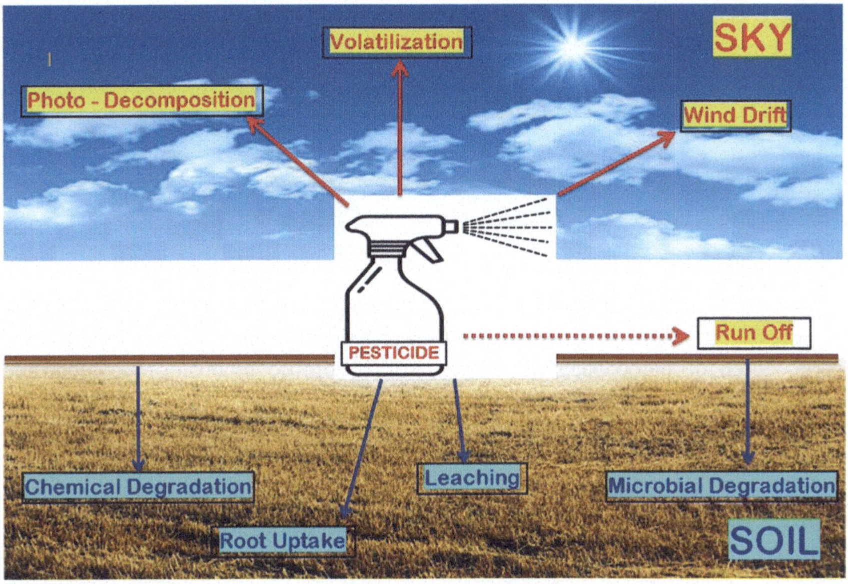

Fig. 13.3 Depicting the behavior of pesticides in environment

components causing ecological imbalance. As per GOI, 2020 in India the use of pesticides progressed geometrically in recent past which is mentioned in Table 13.2.

The most widely used insecticides composed of organophosphates, neonicotinoids and pyrethroids used in cropland. In Rajasthan, Yadav and Dutta (2019) conducted a study on pesticide application and found that in cotton crop, 93.27% of pesticides were applied. It was followed by vegetables (87.2%), wheat (66.4%), millet (52.6%), and mustard (12.6%).

Even though the pesticides specifically act upon the target pest but also have impact on its non-target biota. The persistence of the pesticides in the environment depends of their half-life period which varies in different compounds (Table 13.3).

Water solubility property of pesticide remains low which makes them potential to proliferate in living tissues. The burning example was illegal and extensive use of Chlordecone in France's Caribbean territories and their water resources results in

Table 13.2 Showing increased use of pesticides as per GOI, 2020

Name of the pesticides	Amount used during 2014–15 (in metric ton)	Amount used during 2019–20 (in metric ton)	Hike in use in percentage (%)
Chloropyrifos	471	1431	304
Sulphur	1548	3878	250
2,4-D	1	1067	106,700
Zinc phosphate	65	200	307

Table 13.3 Showing half-life periods of most common pesticides used in India

Name of the pesticides	Half-life period (years)
DDT	22–30
Toxaphene	14
Mirex	12
Dieldrin	7
Chlordecone	30

contaminated agricultural land because of which its highest ever reported concentration was found in both marine and fresh water fishes especially in wild and farmed tilapia, which ultimately became a threats to other wildlife organisms (Bocquené and Franco 2005).

13.3.3 Impact of Pesticides

13.3.3.1 Contamination of Soil

Pesticides contaminate soil by penetrating into it due to spray drift during its application to target plant, wash-off from treated target plants and release from granulates. The persistence, sorption–desorption, volatilization, degradation, absorption by plants, run-off and leaching of pesticides in soil depends not only on the physicochemical status of soil but also on its biological properties. Primarily, soil is contaminated by anthropogenic activities like electroplating, painting, usage of pesticides, herbicides, and fertilizers, which amounts for about 30 million tons of industrial pesticides and other chemicals (Guo et al. 2014). Because of pesticide administration, every year, 3 million people get poisoned and premature death of about 2.5 lakh people occurs (Bourguet and Guillemaud 2016).

Soil organic matter enhances sorption and percolation of pesticides into soil by which they become less procurable to microbial degradation and hence become more persistent. Thus humus enriched soil is more chemically reactive in the presence of pesticides than non-humified soil (Farenhorst 2006). Sorption of pesticide by soil occurs abruptly after immediate pesticide application but slow sorption takes place as along as the time increases. However, by repeated application of pesticides, gradually these get adsorbed to form bound residues. Due to constant variation in soil pH, nitrate fertilizers input, these residues subsequently reach to plants and animals (Gevao et al. 2001).

13.3.3.2 Water Contamination

Water resources are directly contaminated by pesticides by their direct application onto water surface and indirectly by pesticide drift, and runoff leaching from nearby treated land. However, hydrophobic and hydrophilic nature of pesticides, soil properties, weather conditions, landscape type measure the rate of water contamination. Usually, atrazine, metolachlor, diazinon, endosulfan, lindane and aldrin are found in water. Primarily pesticides affect primary productivity of the ecosystem, but subsequently secondary productivity get damaged too. Reductions in producer biomass eventually results in low dissolved oxygen, pH, increased alkalinity and increased conductivity.

13.3.3.3 Effects of Pesticides on Biodiversity

Both direct (short term) and indirect (long term) impacts of pesticides on organisms have been reported in organisms. It is due to alterations caused by the chemicals to the environment mainly to the food sources of the organisms. These effects result in intensive environmental pollution, wild life and biodiversity loss. The negative impacts on ecological parameters like population dynamics, species density, species diversity, biomass, primary productivity, species interaction, and abiotic (physico-chemical) parameters like nutrient level, pH, temperature, turbidity, salinity and food sources etc. result in perturbation to the function of ecosystem.

Eco-toxicologically, pesticides at sub-lethal concentration formulate remarkable changes in individual traits such as changes in neurotransmitters, hormones, enzymes, neuro-signaling, reproduction and behavior. Overall physiological, biochemical, histological, immunological, neurological, histological, genetical, and mutagenic impacts are the major consequences. Also frequency, time and dose of application might damage a particular ecosystem. Effects can be assessed by observing species dominance and their abundance in a particular ecosystem.

The use of broad spectrum pesticides may cause desertification, eutrophication and bio-magnifications. Short life cycle organisms like microbes can recover from pesticide toxicity relatively in small time period. But longer lived organisms like nematodes and earth worms may take somehow longer period to retrieve from its long term impairment.

13.3.3.4 Pesticides on Soil Microorganisms

Soil microorganisms play a vital role for maintenance of soil structure, transformation and mineralization, nutrient cycle, bioturbation and increment of the microbial activities (Lazcano et al. 2008). Somehow, soil microorganisms participate in metabolization and degradation of pollutants, but this may lead to increase its toxicity and persistence. Pesticides cause fluctuation in their population and species diversity drastically influencing the whole soil ecosystem (Hole et al. 2005).

Pencycuron, a fungicide altered the soil microbial metabolic and respiration activities as well, which in turn decreases their population rate, specifically of fungi (Pal et al. 2005). Nitrifying and denitrifying bacteria are negatively affected by fungicides like chlorothalonil and carbendazim (Lang and Cai 2009). However, it was also observed, glyphosate stimulates the population of fungi and actinomycetes with high aggregate microbial activity, even though there is a significant decline in bacterial populations (Araújo et al. 2003).

13.3.3.5 Pesticides on Invertebrates and Vertebrates

Soil invertebrates like nematodes, earthworms, and arthropods (insects, mites, spiders, and springtails) help in decomposition of organic matter, soil aggregation and porosity and hence act as "Ecosystem engineers". But abundant use of various pesticides causes adverse impacts on soil invertebrates (Edwards and Thompson 1973).

Pesticides along with heavy metals have synergistic effects on earthworms at different organizational levels like infra-individual, and individual level, causing DNA damage, disruption in enzymes, low growth and survival rate, and also at population, and community level causing decrease in biomass and density (Uwizeyimana et al. 2017). Chlorpyrifos, Glyphosate and adversely act on cholinesterase activity of earthworm causing DNA damage, reducing feeding activity, decreasing reproductive fecundity, and thus survivability of the organism decreases (Casabé et al. 2007). Some studies reported, in crop land, earthworm population could increase up to 1.5–4.8 times more by decreasing Frequency Index (TFI) of pesticide to half (Pelosi et al. 2021).

Insecticide chlorpyrifos induces behavioural changes and mortality in arthropod communities (spiders and collembola) by reducing their abundance and species diversity. However, juveniles are more sensitive to xenobiotics than adults (Fountain et al. 2007). A drastic decrease can be seen in species richness of bumblebee and butterfly on exposure to pesticide fenitrothion applied farmland while their number increases in organic farming (Feber et al. 2007). These crop protection chemicals not only kill the adults but also hinder the growth and development at larval stages. Adverse and deleterious effects may be declined by inoculation of pesticides into crop field in late summer and early fall, and during larval diapause period (Russell and Schultz 2010).

Glyphosate not only kill plants, but also bring about high rate of mortality in three North American species (*Rana pipiens*, *Bufo americanus*, and *Hyla versicolor*) by killing the population of their tadpoles and juveniles (Relyea and Diecks 2008). Mixture of pesticides could be able to cause severe abnormalities in physiological rates of animals. Laetz et al. (2009) reported negative effects of organophosphate combined with carbamate pesticides on the acetylcholine esterase enzyme in the brains of juvenile Coho salmon. Adoption of no-spray buffer zones can be a recovery measure to benthic macro-invertebrate communities those exposed to pesticide application as reported by a study in UK by Maltby and Hills (2008). This toxicity reduces prey plants and animals, ultimately affecting the farmland bird's population during their breeding. Pesticide DDT induces the decrease in egg shell thickness in Peregrine falcon and Eurasian sparrow-hawks as observed by Ratcliffe (1967). However, mortality, survivability, and reproduction are considered ecologically more relevant parameters to observe the sub-lethal effects of pesticides.

13.3.3.6 Pesticides on Human Being

After the emergence of pesticide use during green revolution, suicide rate due to pesticides severely increased estimating 14 million premature deaths (Karunarathne et al. 2020). The negative impacts of pesticides are associated with various health issues like skin allergies, low immune response, hypersensitivity, congenital defects, genetic manipulation, nervous disorganization, endocrine disruption, tumors, and finally coma and death (Lorenz 2009).

In 2009, an investigation of Wigle et al. (2009) concluded that pesticide exposure to pregnant mother may result in childhood leukemia, and cancer. In another study, pesticide found associated with male sterility declining in sperm count that too pesticide residues found in men's urine and semen (Swan 2006; Hauser 2006). It is also found that, mother who becomes pregnant during April to July gives birth to child with prenatal defects which may be associated with increased level of nitrites, atrazine, and other pesticides during the season (Winchester et al. 2009). Moreover, in spring conceptions, rate of gastroschisis become high due to more use of atrazine in Washington State (Waller et al. 2010). Additionally, organophosphate consumption results in neurodevelopmental abnormalities like hyperactivity disorder (ADHD), neural tube defects, particularly anencephaly during prenatal and postnatal stage of child as dimethyl alkyl phosphate (DMAP) found in urinary metabolites (Bouchard et al. 2010).

A review study by Mendola et al. between 1999 and 2007 summarized pesticide like DDT/DDE cause earlier menarche and puberty as well as menopause in female (Mendola et al. 2008). Pesticides (lindane, atrazine, mancozeb, and maneb) also cause longer menstrual cycles, missed periods and abnormal bleeding consequently give rise to infertility in women. Many case studies revealed a positive correlation between pesticides DDT and toxaphene exposure and breast cancer (Cohn et al. 2007). Particularly in case of breast cancer it was found that the patient had a long exposure to DDT prior to age of 14. It is because; pesticides function as endocrine disruptors affecting male and female reproductive hormones.

13.4 Remediation Techniques for Pesticide-Contaminated Soils

There are mainly 3 categories of remediation, such as physical, chemical and biological remediation as shown in Fig. 13.4. Out of which, physical remediation is the technique of physical removal of contaminants from the soil by involving excavation, soil capping/soil covering, and vitrification by using skimmers, sorbent materials and booms. In case of Chemical remediation, certain forms of chemicals are added to capture the contaminants and remove it from the soil. The chemicals used are phosphate, biochar, clay mineral, salts of aluminium, silico-calcium materials, and sulphide. The mechanisms behind the use of these chemicals include adsorption,

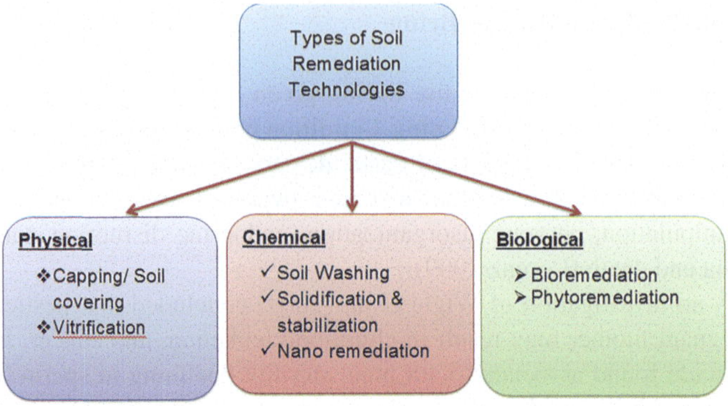

Fig. 13.4 Depicting various types of remediation technologies for contaminated soil

reduction, oxidation, complexation, precipitation and ion exchange. Most commonly used methods of chemical remediation are Soil washing, Solidification and Stabilization, Nano remediation. At last, Biological remediation is the most sustainable and cost effective technique involving microbes or plant named as microbial remediation or phytoremediation. Its sustainability depends on the level of pollution. As this chapter focuses on microbial bioremediation, hence the following lines are fully dedicated to the methods and mechanisms involved in this process.

13.4.1 Microbial Bioremediation

Microbial Bioremediation involves use of microbes for remediation purpose and it is regarded as a safe, inexpensive and sustainable method. However, the microbial bioremediation is worldwide used and trustworthy method that can degrade the contaminants effectively and clean the environment. The conversion of the toxic contaminant to less toxic form or complete degradation to CO_2 or CH_4 involving microbes is called as microbial degradation.

13.4.2 Mechanisms of Microbial Degradation

In this process, pesticides are transformed from highly toxic compounds to less or non-toxic compounds by microorganisms. The microbes (bacteria or fungi) use pesticides as nutrients using various enzymes, like hydrolases, peroxidases, and oxygenases. Hence, the entire biochemical process of pesticides degradation is illustrated in Fig. 13.5.

In first step, the pesticides are converted into more water-soluble and less toxic products by the help of the above said enzymes. It involves biochemical reactions like

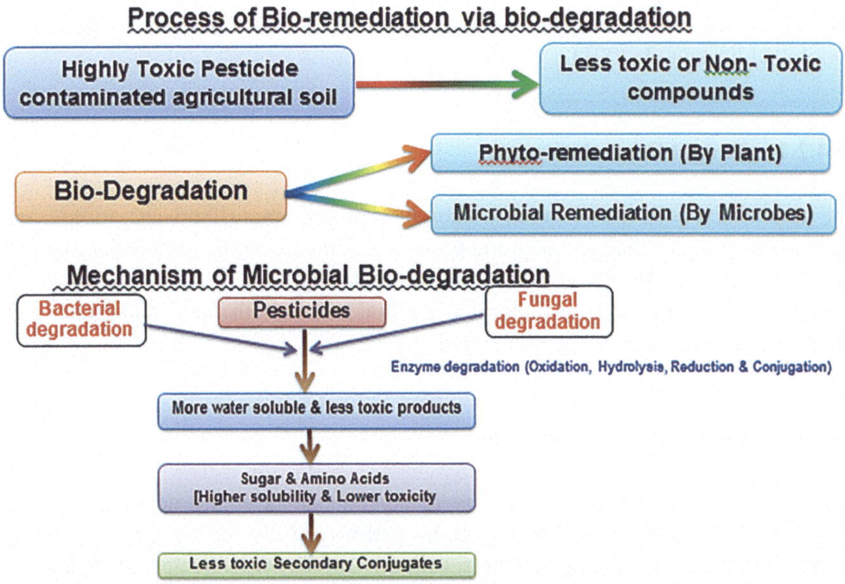

Fig. 13.5 Depicting the mechanism of microbial bio-degradation

oxidation, reduction, and hydrolysis. Then, products formed in first step are converted into smaller compounds having higher solubility in water and low toxicity level such as sugars and amino acids. Finally, the Step-2 metabolites are converted into secondary conjugates which are chemically lesser toxic than the step-1 products. The bioremediation methods are planned as per the time involved in degradation process. This process depends on several parameters such as microbial activity, pesticides concentration, temperature, water content, availability etc. (Soulas and Lagacherie 2001).

13.4.2.1 Bacterial Degradation

Various bacterial strains can be used for degrading the pesticides contaminated soil. The optimum conditions of several factors like pH, temperature, water content, and the nature of pollutants, bacterial strain used. Further, in due course of degradation process, certain byproducts may be formed which hampers the degradation of the original contaminants. It can be understood by the study of Abraham and Silambarasan (2016) that chlorpyrifos is hydrolyzed by microorganisms forming primary degradation product as 3,5,6-trichloro-2-pyridinol (TCP) which has greater water solubility than original contaminants i.e. Chlorpyrifos and causing difficulty in the degradation process and further pollute the aquatic environment. On the other hand, the microbial degradation is more comfortably done by using a bacterial consortium rather using an isolated pure culture because coexistence of various bacteria help to take off the metabolites generated in each steps of pesticide degradation (Doolotkeldieva et al. 2021).

13.4.2.2 Fungal Degradation

The agricultural soils are acting as a habitat for many fungi (like yeast, molds, and filamentous fungi) which can degrade pesticides. Fungal degradation is the involvement of these fungi in biodegradation of pesticides by influencing the permeability of soil. For safe practice, an integrated method to use both fungi and bacteria to increase biodegradation is recommended.

13.4.2.3 Enzymatic Degradation

Enzymatic biodegradation happens by catalysing the pesticide degradation. The enzymes are formed due to the metabolic pathways followed by microbes as by product. It accelerates the rate of the reaction by minimising the activation energy. Main enzymes involved in this process are oxidases, hydrolases and reductases (Scott et al. 2008).

13.4.2.4 Mineralization

Mineralization involves complete breakdown of the highly toxic compounds of pesticides into inorganic compounds (CO_2, water, salts and minerals) without any toxicity which can be further take up by the microbes as nutrient source. The rate of mineralization is directly proportional to the concentration of the microbes. De Souza et al. (2017) observed that chlorothalonil (CTN) being used as fungicide, is degraded into CO_2 in optimum concentration of microbes. In contradiction, with low concentration of microbes certain intermediate metabolites are formed which are more toxic than CTN.

13.4.2.5 Co-metabolism

Co-metabolism is the transformation of an organic compound which is not supporting microbial growth. In the co-metabolic process, the enzymes involved are hydrolases, transferases, oxidases and reductases through a series of reactions. Ma et al. (2014) have studied the co-metabolism of Imidacloprid (IMI) by using different types of substrates used as an energy source like carbohydrates and organic acids while *P. indica* CGMCC 6648 strain degrade IMI into olefin and 5-hydroxy IMI through the hydroxylation pathway.

13.4.3 Strategies of Microbial Bioremediation

The most common bioremediation strategies are in situ and ex situ bioremediations (Das and Dash 2014). In case of in situ bioremediation, bioremediation process is conducted at the contaminated site itself. Here, the biodegradation ability of natural microflora is increased or a group of natural or genetically modified strains is introduced into the contaminated site. While in case of ex situ bioremediation process, contaminated substances are transferred to a distant place where the entire process of bioremediation is applied in optimum conditions of the factors involved in the process (Fig. 13.6).

Intrinsic bioremediation is a natural process where microorganisms indigenous to the contaminated soil degrade pesticides without intervention of human. Biosparging is an in situ bioremediation method involving enhancement of the microbial growth by injecting the air or oxygen with high pressure. It is mainly used to treat the volatile organic compound contaminated ground water. While bioventing is another in situ bioremediation method used to stimulate indigenous microbial degradation of contaminants by increasing oxygen availability. In this process, the unsaturated Vadose zone is aerated where injection of air into the groundwater for its treatment. Similarly, in case of Bio-slurping, a vacuum is applied to the treatment area to remove free phase hydrocarbons along with dissolved phase pollutants. In bio-stimulation, indigenous microorganisms are employed in degrading pollutants such as hydrocarbons; pesticides etc. in soil as well as in groundwater by optimizing all environmental conditions. In addition another method i.e. Bio-augmentation is also employed as an in situ bioremediation technique which is more popular in remediation of pesticide contaminated soils (Pimmata et al. 2013; Cycoń et al. 2017). This technique differs from the above mentioned techniques as where the indigenous microbial populations remain ineffective or less effective in degradation; hence a specific group of natural or genetically engineered microbial strains are introduced to the contaminated site to degrade contaminants (Das and Dash 2014). It can also be used as ex situ method.

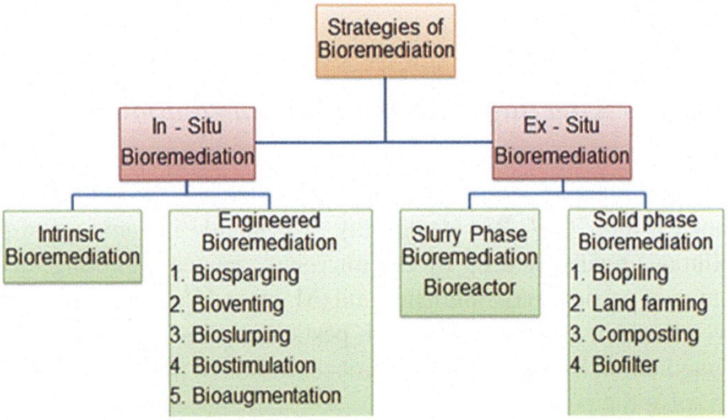

Fig. 13.6 Depicting various strategies of bioremediation

Moreover, many studies inferred that the combination above techniques together increases the effectiveness of bioremediation.

Bioreactor is an ex situ bioremediation practice in which contaminated medium is treated in a container or a vessel called a reactor and slurry or aquatic media are used. It is a controlled environment with optimum conditions required for microbial growth. It is mostly used to treat waste water. A containment apparatus is used where mixture of three phase condition is maintained. It acts as an engineered system where the rate of microbial bioremediation of water soluble pollutants is increased by enhancing the target pollutant's bio availability. This strategy can also be used in removal of the pesticide contaminated soil and water. Similarly, Bio-piling is also another type of In situ bioremediation technique involves piling contaminated material along with soil in a mound or pile and then introducing the microbes to enhance the natural degradation. Moisture, pH, oxygen and nutrients are the environmental factors which remain controlled within the biopile system. This technique has been considered as a promising and competent technique for the removal of pesticide from polluted soils (Parween et al. 2018).

Land farming is a "solid-phase remedial system" used to treat pesticide adulterated soils which includes both ex situ or in situ bioremediation practice. In this method, blended wastes are spread out in a designated area where degradation of contaminants happens by natural microbial metabolism. The contaminants turns to less harmful overtime and the same method again repeated to form such soil layers.

Further, composting is another ex situ bioremediation method where the contaminated soil is excavated and brought to a controlled facility and mixed with organic amendments and bulking agents and also microbes to enhance the bioremediation. Overtime, the composting process helps to break down the contaminants to simpler compounds without any toxic effect. Though composting has been extensively used to remove petroleum hydrocarbon but it has equally potential to remediate pesticides from contaminated soil (Mena et al. 2003).

Bio-filters are designed to create optimal conditions for microbial activity, allowing the microorganisms to break down contaminants into harmless by-products like water and carbon dioxide.

13.4.4 Factors Affecting Microbial Bioremediation

For any microbial reaction, optimum conditions are requiredto get better result and it is illustrated in Fig. 13.7. Bio-availability of pesticide i.e. amount of pesticides (or any pollutant) readiy take up by the microbes, acts as a limiting factor in the bioremediation of pesticide contaminated soil (Maier and Valdes 2000). Accordingly, it is observed that low concentration of th pesticide at low nutrient medium slows down bioremediation. Hence, increase in concentration of pestcide bioavailability in top soil can solve the problem.

Physicochemical factors include various parameters of soil like temperature, pH, moisture content, organic and inorganic content present in soil and texture of the

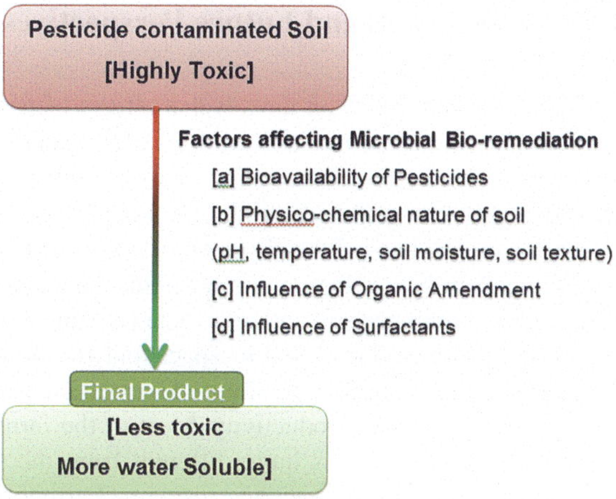

Fig. 13.7 Depicting the factors affecting microbial bioremediation

soil. They affect directly or indirectly on bioavailability and biodegradation. Due to enzyme dependent pathway, at optimum conditions of those physicochemical factors, rate of biodegradation can be accelerated. It can be understood from increase in chlorpyrifos degradation in a bioslurry phase reactor with optimum condition i.e. at pH 7 and temperature 30° (Brajesh et al. 2006). Further, the moisture content of soil cause the increased availability of chemicals required for growth of microorganism involved in biodegradation (Daniel and Timothy 1991). The conversion of solid phase to solution phase of the pesticides is required for inducing the action of microbes and most of the contaminants degrade at this level. As we know that soil has a both organic and inorganic content in a heterogeneous mixture state with various degrees of coat composition and surface activity, hence they may bind with the contaminants i.e. pesticides and either increase or decrease the bioavailability (Torrents and Jayasundera 1997). Similarly, the influence of can be organic amendments may alter the rate of pesticide degradation (Cox et al. 2001). Mukherjee (2012) reported endosulfan degradation can be apprehended to one day with vermicomposting and farmyard manure added soil.

Surfactants are surface-active agents that decrease the surface and interfacial tensions of liquids. Chemically, it is amphiphilic molecule which aggregate to form micelles in solution. Being capable of forming bio-surfactants, many species of bacteria such as *P. aeruginosa*, *Corynebacterium spp.*, *Bacillus subtilis* etc., free energy of the entire system increases in solution phase when the lipophilic part cannot form the hydrogen bond. It alternately absorbs pesticides to stabilise the micelle by decreasing the free energy (Lang 2002).

13.5 Conclusion and Future Perspectives

This review emphasised on the adverse effects of pesticide contamination on soil, environment and moreover on health condition of human and other animals. It is well proved that bioremediation especially the microbial bioremediation effectively remediate pesticide polluted soil. It revealed that the microbial bioremediation techniques are much ahead of the physico-chemical methods due to cost effectiveness as microorganisms have inherent biological mechanisms that enable them to survive in pesticide contaminated soil and remediate the same. Furthermore, it witnessed as an emerging technique due to its low space and simple apparatus requirement also. It can mitigate the potential risk imposed by pesticides on natural ecosystem without hampering agricultural productivity. Most of the farmers independently decide the dose of pesticide which is quite different from the standard recommended dose, causing accumulation of pesticide residues. Furthermore, farmers lack the knowledge about proper storage, handling and application practices regarding pesticides are not proper by which pesticides become hazardous. Strict legal policies can prohibit the misuse/overuse of pesticides. The hands on training on pesticide use may solve most of the problem. However, further studies can be done to identify new microorganisms with such capabilities of bio-remediation of pesticides from agricultural field.

References

Abraham J, Silambarasan S (2016) Biodegradation of chlorpyrifos and its hydrolysis product 3,5,6-trichloro-2-pyridinol using a novel bacterium *Ochrobactrum* sp. JAS2: a proposal of its metabolic pathway. Pestic Biochem Physiol 126:13–21. https://doi.org/10.1016/j.pestbp.2015.07.001

AL-Huqail AA, Kumar P, Eid EM, Adelodun B, AbouFayssal S, Singh J et al (2022) Risk assessment of heavy metals contamination in soil and two rice (*Oryza sativa* L.) varieties irrigated with paper mill effluent. Agriculture 12(11):1864

Alloway BJ (ed) (2012) Heavy metals in soils: trace metals and metalloids in soils and their bioavailability, vol 22. Springer Science & Business Media

Araújo AD, Monteiro RTR, Abarkeli RB (2003) Effect of glyphosate on the microbial activity of two Brazilian soils. Chemosphere 52(5):799–804

Arias-Estévez M, López-Periago E, Martínez-Carballo E, Simal-Gándara J, Mejuto JC, García-Río L (2008) The mobility and degradation of pesticides in soils and the pollution of groundwater resources. Agr Ecosyst Environ 123(4):247–260

Bocquené G, Franco A (2005) Pesticide contamination of the coastline of Martinique. Mar Pollut Bull 51(5–7):612–619

Bouchard MF, Bellinger DC, Wright RO, Weisskopf MG (2010) Attention-deficit/hyperactivity disorder and urinary metabolites of organophosphate pesticides. Pediatrics 125(6):e1270–e1277

Bourguet D, Guillemaud T (2016) The hidden and external costs of pesticide use. Sustain Agric Rev 19:35–120

Brajesh KS, Walker A, Wright DJ (2006) Bioremediation potential of fenamiphos and chlorpyrifos degrading isolates: influence of different environmental conditions. Soil Biol Biochem 2682–2693

Casabé N, Piola L, Fuchs J, Oneto ML, Pamparato L, Basack S et al (2007) Ecotoxicological assessment of the effects of glyphosate and chlorpyrifos in an Argentine soya field. J Soils Sediments 7:232–239

Choudhary S, Yamini NR, Yadav SK, Kamboj M, Sharma A (2018) A review: pesticide residue: cause of many animal health problems. J Entomol Zool Stud 6:330–333

Cohn BA, Wolff MS, Cirillo PM, Sholtz RI (2007) DDT and breast cancer in young women: new data on the significance of age at exposure. Environ Health Perspect 115(10):1406–1414

Conway G (2019) The doubly green revolution: food for all in the twenty-first century. Cornell University Press

Cox L, Cecchi A, Celis R, Hermos M, Koskinen W, Cornejo J (2001) Effect of exogenous carbon on movement of simazine and 2,4-D in soils. Soil Sci Soc Am J 65:1688–1695

Cycoń M, Mrozik A, Piotrowska-Seget Z (2017) Bioaugmentation as a strategy for the remediation of pesticide-polluted soil: a review. Chemosphere 172:52–71

Daniel RS, Timothy BP (1991) Effect of moisture on sorption and biodegradation of carbofuran in soil. J Agric Food Chem 39:2060–2063

Das S, Dash HR (2014) Microbial bioremediation: a potential tool for restoration of contaminated areas. Microbial biodegradation and bioremediation. Elsevier

de Souza AJ, de Andrade PAM, de Araújo Pereira AP, Andreote FD, Tornisielo VL, Regitano JB (2017) The depleted mineralization of the fungicide chlorothalonil derived from loss in soil microbial diversity. Sci Rep 7(1):14646

Doolotkeldieva T, Bobusheva S, Konurbaeva M (2021) The improving conditions for the aerobic bacteria performing the degradation of obsolete pesticides in polluted soils. Air, Soil Water Res 14:1178622120982590

Edwards CA (2002) Assessing the effects of environmental pollutants on soil organisms, communities, processes and ecosystems. Eur J Soil Biol 38(3–4):225–231

Edwards CA, Thompson A (1973) Pesticides and the soil fauna. In: Residue reviews: residues of pesticides and other contaminants in the total environment, pp 1–79

FAO and ITPS (2015) Status of the world's soil resources (SWSR)—main report. Food and Agriculture Organization of the United Nations and Intergovernmental Technical Panel on Soils, Rome, Italy

Farenhorst A (2006) Importance of soil organic matter fractions in soil landscape and regional assessments of pesticide sorption and leaching in soil. Soil Sci Soc Am J 70(3):1005–1012

Feber RE, Johnson PJ, Firbank LG, Hopkins A, Macdonald DW (2007) A comparison of butterfly populations on organically and conventionally managed farmland. J Zool 273(1):30–39

Fountain MT, Brown VK, Gange AC, Symondson WO, Murray PJ (2007) The effects of the insecticide chlorpyrifos on spider and Collembola communities. Pedobiologia 51(2):147–158

Gevao B, Mordaunt C, Semple KT, Piearce TG, Jones KC (2001) Bioavailability of nonextractable (bound) pesticide residues to earthworms. Environ Sci Technol 35(3):501–507

Gollin D, Hansen CW, Wingender A (2018) Two blades of grass: the impact of the green revolution. National Bureau of Economic Research, Cambridge, CA. https://doi.org/10.3386/w24744

Guo H, Jin Y, Cheng Y, Leaderer B, Lin S, Holford TR, Qiu J, Zhang Y, Shi K, Zhu Y, Niu J, Bassig BA, Xu S, Zhang B, Li Y, Hu X, Chen Q, Zheng T (2014) Prenatal exposure to organochlorine pesticides and infant birth weight in China. Chemosphere 110:1–7

Hauser R (2006, July) The environment and male fertility: recent research on emerging chemicals and semen quality. In: Seminars in reproductive medicine, vol 24, No 03. Thieme Medical Publishers, Inc., New York, USA, pp 156–167

Hole DG, Perkins AJ, Wilson JD, Alexander IH, Grice PV, Evans AD (2005) Does organic farming benefit biodiversity? Biol Cons 122(1):113–130

Karunarathne A, Gunnell D, Konradsen F, Eddleston M (2020) How many premature deaths from pesticide suicide have occurred since the agricultural green revolution? Clin Toxicol (Philadelphia, PA) 58(4):227–232

Konradsen F, Van der Hoek, Cole W, Hutchinson DC, Daisley GH, Singh S, Eddleston M (2003) Reducing acute poisoning in developing countries-options for restricting the availability of pesticides. Toxicology 192:249–261

Laetz CA, Baldwin DH, Collier TK, Hebert V, Stark JD, Scholz NL (2009) The synergistic toxicity of pesticide mixtures: implications for risk assessment and the conservation of endangered Pacific salmon. Environ Health Perspect 117(3):348–353

Lang S (2002) Biological amphiphiles (microbial biosurfactants). Curr Opin Colloid Interface Sci 7:12–20

Lang M, Cai Z (2009) Effects of chlorothalonil and carbendazim on nitrification and denitrification in soils. J Environ Sci 21(4):458–467

Lazcano C, Gómez-Brandón M, Domínguez J (2008) Comparison of the effectiveness of composting and vermicomposting for the biological stabilization of cattle manure. Chemosphere 72(7):1013–1019

Lorenz ES (2009) Potential health effects of pesticides. AG Communications and Marketing

Ma Y, Zhai S, Mao SY, Sun SL, Wang Y, Liu ZH et al (2014) Co-metabolic transformation of the neonicotinoid insecticide imidacloprid by the new soil isolate *Pseudoxanthomonas indica* CGMCC 6648. J Environ Sci Health B 49(9):661–670

Maier R, Valdes JJ (2000) Bioavailability and its importance to bioremediation. Int Soc Environ Biotechnol Environ MonitBiodiagration 59–78

Maltby L, Hills L (2008) Spray drift of pesticides and stream macroinvertebrates: experimental evidence of impacts and effectiveness of mitigation measures. Environ Pollut 156(3):1112–1120

Mena E, Garrido A, Hernández T, García C (2003) Bioremediation of sewage sludge by composting. Commun Soil Sci Plant Anal 34(7–8):957–971

Mendola P, Messer LC, Rappazzo K (2008) Science linking environmental contaminant exposures with fertility and reproductive health impacts in the adult female. Fertil Steril 89(2):e81–e94

Mukherjee I (2012) Influence of organic amendments on the degradation of endosulfan. Bull Environ Contam Toxicol

Narayanan J, Sanjeevi V, Rohini U, Trueman P, Viswanathan V (2016) Postprandial glycaemic response of foxtail millet dosa in comparison to a rice dosa in patients with type 2 diabetes. Indian J Med Res 144:712

Nayak S, Sahoo A, Kolanthasamy E, Rao K (2020) Role of pesticide application in environmental degradation and its remediation strategies

Pal R, Chakrabarti K, Chakraborty A, Chowdhury A (2005) Pencycuron application to soils: degradation and effect on microbiological parameters. Chemosphere 60(11):1513–1522

Parween T, Bhandari P, Sharma R, Jan S, Siddiqui ZH, Patanjali P (2018) Bioremediation: a sustainable tool to prevent pesticide pollution. Modern age environmental problems and their remediation. Springer

Pathak C, Mandalia HC (2011) Impact of environmental pollution on human future. World J Environ Pollut (WJEP) 1:8–10

Pelosi C, Bertrand C, Daniele G, Coeurdassier M, Benoit P, Nélieu S et al (2021) Residues of currently used pesticides in soils and earthworms: a silent threat? Agr Ecosyst Environ 305:107167

Pimmata P, Reungsang A, Plangklang P (2013) Comparative bioremediation of carbofuran contaminated soil by natural attenuation, bioaugmentation and biostimulation. Int Biodeterior Biodegradation 85:196–204

Prabagar S et al (2021) Accumulation of heavy metals in grape fruit, leaves, soil and water: a study of influential factors and evaluating ecological risks in Jaffna, Sri Lanka. Environ Sustain Indic 12:100147

Rajaganapathy V, Xavier F, Sreekumar D, Mandal PK (2011) Heavy metal contamination in soil, water and fodder and their presence in livestock and products: a review. J Environ Sci Technol 4:234–249

Ratcliffe DA (1967) Decrease in eggshell weight in certain birds of prey. Nature 215(5097):208–210

Relyea RA, Diecks N (2008) An unforeseen chain of events: lethal effects of pesticides on frogs at sublethal concentrations. Ecol Appl 18(7):1728–1742

Russell C, Schultz CB (2010) Effects of grass-specific herbicides on butterflies: an experimental investigation to advance conservation efforts. J Insect Conserv 14:53–63

Scott C, Pandey G, Hartley CJ, Jackson CJ, Cheesman MJ, Taylor MC et al (2008) The enzymatic basis for pesticide bioremediation. Indian J Microbiol 48:65–79

Shahid N, Kalofolias V, Bresson X et al (2015) Robust principal component analysis on graphs. In: IEEE international conference on computer vision. IEEE Computer Society, pp 2812–2820

Sharma A, Kumar V, Shahzad B, Tanveer M, Sidhu GPS, Handa N, Kohli SK, Yadav P, Bali AS, Parihar RD (2019) Worldwide pesticide usage and its impacts on ecosystem. SN Appl Sci 1:1446

Soulas G, Lagacherie B (2001) Modelling of microbial degradation of pesticides in soils. Biol Fertil Soils 33:551–557

Sun T, Miao J, Saleem M, Zhang H, Yang Y, Zhang Q (2020) Bacterial compatibility and immobilization with biochar improved tebuconazole degradation, soil microbiome composition and functioning. J Hazard Mater 398:122941

Swan SH (2006, July) Does our environment affect our fertility? Some examples to help reframe the question. In: Seminars in reproductive medicine, vol 24, No 03. Thieme Medical Publishers, Inc., New York, USA, pp 142–146

Torrents A, Jayasundera S (1997) The sorption of nonionic pesticides onto clays and the influence of natural organic carbon. Chemosphere 35(7):1549–1565

Uwizeyimana H, Wang M, Chen W, Khan K (2017) The eco-toxic effects of pesticide and heavy metal mixtures towards earthworms in soil. Environ Toxicol Pharmacol 55:20–29

Waller SA, Paul K, Peterson SE, Hitti JE (2010) Agricultural-related chemical exposures, season of conception, and risk of gastroschisis in Washington state. Am J Obstet Gynecol 202(3):241-e1

Wang Y, Shen L, Gong Z, Pan J, Zheng X, Xue J (2019) Analytical methods to analyze pesticides and herbicides. Water Environ Res 91(10):1009–1024

Wigle DT, Turner MC, Krewski D (2009) A systematic review and meta-analysis of childhood leukemia and parental occupational pesticide exposure. Environ Health Perspect 117(10):1505–1513

Winchester PD, Huskins J, Ying J (2009) Agrichemicals in surface water and birth defects in the United States. Acta Paediatr 98(4):664–669

Yadav IC, Devi NL (2017) Pesticides classification and its impact on human and environment. Environ Sci Eng 6:140–158

Yadav S, Dutta S (2019) A study of pesticide consumption pattern and farmer's perceptions towards pesticides: a case of Tijara Tehsil, Alwar (Rajasthan). Int J Curr Microbiol Appl Sci 8(04):96–104

Chapter 14
Phytoremediation Mechanisms for Emerging Contaminants: Approaches Towards Sustainable Environment

Pranjal Roy, Esha Dutta, Subhra Satahrada, Subrat Kumar, and Ritesh Pattnaik

Abstract The rapid urbanization and increase global population driven activities result in release of toxins from various anthropogenic activities, which can antagonistically influence untamed life, affecting both human and the surrounding environment. Organic pollutants may pose environmental risk and have xenobiotic effect on plants. Various types of emerging contaminants including heavy metals like lead, zinc, cadmium, arsenic are refractory in nature and even if available in low concentration levels represent a potential risk of bioaccumulating at higher tropic levels in both terrestrial and aquatic ecosystem. Heavy metal pollution being mutagenic, endocrine, cancer-causing and teratogenic, may cause apprehensive medical disorders. The concentration of these xenobiotics can be potentially limited by adapting various remediation techniques. Phytoremediation can offer an efficient biocontrol strategy, involving wetland, aquatic plants and soil microbiota to mitigate pollutants into less toxic components. The phenomenon of phytoremediation incorporates sequestration of heavy metals by either bioaccumulating in plants tissues, or by immobilizing the contaminants and detoxification in the roots. Studies on biocontrol suggest that different plant species like *Pteris vittata, Helianthus annus, Arabidopsis haller*i can potentially be used for bioremediation of emerging contaminants available in terrestrial and aquatic environments. Activities like corrosive downpours, modern and local waste waters result in excessive disposal of inorganic contaminants and toxins in such natural ecosystems. However, macrophytic species and rhizospheric flora including *Azolla, Eichhornia, Spirodela* serve as natural occurring phytoremediators for bioaccumulation and subsequent reduction of toxic chemicals. Besides, other effective techniques like phytoextraction, phytostabilization, rhizofiltration methods are being adopted globally for successful mitigation of heavy metals. Recent advances include adopt of agronomic strategies, green technology, artificial intelligence and control biotoxicity by PGPR induced transgenic plants for effective bioremediation

Pranjal Roy and Esha Dutta are Equal contributing authors.

P. Roy · E. Dutta · S. Satahrada · S. Kumar · R. Pattnaik (✉)
KIIT School of Biotechnology, KIIT University, Bhubaneswar, Odisha, India
e-mail: rpattnaik@kiitbiotech.ac.in

A. Kuanar et al. (eds.), *Recent Advances in Bioremediation and Phytoremediation*, Environmental Science and Engineering,
https://doi.org/10.1007/978-3-031-77884-1_14

of aquatic environments and restoration of polluted environments. Besides, various limitations and research gaps are required to be addressed for effective and enhanced mitigation of emerging contaminants. In the present chapter emphasis on various emerging contaminants and its potential mitigation through various phytoremediation techniques towards sustainable environment are highlighted. Besides, extensive mechanisms of phytoremediation and its limitations are also outlined for improving the efficacy of the strategies.

Keywords Emerging contaminants · Toxicity · Phytoremediation · Mitigation · Sustainable environment

14.1 Introduction

The rise in industrial activities contributing towards environmental pollution has been raising concern globally. In order to control the anthropogenic activities, various strategies have been adopted and potential eco-friendly, cost-effective methods are being developed to mitigate the risk associated with environmental pollutants (Bhatt et al. 2023). Bioremediation has become the indispensable strategy to prevent, eliminate, and to restore the adverse effects of pollutants, toxins released into the environment (Khan et al. 2022). Influential and recyclable approaches for eliminating contaminants include biomass adsorbents and plant-based environment friendly approaches. Phytoremediation is an emerging content proposed as a plant-based approach to detect, degrade and foster environmental remediation. For example: garden plants like sunflowers help in the removal of soil pollutants. Phytoremediation is a bioremediation process because it uses various plants to remove/destroy secondary pollutants by locking them in their roots, shoots or by breaking them into less toxic components, simpler forms. The term comes from *phyto* (Greek term for plant) and *remedium* (*Latin word* which means to correct or restore). Over the decades, phytoremediation emphasizes on the importance of plants in detoxifying wastewater through the elimination of pollutants, heavy metals, dyes, inorganic substances, pesticides, hydrocarbons, and pharmaceutical contaminants. Phytoremediation is the organic phenomenon where living plants biologically degrade xenobiotics and recalcitrant substances and explores the potentiality of plants (especially roots) to purify soil and water contaminated with heavy metals. The term Phytoremediation has been accepted worldwide as the Superfund Program of the US Environmental Protection Agency (EPA) to detoxify a range of pollutants from lead to arsenic and instill genetically engineered techniques (advanced genomics, transcriptomics and proteomics) to enhance phytoremediation strategies (Alves et al. 2022). Thus, the environment naturally benefits from spontaneously occurring plants which are then modified using transgenic means viable in removing contamination, provide habitat for an ecosystem and lead to an increase in soil aeration and microbial activity.

14.2 Background Information

The term 'Phytoremediation' was first introduced in the year 1883 by R. L Chaney. The basic information of phytoremediation was introduced and showcased variety of research which includes constructed wetlands, oil spills, and unique ability of some agricultural plants to accumulate heavy metals and tolerate environmental pollutants. These plants were defined as hyperaccumulators because they actively take up metals into their biomass, some extract metals and secures into the rhizosphere by precipitation to minimize erosion and metal contaminant transport. In earlier days, the concept of disposal, reuse and recycling of industrial waste was regarded as an unproductive function to be achieved at least possible cost, that led to extensive pollution affecting progress and prosperity (Glenna et al. 2022). Recognition of the thread imposed by pollution raised national awareness and transformed disposal of industrial waste into a productive venture, thereby promoting the emerging alternative- use of green technology, phytoremediation. The evolution of phytoremediation has been observed for the past 300 years which originated from unusual interaction between plants and environment reducing pollutant concentration in vegetated soils, and is further progressing under the constant changing paradigms of technology (Stephenson and Black 2014). Thus, phytoremediation relates to the overall idea of utilizing plant-based environmental approaches in this report, instead of a specific application.

14.3 Emerging Pollutants: Characteristics and Toxicity

Certain substances that have very recently emerged to be potential environmental hazards have been identified as emerging pollutants. Emerging Pollutants (EP) also known as micropollutants are defined as synthetic or naturally occurring chemical compounds that pose a danger and cause adverse effects on the environment, ecosystems, including human health and increase toxicity, persistence, and constrains productivity. These pollutants encompass a diverse group of compounds such as pharmaceutical sector pollutants including antibiotics, drugs and steroids, endocrine disruptors, hormones, toxins, industrial additives and biological contaminants, such as microplastics, microbeads and nanoparticles (Geissen et al. 2015). Emerging pollutants are persistent is nature and also includes heavy metal, excessive nutrients, petroleum, pesticides, personal care products, PCB (Polychlorinated Biphenyl), drinking water contaminants (Chromium, PFOS- Perfluorooctane sulfonic acid, PFOA- Perfluorooctanoic acid). On the basis of their chemical nature, these pollutants may possess distinct levels of toxicity and they frequently exhibit negative effects on the ecosystem, human health, and aquatic life (Castro et al. 2021). Emerging pollutants were previously found in lower concentrations but due to the increased anthropogenic global production, multiple reduction measures are being borne by the researchers. These compounds if not regulated or monitored they will create a high risk to our livelihood. Recognition and minimization of such pollutants extremely unfavorable for the environment has maximized the intensity of research and monitoring.

14.4 Consequences of Emerging Pollutants

Numerous minerals and chemicals have discharged and integrated in the environment in recent decades as a result of fast industrialization, urbanization, and technological breakthroughs. Of them, a subset referred to as "emerging pollutants" has attracted more attention because of the possible risk they might induce in environment and human health. A wide variety of compounds that are not often monitored or controlled, such as industrial chemicals, medicines, personal care items, microplastics, and nanomaterials, are considered emerging contaminants (Fig. 14.1). The emergence of new contaminants presents a variety of issues that need to be addressed immediately by communities, businesses, scientists, and governments. It will need a coordinated effort to improve monitoring capacities, create cutting-edge mitigation technologies, fortify regulatory frameworks, and encourage sustainable habits in order to address the effects of these contaminants. We can protect the environment and public health for both present and future generations by recognizing the hazards associated with new contaminants and acting proactively to minimize them. The following section provides a detailed overview of the various emerging contaminants and their various impacts on our environment.

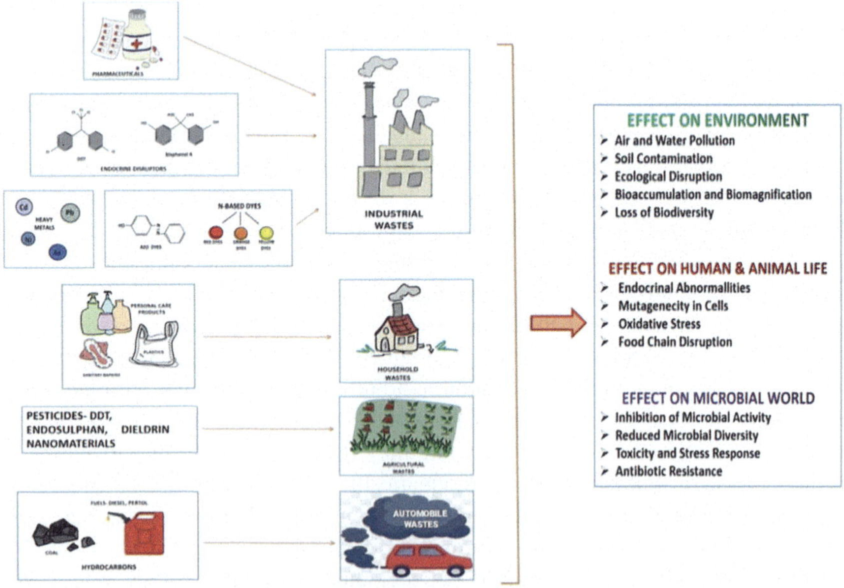

Fig. 14.1 An overview of emerging contaminants and their detrimental impact on environment

14.4.1 Pharmaceutical and Personal Care Products

An ecosystem can be impacted via the presence of emerging pollutants, especially from personal care products (consumer goods like cosmetics) and pharmaceutical products (like therapeutic drugs) (PPCPs), for improper disposal leading to presence of active metabolites of PPCPs in the surface water, groundwater and wastewater treatment plants (Birch et al. 2015; Huerta et al. 2016). The consequences are as follows:

Ecological Impacts: PPCPs possess the ability to interrupt the endocrine systems of aquatic animals, accumulates on the tissues of invertebrates, fishes, higher vertebrates, affects the algal community, interferes with the metabolic activities, structural and functional imbalance. These chemicals might have a subtle effect including antimicrobial resistance but causes mutation on reproduction, targets the mitochondrial respiratory proteins and cytochrome enzymes (Birch et al. 2015), affects the offsprings produced thereby posing a threat to the ecosystem.

(a) **Concerns Regarding Drinking Water Quality**: PPCPs having a possibility to impair water supplies and endanger public health. Drinking water supplies might have traces of these compounds due to inadequate elimination during water treatment operations (Huerta et al. 2016).
(b) **Antimicrobial Resistance**: The use of antibiotics can turn less effective in treating disorders when drugs accumulate in the environment may promote microorganisms to become resistant to antibiotics (Guanqun et al. 2021).
(c) **Bioaccumulation**: Varying concentration of these acute or chronic toxins gradually leads specific PPCPs to enter and accumulate inside the tissues of living organisms, causing increased levels of physiological ailments. The intensity of bioaccumulation is popular among higher-ranking organisms in the food chain. Ecosystem could possibly be susceptible to a sudden ecological chain reaction resulting from this bioaccumulation over a long period of time.
(d) **Heath Care Hazards**: Although the exact cause of these threats remains under investigation, drinking water or consuming food exposed to relatively small amounts of PPCPs gradually could result in long-term detrimental health effects (Rego Richelle et al. 2021).
(e) **Enforcement Obstacles**: The increased abundance of PPCPs contamination has lead to change in biota and climate. More warranted, eco-friendly and cost-effective techniques are the upcoming challenges for detection, degradation and monitoring (Mishra et al. 2021; Mishra et al. 2023). Comprehensive techniques and legal structures are essential for the surveillance and management of these chemical compounds.
(f) **Challenges In Sewage Purifications**: Several PPCPs may not be entirely eliminated by conventional wastewater treatment facilities and end up in release of them into the environment. It may take highly sophisticated therapies to overcome the issues (Guanqun et al 2021).

(g) **Influence on Soil Purity**: Biological solids (sewage effluents) containing PPCPs when utilized as fertilizers, the compounds can penetrate the soil and might have an impact on crop production and soil fertility (Smith et al. 2006).

With the goal to restrict the consequences of developing contaminants (PPCPs) in the environment and to prioritize the well-being of humans, addressing these repercussions requires an integrated strategy involving studies, policy formulation, increased wastewater management, and raising awareness.

14.4.2 Endocrine Disputors

Emerging contaminants, notably endocrine disruptors, may result in serious consequences. Chemicals like DDT, bisphenol A, dioxin, organochlorine compounds cause disturbances to the endocrine system by imitating regular hormone operation, opposing hormone synthesis or action which adversely affects the metabolism of hormones in humans as well as animals. Hormonal imbalances, development of abnormalities, imperative issues related to reproduction, and an increased risk of specific infirmity are the highly possible potential outcomes (Huerta et al. 2016). Furthermore, these contaminants may affect biodiversity and ecosystem services through a chain reaction of repercussions. Endocrine disruptors need to be addressed and controlled if human health and ecological balance are to be conserved.

EDCs produces hormones, which regulates many important autonomic and anatomical functions, which includes:

- Development of growth
- Reproduction and Fertility
- Metabolism
- Mood and Behavior
- Immune system Function

Natural hormones can be imitated or blocked by EDCs, which may result in a wide range of adverse impacts on both humans and wildlife (Lauretta et al. 2019). Effect of EDCs can be from moderate to severe and may occur at any point in lifetime.

In Humans

Reproductive Health Issues: EDCs are associated with a plenty of human reproductive disorders, such as testicular cancer, endometriosis, polycystic ovary syndrome (PCOS) and decreased fertility.

Developing Issues: EDCs can hamper a fetus's normal development, leading to low birth weight, learning disabilities, and birth defects.

Cancer: EDCs has been associated to posing a greater probability for developing malignancy in the thyroid, prostate, and breast regions of the body.

Besides medical conditions: EDC's has also been identified as the host to many other additional medical conditions, such as immune system disorders, diabetes, obesity, and thyroid-related problems.

In Wildlife

Population decline: Fishes, amphibians, birds, alongside other wildlife species are undergoing demographic decrease as a result of increased EDC in the environment.

Deformations and abnormalities: EDCs have been scientifically associated with an array of physical deformities in wildlife, including masculinized and feminized males and females. The reproduction process is impaired since EDC's may hinder the ability of organisms to reproduce, thus lowering fertility and birth rates and leading to fall in population.

14.4.3 Agricultural Pesticicdes

Agricultural pesticides are one of the most elemental constituents of the present agricultural practices and is a vital sector in enhancing economic wealth of any nation. Pesticides are extensively used to shield crops from insects and pests, increase crop production and yield development in the agricultural field (Prem et al. 2023; Mishra et al. 2021). Unfortunately, studies aim to unravel the potential risks imposed by agricultural pesticides to the biosphere and have unexpected repercussions from the release of emerging pollutants, detrimental chemicals present in its residues into the natural environment (Tudi et al. 2021). The presence of these pollutants in soil, water, and food can have a variety of associated effects and are lethal to the ecosystems especially human health. These pollutants tend to be man-made which requires thorough regulation and is a major global concern to maintain sustainability. The following are the main effects of emerging contaminants in agricultural pesticides:

Environmental Impacts:

Leaching of Soil: Emerging pollutants have the potential to linger in soil for extended periods of time, infecting crops, influencing microbial activity and disrupting soil fertility. Decrease in agriculture yields and ecological hazards might occur as a result of leaching (Das et al. 2023).

Water Pollution: Water pollution may affect aquatic life and contaminate drinking water supplies whenever pesticidal runoff from agricultural fields floods into rivers, lakes, and groundwater (Mishra et al. 2022).

Loss of Biodiversity: The pollinating insects such as bees and butterflies can suffer from exposure to emerging contaminants, adversely counter effect the food webs and cause a decrease in diversification (Bhuyan and Giri 2020).

Human Health Impacts:

Exposure through water and food: Food and water contamination can expose people to new harmful substances, which can cause a variety of health issues such as cancer (Tudi et al. 2021), endocrine disruption, and developmental abnormalities.

Workplace exposure: Workers managing pesticides including farmworkers are more likely to be exposed to new pollutants which can cause skin irritation, respiratory problems and other health issues (Enrico et al. 2015; Mishra et al. 2019).

Difficulties in Handling New Pollutants:

Little Data and Research: There is not much data available about the effects of many new toxins on the environment and human health.

Combined Toughness and Mobility: Emerging pollutants have the potential to be extremely mobile and persistent, which allows pollutants to travel great distances and stay in the environment for extended periods of time.

Absence of Guidelines: There are currently no restrictions on the consumption of several novel contaminants or their emission into the environment.

14.4.4 Hydrocarbons

Hydrocarbons are the major element present in petroleum-based products and an intrinsic component of several industrial goods widely available in our natural surroundings (Lahiri et al. 2021). Despite being necessary for our modern industrial and energy systems, hydrocarbons pose multitude of adverse effects to human health and innumerable environmental consequences (Pollard 1977). Incomplete combustion of organic wastes, concomitant poor disposal of waste, inadequate hydrocarbon extraction, processing and usage releases a vast array of toxic group of contaminants. Hydrocarbons imposes toxicological effects and is a raising area of concern which requires proper regulation and control (Behera et al. 2020).

Principal effects of recently discovered contaminants in hydrocarbons:

Effect on the Environment:

Alterations to Ecosystems: Polycyclic Aromatic Hydrocarbons (PAHs) and alkylphenols are two examples of emerging contaminants that can be hazardous for both marine and terrestrial individuals, induce alterations in food chains thereby causing large-scale changes on the existing ecosystems.

Environmental Damage of Soil and Groundwater: Leakage from the transportation and storage of hydrocarbons can contaminate groundwater as well as soil with dangerous substituents affecting the quality of drinking water and endangering plant and animal life.

Changes in the Climate and Pollution in the Atmosphere: During combustion of hydrocarbons, airborne contaminants including nitrogen oxides and volatile organic compounds (VOCs) are released into the atmosphere, which in turn causes smog, acid rain, and climate change (David et al. 2018).

Impact on Human Health:

Chronic Disorders (including Cancer): Evidences incur exposure to particular hydrocarbons and the pollutants they are linked to (especially PAHs) emanate a greater probability of cancer, respiratory problems, and neurological impairments.

Hormonal Imbalance and Reproductive Health: Certain hydrocarbons include endocrine disrupting chemicals (EDCs), which may cause disruption with hormonal processes and affect immunological networks, development, and fertility.

Water-borne Infections: Drinking water contaminated by hydrocarbons and related pollutants can result in waterborne diseases that affect sanitation and overall wellness.

Problems in Dealing with Newly Emerged Pollutants:

Limited data and information: Continuous studies and research are being done to determine the long-term health and environmental effects of many newly discovered contaminants, many of which are poorly understood.

Tenacity and Adaptability: Certain hydrocarbon pollutants are mobile, travels great distances (Kurt et al. 2005) environment for decades because of their high persistence, motility and multiple adaptive capabilities (such as PAHs).

Absence of Control and Regulation: In distinct parts of certain developing nations, absence of strict regulative laws restricting the anarchic use and improper discharge of emerging contaminants in hydrocarbons leads to rise in pollution.

Scientific and Technological Expenditures: In order to have a complete idea about the range of emerging contaminants present in the hydrocarbons, enunciate efficient mitigation plans (Bruno et al. 2019), promote further investigation and adapt remediate technologies expenditure and proper provision is an indispensable factor affecting the scientific junctures.

Adaption of more Environmentally Friendly Technologies: The production and discharge of emerging pollutants minimization requires proper extraction of hydrocarbons, eco-friendly and cost-effective ventures especially biocontrol methods to mitigate and reduce hydrocarbon pollution and enforce more sustainability and healthy future by addressing the issues and putting workable solutions in place.

14.4.5 Heavy Metals and Dyes

Heavy metals toxicity from air pollution, sewage sludge, anthropogenic activities (Bruno et al. 2019) and metal-containing dyes from textile wastewater contains large quantities of chemical effluents that significantly increases the biological and chemical oxygen demand (BOD and COD) stands as one of the major global polluters. Dyes and heavy metals are two major groups of systemic toxicants (Tchounwou et al. 2009) and natural byproducts of industrial activities which unitedly corrode the environment extensively (Mishra et al. 2022).

Environmental Impact:

Biological Accumulation and Harmful Effects: Hazardous concentrations of heavy metals, such as arsenic, lead, and mercury, can bioaccumulate in aquatic organisms and propagate toxicity through the food chain and possess the potential to impair essential biological metabolic processes (Nagajyoti et al. 2010) in both plants and animals, diminish biodiversity and affect the ecosystems subsequently.

Contamination of water and soil: The excess application of multiple industrial activities, mining (Tchounwou et al. 2009), and inappropriate waste disposal causes heavy metals contamination of soil and water. This can alter aquatic ecosystems adversely, destroy plant development, and contaminate supplies of drinking water.

Atmospheric Contamination: Contamination from toxic release of industrial exhaust and the burning of fossil fuels leads to air pollution enhancing respiratory problems in both human and animals (Nagajyoti et al. 2010).

Influence on Human Health:

Neurotoxicological Impact: Lead and mercury exposure predominantly attack the nervous system, results in serious medical impairments, memory loss, and possibly death.

Cancer and Other Long-term Illnesses: Chronic exposure to heavy metals has been associated with a higher chance of acquiring cardiovascular diseases, gastrointestinal and kidney malfunction, risk of developing cancer, and other chronic health issues (Velusamy et al. 2021).

Reproductive issues and neurodevelopmental delays: Heavy metals have the ability to interfere with fetal growth and development and disturb the hormonal balance, which may result in birth abnormalities and developmental delays in offspring.

14.5 Biocontrol Measures for Mitigating Emerging Pollutants

The majority of researchers are adopting bioremediative methods to eliminate both organic and inorganic contaminants. In order to reduce environmental pollution, bioremediation plans for environmentally friendly detoxification and the efficient breakdown of several organic and inorganic pollutants. "Bioremediation" is the term used to describe the biochemical degradation of organic wastes under regulated conditions. Through the use of bioremediation, toxic compounds can be broken down or detoxified by giving organisms the nutrients and other chemicals they require to function at their best.

The various Bioremediation methods include:

1. **Biopiling**: The piled-up polluted soil above-ground microbial metabolic activity is boosted by aeration and nutrient supplementation. This process includes the use of aeration, fertilizers, irrigation, leachate collection, and treatment bed systems. Ex situ biodegradation is a process that is gaining popularity due to its affordability and helpful features, such as pH and nutrient management. It is possible to use biopile method to treat low-molecular-weight volatile contaminants and clean up polluted cold environments.
2. **Windrows**: It spins the piled contaminated soils and accelerates the biodegradation of native and transient hydrocarbon plastic present in the contaminated soils. Through acclimation, biological therapy, and mineralization, hazardous soil can be aerated, mineralized, and bio transformed (De Silva et al. 2019). Due to the anaerobic environment created inside the piled-up polluted soil, windrow treatment is a source of greenhouse gas (CH_4).
3. **Land Farming**: Due to its cheap operating costs and absence of specialist equipment, land farming is the most important and straightforward bioremediation method. In order to allow native microorganisms to aerobically breakdown toxins, extracted contaminated soils are typically spread out over a permanent layer of substrate that is considerably above the Earth's surface. With the use of land farming bioremediation technologies, toxic soil may be cleaned up in a relatively quick and easy manner that leaves little of an environmental or energy-intensive imprint (Ghosh and Das 2020).
4. **Bioreactor**: Bioreactors transform raw materials into specific products. Bioremediation thrives in a bioreactor, which provides the ideal conditions for growth. As an efficient bioremediation process based on bioreactors that can precisely regulate (Sharma et al. 2019) pH, agitation, temperature, aeration, substrate concentration, and inoculum concentration significantly reduce the time required for bioremediation (Nag et al. 2023).
5. **Bioventing**: This method that delivers oxygen to the unsaturated zone using regulated airflow to boost the activity of local bacteria for bioremediation. The addition of nutrients and moisture during the bioventing (Abhilash et al. 2009) process helps the bioremediation process. This will result in the microbial conversion of contaminants into safe materials.

6. **Bioslurping**: This approach is carried out to reach soil and groundwater levels for restoration requires the utilization of a direct oxygen supply, stimulation of pollutant biodegradation, vacuum-assisted pumping, bioventing, and soil vapour extraction (SVE) (Slama et al. 2021). Light non-aqueous phase liquids (LNAPLs) and saturated and unsaturated zones can both be recovered using this method. This approach can be used to clean up flammable and somewhat flammable organic substances-contaminated soils.
7. **Biosparging**: This techniqueissimilar to bioventing, air is injected into the soil core to promote microbiological activity, which eliminates contaminants from polluted places. By infusing air into a saturated zone (Slama et al. 2021; Singh et al. 2011) to assist the flow of combustible organic compounds upward to a nearby unsaturated zone, bioventing is an alternative to conventional biodegradation procedures.
8. **Phytoremediation**: phytoremediation is an emerging content where plants are used to eliminate pollutants from contaminated sites (soil, ground water, aquatic environments), such as heavy metals, excessive nutrients, petroleum, pharmaceuticals, pesticides, etc. For example: garden plants like sunflowers help in the removal of soil pollutants. Phytoremediation is a bioremediation (Russell et al. 1999) process because it uses various plants to remove/destroy secondary pollutants by locking them in their roots, shoots or by breaking them into less toxic components, simpler forms.
9. **Bioaugmentation**: It is doneusing carefully selected microorganisms, which includes fungi or bacteria, in a contaminated environment to hasten the breakdown of contaminants is known as bioaugmentation. These microorganisms accelerate (Chunping et al. 2010) the process of natural biodegradation because they have the enzymes and metabolic pathways required to address particular pollutants.
10. **Constructed Wetlands**: This approach is an artificial wetland is a naturally occurring organic wastewater treatment system that replicates and enhances the procedures that aid in water purification. Water, aquatic plants (such as reeds and duckweed), naturally occurring microorganisms, and a filter bed (often made of sand, soil, or gravel) are the components of the system. Wastewater treatment using constructed wetlands might be secondary or tertiary. There are many various designs available, such as vertical (Sun et al. 2018) wetlands, which use less space than horizontally wetlands but require more energy for pumping or siphoning due to topography and gravity.
11. **Microbial Fuel Cells (MFCs)**: Microbial Fuel Cells (MFCs) are bioelectrochemical technologies that use organic compound oxidation (metabolic oxidation) of microorganisms to produce power source. The basic idea behind MFCs is based on the specific capacity of a bacteria to move electrons generated during the breakdown of organic matter into an electrode to produce an electric current.
12. **Biofiltration**: utilizing microorganisms like fungi and bacteria, biofiltration is a method for controlling pollution in water and air. The procedure depends on the capacity of microbes to metabolize and decompose organic and inorganic chemicals (Arshad et al. 2021) into less toxic forms. Biofilters are widely used

as an environmentally acceptable and sustainable method of pollution control for wastewater and air pollutants.

13. **Algal Remediation**: This strategy isa green technique that makes uses the inherent ability of algae to absorb and eliminate contaminants (Ons et al. 2020) from a variety of environments, particularly water. Algae are photosynthetic organisms that have the ability to absorb and accumulate pollutants. As such, they provide a potential remedy for cleaning up contaminated waters. This includes both macro and microalgae.
14. **Biochar Amendment**: To enhance soil fertility, structure, and environmental sustainability, biochar amendments entail adding carbon-rich material called biocharwhich is obtained from the pyrolysis of organic biomass (Sharma et al. 2023), useful tool for improving soil qualities. It is a stable form of carbon that can linger in soils for hundreds to thousands of years.
15. **Integrated Biocontrol Strategies**: In order to manage pests, diseases, and other unwanted organisms in agricultural and environmental contexts, integrated biocontrol strategies (Barathi et al. 2023) entail the coordinated employment of several biological control techniques. By reducing the need for chemical pesticides, integrated biocontrol aims to improve pest management's efficiency, viability, and economic viability. Several biocontrol techniques are combined in integrated biocontrol, each of which adds to the overall control plan (Mishra et al. 2022; Tripathy et al. 2022).

14.6 Different Types of Potential Phyto-Technologies

Phytoremediation is an emerging content where plants are used to eliminate pollutants from contaminated sites (soil, ground water, aquatic environments), such as heavy metals, excessive nutrients, petroleum, pharmaceuticals, pesticides, etc. For example: garden plants like sunflower helps in the removal of soil pollutants. Phytoremediation was coined in the year 1891, which states phyto = plant and remediation = detoxify pollutant, and it provides an overall idea of plant-based environmental technologies. Phytoremediation is a Bioremediation process because it uses various plants to remove/destroy secondary pollutants by locking them in their roots, shoots or by breaking them into less toxic components, simpler forms. Depending on the pollutant to be remediated and site-specific condition, there are various kinds of potential phytoremediative technologies (Fig. 14.2) as highlighted below:

14.6.1 Phytoaccumulation

Phytoaccumulation is a shoot area specific process which uses hyperaccumulating plants (naturally or assisted) to sequester large amounts of heavy metals in their tissue. The process of phytoaccumulation involves complex mechanisms ranging

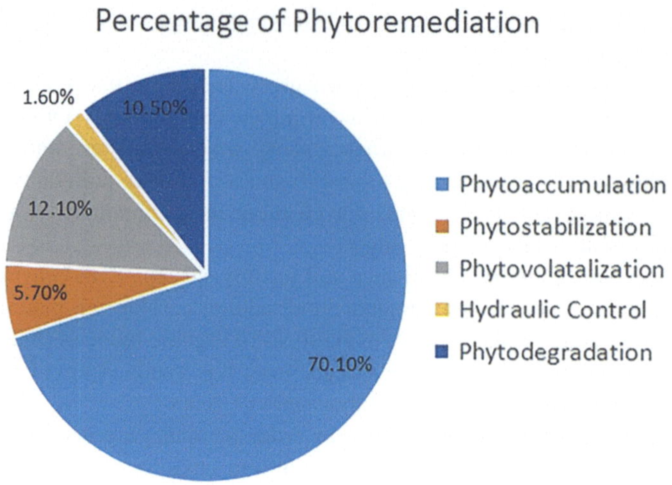

Fig. 14.2 Strategic phytoremediation approaches for effective biocontrol of emerging pollutants

from mobilization, bioaccumulation, translocation, sequestration and compartmentalization. The heavy metals (HM) present in the soil habitat is insoluble in nature and is not bioavailable (Li et al. 2010) for uptake by plant roots. Therefore, the plants release root exudates which in turn alters pH of the rhizosphere allowing the heavy metals to cationize and increase solubility. The heavy metals convert into a soluble form, mobile enough for surface absorption via root hairs and subsequently follows apoplastic (passive transport) methods, symplastic (active transport) methods to enter the root cells. Apoplastic pathway is inherently energy dependent in nature and requires an active metal ion carrier for transport. The process of bioaccumulation includes chelators (organic acids- citric acid) acting as ligands binds to the soluble metal, is a rapid procedure entailing the formation of metal–ligand complex or metal-phytochelatin complex (MPCs) (Nnaji et al. 2023). The MPCs inturn sequesters the metal ion to ascend into the extracellular (cell wall) and intracellular (vacuoles) spaces of the plant cell (Pankaj and Jugade 2023). Channeled transporters (H^+ coupled transporters) present in the transmembrane via cell-signalling and specified proteins (ZIP proteins; ZRT-IRT) translocate the heavy metals (HM) from roots to the aerial parts of the plant. In the shoot area, metal ions are mediated to the tissues where they are sequestered and compartmentalized. Thus, phytoaccumulation serves as a potential permanent solution which is also cost-effective, befitting commercial application. Around 44 families of angiosperms entailing more than 440 species extending from annual herbs to perennial shrubs and trees are identified as potential heavy metal hyperaccumulators. Napier grass cultivars (*Pennisetum purpureum*) helps in the absorption of zinc and cadmium metals (Lacalle et al. 2023). Fabaceae, Brassicaceae, Asteraceae, Euphorbiaceae and Lamiaceae are some other plant families with potential bioaccumulating capabilities. Researchers have demonstrated certain plants like *A. retroflexus, A. bettzickiana*can bioaccumulate more amounts of copper, nickel and molybdenum in their shoot areas.

14.6.2 Phytostabilization

Phyto-stabilization is one of the most specifically used plant-based method for eliminating toxins from highly polluted areas like industries, agricultural effluents and anthropogenic activities (Li et al. 2010; Lacalle et al. 2023). It involves immobilization of pollutants by plant roots to remove contaminants from surface and ground water. The phytoremediation technique of immobilization helps to reduce the bioavailability of heavy metals by reacting with soil for longer periods of time by natural attenuation or by addition of soil amendments. The bioavailability of heavy metals in the soil is directly related to the distinct physical, biological and chemical changes occurring in the rhizosphere. The excessive bioavailability of these toxic metalloids can be repressed by adapting phytostabilizing techniques. Selection of appropriate plant serves as an important step towards effective phyto-stabilization, as the plant must be able to withstand extremes of physiological conditions, bear heavy metal toxicity with minimal translocation to shoot areas and indigenous to the polluted areas (Mocek-Płóciniak et al. 2023). Phytostabilizing plant uses root exudates comprising of enzymes and polysaccharides to immobilize the heavy metals like arsenic, cadmium by the action of organic acids and siderophores. Protein complexes like phytosiderophore can bind to iron, cadmium, zinc and copper to cause immobilization or inactivation of contaminants. For bioaccumulation of the contaminants in the plant cell, root-metal molecular binding occurs with the help of ligands- phytochelatins (PCs) and metallothiones (MTs) which are activated due to the presence of heavy metals (Limmer and Burken 2016). Cell wall response in stimulated and through passive or active transport heavy metals are deposited in the root cells with the help of specified metal ion transporters (cell-signalling occurs) (Mocek-Płóciniak et al. 2023). The following leads to phytotoxic effects in roots of the plant and provides carbon source to enhance the growth of soil bacteria and mycorrhizae to mitigate toxicity of contaminated soil. Thus, phytostabilization is the possible remediative method for application in vast scale to foster improvement in soil growth (Limmer and Burken 2016).

14.6.3 Phytovolatilization

Phytovolatilization involves the conversion of heavy metals into volatile form which are dispersed from the evaporating sites of the plants. The mechanism involves compartmentalization of the contaminant into the vacuole spaces of the plant followed by subsequent diffusion in the atmosphere and is considered as one of the major contamination loss mechanisms (Sakakibara et al. 2010). There are two major forms of phytovolatilization are direct and indirect forms. Direct method is essential for uptake of pollutants from roots to upper parts via translocation and is converted into volatilize compounds which is emitted out by transpiration openings (stomata, cuticle, lenticels and periderm) of the plants. Indirect method involves absorption of advected contaminant and increase in volatile contaminant (gas) flux as a result of plant metabolic activities. The entire process of translocation and volatilization is regulated by the principle of passive transport, diffusion. It removes the contaminants (like Se, Hg, As) (Terry et al. 1995; Cheng et al. 2017) from its place of origin as it is dispersed in gaseous, less toxic form into the atmosphere. *Pteris vitata*is an example of phytovolatilizing plant that helps in neutralizing the toxicity of arsenic. Genetic engineering techniques are allowing model plants like hybrid poplars to volatilize certain specific contaminants (methyl-Hg) from the rhizosphere into the atmosphere in form of HgO, chlorinated acetates and CO_2. Researches are being constantly conducted to achieve phytovolatilization as an effective and safe technique for removal of contaminated agricultural soils and industrial wastewaters from the ecosystem (Agarwal et al. 2022).

14.6.4 Hydraulic Control

The term "hydraulic control" describes the process of adjusting subsurface water flow with plants and constructed systems. This exciting new technology has the ability to solve a number of environmental issues, including remediation of contaminated soil and groundwater plants can be used to extract and degrade contaminants from soil and groundwater through a process called phytoremediation. Hydraulic control can be used to direct the flow of contaminated water towards plant roots (Colares et al. 2020), enhancing the efficiency of phytoremediation. Erosion control since their roots are anchored in the earth, plants help stabilize slopes and stop soil erosion. Artificial wetlands and riparian buffers, which effectively prevent erosion and preserve water quality, can be created using hydraulic management. Leachate administration in landfills leachate, a contaminated liquid that can affect groundwater, is produced by landfills. Leachate can be diverted via hydraulic control from sensitive areas to treatment wetlands, where plants can aid in the removal of pollutants. Drought mitigation plants can help mitigate drought by drawing water from the soil and storing it there. Plants can be made more drought-resistant by optimizing their water usage through hydraulic management. Phyto-technologies can be employed with a variety of hydraulic control strategies (Sharma et al. 2021), which include pumping and treating: before being released or used again, contaminated water can be extracted from below the surface and disinfected. Drainage systems: you may regulate the water flow in the subsurface with trenches, subsurface drains, and ditches, wells for phytoremediation can be built to draw contaminated water out of the ground and supply it to plant roots. However, using hydraulic control exhibit numerous challenges including limited applicability, appropriate public education and economical affordability.

14.6.5 Phytodegradation

Phytodegradation (or phytotransformation) involves the degradation of organic contaminants and heavy metals by the action of enzymes secreted by the plants to enhance breakdown of toxic components into simpler molecules. These simpler molecules inturn helps in plant growth by incorporation of its fragments to the newly developed plant tissues. A transformation process to detoxify the pollutants with the help of plant–microbe interaction and primary and secondary metabolites produced in plants initiates plant growth, which in turn alleviates heavy metal toxicity. Depending on the stress conditions in the environment, root exudates serve as rich source of nutrients for PGPMs (Plant Growth Promoting Microorganisms) enriched with various organic acids which forms complexes with heavy metals (Li et al. 2010). Rhizosphere microorganisms in a favourable environment perform the most of the phytodegrading activities and plays a significant role in the degradation of pollutants like heavy metals, pesticides, herbicides, industrial chemicals and wastes (Sumira et al. 2016). The rate of degradation depends on the bioavailability of emerging pollutants, rhizospheric environment, plant specificity. The initial steps involve uptake of pollutants through the root system and metal-binding chelation occurs for translocation to specific tissues in the plant bodies. Metabolic activities occur followed by biochemical transformation catalyzed by the action of specific enzymes. These transformations lead to conversion of harmful pollutants to a certain less toxic form which are either volatilized or exudated. Hybrid poplars (*Populus spp.*) secretes dehalogenase, nitro-reductase enzyme to target chlorinated solvents and nitroaromatic compounds. Other phytodegrading plants like parrot feather (*Myriophyllum aquaticum*), willow (*Salix spp.*), giant duckweed (*Spirodelapolyrhiza*) secretes laccase, nitrilase, peroxidase phosphatase to catalyze cyanide groups, TNT, phenol phosphates and organophosphate pesticides into less toxic and more easily degradable form (Cherian and Margarida Oliveira 2005).

14.7 Phytoremediation—Mechanisms and Critical Factors

A plant-based approach to environmental remediation known as phytoremediation employs plants to absorb and remove toxins from the soil, water, and air. This process involves the capacity of some plants, referred to as hyperaccumulators, to take up, retain, and occasionally change contaminants. Compared to conventional remediation techniques, phytoremediation can be a more economical and environmentally beneficial strategy. The performance of phytoremediation initiatives is heavily influenced by a number of critical parameters pertaining to the pollutants and plants involved. The biomass of plants and the bioavailability of heavy metals in soil are the key factors affecting the efficiency of phytoremediation. It's worth noting that the low remediation efficiency and the lack of effective disposal methods for contaminated biomass have limited its development and application (Table 14.1).

Table 14.1 Various strategies of phytoremediation for effective bioremediation of pollutants in environments

Sl No	Types of degradation	Mechanism of degradation	Critical factors
1	Phytoaccumulation	1. Absorption and accumulation 2. Enhanced phytodegradation 3. Stabilization of breakdown products	1. Selection of proper plants 2. Contaminants uptake efficiency 3. Mobility of contaminants 4. Harvesting and disposal 5. Time taking procedure
2	Phytostabilization	1. Biotransformation 2. Bioaccumulation by roots 3. Immobilization	1. Specific plant selection 2. Uptake & accumulation by Roots 3. Soil amendments 4. pH levels 5. Environmental factors
3	Phytovolatilization	1. Uptake 2. Translocation 3. Transformation 4. Volatilization	1. Type of contamination 2. Availability of nutrients 3. Interaction of the microbe 4. Condition of the environment 5. Plant growth stage
4	Hydraulic control	1. Plant selection 2. Root growth 3. Water uptake 4. Contaminant control 5. Secondary mechanisms	1. Properties of the fluid 2. Pressure and flow rate 3. Valve properties 4. Response time 5. Environmental factors
5	Phytodegradation	1. Root growth 2. Water uptake 3. Contaminant interception 4. Rhizosphere 5. Biodegradation	1. Selection of plant species 2. Root exudates 3. Condition of soil 4. Temperature and climate factors 5. Phytodegradation pathways

14.8 Advanced Modifications in Phytoremediation

Phytoremediation is one of the most interesting field of ongoing research. Many phyto-technologies are regularly practiced to mitigate the increasing concentration of emerging pollutants from soil. Phytoextraction, phytodegradation, phytovolatization and phytostablization (as highlighted in Fig. 14.3) are the most practiced and common plant-based methods to decontaminate the soil (Conte et al. 2021). Recent advancements in this field of research includes genetic engineering methods to design transgenic plants that degrades and detoxifies xenobiotic synthetics. These plants are specifically designed to express specific genes that has increased ability to absorb and accumulate emerging contaminants (Adeoye et al. 2022).

Genes are encoded with stronger metal binding proteins that quickly degrades contaminants (Azhar et al. 2022; Latif et al. 2023). The over-expression of transport

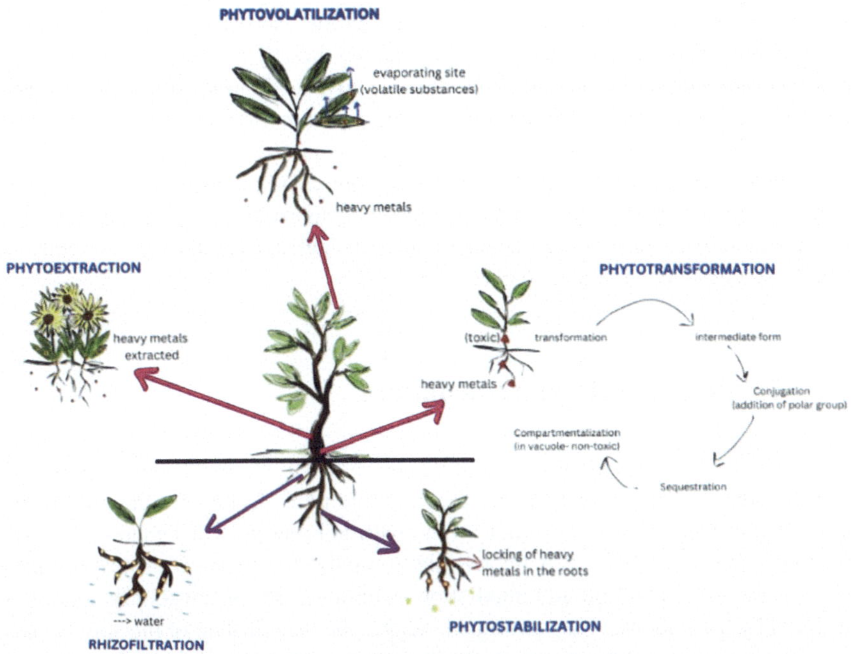

Fig. 14.3 A summary of various phytoremediation approaches for emerging contaminants

proteins in the roots enhances uptake of pollutants and increases bioavailability. The use of synthetic biology, induction of gene of interest and genome editing (CRISPR/Cas9 techniques) can help in improvement of stress tolerance enabling better metabolic uptake from soil. Transgenic endophytic organisms like bacteria and fungi when introduced into the plant body enhances the remediating capability by better uptake through roots to mitigate the pollutants. Facilitated Plant–Microbe interaction can establish symbiotic positive interaction of plant with beneficial microorganisms to increase the processes of accumulation and extraction. Bioinformatics and systemic biological studies helps to focus on understanding the complex interactions between plant and microbes and create more effective techniques (Mocek-Płóciniak et al. 2023).

A current advancement in phytoremediation includes development of designer plants with customized and specific trait of interest. Every plant species that contains unique beneficial characteristics can be crossed or genetically combined for an enhanced approach towards phytoremediation enforcing the advancements for production of more chimeric plants (Latif et al. 2023). Improvement in the rhizosphere and translocation in roots which includes root exudates modification for better soil structure and metal-binding chelation, enhanced rhizospheric processes to increase microbial activity is one of the essential field of research nowadays to expand the use of phytoextraction. The use of nano particles or nanotubes are categorical under the recent advancements. These nano materials improves uptake and absorption by plants increase nutrient availability, uptake thereby promoting plant growth and phytoremediation. Implementation of these advanced modifications, integrated automated control systems for appropriate optimization of phytoremediation technologies, advanced remote sensing technologies to monitor plant growth, tolerance, contamination concentration levels and health requires proper assessments ensuring sustainability of the advanced practices. Futuristic approaches towards phytoremediation would likely involve addressing of current challenges and more environment friendly potential technologies (Conte et al. 2021).

14.9 Limitation of Phytoremediation

Toxic heavy metals are only relocated by phytoremediation; they are not eliminated from the area. The depth and surface area that the roots occupy are the only areas where phytoremediation may occur. Low biomass and slow growth demand sustained effort (Houida et al. 2023). It is impossible to totally stop toxins from seeping into groundwater using plant-based remediation techniques. The survival of the plants is impacted by the toxicity of the polluted land and the overall state of the soil. Sometimes heavy metals are incorporated into the organic substance of the soil, rendering them inaccessible to plants. Phytoremediation is a quite a time taking process and it depends on many factors, like the age of the plants, the depth of the rootsfor effectiveness the roots of the plants must be in contact with the contaminated soil), the climatic condition and chemistry of the soil. The plants promote phytoremediation, always suffers from shunted growth and the contamination is always very high (Dickinson et al. 2009; Witters et al. 2012). And most importantly, when such plants are consumed, it is a great threat to human and animal life.

1. **Slow Process**: It achieves appreciable reductions in pollutant levels through phytoremediation might take years because it is sometimes a lengthy process This could be a drawback if quick correction is needed.
2. **Limited Plant Species**: There are a restricted number of hyperaccumulator species—plants that can accumulate high quantities of contaminants—and not all plants are appropriate for phytoremediation. It's possible that certain pollutants lack appropriate hyperaccumulator plants.
3. **Depth of Contamination**: In general phytoremediation works well for pollutants found in the top soil layers. It might not be appropriate to treat groundwater contamination or deeply ingrained contamination.
4. **Sensitivity towards Changes in the Climate**: The climate can have an impact on the effectiveness of phytoremediation. Drastic weather, high temperatures, or other unfavorable meteorological circumstances might impact on plant growth and, in turn, the restoration process.

5. **Contaminant Mobility**: Phytoremediation may not be able to stop highly mobile contaminants—like some heavy metals—from migrating to deeper soil layers or groundwater (Vangronsveld et al. 2009).
6. **Biomass Management**: After cleanup, harvesting and getting rid of the plant biomass can be logistically difficult. The harvested biomass may need to be properly disposed of in order to prevent recontamination, depending on the contaminants present.
7. **Needs for Nutrients:** Nutrients or soil amendments may be needed for some phytoremediation procedures in order to promote plant development and pollutant uptake. This may increase the cleanup effort's total cost and complexity (Witters et al. 2012).
8. **Danger of Release of Contaminants**: contaminants from the soil may occasionally be mobilized (Williams et al. 2002) into the plant tissues during the phytoremediation process. Contaminants may be released into the environment if this is not well handled.
9. **Site-Specific Environment**: The efficacy of phytoremediation greatly depends on the site. The particular pollutants present, the state of the soil, and the plant species chosen all affect how well the approach works. What is appropriate in one place could not be in another (Krishna et al. 2018).
10. **Regulatory Acceptance and Public Opinion**: It might be difficult to get public support and regulatory permission for phytoremediation projects. Concerns concerning the safety and efficacy of the approach may limit its wider implementation.

Despite these drawbacks, continuing studies and technological developments in phytoremediation seek to alleviate some of these issues. Furthermore, phytoextraction, sometimes called as hybrid systems, is a complete strategy that combines phytoremediation with other remediation processes to improve overall remediation efficiency.

14.10 Conclusion and Future Prospects

A green and ecologically friendly method of cleaning contaminated soil and water is called phytoremediation, which uses flora to eliminate, detoxify, or stabilize persistent contaminants. Plants are the most effective means of on-site heavy metal remediation because of their high biomass productivity and ease of removal. Heavy metals can enter a plant's body through its roots, stems, or leaves, and then accumulate in its organs. Elements are preferentially absorbed by plants. Soil pollution by metals is a global issue that varies in severity and extent depending on the location. Numerous remediation technologies have previously been thoroughly covered in other places, and each has a variety of benefits and drawbacks. Every method aims to use metal-contaminated soils safely for safe food chains and environmental quality. Phytoremediation is said to be the least expensive, least disruptive, and most ecologically benign cleanup technique available. However, the slow and labor-intensive nature of phytoremediation as well as the long-term viability of the plants involved must be carefully considered before the method is used on a commercial basis. There are many different types of plants that can accumulate large amounts of metals in their aerial parts (phytoextraction), hold the metals in their roots or stabilize them in the soil to prevent them from moving to the shoots (phytostabilization), or remove the metals from the soil by generating volatile compounds (phytovolatilization). Different mechanisms are involved in each of the aforementioned technologies, which have previously been thoroughly discussed. The type of soil, the kind of metal, the degree and extent of contamination, and the degree of environmental disturbance involved all influence the choice of phytoremediation technique to be used for cleaning up metal-contaminated locations. A thorough comprehension of the many dynamics at play will greatly enhance the decision-making process when it comes to adopting a certain technology. Phytoextraction is one of the fastest growing phytoremediation techniques, and a variety of hyperaccumulator plants that can accumulate high metal concentrations have been found. However, the majority of hyperaccumulators are low-biomass generating, slowly growing plants, which adds to the sluggish pace of this technology's development. Alternative method is to use fast-growing, high-biomass plants for chelant-assisted phytoextraction; nevertheless, leaching or run-off of solubilized metals into surface and subsurface water bodies is a major concern. Identification and induction of genes responsible for hyperaccumulation in hyperaccumulator plants into those plants which are capable of accumulating metals and producing high biomass might change the phytoremediation technique.

The efficiency of phytoextraction, the final use of its byproducts, and its overall economic feasibility all play major roles in its acceptability. The assumption that site rehabilitation should be completed in a stipulated time frame similar to other clean-up methods has limited commercial phytoextraction to this point. The majority of phytoremediation studies have been carried out in lab settings using heavy metal diets given to plants cultivated in hydroponic environments. While these results are encouraging, experts readily concede that the solution culture is substantially different from that of soil. The main obstacle faced in removing heavy metals (HM's) from actual soil is that many of them are bound up in insoluble forms that reduce their availability. Since phytoremediation is still in its early stages of study and development, there are still a number of technological issues that need to be resolved.

Acknowledgements The authors acknowledge their department and Institute for providing essential facilities and support for the preparation of the draft.

Conflict of Interest There is no potential conflict of interest between authors.

References

Abhilash PC, Jamil S, Singh N (2009) Transgenic plants for enhanced biodegradation and phytoremediation of organic xenobiotics. Biotechnol Adv 27(4):474–488

Adeoye AO, Adebayo IA, Afodun AM, Ajijolakewu KA (2022) Benefits and limitations of phytoremediation: heavy metal remediation review. In: Bhat RA, Tonelli FMP, Dar GH, Hakeem K (eds) Phytoremediation, pp 227–238. Academic Press

Agarwal S, Pramanick P, Mitra A (2022) Potential role of blue carbon in phytoremediation of heavy metals. In: Madhav S, Singh P, Mishra V, Ahmed S, Mishra PK (eds) Recent trends in wastewater treatment

Alves ARA, Yin Q, Oliveira RS, Silva EF, Novo LAB (2022) Plant growth-promoting bacteria in phytoremediation of metal-polluted soils: Current knowledge and future directions. Sci Total Environ 838(4)

Arshad U, Azeem F, Mustafa G, Bakhsh A, Toktay H, McGiffen M, Nawaz MA, Naveed M, Ali MA (2021) Combined application of biochar and biocontrol agents enhances plant growth and activates resistance against meloidogyne incognita in tomato. J Crop Health

Azhar U, Ahmad H, Shafqat H, Babar M, Munir HMS, Sagir M, Arif M, Hassan A, Rachmadona N, Rajendran S, Mubashir M, Khoo KS (2022) Remediation techniques for elimination of heavy metal pollutants from soil: a review. Environ Res 214(4)

Barathi S, Lee J, Venkatesan R, Vetcher AA (2023). Current status of biotechnological approaches to enhance the phytoremediation of heavy metals in India-a review. Plants (Basel)

Behera ID, Basak G, Kumar RR, Sen R, Meikap BC (2020) Treatment of petroleum refinery sludge by petroleum degrading bacterium Stenotrophomonas pavanii IRB19 as an efficient novel technology. J Environ Sci Health Part A Toxic/Hazard Subst Environ Eng 56(2). https://doi.org/10.1080/10934529.2020.1866924

Bhatt P, Ganesan S, Santhose I, Durairaj T (2023) Phytoremediation as an effective tool to handle emerging contaminants. Phys Sci Rev 8(9):2321–2340

Bhuyan K, Giri A (2020) Polycyclic aromatic hydrocarbon compounds as emerging water pollutants: toxicological aspects of phenanthrene on aquatic animals. In: Shikuku V (ed) Effects of emerging chemical contaminants on water resources and environmental health, pp 45–67. IGI Global

Birch GF, Drage DS, Thompson K, Eaglesham G, Mueller JF (2015) Emerging contaminants (pharmaceuticals, personal care products, a food additive and pesticides) in waters of Sydney estuary, Australia. Mar Pollut Bull 97(1–2):56–66

Bruno L, Fávaro-Polonio CZ, Pamphile JA, Polonio JC (2019) Effects of textile dyes on health and the environment and bioremediation potential of living organisms. Biotechnol Res Innovat 3(2):275–290

Castro G, Ramil M, Cela R, Rodríguez I (2021) Identification and determination of emerging pollutants in sewage sludge driven by UPLC-QTOF-MS data mining. Sci Total Environ 778

Cheng L, Wang Y, Cai Z, Liu J, Yu B, Zhou Q (2017) Phytoremediation of petroleum hydrocarbon-contaminated saline-alkali soil by wild ornamental Iridaceae species. Int J Phytoremediation 19(3):300–308

Cherian S, Margarida Oliveira M (2005) Environmental science & technology 39(24):9377–9390

Chunping Y, Chen H, Zeng G, Yu G, Luo S (2010) Biomass accumulation and control strategies in gas biofiltration. Biotechnol Adv 28(4):531–540

Colares GS, Dell'Osbel N, Wiesel PG, Oliveira GA, Lemos PHZ, da Silva FP, Lutterbeck CA, Kist LT, Machado EL (2020) Floating treatment wetlands: a review and bibliometric analysis. Sci Total Environ 714

Conte A, Chiaberge S, Pedron F, Barbafieri M, Petruzzelli G, Vocciante M, Franchi E, Pietrini I (2021) Dealing with complex contamination: a novel approach with a combined bio-phytoremediation strategy and effective analytical techniques. J Environ Manag 288

Cray TYJ, Connor JA, Mairéad C, Stevenson A, Houghton JDR, Rangel DEN, Cooke LR, Hallsworth JE (2016) Biocontrol agents promote growth of potato pathogens, depending on environmental conditions. Microb Biotechnol J Microb Biotechnol 9:330–354

Das AP, Dutta K, Khatun R, Behera ID, Singh S, Mishra S (2023) Microfiber pollution and its microbial mitigation: a review on current trends and future prospects. J Taiw Inst Chem Eng 105104. https://doi.org/10.1016/J.JTICE.2023.105104

David GS, Ramírez-Restrepo S, López-Luján L, Peláez-Jaramillo CA, Bedoya-Pérez JC (2018) Screening of a biological control bacterium to fight avocado diseases: from agroecosystem to bioreactor. Biocatal Agric Biotechnol 14:109–115

Dickinson NM, Baker AJM, Doronila A, Laidlaw S, Reeves RD (2009) Phytoremediation of inorganics: realism and synergies. Int J Phytorem 11(2):97–114

Dietz AC, Schnoor JL (2001) Advances in phytoremediation. Environ Health Perspect 109 Suppl1(Suppl 1):163–168

Enrico B, Basaglia M, Fontana F, Shapleigh JP, Casella S (2015) Development, assessment and evaluation of a biopile for hydrocarbons soil remediation. Int Biodeterior Biodegradation 98:66–72

Geissen V, Mol H, Klumpp E, Umlauf G, Nadal M, van der Ploeg M, Sjoerd EATM, van de Zee C, Ritsema J (2015). Emerging pollutants in the environment: a challenge for water resource management. Int Soil Water Conserv Res 3(1):57–65

Ghosh S, Das AP (2020) Microbial metagenomics: current advances in investigating microbial ecology and population dynamics. In: Nayak SK, Mishra BB (eds) Frontiers in soil and environmental microbiology, 1st ed, pp 247–254. CRC Press

Glenna T, Sheridan C, Holm PE (2022) A critical review of phytoremediation for acid mine drainage-impacted environments. Sci Total Environ 811

Guanqun F, Huang H, Chen Y (2021) Effects of emerging pollutants on the occurrence and transfer of antibiotic resistance genes: a review. J Hazard Mater 420

Houida S, Yakkou L, Kaya LO, Bilen S, Raouane M, El Harti A, Amghar S (2023) Plant growth-promoting bacteria isolated from earthworms enhance spinach growth and its phytoremediation potential in metal-contaminated soils. Int Microbiol

Huerta B, Rodriguez-Mozaz S, Nannou C, Nakis L, Ruhí A, Acuña V, Sabater S, Barcelo D (2016) Determination of a broad spectrum of pharmaceuticals and endocrine disruptors in biofilm from a waste water treatment plant-impacted river. Sci Total Environ 540:241–249

Khan AU, Khan AN, Waris A, Ilyas M, Zamel D (2022) Phytoremediation of pollutants from wastewater: a concise review. Open Life Sci 13 16(1):488–496

Khandare RV, Govindwar SP (2015) Phytoremediation of textile dyes and effluents: current scenario and future prospects. Biotechnol Adv 33(8):1697–1714

Krishna KY, Gupta N, Kumar A, Reece LM, Singh N, Rezania S, Khan SA (2018) Mechanistic understanding and holistic approach of phytoremediation: a review on application and future prospects. Ecol Eng 10:274–298

Kurt S, Baan R, Grosse Y, Secretan B, El Ghissassi F, Cogliano V (2005) Carcinogenicity of polycyclic aromatic hydrocarbons. Lancet Oncol 6(12):931–932

Lacalle RG, Bernal MP, Álvarez-Robles MJ, Clemente R (2023) Phytostabilization of soils contaminated with As, Cd, Cu, Pb and Zn: Physicochemical, toxicological and biological evaluations. Soil Environ Health 1(2)

Lahiri D, Nag M, Dey A, Sarkar T, Joshi S, Pandit S, Das AP, Pati S, Pattanaik S, Tilak VK, Ray RR (2021) Biofilm mediated degradation of petroleum products. Geomicrobiol J 1–10.https://doi.org/10.1080/01490451.2021.1968979

Latif A, Abbas A, Iqbal J et al (2023) Remediation of environmental contaminants through phytotechnology. Water Air Soil Pollut 234, 139

Lauretta R, Andrea S, Massimiliano S, Francesco R, Marialuisa A (2019) Endocrine disrupting chemicals: effects on endocrine glands. Front Endocrinol 10

Li JT, Liao B, Lan CY, Ye ZH, Baker AJM, Shu WS (2010) Cadmium tolerance and accumulation in cultivars of a high-biomass tropical tree (*Averrhoa carambola*) and its potential for phytoextraction

Limmer M, Burken JG (2016) Phytovolatilization of organic contaminants. Environ Sci Technol 50(13):6632–6643. American Chemical Society (ACS)

Limmer M, Burken J (2016) Environ Sci Technol 50(13):6632–6643. https://doi.org/10.1021/acs.est.5b04113
Mishra S, Swain S, Sahoo M, Mishra S, Das AP (2021) Microbial colonization and degradation of microplastics in aquatic ecosystem: a review. Geomicrobiol J. https://doi.org/10.1080/01490451.2021.1983670
Mishra S, Dash D, Al-Tawaha ARMS, Das AP (2022) A review on heavy metal ion adsorption on synthetic microfiber surface in aquatic environments. Appl Biochem Biotechnol 194(10):4639–4654. https://doi.org/10.1007/S12010-022-04029-W/METRICS
Mishra S, charan Rath C, Das AP (2019) Marine microfiber pollution: a review on present status and future challenges. Mar Pollut Bull 140:188–197.https://doi.org/10.1016/J.MARPOLBUL.2019.01.039
Mishra S, Rout PK, Das AP (2021) Emerging microfiber pollution and its remediation. In Prasad R (ed) Environmental and microbial biotechnology, pp 247–266. Springer, Singapore. https://doi.org/10.1007/978-981-15-5499-5_9
Mishra S, Singh RP, Rout PK, Das AP (2022) Membrane bioreactor (MBR) as an advanced wastewater treatment technology for removal of synthetic microplastics. Dev Wastewater Treat Res Process 45–60.https://doi.org/10.1016/B978-0-323-85583-9.00022-3
Mishra S, Dash D, Das AP (2023) Aquatic microbial diversity on plastisphere: colonization and potential role in microplastic biodegradation. Geomicrobiol J 1–12.https://doi.org/10.1080/01490451.2023.2209750
Mocek-Płóciniak A, Mencel J, Zakrzewski W, Roszkowski S (2023) Phytoremediation as an effective remedy for removing trace elements from ecosystems. Plants 12:1653
Nag M, Lahiri D, Ghosh S, Sarkar T, Pati S, Das AP, Ram DK, Bhattacharya D, Ray RR (2023) Application of microorganisms in biotransformation and bioremediation of environmental contaminant: a review. Geomicrobiol J. https://doi.org/10.1080/01490451.2023.2261443
Nagajyoti PC, Lee KD, Sreekanth TVM (2010) Heavy metals, occurrence and toxicity for plants: a review. Environ Chem Lett 8:199–216
Nnaji ND, Onyeaka H, Miri T et al (2023) Bioaccumulation for heavy metal removal: a review. SN Appl Sci **5**:125
Ons L, Bylemans D, Thevissen K, Cammue BPA (2020) Combining biocontrol agents with chemical fungicides for integrated plant fungal disease control. Microorganisms 8:1930
Pankaj B, Jugade R (2023) Phytostabilization and rhizofiltration of toxic heavy metals by heavy metal accumulator plants for sustainable management of contaminated industrial sites: a comprehensive review. J Hazard Mater Adv 10
Pollard RT (1977) Chapter 2 hydrocarbons. In: Bamford CH, Tipper CFH (ed) Comprehensive chemical kinetics, vol 16, pp 249–367. Elsevier
Prem R, Roy S, Ganguly A, Mandi M, Dutta A, Das K, Nanda S, Ghanty S, Biswas G (2023) Agricultural pesticides–friends or foes to biosphere. J Hazard Mater Adv 10
Rego Richelle M, Kuriya G, Kurkuri MD, Kigga M (2021) MOF based engineered materials in water remediation: recent trends. J Hazard Mater 403
Russell RC (1999) Constructed wetlands and mosquitoes: health hazards and management options—an Australian perspective. Ecol Eng 12(1–2):107–124
Sakakibara M, Watanabe A, Inoue M, Sano S, Kaise T (2010) Phytoextraction and phytovolatilization of arsenic from as-contaminated soils by Pteris vittata. In: Proceedings of the Annual International Conference on Soils, Sediments, Water and Energy, vol 12, Article 26
Sandermann H Jr (1994) Higher plant metabolism of xenobiotics: the 'green liver' concept. Pharmacogenetics 4(5):225–241
Sharma R, Vymazal J, Malaviya P (2021) Application of floating treatment wetlands for stormwater runoff: a critical review of the recent developments with emphasis on heavy metals and nutrient removal. Sci Total Environ 777
Sharma RS, Karmakar SK, Mishra P, Vandana (2019) Application of filamentous phages in environment: a tectonic shift in the science and practice of eco-restoration. Ecol Evolut Ecol 9

Sharma JK, Nitish K, Singh NP, Rani SA (2023) Phytoremediation technologies and their mechanism for removal of heavy metal from contaminated soil: an approach for a sustainable environment. Front Plant Sci 13

De Silva Nimali I, Brooks S, Lumyong S, Hyde KD (2019) Use of endophytes as biocontrol agents. Fungal Biol Rev 33(2):133–148

Singh A, Parmar N, Kuhad RC, Ward OP (2011) Bioaugmentation, biostimulation, and biocontrol in soil biology. In: Singh A, Parmar N, Kuhad R (eds) Bioaugmentation, biostimulation and biocontrol. Soil biology, vol 108. Springer

Slama HB et al (2021) Biotechnology and bioinformatics of endophytes in biocontrol, bioremediation, and plant growth promotion. In: Maheshwari DK, Dheeman S (eds) Endophytes: mineral nutrient management, vol 3. Sustainable development and biodiversity, vol 26. Springer, Cham

Smith MJ, Flowers TH, Duncan HJ, Alder J (2006) Effects of polycyclic aromatic hydrocarbons on germination and subsequent growth of grasses and legumes in freshly contaminated soil and soil with aged PAHs residues, \qewzx Environ Pollut 141(3):519–525

Stephenson C, Black CR (2014). One step forward, two steps back: the evolution of phytoremediation into commercial technologies. Biosci Horiz Int J Stud Res 7

Sumira J, Rashid B, Azooz MM, Hossain MA, Ahmad P (2016) Genetic strategies for advancing phytoremediation potential in plants: a recent update, Parvaiz Ahmad. Plant metal interaction. Elsevier, pp 431–454

Sun R, Sun P, Zhang J, Esquivel-Elizondo S, Wu Y (2018) Microorganisms-based methods for harmful algal blooms control: a review. Bioresour Technol 248(B):12–20

Tchounwou PB, Yedjou CG, Patlolla AK, Sutton DJ (2009) Heavy metal toxicity and the environment. In: Wang J, Chen C (2009) Biosorbents for heavy metals removal and their future. Biotechnol Adv 27(2):195–226

Terry N, Zayed A, Pilon-Smits E (1995) Bioremediation of selenium by plant volatilization. United States

Thijs S, Sillen W, Weyens N, Vangronsveld J (2017) Phytoremediation: state-of-the-art and a key role for the plant microbiome in future trends and research prospects. Int J Phytorem 19(1):23–38

Tripathy B, Dash A, Das AP (2022) Detection of environmental microfiber pollutants through vibrational spectroscopic techniques: recent advances of environmental monitoring and future prospects. Crit Rev Anal Chem 1–11.https://doi.org/10.1080/10408347.2022.2144994

Tudi M, Daniel Ruan H, Wang L, Lyu J, Sadler R, Connell D, Chu C, Phung DT (2021) Agriculture development, pesticide application and its impact on the environment. Int J Environ Res Public Health 27

Vangronsveld J, Herzig R, Weyens N et al (2009) Phytoremediation of contaminated soils and groundwater: lessons from the field. Environ Sci Pollut Res 16:765–794

Velusamy SR, Sundaram A, Senthilarasu KM (2021) A review on heavy metal ions and containing dyes removal through graphene oxide-based adsorption strategies for textile wastewater treatment. Chem Record 21:1570–1610

Williams JB (2002) Phytoremediation in wetland ecosystems: progress, problems, and potential. Crit Rev Plant Sci 21(6):607–635

Witters N, Mendelsohn RO, Van Slycken S, Weyens N, Schreurs E, Meers E, Tack F, Carleer R, Vangronsveld J (2012) Phytoremediation, a sustainable remediation technology? Conclusions from a case study. I: energy production and carbon dioxide abatement. Biomass Bioenergy 39:454–469

Chapter 15
Green Technology for Sustainable Environmental Waste Management Through Phytoremediation

Prajna Priyadarshini Jena, Sailaja Priyadarsini, Jyotirmayee Mohanty, and Alok Prasad Das

Abstract The rapid expansion of industrialization and progress in technology can lead to the creation of novel products and materials, some of which might contain harmful elements. The quick disposal of electronic devices, for instance, contributes to electronic waste, often containing hazardous substances. Insufficient or improper waste management methods, illegal dumping, inadequate storage, and the absence of appropriate disposal facilities can result in the emission of hazardous substances into the environment, worsening the issue. The Energy and Resources Institute (TERI) reported that India generates over 62 million tons (MT) of waste each year. Environmental biotechnology emerges as a nascent discipline that incorporates living materials, predominantly plants, along with small organisms such as earthworms and microorganisms to tackle issues related to environmental management and sustainable development. Phytoremediation, an eco-friendly and sustainable approach, harnesses the unique capabilities of plants to address diverse types of waste materials, encompassing heavy metals, organic pollutants, and contaminants in soil, water, and air. T(Dashis method entails the utilization of specially chosen plants, recognized as hyperaccumulators, which can assimilate, accumulate, and detoxify pollutants through various physiological and biochemical mechanisms. As the global demand for sustainable environmental solutions increases, phytoremediation emerges as a promising technology with the potential to alleviate the repercussions of industrial activities and pollution. Further exploration and application of phytoremediation strategies can contribute to fostering a cleaner and healthier environment, fostering a balanced coexistence between nature and human endeavors. This review discusses various human-caused origins of environmental waste and their effects on the environment. Additionally, it examines diverse phytoremediation techniques and the involvement of mycorrhiza in this process.

P. P. Jena · S. Priyadarsini · J. Mohanty · A. P. Das (✉)
Department of Life Sciences, Rama Devi Women's University, Vidya Vihar, Bhubaneswar, Odisha, India
e-mail: alok1503@gmail.com

A. Kuanar et al. (eds.), *Recent Advances in Bioremediation and Phytoremediation*, Environmental Science and Engineering,
https://doi.org/10.1007/978-3-031-77884-1_15

Keywords Microorganisms · Heavy metals · Phytoremediation · Pollutants · Toxicity

15.1 Introduction

Sustainable development is a fundamental principle of environmentally friendly technology (Mohanty et al. 2024). The prevailing assumption posits that over the past two decades, this field will yield both significant disruptions and advancements in everyday life because of the explosion of technology which generates a significant amount of waste materials like plastics (Das et al. 2023; Mishra et al. 2023; Mishra and Das 2023), heavy metal (Das et al. 2011; Das and Mishra 2008) (Swain et al. 2025), and biomedical waste (Khatun et al. 2023; Sadasivuni et al. 2019). At this nascent stage, it is challenging to anticipate the ultimate scope of Green Technology. The ultimate objective of Green Technology is to meet society's needs sustainably, without exhausting natural resources. The quest for environmentally friendly technologies aims to replace energy production methods that are both sustainable and cost-effective (Yousuf et al. 2025). This concern is specifically focused on meeting future energy requirements through renewable energy production and reducing greenhouse gas emissions (Mishra et al. 2022) (Das et al. 2025).

The Energy and Resources Institute (TERI) estimates that India produces more than 62 MTs of garbage annually. Out of this total waste, only 43 MTsare collected, while the residual 31 MTsare directly dumped inyards, with 12 MTs undergoing processing before disposal. According to data published in the Journal of Urban Management (December 2021), there are 62 MT of rubbish created annually, of which 0.17 MT are biological waste, 7.9 MT are hazardous waste, 1.5 MT are e-waste, and 5.6 MT are plastic waste. The Central Pollution Control Board (CPCB) recently projected that by 2030, India's annual waste generation would amount to 165 MT. It is anticipated that the amount of hazardous, plastic, e-waste, and biomedical waste produced will rise correspondingly. Owing to the ever-increasing process of industrialization coupled with the agricultural process, the pollution of soil with heavy metals has emerged as a significant environmental fearon a global scale (Das and Ghosh 2023a, b; Mishra et al. 2023). The heavy metals are additionally divided into two groups: essential and non-essential. The micronutrients known as essential heavy metals, such as manganese (Mn), iron (Fe), nickel (Ni), chromium (Cr), and zinc (Zn), cobalt (Co),copper (Cu) and lead (Pb) are necessary for many biological functions, but when ingested in excess, they can become poisonous (Das and Ghosh 2023g; Prabhakar et al. 2019; Raj and Das 2023). However, non-essential heavy metals are extremely harmful to living things (Priyadarsini and Das 2024, 2025). Examples of these are lead (Pb), cadmium (Cd), and mercury (Hg). Though naturally occurring in rocks, human activity has significantly increased the accumulation of heavy metals. Heavy metals in soil can come from the over-employment of manure, wastewater, sewage sludge, biosolids, and, agrochemicals among other things (Khatun et al. 2023b; Kumar et al. 2023). It has been discovered that the buildup of these heavy metals in the soil has detrimental effects on human, animal, and plant health. Based on extant research, it has been approximated that 10 million individuals globally have been affected by heavy metal-contaminated soil (Mishra et al. 2023b; Nag et al. 2023a). Heavy metal accumulation in crops due to soil-root interaction carries a considerable risk. The two most crucial natural resources for soil and water for both civilization's long-term survival and the sustainability of agriculture are generally accepted. Additionally, both human and natural activity have contributed to soil and water contamination (Das and Ghosh 2023h, e; Ghosh et al. 2023). The ecology, natural habitats, and public health are all seriously endangered by the garbage that these activities produce. Waste's organic components decompose naturally, but heavy metals and metalloids pose a different kind of problem because of their enduring presence in the environment (Bal et al. 2020; Behera et al. 2020; Kanungo et al. 2023; Nag et al. 2023b). Some of these heavy metals are useful to biological systems, therefore they need to be present in certain concentration ranges. Humans require heavy metals such as Fe, Co, Mn, Mo, and Znfor various physiological and biochemical functions, but excessive amounts can have detrimental effects (Das and Ghosh 2023b, c, d). However, certain heavy metals such as Hg, Cd, Pu, and Pb do not have any beneficial impact on organisms (Das and Mishra 2009; Lahiri et al. 2021; Mishra et al. 2009).

From the standpoint of plant-based remediation, phytoremediation entails either extracting and removing contaminants or lowering their mobility and bioavailability in soil. Because the plants have vast root systems, these techniques work even in low-contamination situations. Phytoremediation offers an economically viable, environmentally friendly approach that can be applied on a large scale. This procedure also introduces various organic compounds into the ground, enhancing soil fertility by inhibiting erosion and the leaching of metals. While phytoremediation holds promise, it faces challenges in effectively purifying polluted land. The management and safe disposal of plant biomass pose a significant hurdle due to the potential release of pollutants into the nearby environment (Mohanty et al. 2025). Over the years, researchers have been focusing on various technologies, such as compaction, combustion, composting, and gasification, to handle and dispose of contaminated biomass safely. Recent research has expanded the scope of phytoremediation, introducing novel technologies for addressing metal-contaminated soil, such as phytoextraction, Phyto-stabilization, and phytovolatilization. Within the field of phytoremediation, plants with the ability to accumulate metals play a crucial role. These plants possess the remarkable capability to absorb metals at concentrations up to 100–1000 times higher compared to non-accumulating plants. It is noteworthy that metal-accumulating species have been identified across at least 45 plant families, with individual species demonstrating the capacity to accumulate various metals including but not limited to Se, As, Cu, Cd, Co, Ni, and Zn, either individually or in specific combinations.

15.2 Sources of Environmental Waste

15.2.1 Heavy Metals

Heavy metal pollution has drawn global attention due to its harmful effects on living organisms via the accumulation process in biological systems. Heavy metals, characterized by their higher elemental density, are commonly categorized in this regard (Mishra et al. 2009, 2023; Das and Mishra 2010; Das and Bissoyi 2011). These metals are further classified based on their density, weight, or atomic properties. A proposed new standard recommends that heavy metals should possess an atomic density exceeding 41 g/cm^3, with examples including Pb, Zn, Co, Cr, Hg, Cu, Cd, and other metals. Even minimal amounts of heavy metals within a system can profoundly affect ecosystems, both on land and in water. The origins of this impact are often linked to diffuse factors, such as air emissions, identified as the primary culprits for soil and water pollution. Heavy metals stand out as particularly perilous and potent water contaminants due to their thoughtful threats to both human health and biological habitats (Das and Ghosh 2018; Ghosh et al. 2020; Mohanty et al. 2018). Although soil erosion, volcanic eruptions, and rock fragmentation are among the natural processes that release heavy metals into the environment, human activity also contributes significantly to their existence (Das et al. 2012, 2015b; Mohanty et al. 2016). The incomplete combustion of fossil fuels, mineral mining, landfilling, metal refining, the production of electronic goods, the production of dyes, the use of agricultural chemicals, military operations, and automobile emissions are all included in this list of activities. Toxic heavy metals like Cu, As, Pb, Cr, Cd, and Hg, and are frequently found in wastewater from mining activities, sewage, dyes, and alloys. However, compared to natural sources, human-made activities contribute significantly more to environmental contamination. Many heavy metals are extremely harmful due to their reactivity, which has a deleterious effect on the environment and human health (Das et al. 2012, 2015a; Ghosh and Das 2017).

15.2.2 Wastewater

Wastewater, a significant waste source originating from human activities, encompasses both household dirty water and industrial polluted water (Mishra and Das 2023b; Pradhan et al. 2023). Consequently, governmental attention has consistently focused on wastewater treatment due to its detrimental impact, including the direct emission of wastewater that harms aquatic organisms, alters the pH of the surrounding soil, and poses health risks. There are several ways to treat wastewater, including chemical, biological, and physical methods. Scientists have recently argued that biological wastewater treatment is preferable to chemical techniques. The fastest method of purification is chemical treatment; nevertheless, the presence of synthetic chemical compounds in the treated water might have unanticipated ecological and environmental effects. Furthermore, it's thought that the biological method of treating wastewater is the most eco-friendly. Microorganisms have been used for biodegradation—the breakdown of big particles and removal of contaminants from wastewater—over the past few decades. Among the various types of biodegradations, aerobic biodegradation, which requires oxygen, is the most common. In the realm of biological wastewater treatment, microorganisms have been widely used to remove pollutants such as nitrates, phosphorous, and sulfates (Zheng et al. 2013) (Mohanty et al. 2025). Additionally, microalgae have gained popularity in the field of wastewater treatment. Notably, microalgae are effective in removing emerging contaminants found in municipal wastewater, as well as waste streams from pharmaceutical and personal care products (Yadav et al. 2019). It has been reported that the quantity of dairy wastewater produced is approximately three times that of the milk treated. Dairy effluent has high levels of BOD and COD and is made up of complex inorganic nutrients like phosphates and nitrogen. High concentrations of fats, oils, and dissolved particles also make the treatment procedure more difficult. To address this challenge, new technologies have been developed for the treatment of dairy wastewater.

15.2.3 Waste from Aquaculture

The primary contributors to solid waste comprise unconsumed feed and excrement from farmed fish, as indicated by (Akinwole et al. 2016). Occasionally, this category may encompass fish that do not complete the cultivation process. Solid wastes can be further categorized into suspended solids and settled solids. Dissolved wastes are constituted by the byproducts of fish food metabolism or the decomposition of uneaten feed. The two main elements of dissolved waste that need to be taken into consideration are phosphorus (P) products and nitrogen (N) (Boyd and Massaut 1999).

15.3 Phytoremediation

Phytoremediation utilizes metal-accumulating plants to restore damaged primary sources like soil and water (Muthusaravanan et al. 2018). This technology employs various mechanisms such as phytostabilization, phytovolatilization, phytodegradation, rhizodegradation, phytoextraction, and phytoaccumulation for heavy metal absorption (Tangahu et al. 2011). Selected plant species demonstrate the ability to absorb significant quantities of heavy metals into their roots, withstand different metal types, and adapt physiologically to diverse environments (Prieto et al. 2018). Aquatic hydrophytes like E. crassipes, E. stagnina, L. stolonifera, and P. australis show promise in accumulating Cd, Pb, and Ni, particularly in polluted wetlands (Eid et al. 2020). Phragmites australis exhibits the highest capacity for Cd and Ni accumulation, while Echinochloa stagnina shows significant Pb accumulation. Except Ni in

Echinochloa stagnina and Cd in Ludwigia stolonifera, these species are suitable for phytoextraction and phytostabilization of the tested heavy metals, making phytoremediation an economically viable and effective technique for mitigating heavy metals in water and soil.

15.4 Role of Mycorrhiza in Phytoremediation

15.4.1 Nutrient Acquisition

Apart from aiding plants in absorbing detrimental metals from their surroundings, mycorrhizae also enhance the supply of essential nutrients. Studies indicate that arbuscular mycorrhizal (AM) fungus can enhance the fungus's tolerance to metals and can bind metals to polyphosphates within the fungal hyphae. This binding mechanism can fortify the resistance of certain plants to harmful metal pollution (Barea et al. 2005; Morgan et al. 2005). The establishment of the mycorrhizal network offers various substantial advantages for the host plant's acquisition of mineral nutrients.

(i) The fungal hyphae extend beyond the zone depleted of nutrients surrounding the plant roots.
(ii) The fungal hyphae substantially increase the available surface area for absorbing nutrients.
(iii) Plant roots are unable to reach soil pores that hyphae can enter.
(iv) Certain mycorrhizal fungus, particularly those found in organic kinds, can provide plants with forms of phosphorus (P) and nitrogen (N) that are unavailable to plants that do not have mycorrhizal relationships.

The impact of AM fungus and refuse compost on phytoremediation using vetiver grass slips was assessed in a field study conducted on Pb/Zn mining tailings (Morgan et al. 2005; Wu et al. 2011). Following mycorrhizal treatments, the shoot contents of N and P were notably higher (N: 13.6 mg kg^{-1}; P: 1.26 mg kg^{-1}) compared to shoots without AM fungus inoculation (N: 11.7 mg kg^{-1}; P: 0.91 mg kg^{-1}). Additionally, AM fungus produces glomalin, an insoluble glycoprotein that forms a complex, sequestering trace element. This contributes to enhancing the remediation of contaminated soils (Lovelock et al. 2004).

15.4.2 Plant Protection

Another crucial role of mycorrhizal symbiosis is to shield roots from the harmful impacts of metal toxicity by regulating communicationsamong metals and plant roots. Various types of mycorrhizae, such as ectomycorrhizae, arbuscular mycorrhizae (Sharples et al. 2000; Tian et al. 2011), and ericoid mycorrhizae (Sharples et al. 2000), have demonstrated the ability to remove the toxic effects of metals on higher plants. In some cases, natural mycorrhizae can precipitate metal oxalates within the intracellular compartments of the host plant or fungus when exposed to contaminated environments. This precipitation restricts the apoplastic transport of metals through the Casparian strips. As an alternative, mycorrhizal fungi can actively defend the plant by stopping toxic levels of contaminants from building up. They do this by binding the contaminants inside the fungal tissues connected to the roots or by secreting certain detoxifying substances, including organic acids. By doing this, a physical barrier is established that stops harmful metals from moving to the plant itself. Studies on the ectomycorrhizal defense against copper toxicity offered by Pinus sylvestris were carried out by Vosatka et al. When S. bovinus and T. terrestris mycorrhizae were present, the average copper levels in the roots of plants grown in copper-contaminated soil were found to be much lower (433 and 443 mg Cu g^{-1} dry weight) than when these mycorrhizae were absent (3,630 mg Cu g^{-1} dry weight).

Fungi can amass micronutrients and toxic metals over their nutritional needs, giving rise to the acknowledgment of the biosorption capacity of fungal biomass in eliminating toxic metals. The absorption of metal ions by both living and deceased cells can transpire through two mechanisms. In the physical process, complexation interactions with functional groups like carboxyls, sulphydryls, and amineslead to the biosorption of harmful metal ions on the surface of cells through ion exchange. Metal ions enter the cytosol during the biological process by passing through the cell wall and cell membrane. According to the roots of *Nicotiana tabacum* colonized with *G. intraradices* (120 mg kg^{-1}) absorbed Cd 20 times less than the extraradical mycelium of *Glomus intraradices* treated to Cd (2,592 mg kg^{-1}). They found that by improving Cd sorption in the arbuscular mycorrhizal (AM) fungus, mycorrhizal symbiosis may help reduce Cd toxicity in plants. Consequently, their results point to a possible function for AM fungal symbiosis in the phytostabilization of contaminated soils.

15.5 Phytoremediation Techniques

15.5.1 Phytoextraction

Toxins are absorbed and transported from plant roots to shoots by a process called phytoextraction, which is sometimes referred to as phytoaccumulation, photoabsorption, or phytosequestration. To effectively eliminate soil contaminants, it is essential to harvest both the roots and shoots. Chandra and Kumar (2017) conducted a study revealing that twelve naturally occurring plants cultivated on sludge demonstrated the ability to absorb iron. Among them, *A. spinosus L.* emerged as an accumulator of Zn and Mn in the shoots. It was found that the leaves of all plants—aside from A. spinosus L. and Ricinus communis—accumulated Fe, Mn, and Zn. The research conducted by Lajayer and colleagues showcased the auspicious capacity of ornamental plants in phytoextracting heavy metals (HMs) from contaminated surroundings. This approach appears to be more environmentally friendly compared to traditional cleansing methods. crops produced in agricultural soils irrigated with refinery wastewater may have health risks, so Ugya et al. (2019) investigated into the phytoextraction capacity of several crops. A notable build-up of mercury and cadmium in the veggies was found by the study. Using Trifolium repens, the phytoextraction method for Cd, Ni, and Cu was investigated for potential improvement in a study by Makarova et al. The outcomes showed that potassium salt of K2HEDP, iron chelate, and plant development regulators were capable of extracting copper using phytoextraction.

15.5.2 *Phytostabilization*

However, the process involves immobilizing heavy metals (HMs) in polluted soils using plants that possess resistance to these metals, thereby reducing their availability in the environment (Yan et al. 2020). This prevents their entry into the food chain or migration into groundwater. investigated the effectiveness of four woody plants combined with organic additives for stabilizing zinc smelting residue. Results showed that planting these plants directly in the zinc smelting residue increased nutrient accumulation and decreased the availability of lead, copper, and zinc. In contrast, Saran et al. (2020) conducted a demonstration of phytostabilization using Helianthus petiolaris, an aromatic plant. H. petiolaris effectively stabilized soil with concentrations of up to 500 mg/Kg of Pb^2 and 50 mg/Kg of Cd^2. It also exhibited the capability to transfer significant amounts of Pb to its aboveground parts, accumulating nearly three times the level of Cd present in the soil in its aboveground parts.

15.5.3 *Rhizofiltration*

Rhizofiltration, a method utilizing plant roots to purify wastewater from contaminants, involves the absorption of heavy metals (HMs) by root exudates, leading to changes in the pH of the rhizosphere (Yan et al. 2020). In accordance with a study by Benavides et al. (2018), Zea mays has a greater capacity for absorbing and accumulating mercury. Two distinct species, *Phragmites australis* and *Kyllinga nemoralis*, were used to evaluate the efficacy of a rhizofiltration device in the removal of heavy metals (HMs) from municipal wastewater in the setting of a sewage treatment plant in KwaZulu-Natal, a province in South Africa. There were differences in the HM concentrations between the rhizofilter's planted and reference portions. Additionally, the root system of *K. nemoralis* exhibited a 33% increase in Cd levels and a 21% increase, indicating the system's effectiveness in extracting HMs from wastewater (Odinga et al. 2019).

15.5.4 Phytovolatilization

With the use of plants, different pollutants are converted into less dangerous volatiles using biofiltration technology, and these volatiles are then released into the atmosphere through the foliar system. Yan and colleagues state that this method can be applied to remove organic pollutants as well as heavy metals as arsenic (As), mercury (Hg), and selenium (Se). Dimethyl diselenide is produced when *Astragalus racemosus* phytovolatilizes selenium, and mercury's volatility is increased when Arabidopsis thaliana converts Hg^{2+} to Hg0. The results align with those of earlier research. Additionally, Sakakibara et al. (2010) demonstrated that Pteris vittata can convert arsenic into volatile forms. During the phytovolatilization process, the removal or disposal of heavy metal (HM) contaminants involves extracting them from the soil and allowing them to disperse as gaseous molecules (Yan et al. 2020).

15.5.5 Phytodegradation

Phytodegradation refers to the process where bacteria and plants jointly uptake, digest, and break down organic pollutants. This method employs plant roots along with microorganisms to cleanse soil tainted with organic contaminants (Garbisu and Alkorta 2001), also referred to as Phyto-transformation. Specific plants can detoxify soil, sludge, silt, surface water, and groundwater by generating enzymes. This technique addresses organic substances like herbicides, insecticides, chlorinated solvents, and inorganic pollutants (Pivetz 2001) (Fig. 15.1 and Table 15.1).

15.6 Enhancement of Phytoremediation Efficiency with Natural Amendments

15.6.1 Application of Organic Soil Amendments

The efficiency of phytoremediation is significantly impacted by the mobility and availability of trace elements in the soil, which can be modified by adding either organic or inorganic supplements. The presence of organo-metallic compounds can either increase or decrease the availability of metals in the soil solution (Kumpiene et al. 2008). According to Wang et al. (2012), incorporating organic supplements into the soil prompts the production of biochelators, resulting in increased availability and mobility of soil contaminants, thus improving the effectiveness of phytoextraction. Moreover, the influence of soil supplements on plants is pivotal in enhancing various aspects of plant growth. Commonly used methods include the addition of biochar, utilization of organic acid exudates from plant materials, and integration of organic substances such as humic compounds.

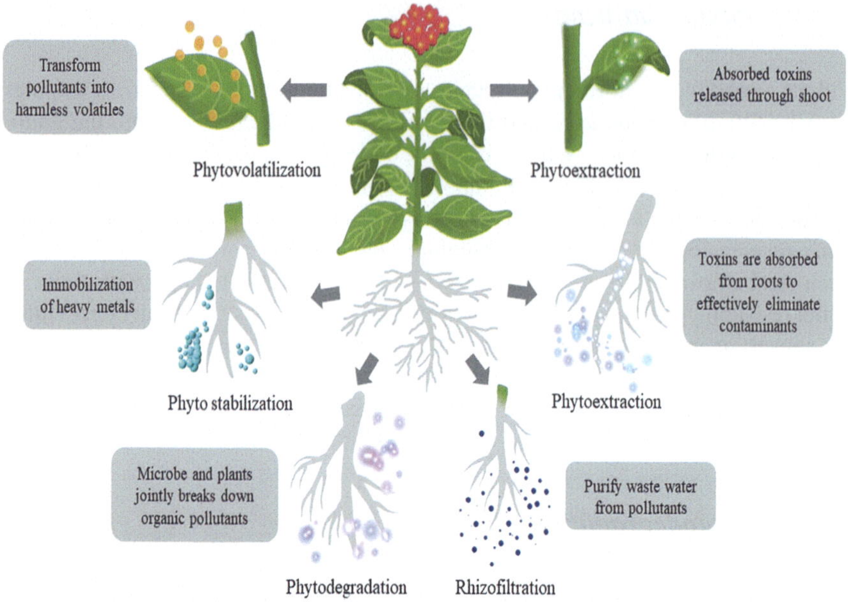

Fig. 15.1 Describes different phytoremediation processes for bioremediation of contaminants

15.6.2 Addition of Biochar

The carbon matrix structure inherent in biochar provides a notable degree of microporosity and an expanded surface area, making it an excellent surface adsorbent. Its advantageous qualities include high pH and cation exchange capacity, which contribute to enhancing the effectiveness of phytoremediation (Jiang et al. 2012). Biochar can be produced by subjecting waste biomass to pyrolysis in low oxygen environments. Recent studies have predominantly focused on immobilizing soil pollutants with biochar to decrease their bioavailability and subsequent phytotoxicity. The incorporation of biochar materials substantially elevates soil pH, resulting in the precipitation of contaminants. Additionally, biochar provides noteworthy agricultural advantages by delivering and retaining nutrients, enhancing soil fertility, and ultimately boosting biomass production (Zhang et al. 2013). The recommended amount of biochar to add typically ranges from 1 to 5%, dependent on soil texture and the cultivated crop. However, in the context of bioremediation, the pyrolysis temperature and the type of pollutant are also crucial considerations (Zhang et al. 2013). Zhang et al. (2013) financed the beneficial impacts of biochar on increased plant development to its effects on the physical and biochemical features of the soil, which include nutrient and water retention, biological activity, cation exchange capacity, and nutrient cycling. However, it has been observed that adding biochar may cause the amount of organic matter in the soil to decrease and introduce xenobiotics (Jiang et al. 2012). Long-term research is therefore essential to assess the advantages and

Table 15.1 Describe different plant species used in the phytoremediation of different heavy metal contaminants

Sl no	Plants species	Heavy metals	Concentration	References
1	*Sinapis arvensis*	Zn	26.4 mg/kg	Drozdova et al. (2019)
	Brassica campestris	Cu	1.5 mg/kg	
	Brassica juncea	Ni	0.9 mg/kg	
2	*Tagetes erecta L.*	Cd Zn	10–160 mg/kg	Madanan et al. (2021)
3	*Plectramthus sp.*	Hg	230–6,320 mg/g	Marrugo-Negrete et al. (2016)
	Clidemia sp.			
	Capsicum annuum			
	Phyllanthus niruri			
	Inga edulis			
4	*Brachiaria mutica*	Cr	100 mg/kg	Ullah et al. (2021)
	Leptochloa fusca			
5	*Typha angustifolia L.*	Cr	50 mg/L	Taufikurahman et al. (2019)
	Canna indica L.			
	Hydrocotyle umbellata L.			
6	*Typha angustifolia*	Ni	44.4 μg/L	Pandey et al. (2019)
	Echhornia crassipus	Pb	114.6 μg/L	
7	*Pterocypsela laciniata*	Cd	5, 10, and 25 mg/kg	Zhong et al. (2019)
8	*Malva rotundifolia*	Cd	5–100 mg/kg	Wu et al. (2018)
	Abelmoschus manihot			
9	*Pinus sylvestris*	Pb	....	Andráš et al. (2016)
	Quercus robur	Zn		
10	*Lantana camara L.*	Cd	100 mg/kg	Liu et al. (2019)

effects of biochar on the environment. This highlights the need for more investigation into the various plant materials and their applicability (Jiang et al. 2012).

15.6.3 Addition of Compost to the Soil Medium

During the composting process, diverse bacteria work on decomposing complex organic wastes, yielding simpler products like humus, micronutrients, and inorganic components. This enriches the soil with organic matter, promoting plant growth and mitigating adverse effects (Das and Ghosh 2022). Compost can alter the mobility and bioavailability of heavy metals in soil through redox reactions, complexation with other minerals, adsorption to soil particles, and precipitation of highly mobile elements.

15.6.4 Phytochelators

The amount of different organic acids present in plant root exudates has a big impact on what elements are available in the soil. These acids can chelate metals, are readily degradable, and generally have low molecular weights, as mentioned by Jiang et al. (2012). Chelators that are known to improve the mobilization of these elements include EDTA, vanillic acid, gallic acid, and citric acid (Ehsan et al. 2014). According to research by Hofrichter and Steinbuchel (2001), root exudates are useful in reducing the symptoms of pollution-induced poisoning as well as helping to release metals into the soil for plant uptake.

15.6.5 Humic Substances

Naturally occurring organic compounds are present in the upper soil layer due to the breakdown and decomposition of plant and animal wastes. Specific materials can efficiently mobilize or immobilize metal ions by creating stable complexes. Shahid and colleagues (2012) identified the three primary humic compounds in soil: humic acid, fulvic acid, and humine. Unlike humine, which precipitates under strong acidity or basicity, humic and fulvic acids can be easily extracted from soil and other solid-phase sources using a strong alkaline solution. However, it has been observed that various metal species exhibit varying degrees of humic acid's ability to facilitate phytoextraction. This variability is attributed to the fact that certain metals, such as Pb and Cu, display a strong affinity for organic materials, reducing their bioavailability. Additionally, the accumulation of organic matter in the soil decreases the soil pH, transforming metals into soluble forms.

15.6.6 Selection of Potential Hyperaccumulators

The significance of hyperaccumulators in achieving effective phytoremediation cannot be overstated. It is advisable to choose hyperaccumulators capable of both accumulating and tolerating higher concentrations of contaminants by forming intracellular complexes. Detailed research has been conducted on potential hyperaccumulators for various contaminants, as outlined below. Additionally, there has been exploration into the use of soil amendments, including plant exudates, humic substances, compost, and biochar, derived from secondary biomass produced by plants in contaminated sites. While some studies have focused on utilizing biomass accumulated with heavy metals (HMs), investigations into the utilization of biomass exposed to radioactive nuclides (RNs) are limited. This limitation arises from the fact that RN contamination is primarily restricted to specific regions of the world, such as Ukraine and Russia, unlike HMs, which are more widely generated because of industrialization.

15.7 Limitation of Phytoremediation Techniques

Phytoremediation, despite its many advantages, is subject to certain limitations that can influence its effectiveness and practicality in certain situations. Some of the key limitations include:

- The extent of this phenomenon is contingent upon the depth of the roots and the surface area covered by the plants.
- The process of slow growth is a long-term one.
- While it is not feasible to eliminate the leaching of pollutants into the groundwater, it is possible to partially degrade or stabilize them within the delimited contaminated area.
- The viability of the plant is contingent upon the state of the soil and the conditions of growth.
- Owing to their bioavailability, they have the potential to become assimilated into the food chain and pose a threat to plants, animals, and humans.

15.8 Conclusion

Despite the presence and utilization of various methods for addressing pollutants such as heavy metals and radioactive nuclides, phytoremediation technology is considered a readily accessible, effective, easily monitored, and environmentally friendly approach with minimal constraints. The foremost consideration should be the selection of appropriate plant species tailored to the specific polluted area. To expedite the phytoremediation process, the addition of soil amendments like biochar, humic materials, and organic acid exudates to polluted areas can be beneficial. Post-remediation, the primary focus should shift towards the utilization and disposal of PPBs. The danger posed by heavy metals to soil, water, and human health underscores the need for investigating cost-effective and environmentally friendly methods to remediate these pollutants. Consequently, phytoremediation has emerged as a widely adopted and efficient plant-based strategy for removing contaminants from polluted areas, to minimize adverse effects on water bodies and soil structure. The impetus to enhance and expedite phytoremediation arises from both research findings and practical applications. This drive is fuelled not only by ecological advantages but also by the drawbacks linked to the extended time needed to eliminate heavy metals and the potential toxicity to plant well-being. This study delves into the sources of heavy metal pollution, the contamination of soil and water by heavy metals, and the adverse effects on human health in such environments. It aids in selecting appropriate techniques for treating metal-bearing effluents based on time efficiency and effectiveness while identifying commonly used plants for phytoremediation. The study also explores approaches to enhance phytoremediation, encompassing various methods, biotechnological strategies, the utilization of different plant species, and factors influencing the process. As an alternative to conventional means of extracting heavy metals from the environment, employing metal-accumulating plants in phytoremediation proves to be a cost-effective, efficient, sustainable, and ecologically beneficial solution. The improvement of the procedure involves introducing new plant species capable of extracting heavy metals from contaminated soils. Techniques that enhance the bioavailability of metals, such as soil pH reduction, and methods to enhance plant uptake of heavy metals, such as utilizing bio-augmented acidified manure, can be employed. Additionally, genetic engineering and molecular research may play a vital role in enhancing the effectiveness of phytoremediation and recovering precious metals from extensively contaminated areas. This approach can be combined with chelating chemicals and microorganisms to increase heavy metal bioavailability, facilitating higher accumulation in plants, and easing soil remediation. However, further research is necessary to fully understand the impact of different catalysts on the effectiveness of phytoremediation, enhancing its practicality for environmental restoration.

References

Abbas J, Sagsan ˘ M (2019) Impact of knowledge management practices on green innovation and corporate sustainable development: a structural analysis. J Clean Prod 229. https://doi.org/10.1016/j.jclepro.2019.05.024

Akinwole AO, Dauda AB, Ololade AO (2016) Haematological response of Clarias gariepinus juveniles reared in treated wastewater after waste solids removal using alum or Moringa oleifera seed powder. Int J Acarology 6(11):1–8

András P, Turisova˘´ I, Buccheri G, de Matos JMX, Dirner V (2016) Comparison of heavy-metal bioaccumulation properties in pinus sp. and quercus sp. in selected European Cu deposits. Web Ecol 16:81–87. https://doi.org/10.5194/we-16-81-2016

Asati A, Pichhode M, Nikhil K (2016) Effect of heavy metals on plants: an overview. Int J Appl Innov Eng Manag 5(3):56–66

Awa SH, Hadibarata T (2020) Removal of heavy metals in contaminated soil by phytoremediation mechanism: a review. Water Air Soil Pollut 231(2):1–15

Bal M, Behera ID, Kumari U, Biswas S, Meikap BC (2020) Hydrodynamic study and particulate matter removal in a self priming venturi scrubber. Environ Technol Innovat 20. https://doi.org/10.1016/j.eti.2020.101167

Barea JM, Pozo MJ, Azc´on R, Azc´on-Aguilor C (2005) Microbial co-operation in the rhizosphere. J Exp Bot 56:1761–1778

Behera ID, Basak G, Kumar RR, Sen R, Meikap BC (2020) Treatment of petroleum refinery sludge by petroleum degrading bacterium Stenotrophomonas pavanii IRB19 as an efficient novel technology. J Environ Sci Health Part A Toxic Hazard Subst Environ Eng 56(2). https://doi.org/10.1080/10934529.2020.1866924

Benavides LCL, Pinilla LAC, Serrezuela RR, Serrezuela WFR (2018) Extraction in laboratory of heavy metals through rhizofiltration using the plant Zea mays (maize). Int J Appl Environ Sci 13(1):9–26

Boyd CE, Massaut L (1999) Risks associated with the use of chemicals in pond aquaculture. Aquacult Eng 20:113–132

Chandra R, Kumar V (2017) Phytoextraction of heavy metals by potential native plants and their microscopic observation of root growing on stabilised distillery sludge as a prospective tool for in situ phytoremediation of industrial waste. Environ Sci Pollut Control Ser 24(3):2605–2619

Chirakkara RA, Reddy KR (2015) Biomass and chemical amendments for enhanced phytoremediation of mixed contaminated soils. Ecol Eng 85:265–274

Colpaert JV, Vandenkoornhuyse P, Adriaensen K, Vangronsveld J (2000) Genetic variation and heavy metal tolerance in the ectomycorrhizal basidiomycete Suillus luteus. New Phytol 147:367–369

Das AP, Ghosh S (2018) Bioleaching of manganese from mining waste materials. Mater Today Proceed 5(1):2381–2390. https://doi.org/10.1016/J.MATPR.2017.11.459

Das A, Ghosh S (2022) Role of microorganisms in extenuation of mining and industrial wastes. Geomicrobiol J 39:173–175. https://doi.org/10.1080/01490451.2022.2038953

Das AP, Ghosh S (2023a) Cost-effectiveness of the biomining process. Manganese Min Microorg 143–151. https://doi.org/10.1016/B978-0-12-822147-1.00008-X

Das AP, Ghosh S (2023b) Factors affecting bioleaching. Manganese Min Microorg 77–86b. https://doi.org/10.1016/B978-0-12-822147-1.00004-2

Das AP, Ghosh S (2023c) Manganese and its application. Manganese Min Microorg 29–40. https://doi.org/10.1016/B978-0-12-822147-1.00007-8

Das AP, Ghosh S (2023d) Mechanism of manganese bioleaching. Manganese Min Microorg 87–98. https://doi.org/10.1016/B978-0-12-822147-1.00014-5

Das AP, Ghosh S (2023e) Metagenomic insights into unculturable microbial diversity. Manganese Min Microorg 63–76. https://doi.org/10.1016/B978-0-12-822147-1.00002-9

Das AP, Ghosh S (2023f) Microorganisms in manganese biomining processes. Manganese Min Microorg 55–61. https://doi.org/10.1016/B978-0-12-822147-1.00013-3

Das AP, Ghosh S (2023g) Mixed culture bioleaching: an insight into manganese biomining process and efficacy. Manganese Min Microorg 129–141. https://doi.org/10.1016/B978-0-12-822147-1.00001-7

Das AP, Ghosh S (2023h) Role of metabolites and metabolomics for understanding of manganese bioleaching. Manganese Min Microorg 117–127. https://doi.org/10.1016/B978-0-12-822147-1.00011-X

Das AP, Mishra S (2008) Hexavalent chromium (VI): health hazards and environmental pollutant. J Environ Res Dev 2(3):386–392

Das AP, Sukla LB, Pradhan N, Nayak S (2011) Manganese biomining: a review. Biores Technol 102(16):7381–7387. https://doi.org/10.1016/J.BIORTECH.2011.05.018

Das AP, Swain S, Panda S, Pradhan N, Sukla LB (2012) Reductive acid leaching of low grade manganese ores. Geomaterials 2012(04):70–72. https://doi.org/10.4236/GM.2012.24011

Das AP, Ghosh S, Mohanty S, Sukla LB (2015a) Advances in manganese pollution and its bioremediation. In Sukla LB, Pradhan N, Panda S, Mishra BK (eds) Environmental microbe biotechnology, 1st ed. Springer, Cham, pp 313–328. https://doi.org/10.1007/978-3-319-19018-1_16

Das AP, Ghosh S, Mohanty S, Sukla LB (2015b) Consequences of manganese compounds: a review. Toxicol Environ Chem 96(7):981–997. https://doi.org/10.1080/02772248.2015.1005428

Das AP, Dutta K, Khatun R, Behera ID, Singh S, Mishra S (2023) Microfiber pollution and its microbial mitigation: a review on current trends and future prospects. J Taiwan Inst Chem Eng 105–104. https://doi.org/10.1016/J.JTICE.2023.105104

Das AP, Mishra S (2009) Hexavalent chromium [Cr (VI)]: yellow water pollution and its remediation. Sarovar Saurabh ENVIS Newsl Wetl Ecosyst 5(2)

Das A, Mishra S (2010) Biodegradation of the metallic carcinogen hexavalent chromium Cr(VI) by an indigenously isolated bacterial strain. J Carcinog 9. https://doi.org/10.4103/1477-3163.63584

Das AP, Bissoyi A (2011) Computational approach for comparative phylogenetic analysis of isolated chromium resistant strain Brevibacterium casei. J Eng Technol Res 3(3), 82–87. https://doi.org/10.5897/JETR.9000020

Das A, Dey S, Gacem A, Yadav KK, Nath H, Ram DK, Das AP (2025) Revolutionizing Solid Hydrocarbon Sourcing: Plastic's Journey from Waste to Upcycled Treasure. 331–351. https://doi.org/10.1007/978-3-031-81615-4_14

Das AP, Sukla LB, Pradhan N (2012) Microbial recovery of manganese using *Staphylococcus Epidermidis*. Int J Nonferrous Metall (02):9–12. https://doi.org/10.4236/IJNM.2012.12002

Drozdova I, Alekseeva-Popova N, Dorofeyev V, Bech J, Belyaeva A, Roca N (2019) A comparative study of the accumulation of trace elements in Brassicaceae plant species with phytoremediation potential. Appl Geochem 108. https://doi.org/10.1016/j.apgeochem.2019.104377

Ehsan S, Ali S, Noureen S, Mahmood K, Farid M, Ishaque W, Shakoor MB, Rizwan M (2014) Citric acid assisted phytoremediation of cadmium by Brassica napus L. Ecotoxicol Environ Saf 106:164–172

Eid EM, Galal TM, Sewelam NA, Talha NI, Abdallah SM (2020) Phytoremediation of heavy metals by four aquatic macrophytes and their potential use as contamination indicators: a comparative assessment. Environ Sci Pollut Control Ser 27(11):12138–12151

Ekmekyapar F, Sabudak T, Seren G (2012) Assessment of heavy metal contamination in soil and wheat (Triticum aestivum L.) plant around the Çorlu–Çerkezkoy highway in Thrace region. Glob NEST J 14(4):496–504

Galal TM, Gharib FA, Ghazi SM, Mansour KH (2017) Phytostabilization of heavy metals by the emergent macrophyte Vossia cuspidata (Roxb.) Griff.: a phytoremediation approach. Int J Phytoremediation 19(11):992–999

Garbisu C, Alkorta I (2001) Phytoextraction: a cost-effective plantbased technology for the removal of metals from the environment. Bioresour Technol 77(3):229–236

Ghosh S, Das AP (2017) Bioleaching of manganese from mining waste residues using Acinetobacter sp. Geol Ecol Landsc 1(2):77–83. https://doi.org/10.1080/24749508.2017.1332847

Ghosh S, Tripathy B, Dey S, Das AP (2023) Comparative investigation of fungal and bacterial manganese biomining mechanisms. Manganese Min Microorg 99–115. https://doi.org/10.1016/b978-0-12-822147-1.00016-9

Ghosh S, Gandhi M, van Hullebusch ED, Das AP (2020) Proteomic insights into Lysinibacillus sp.-mediated biosolubilization of manganese. Environ Sci Pollut Res 28(30):40249–40263. https://doi.org/10.1007/S11356-020-10863-4

Hadia-e-Fatima AA (2018) Heavy metal pollution–a mini review. J Bacteriol Mycol Open Access 6(3):179–181

Hofrichter M, Steinbuchel A (2001) Biopolymers, 1, Lignin, humic substances and coal. Wiley-VCH, Weinheim

IEEE technical subcommittee on green communications and computing (2011) https://tcgcc.committees.comsoc.org/history/

Jiang J, Xu RK, Jiang TY, Li Z (2012) Immobilization of Cu(II), Pb(II) and Cd(II) by the addition of rice straw derived biochar to a simulated polluted Ultisol. J Hazard Mater 229–230:145–150

Kanungo J, Sahoo T, Bal M, Behera ID (2023) Performance of bioremediation strategy in waste lubricating oil pollutants: a review. Geomicrobiol J. https://doi.org/10.1080/01490451.2023.2245395

Kapoor D, Singh MP (2021) Heavy metal contamination in water and its possible sources. Heavy metals in the environment. Elsevier, pp 179–189

Khatun R, Singh S, Dubey NK, Prasad Das A (2023) A review on marine-based phytochemicals and their application in biomedical research. Recent Front Phytochem Appl Food Pharm Cosmet Biotechnol 383–395. https://doi.org/10.1016/B978-0-443-19143-5.00013-X

Khatun R, Singh S, Dubey NK, Prasad Das A (2023b) A review on marine-based phytochemicals and their application in biomedical research. In: Recent frontiers of phytochemicals: applications in food, pharmacy, cosmetics, and biotechnology. Elsevier, pp 383–395. https://doi.org/10.1016/B978-0-443-19143-5.00013-X

Kumar M, Pandit S, Patel V, Khanna N, Nag M, Lahiri D, Ray RR, Das A, Das D (2023) Maximization of energy recovery from starch processing wastewater by thermophilic dark fermentation coupled with microbial fuel cell technology. Geomicrobiol J 1–12. https://doi.org/10.1080/01490451.2023.2209555

Kumar SS, Kadier A, Malyan SK, Ahmad A, Bishnoi NR (2017) Phytoremediation and rhizoremediation: uptake, mobilization and sequestration of heavy metals by plants. Plant-microbe interactions in agro-ecological perspectives, pp 367–394

Kumpiene J, Lagerkvist A, Maurice C (2008) Stabilization of As, Cr, Cu, Pb and Zn in soil using amendments: a review. Waste Manag 28:215–225

Lahiri D, Nag M, Dey A, Sarkar T, Joshi S, Pandit S, Das AP, Pati S, Pattanaik S, Tilak VK, Ray RR (2021) Biofilm mediated degradation of petroleum products. Geomicrobiol J 1–10. https://doi.org/10.1080/01490451.2021.1968979

Leung HM, Wu FY, Cheung KC, Ye ZH, Wong MH (2010a) The effect of arbuscular mycorrhizal fungi and phosphate amendment on arsenic uptake, accumulation, and growth of Pteris Vittata in As-contaminated soil. Int J Phytor

Liu S, Ali S, Yang R, Tao J, Ren B (2019) A newly discovered Cd-hyperaccumulator Lantana camara L. J Hazard Mater 371:233–242. https://doi.org/10.1016/j.jhazmat.2019.03.016

Liu S, Ali S, Yang R, Tao J, Ren B (2019) A newly discovered Cd-hyperaccumulator Lantana camara L. J Hazard Mater 371:233–242. https://doi.org/10.1016/j.jhazmat.2019.03.016

Lovelock CE, Wright SF, Clark DA, Ruess RW (2004) Soil stocks of glomalin produced by arbuscular mycorrhizal fungi across a tropical rain forest landscape. J Ecol 92:278–287

Madanan MT, Shah IK, Varghese GK, Kaushal RK (2021) Application of Aztec Marigold (Tagetes erecta L.) for phytoremediation of heavy metal polluted lateritic soil. Environ Chem Ecotoxicol 3:17–22. https://doi.org/10.1016/j.enceco.2020.10.007

Makarova A, Nikulina E, Avdeenkova T, Pishaeva K (2021) The improved phytoextraction of heavy metals and the growth of Trifolium repens L.: the role of K2HEDP and plant growth regulators alone and in combination. Sustainability 13(5):2432

Marrugo-Negrete J, Marrugo-Madrid S, Pinedo-Hernández J, Durango-Hernández J, Díez S (2016) Screening of native plant species for phytoremediation potential at a Hg-contaminated mining site. Sci Total Environ 542:809–816. https://doi.org/10.1016/j.scitotenv.2015.10.117

Mishra S, Das AP (2023) Microbial remediation of synthetic microfiber contaminated wastewater. Microbial Technol Ind Wastewater Treat 337–350. https://doi.org/10.1007/978-981-99-2435-6_17/COVER

Mishra S, Dash D, Al-Tawaha ARMS, Das AP (2022) A review on heavy metal ion adsorption on synthetic microfiber surface in aquatic environments. Appl Biochem Biotechnol 194(10):4639–4654. https://doi.org/10.1007/S12010-022-04029-W/METRICS

Mishra S, Ghosh S, Hullebusch van ED, Singh S, Das AP (2023a) A critical review on the recovery of base and critical elements from electronic waste-contaminated streams using microbial biotechnology. Appl Biochem Biotechnol 1–30. https://doi.org/10.1007/S12010-023-04440-X/METRICS

Mishra S, Ghosh S, van Hullebusch ED, Singh S, Das AP (2023b) A critical review on the recovery of base and critical elements from electronic waste-contaminated streams using microbial biotechnology. Appl Biochem Biotechnol 1–30. https://doi.org/10.1007/S12010-023-04440-X/METRICS

Mishra S, Dash D, Das AP (2023) Aquatic microbial diversity on plastisphere: colonization and potential role in microplastic biodegradation. Geomicrobiol J 1–12. https://doi.org/10.1080/01490451.2023.2209750

Mishra S, Das AP (2023b) Microbial remediation of synthetic microfiber contaminated wastewater. In: Shah MP (ed) Microbial technologies in industrial wastewater treatment. Springer, Singapore, pp 337–350. https://doi.org/10.1007/978-981-99-2435-6_17

Mishra S, Das AP, Seragadam P (2009) Microbial remediation of hexavalent chromium from chromite contaminated mines of Sukinda Valley, Orissa (India). J Environ Res Dev 1122–1127

Mohanty S, Ghosh S, Nayak S, Das AP (2016) Isolation, identification and screening of manganese solubilizing fungi from low-grade manganese ore deposits. Geomicrobiol J 34(4):309–316. https://doi.org/10.1080/01490451.2016.1189016

Mohanty M, Mohanty J, Dey S, Dutta K, Shah MP, Das AP (2024) The face mask: a tale from protection to pollution and demanding sustainable solution. Emerg Contam 10(2):100298. https://doi.org/10.1016/J.EMCON.2023.100298

Mohanty S, Ghosh S, Bal B, Das AP (2018) A review of biotechnology processes applied for manganese recovery from wastes. Rev Environ Sci Bio/Technol 17(4):791–811. https://doi.org/10.1007/S11157-018-9482-1

Morgan JAW, Bending GD, White PJ (2005) Biological cost and benefits to plant-microbe interactions in the rhizosphere. J Exp Bot 56:1729–1739

Muthusaravanan S, Sivarajasekar N, Vivek J, Paramasivan T, Naushad M, Prakashmaran J, Gayathri V, Al-Duaij OK (2018) Phytoremediation of heavy metals: mechanisms, methods and enhancements. Environ Chem Lett 16(4):1339–1359

Nag M, Lahiri D, Ghosh S, Sarkar T, Pati S, Das AP, Ram DK, Bhattacharya D, Ray RR (2023a) Application of microorganisms in biotransformation and bioremediation of environmental contaminant: a review. Geomicrobiol J 1–18. https://doi.org/10.1080/01490451.2023.2261443

Nag M, Lahiri D, Ghosh S, Sarkar T, Pati S, Das AP, Ram DK, Bhattacharya D, Ray RR (2023b) Application of microorganisms in biotransformation and bioremediation of environmental contaminant: a review. Geomicrobiol J. https://doi.org/10.1080/01490451.2023.2261443

Odinga CA, Kumar A, Mthembu MS, Bux F, Swalaha FM (2019) Rhizofiltration system consisting of Phragmites australis and Kyllinga nemoralis: evaluation of efficient removal of metals and pathogenic microorganisms. Desalin Water Treat 169:120–132

Pandey SK, Upadhyay RK, Gupta VK, Worku K, Lamba D (2019) Phytoremediation potential of macrophytes of urban waterbodies in central India. J Health Pollut 9. https://doi.org/10.5696/2156-9614-9.24.191206

Pivetz BE (2001) Ground water issue: phytoremediation of contaminated soil and ground water at hazardous waste sites, pp 1–36

Prabhakar A, Mishra S, Das AP (2019) Isolation and identification of lead (Pb) solubilizing bacteria from automobile waste and its potential for recovery of lead from end of life waste batteries. Geomicrobiol J 36(10):894–903. https://doi.org/10.1080/01490451.2019.1654044

Pradhan G, Tripathy B, Ram DK, Digal AK, Das AP (2023) Bauxite mining waste pollution and its sustainable management through bioremediation. Geomicrobiol J 1–10. https://doi.org/10.1080/01490451.2023.2235353

Prieto M, Acevedo S, Prieto G, Gonz´ alez N (2018) Phytoremediation of soils contaminated with heavy metals. Biodiversity Int J 2:362–376

Prieto M, Acevedo S, Prieto G, Gonz´ alez N (2018) Phytoremediation of soils contaminated with heavy metals. Biodiversity Int J 2:362–376

Priyadarsini S, Das AP (2024) Analytical and structural characterization of waste lithium-ion batteries for their effective recycling strategy. Environ Chem Ecotoxicol. https://doi.org/10.1016/j.enceco.2024.12.004

Priyadarsini S, Das AP (2025) Lithium bioleaching: A review on microbial-assisted sustainable technology for lithium bio-circularity. J Water Proc Eng 69, 106744. https://doi.org/10.1016/j.jwpe.2024.106744

Raj K, Das AP (2023) Lead pollution: impact on environment and human health and approach for a sustainable solution. Environ Chem Ecotoxicol 5:79–85. https://doi.org/10.1016/J.ENCECO.2023.02.001

Sadasivuni KK, Rattan S, Waseem S, Brahme SK, Kondawar SB, Ghosh S, Das AP, Chakraborty PK, Adhikari J, Saha P, Mazumdar P (2019) Silver nanoparticles and its polymer nanocomposites—synthesis, optimization, biomedical usage, and its various applications. Lecture Notes Bioeng 331–373. https://doi.org/10.1007/978-3-030-04741-2_11

Saif S, Khan M (2017) Assessment of heavy metals toxicity on plant growth promoting rhizobacteria and seedling characteristics of Pseudomonas putida SFB3 inoculated greengram. Acta Sci Agric 1:47–56

Sakakibara M, Watanabe A, Inoue M, Sano S, Kaise T (2010) Phytoextraction and phytovolatilization of arsenic from As-contaminated soils by Pteris vittata. In: Proceedings of the annual international conference on soils, sediments, water and energy

Saran A, Fernandez L, Cora F, Savio M, Thijs S, Vangronsveld J, Merini LJ (2020) Phytostabilization of Pb and Cd polluted soils using Helianthus petiolaris as pioneer aromatic plant species. Int J Phytoremediation 22(5):459–467

Shahid M, Pinelli E, Dumat C (2012) Review of Pb availability and toxicity to plants in relation with metal speciation: role of synthetic and natural organic ligands. J Hazard Mater 219:1–12

Sharples JM, Meharg AA, Chambers SM, Cairney JWG (2000) Mechanism of arsenate resistance in the ericoid mycorrhizal fungus Hymenoscyphus ericae. Plant Physiol 124:1327–1334

Swain S, Purnima Sudha M, Dey S, Badu Y, Bhoi SR, Alreshidi MA, Qasim MT, Nath H, Ram DK, Das AP (2025) Sources of micro and nano plastic hydrocarbon pollutants in marine bodies and their treatment 45–67. https://doi.org/10.1007/978-3-031-81615-4_3

Tangahu BV, Sheikh Abdullah SR, Basri H, Idris M, Anuar N, Mukhlisin M (2011) A review on heavy metals (As, Pb, and Hg) uptake by plants through phytoremediation. Int J Chem Eng (2011)

Taufikurahman T, Pradisa MAS, Amalia SG, Hutahaean GEM (2019) Phytoremediation of chromium (Cr) using Typha angustifolia L., Canna indica L. and Hydrocotyle umbellata L. in surface flow system of constructed wetland. IOP conference series: earth and environmental science, vol 308, no 1. https://doi.org/10.1088/1755-1315/308/1/012020

Tian W, Zhang CQ, Qiao P, Milne R (2011) Diversity of culturable ericoid mycorrhizal fungi of Rhododendron decorum in Yunnan, China. Mycologia 103:703–709

Ugya A, Ahmad A, Adamu H, Giwa S, Imam T (2019) Phytoextraction of heavy metals and risk associated with vegetables grown from soil irrigated with refinery wastewater. J Appl Biol Biotechnol 7(2):14–19

Ullah S, Mahmood S, Ali R, Khan MR, Akhtar K, Depar N (2021) Comparing chromium phyto-assessment in Brachiaria mutica and Leptochloa fusca growing on chromium polluted soil. Chemosphere, 269. https://doi.org/10.1016/j.chemosphere.2020.128728

Wang J, Feng X, Anderson CW, Xing Y, Shang L (2012) Remediation of mercury contaminated sites–a review. J Hazard Mater 221:1–18. https://doi.org/10.1016/j.jhazmat.2012.04.035

Wu J, Guo S, Li J, Zeng D (2016) Big data meet green challenges: greening big data. IEEE Syst J 10(3):873–887. https://doi.org/10.1109/JSYST.2016.2550538

Wu M, Luo Q, Zhao Y, Long Y, Liu S, Pan Y (2018) Physiological and biochemical mechanisms preventing Cd toxicity in the new hyperaccumulator abelmoschus manihot. J Plant Growth Regul 37(3):709–718. https://doi.org/10.1007/s00344-017-9765-8

Wu SC, Wong CC, Khan AG, Wong MH (2011)

Wu J, Guo S, Huang H, Liu W, Xiang Y (2018) Information and communications technologies for sustainable development goals: state-of-the-art, needs and perspectives. IEEE Commun Surv Tutor 20(3):2389–2406. https://doi.org/10.1109/COMST.2018.2812301

Yadav G, Dash SK, Sen R (2019) A biorefinery for valorization of industrial wastewater and flue gas by microalgae for waste mitigation, carbon-dioxide sequestration and algal biomass production. Sci Total Environ 688:129–135. https://doi.org/10.1016/j.scitotenv.2019.06.024.

Yan A, Wang Y, Tan SN, Yusof MLM, Ghosh S, Chen Z (2020) Phytoremediation: a promising approach for revegetation of heavy metal-polluted land. Front Plant Sci 11

Yousuf M, Das AP, Masurkar P, Bag MK, Mitra D, Ray A (2025) Recent advances and research in biodiesel as a sustainable, eco-friendly approach toward the Green Economy. 305–329. https://doi.org/10.1007/978-3-031-81615-4_13

Zhang XK, Wang HL, He LZ, Lu KP, Sarmah A, Li JW, Bolan NS, Pei JC, Huang HG (2013) Using biochar for remediation of soils contaminated with heavy metals and organic pollutants. Environ Sci Pollut Res 20:8472–8483

Zheng C, Zhao L, Zhou X, Fu Z, Li A (2013) Treatment technologies for organic wastewater. In: Water Treatment. InTech. https://doi.org/10.5772/52665

Zhong L, Lin L, Liao M, Wang J, Tang Y, Sun G, Liang D, Xia H, Wang X, Zhang H, Ren W (2019) Phytoremediation potential of pterocypsela laciniata as a cadmium hyperaccumulator. Environ Sci Pollut Res 26(13):13311–13319. https://doi.org/10.1007/s11356-019-04702-4

Chapter 16
Nanomodified Biochar for Heavy Metal Removal from Wastewaters: An Overview

Sourav Chattaraj, Hrudayanath Thatoi, and Pradeep K. Das Mohapatra

Abstract Biochar (BC), a carbon-rich material derived from biomass through thermal combustion in an oxygen-limited environment, offers a versatile solution with various applications. Biomass waste sources suitable for biochar production encompass agricultural crop residues, forestry waste, municipal solid waste, and animal manures. The distinctive characteristics of biochar, including its large surface area, high porosity, functional groups, and cation exchange capacity, render it suitable for diverse applications. Notably, the pyrolysis method significantly influences biochar yield, with slow pyrolysis yielding 30% more charcoal compared to fast pyrolysis or gasification methods. Biochar has garnered considerable attention for its efficacy in removing contaminants, particularly heavy metals, from wastewater, offering a cost-effective and environmentally friendly solution. However, the original biochar's capacity for heavy metal sorption is limited. This review discusses various feedstock materials and production methods for biochar, spanning traditional and modern techniques. Furthermore, the potential application of nanomodified biochar for the removal of toxic heavy metals from wastewater is explored, highlighting future prospects for sustainable environmental remediation.

Keywords Nanobiochar · Biochar · Nanotechnology · Environmental pollution · Remediation

S. Chattaraj · H. Thatoi (✉)
Centre for Industrial Biotechnology Research, Siksha 'O' Anusandhan, Deemed to be University, Kalinganagar, Ghatikia, Bhubaneswar, Odisha 751003, India
e-mail: director.cibr@soa.ac.in

P. K. Das Mohapatra
Department of Microbiology, Raiganj University, Raiganj, West Bengal, India

A. Kuanar et al. (eds.), *Recent Advances in Bioremediation and Phytoremediation*, Environmental Science and Engineering,
https://doi.org/10.1007/978-3-031-77884-1_16

16.1 Introduction

In recent years, the increasing contamination of water bodies with heavy metals has become a pressing environmental concern worldwide. Heavy metals, such as lead, cadmium, arsenic, manganese, chromium, and mercury, pose significant risks to human health and the environment due to their persistence, toxicity, and ability to bioaccumulate in living organisms (Das et al. 2011; Raj and Das 2023). Industrial activities, mining operations, and urbanization are among the primary sources of heavy metal pollution in aquatic ecosystems, threatening both freshwater resources and marine habitats (Mishra and Das 2008; Prabhakar et al. 2019). Addressing this challenge requires effective and sustainable remediation strategies capable of mitigating heavy metal contamination in wastewater (Das et al. 2015; Dey et al. 2023). Among the various remediation approaches, biochar has emerged as a promising candidate for heavy metal removal due to its high surface area, presence of functional groups at the surface, porous structure, and strong affinity for metal ions. Biochar, a carbon-rich material produced through the pyrolysis of biomass under oxygen-limited conditions, offers several advantages for wastewater treatment, including cost-effectiveness, abundance of raw materials, and environmental sustainability (Kapoor 2021). However, conventional biochar materials often exhibit limited adsorption capacity and selectivity, necessitating the exploration of innovative approaches to enhance their performance.

Biochar, derived from pyrolysis of biomass materials, possesses high affinity for heavy metal ions due to its porous structure and various functional groups. Biochar presents a cost-effective and sustainable solution for addressing heavy metal contamination in wastewater, distinguished by its ability to be regenerated and reused, or repurposed into electrodes for supercapacitors or stable catalyst materials, thus attracting considerable attention in biomass thermochemical conversion (Lima et al. 2022). Its production, primarily through pyrolysis of various feedstocks like agricultural and forestry residues or municipal solid waste, yields a highly porous, carbon-rich material whose physicochemical properties vary with different pyrolysis conditions, impacting heavy metal adsorption capacity (Pariyar et al. 2020). However, optimizing biochar production and modification processes is crucial to enhancing its adsorption properties, with nano-modified biochar showing significant promise, particularly when derived from agro-wastes. By combining biochar with nanoparticles or metal oxides, hybrid adsorbents can be created, overcoming limitations and enhancing heavy metal adsorption capacity and selectivity, thus establishing biochar-nanocomposites as a potent strategy for remediating toxic heavy metals from polluted environments, with potential for regeneration and reuse (Li et al. 2018). Hence, the current review aims to explore a range of feedstock materials and the methods employed for their production of biochar, encompassing both traditional and modern techniques. Additionally, it looks into the potential applications of nanomodified biochar in addressing the issue of heavy metal removal from wastewater. This discussion offers insights into how advancements in biochar technology can contribute to environmental remediation efforts, particularly in managing water pollution caused by heavy metals.

16.2 Feed Stock

Biochar production relies on various feedstocks sourced from organic materials, offering a diverse array of options for its creation. Agricultural residues stand out prominently as viable feedstocks, encompassing materials like rice and wheat stalks, fruit peels, and corn stalks (Chaturvedi et al. 2021). These residues, often abundant and readily available, provide an excellent resource for biochar production. Additionally, forest wastes contribute significantly to the feedstock pool, including wood residues and other byproducts from forestry operations. Livestock waste, such as that from pigs and cattle, adds another dimension, offering a valuable source of organic matter for biochar production. Municipal solid waste (MSW) emerges as a unique yet potent feedstock category, encompassing both residential and non-residential waste streams. This includes organic waste from households, businesses, and industries, as well as sewage sludge and other solid wastes. Categorically, these feedstocks fall into three primary groups: virgin resources, residues, and MSW, each with its own characteristics and potential for biochar production (Chaturvedi et al. 2021). While virgin resources like forest materials provide a foundational base, residues from agriculture and livestock add depth, and MSW introduces a recycling aspect to the process. Biochar, resulting from diverse organic materials such as agricultural and forest residues, along with municipal solid waste, holds significant promise as an energy resource. These materials fall into three main categories: virgin resources, residues, and municipal solid waste, each offering unique opportunities for conversion. Examples of these include forest resources, agricultural byproducts, and residential or non-residential waste. With appropriate conversion methods, these biomass sources can play a crucial role in future energy solutions (Pariyar et al. 2020). Overall, the diversity of feedstocks available for biochar production underscores its versatility and potential as a sustainable solution for waste management and renewable energy production.

16.3 Biochar Preparation

Biochar (BC) derives from a variety of biomasses with diverse chemical and physical characteristics, crucial in thermal conversion methods. Common biomass sources for BC production include crops, forest residues, organic and agricultural waste, kitchen scraps, and sewage sludge. Among thermochemical conversion techniques like pyrolysis, hydrothermal carbonization, gasification, and torrefaction, pyrolysis stands out as the most prevalent, involving heating biomass in low-oxygen conditions at temperatures from 400 to 1000 °C, leading to decomposition of organic compounds

(Ercan et al. 2023). The resulting biochar's properties are primarily influenced by feedstock type and pyrolysis conditions such as temperature, residence time, reactor type, and heating rate, with various thermochemical processes yielding different biochar quantities (Vieira et al. 2020). Recent research has extensively explored the use of lignocellulosic feedstocks like wood, tea waste, coffee hulls, and agricultural residues, demonstrating their potential for biochar production through pyrolysis.

16.3.1 Conventional Pyrolysis

Biochar production occurs in anaerobic conditions and is categorized into traditional and modern methods based on advancements. Conventional pyrolysis, also known as slow pyrolysis, is a widely employed method for biochar production, characterized by its gradual heating of biomass at temperatures typically ranging from 300 °C to 600 °C with a heating rate of 5–7 °C min^{-1}. This approach relies on the transfer of heat from an external source to the biomass through mechanisms such as conduction, radiation, and convection. Slow pyrolysis entails a longer processing time, often spanning several hours, allowing for thorough decomposition of organic compounds within the biomass. Typically, slow pyrolysis of wood results in biochar, biooil, and syngas yields accounting for approximately 35%, 30%, and 35% of the dry feed mass, respectively (Guo et al. 2019). With its relatively lower temperatures and longer residence times, slow pyrolysis offers greater control over the final product composition and properties, making it well-suited for small-scale and specialized biochar production applications.

16.3.2 Modern Approaches

At the end of the twentieth century, several modifications and changes in biochar production methods were surfaced. Eventually, various approaches have been developed for biochar production, such as modern pyrolysis (flash-, vacuum-, and microwave-pyrolysis), gasification, torrefactions, hydrothermal carbonization, electro-modified techniques, etc. where several advanced technologies were employed.

16.3.2.1 Flash Pyrolysis

Flash pyrolysis is a rapid thermal decomposition process that operates at extremely high temperatures for a very short duration, typically in the range of milliseconds to seconds. In this method, biomass is subjected to intense heating at temperatures ranging from 500 to 1200 °C, causing instantaneous vaporization and decomposition of organic materials (Piskorz et al. 2000). The process occurs in the absence of oxygen, preventing combustion and ensuring the production of biochar rather than ash. Flash pyrolysis is characterized by its ability to quickly convert biomass into biochar, bio-oil, and syngas. The high heating rates and short residence times result in bio-oil being the primary product, with biochar produced in lower quantities compared to slow pyrolysis. While flash pyrolysis offers rapid processing and high bio-oil yields, it requires sophisticated equipment and precise control over operating conditions (Okoroanyanwu et al. 2021).

16.3.2.2 Vacuum Pyrolysis

Vacuum pyrolysis is a variation of traditional pyrolysis that operates under reduced pressure conditions (Zhang et al. 2021). By subjecting the biomass to vacuum, typically ranging from a few millibars to a few torrs, the boiling point of the pyrolysis products is lowered, allowing for thermal decomposition to occur at lower temperatures. This method is particularly advantageous for the production of high-quality biochar with enhanced porosity and surface area. Vacuum pyrolysis reduces the risk of thermal degradation and promotes the retention of volatile compounds, resulting in biochar with improved chemical and physical properties. Additionally, vacuum pyrolysis offers better control over the distribution of pyrolysis products and minimizes the formation of undesired byproducts. However, vacuum pyrolysis requires specialized equipment capable of maintaining vacuum conditions, making it more complex and costly compared to conventional pyrolysis methods (Lopez et al. 2009).

16.3.2.3 Microwave Pyrolysis

Microwave pyrolysis utilizes electromagnetic radiation in the microwave frequency range to induce thermal decomposition of biomass materials. In this method, biomass is exposed to microwave irradiation, causing rapid heating and vaporization of moisture and volatile organic compounds. Microwave pyrolysis offers several advantages over conventional heating methods, including uniform heating, precise temperature control, and shorter processing times (Robinson et al. 2015). By selectively heating the biomass, microwave pyrolysis reduces energy consumption and improves process efficiency. Moreover, microwave pyrolysis can be tailored to specific feedstocks and desired biochar properties, allowing for greater flexibility in biochar production. However, microwave pyrolysis systems require specialized equipment capable of generating and controlling microwave energy, which may entail higher initial investment costs. Despite these challenges, microwave pyrolysis shows promise for scalable and efficient biochar production, particularly for niche applications requiring rapid processing and precise control over product quality.

16.3.2.4 Gasification

Gasification pyrolysis is a thermochemical process that converts biomass into a combustible gas mixture known as syngas, along with char residue, through partial oxidation in a controlled environment. Unlike conventional pyrolysis, which occurs in the absence of oxygen, gasification pyrolysis involves the introduction of a limited amount of oxygen or steam to facilitate the partial oxidation of biomass. This process typically occurs at higher temperatures ranging from 700 to 1000 °C, depending on the feedstock and desired syngas composition (Hansen et al. 2015). The gasification pyrolysis process begins with the drying and heating of the biomass feedstock, followed by pyrolysis, during which organic materials decompose into volatile compounds such as methane, carbon monoxide, hydrogen, and various hydrocarbons. As the temperature increases, these volatile compounds undergo further reactions with oxygen or steam, leading to gasification reactions that produce syngas (Couto et al. 2013). The syngas consists mainly of carbon monoxide and hydrogen, along with smaller amounts of carbon dioxide and other trace gases (Abubackar et al. 2011). The char residue left behind after gasification contains carbonaceous materials and ash, which can be further processed into biochar or used as a solid fuel. Gasification pyrolysis offers several advantages over conventional pyrolysis methods, including higher syngas yields, improved energy efficiency, and greater flexibility in feedstock selection. Additionally, the syngas produced can be utilized as a renewable energy source for heat and power generation, as well as for the production of biofuels and chemicals through further processing. Gasification pyrolysis can be implemented using various reactor configurations, including fixed-bed, fluidized bed, and entrained-flow reactors, each offering unique advantages in terms of operating conditions, efficiency, and scalability. Fixed-bed gasifiers are commonly used for small-scale applications, while fluidized bed and entrained-flow gasifiers are preferred for larger-scale operations due to their ability to handle a wide range of feedstocks and produce syngas with high energy content (Thomson et al. 2020). Gasification pyrolysis represents a promising technology for converting biomass into valuable syngas and biochar, offering opportunities for sustainable energy production and waste management. Continued research and development efforts aimed at optimizing process parameters, improving reactor designs, and integrating gasification pyrolysis with other renewable energy technologies will further enhance its economic viability and environmental sustainability.

16.3.2.5 Hydrothermal Carbonization

Hydrothermal carbonization (HTC) is an innovative thermochemical conversion process that transforms biomass into a carbonaceous material known as hydrochar, in the presence of water and at elevated temperatures and pressures. Unlike traditional pyrolysis methods that require high temperatures, HTC operates at moderate temperatures ranging from 180 to 250 °C, making it suitable for processing wet biomass feedstocks such as sewage sludge, organic waste, algae, and lignocellulosic materials (Marzbali et al. 2021). The process begins with the mixing of biomass with water in a pressurized reactor, followed by heating to the desired temperature range. Under these conditions, biomass undergoes hydrolysis, dehydration, decarboxylation, and polymerization reactions, leading to the formation of hydrochar. Hydrothermal carbonization offers several advantages over conventional thermochemical conversion methods, including the ability to process wet biomass without the need for drying, reduced energy consumption, and the production of hydrochar with enhanced energy density and stability. The hydrochar produced through HTC exhibits unique properties, such as high carbon content, porous structure, and low ash content, making it suitable for various applications including soil amendment, carbon sequestration, and renewable energy production (Kambo and Dutta 2015). In addition to its environmental benefits, HTC has garnered interest for its potential in valorizing waste biomass and addressing challenges related to waste management and resource recovery. By converting organic waste streams into value-added products like hydrochar, HTC contributes to the circular economy by closing the loop on nutrient and carbon cycles and reducing reliance on fossil fuels. The HTC process can be tailored to optimize the properties of the hydrochar by adjusting various parameters such as reaction temperature, residence time, biomass feedstock composition, and pH of the reaction medium. These parameters influence the yield, quality, and properties of the hydrochar, allowing for the production of customized materials for specific applications (Wang et al. 2018). Hydrothermal carbonization technology is still in the early stages of development, and ongoing research efforts are focused on improving process efficiency, scalability, and economic viability. Future advancements in HTC technology may involve the integration of novel reactor designs, catalysts, and process optimization strategies to enhance the overall performance and competitiveness of hydrochar production.

16.3.2.6 Torrefaction

Torrefaction is a thermal conversion process that involves heating biomass in the absence of oxygen at temperatures typically ranging from 200 to 300 °C (Ru et al. 2015). This process aims to remove moisture and volatile organic compounds from biomass, resulting in a dry, energy-dense solid material known as torrefied biomass or biochar. Torrefaction is often considered a pretreatment step for biomass prior to combustion, gasification, or pyrolysis, as it improves the physical and chemical properties of the biomass, making it more suitable for energy conversion processes (Chew and Doshi 2011). During torrefaction, biomass undergoes several chemical and physical transformations. First, moisture is removed from the biomass, reducing its moisture content to less than 5%. Next, volatile organic compounds such as hemicellulose, cellulose, and lignin degrade and evaporate, resulting in the release of gases such as water vapor, carbon dioxide, methane, and volatile organic compounds. As a result, the biomass becomes darker in color, lighter in weight, and more brittle. The torrefied biomass produced through this process exhibits several desirable properties for energy applications. It has a higher energy density compared to raw biomass due to the removal of moisture and volatile compounds. Additionally, torrefied biomass has improved grindability and hydrophobicity, making it easier to handle, transport, and store. Furthermore, torrefaction reduces the reactivity of biomass, resulting in improved combustion characteristics such as increased ignition temperature and reduced emissions of pollutants like nitrogen oxides and volatile organic compounds. Torrefaction technology offers several advantages for biomass utilization and energy production. It can be applied to a wide range of biomass feedstocks, including wood chips, agricultural residues, energy crops, and municipal solid waste. Torrefied biomass can be used as a renewable fuel for heat and power generation in industrial boilers, power plants, and residential heating systems. It can also be co-fired with coal to reduce greenhouse gas emissions and improve the sustainability of fossil fuel-based energy production (Sarker et al. 2021). In addition to energy applications, torrefied biomass has potential uses in other industries such as metallurgy, agriculture, and wastewater treatment. It can be used as a carbon source for iron ore reduction in steel production, as a soil amendment to improve soil quality and fertility, or as an adsorbent for removing contaminants from wastewater streams.

16.4 Electro-Modified Biochar

Electro-modified biochar refers to biochar materials that have been subjected to electrochemical treatments or modifications to enhance their properties and functionalities for various applications. This emerging field combines principles of electrochemistry with the unique characteristics of biochar, such as its high surface area, porosity, and chemical reactivity, to create tailored materials with improved performance in areas such as environmental remediation, energy storage, and catalysis. One of the key applications of electro-modified biochar is in environmental remediation, particularly in the removal of contaminants from water and soil. Electrochemical techniques such as electrocoagulation, electrooxidation, and electrochemical reduction can be used to enhance the adsorption, oxidation, or reduction of pollutants by biochar. For example, electrochemical activation of biochar can increase its surface area and the number of surface functional groups, thereby improving its adsorption capacity for heavy metals, organic pollutants, and emerging contaminants. In addition to environmental remediation, electro-modified biochar has shown promise in energy storage applications, particularly as electrode materials for supercapacitors and batteries. By modifying the surface chemistry and structure of biochar through electrochemical treatments, researchers can enhance its conductivity, capacitance, and cycling stability, making it suitable for use in energy storage devices. Furthermore, biochar-derived materials have been investigated as catalyst supports in electrochemical reactions for energy conversion processes such as oxygen reduction reaction (ORR) and hydrogen evolution reaction (HER) (Maliutina et al. 2023). The electrochemical modification of biochar can be achieved through various methods, including direct electrochemical treatment, electrochemical deposition, and electrodeposition of metal nanoparticles or conducting polymers onto the biochar surface. These methods allow for precise control over the modification process, enabling the tuning of biochar properties such as surface chemistry, pore structure, and electrical conductivity to meet specific application requirements. Research in the field of electro-modified biochar is still in its early stages, and ongoing efforts are focused on understanding the underlying mechanisms of electrochemical modification, optimizing synthesis protocols, and exploring new applications. Challenges remain in terms of scalability, cost-effectiveness, and the environmental impact of electrochemical treatments on biochar production. However, the potential of electro-modified biochar to address pressing environmental and energy challenges makes it an exciting area of research with significant promise for future technological developments (Table 16.1).

16.5 Heavy Metal Removal by Biochar from Wastewater

Heavy metal contamination in wastewater is a significant issue. Heavy metal removal by biochar from wastewater involves a multifaceted process wherein biochar, a highly porous carbonaceous material derived from the pyrolysis of organic biomass, acts as an effective adsorbent for heavy metal ions present in wastewater. The high surface area and numerous functional groups of biochar facilitate the adsorption of heavy metals through mechanisms such as ion exchange, complexation, and surface precipitation. Additionally, the presence of specific functional groups on biochar surfaces can enhance selectivity towards certain metal ions, thereby offering a tailored approach for wastewater treatment. Furthermore, the physicochemical properties of biochar, such as pore size distribution and surface area, can be optimized through various production techniques, enabling improved removal efficiency. Overall, the utilization of biochar for heavy metal removal presents a promising and sustainable solution for mitigating water pollution, offering an environmentally friendly alternative to conventional treatment methods. Researchers have explored biochar made from various sources, such as rice husk and wood bark, under different conditions to remove metals like cadmium, lead, nickel, arsenic, mercury, and chromium. The removal mechanisms include electrostatic interactions, cation exchange, precipitation, reduction, sorption, and complexation, which vary depending on the metal's valence state and the solution's pH (Li et al. 2017). Biochar is extensively utilized as an adsorbent for the extraction of heavy metals from wastewater (Gao et al. 2022). The sorption mechanisms exhibit significant variability based on the feedstock, production method, and target metal. Pyrolytic methods, conditions, and feedstock composition also exert an influence on the physical and chemical characteristics of biochar, thereby affecting its metal removal capacity. Generally, high-temperature pyrolysis induces the degradation of active functional groups present on the biochar surface, resulting in a decrease in its sorption capacity. Removal mechanisms vary with the target metal's valence state across different solution pH levels (Li et al. 2017). Four primary mechanisms dominate the removal of heavy metals from water by biochar (i) electrostatic attraction between heavy metals and the biochar surface; (ii) ion exchange between heavy metals and alkali or alkaline earth metals or protons on the

Table 16.1 Electrochemical modification strategies of biochar derived from diverse waste sources

Electro modification	Material type	Mode of action and application	References
Electrochemical activation	Wood-based biochar	Electrochemical activation involves applying an electrical current to wood-based biochar, enhancing its surface area and pore structure. This modified biochar is highly effective in the removal of heavy metals from wastewater due to its increased adsorption capacity and improved surface reactivity. It finds application in water treatment plants and industrial effluent treatment facilities. It is also used as supercapacitors	Yan et al. (2022)
Electrochemical deposition of metal nanoparticles	Agricultural residue-derived biochar	In this process, metal nanoparticles are deposited onto the surface of agricultural residue-derived biochar through electrochemical deposition. The resulting composite material exhibits enhanced conductivity and catalytic activity, making it suitable for use in energy storage devices such as supercapacitors. It enables efficient energy storage and release, contributing to renewable energy systems and portable electronic devices	Li et al. (2022)
Electrodeposition of conducting polymers	Municipal solid waste-derived biochar	Conducting polymers are electrodeposited onto the surface of municipal solid waste-derived biochar, improving its electrical conductivity and stability. This modified biochar serves as an efficient catalyst support in organic synthesis reactions, facilitating the production of value-added chemicals and pharmaceuticals. It finds applications in chemical manufacturing and pharmaceutical industries	Abdelbasir et al. (2020)
Direct electrochemical treatment	Algae-derived biochar	Direct electrochemical treatment involves subjecting algae-derived biochar to an electrical current in an electrolyte solution. This process enhances the electrochemical properties of biochar, enabling its use in electrochemical water treatment systems for the removal of contaminants such as heavy metals and organic pollutants. It provides a sustainable and cost-effective solution for water purification in both industrial and municipal settings	Gong et al. (2022)

(continued)

Table 16.1 (continued)

Electro modification	Material type	Mode of action and application	References
Electrochemical reduction	Sewage sludge-derived biochar	Sewage sludge-derived biochar undergoes electrochemical reduction, resulting in the modification of its surface chemistry and reactivity. The modified biochar serves as an effective electrocatalyst for the oxygen reduction reaction (ORR) in fuel cells and metal-air batteries, enhancing their energy conversion efficiency and durability. It contributes to the development of clean energy technologies for transportation and stationary power generation	Ren et al. (2021)
Electrooxidation	Food waste-derived biochar	Electrooxidation treatment modifies the surface of food waste-derived biochar, enhancing its oxidative properties and surface reactivity. The modified biochar exhibits improved adsorption and degradation capabilities, making it suitable for the removal of organic pollutants from water bodies and wastewater streams. It offers a sustainable solution for environmental remediation and pollution control in industries such as food processing and wastewater treatment	Chakraborty et al. (2020)
Electrocoagulation	Lignocellulosic biochar	Lignocellulosic biochar is subjected to electrocoagulation, resulting in the formation of metal hydroxide coagulants on its surface. The modified biochar acts as an effective soil amendment, enhancing soil fertility, water retention, and nutrient availability. It finds applications in agriculture and land reclamation projects, promoting sustainable soil management practices and improving crop yields	Othmani et al. (2021), Tian et al. (2024)

biochar surface; (iii) complexation with π electron-rich domains or surface functional groups; (iv) co-precipitation to form insoluble compounds (Tan et al. 2016; Li et al. 2017; Qian et al. 2020).

16.5.1 Nanomodified Biochar for HM Removal

Nanomodified biochar emerges as a promising avenue for heavy metal removal from wastewater, leveraging the synergistic benefits of biochar's porous structure and the unique properties of nanomaterials. Through various nanomodification techniques, such as impregnation, in situ synthesis, or coating, biochar can be enhanced with nanoparticles to augment its adsorption capacity and selectivity towards heavy metal ions. For instance, the incorporation of zero-valent iron nanoparticles onto biochar surfaces has demonstrated significant improvements in the removal of contaminants like arsenic, chromium, and lead, owing to enhanced redox reactions and surface area. Similarly, the integration of graphene oxide nanoparticles with biochar has shown remarkable efficiency in sequestering heavy metals due to increased surface functional groups and π-π interactions. Moreover, the utilization of nanomodified biochar can also address challenges associated with regeneration and reusability, ensuring its practical applicability in wastewater treatment systems. Overall, nanomodified biochar presents a versatile and sustainable approach for heavy metal removal, offering tailored solutions that hold considerable promise for addressing water pollution challenges. Research indicates limited adsorption capacity of untreated biochar, necessitating modification for better performance. Surface modification with nanomaterials is promising for enhancing sorption capacity. Various methods, including metal, ultrasonic, acid–base, and thiol modification, have been explored (Ndirangu et al. 2019; Wang et al. 2021). For instance, Nath et al. (2019) improved rice husk nanobiochar's adsorption capacity by adding iron oxides. Modified nanobiochars consistently show superior pollutant removal capabilities.

The synthesis of tri-metallic surface-engineered nanobiochar from *Cynodon-dactylon* residues, accomplished in a single-step process, has been investigated for its efficacy in removing copper and lead ions. The study encompassed both batch and column experiments to assess its performance (Vishnu et al. 2022).

16.5.2 Adsorption of Heavy Metals by Nanobiochar

Nanobiochar effectively immobilizes heavy metals like Cd, Pb, and Cu in water, with absorption rates increasing with nanobiochar concentration due to its exchangeable sites and larger surface area (Li et al. 2017; Yue et al. 2019; Vishnu et al. 2022). However, high concentrations may cause agglomeration, reducing effectiveness, especially in acidic conditions (Guo et al. 2017). Lyu et al. (2018) found that the adsorption rate of Nickel (Ni) increased with low nanobiochar application rates in soils but decreased with higher rates. Additionally, heteroaggregation between nanobiochar and soil minerals is common, driven by electrostatic interactions between positively charged soil minerals and negatively charged nanobiochar surfaces (Zhang et al. 2022). Modified nanobiochars show improved pollutant adsorption, like enhanced chromium removal compared to pristine nanobiochar (Ahuja et al. 2022). Nanobiochars with surface functional groups or introduced negative charges demonstrate enhanced adsorption, as seen with thiol-modified nanobiochar for mercury removal (Pratap et al. 2022). Iron oxide-modified rice husk nanobiochar removes up to 95% of arsenic (Nath et al. 2019). Nanobiochar derived from pine wood effectively eliminated micropollutants from water, with 95% removal of carbamazepine (CBZ) achieved within 3 h using nanobiochar particles averaging 60 nm (Naghdi et al. 2019). Liu et al. (2020a) demonstrated that nanobiochar application enhanced plant biomass and seed germination while stabilizing cadmium (Cd) ions in soil, promoting the growth of Actinobacteria and Bacteroidetes, and decreasing available Cd levels. Ramanayaka et al. (2020) prepared nanobiochar using a mechanized grinding and top-down approach, derived from dendro biochar, to remove various contaminants from water, including antibiotics, agrochemicals, and hazardous elements, showing efficacy in removing glyphosate (GL), Cr(VI), oxytetracycline (OTC), and Cd(II). Li et al. (2018) found that ZnO/ZnS-modified nanobiochar exhibited superior adsorption capacities for Cu(II), Pb(II), and Cr(VI) compared to normal biochar, indicating potential for efficient removal of heavy metal contaminants. Furthermore, numerous studies have confirmed the ability of biochar nanocomposites to effectively remove various pollutants such as arsenic, lead, copper, chromium, cadmium, as well as toxic chemicals, dyes, drugs, and antibiotics (Chausali et al. 2021).

High hemicellulose biomass typically yields nanobiochar with distinctive properties, characterized by low carbon content and elevated oxygen levels, contrasting with the strong aggregation tendencies observed in nanobiochar derived from lignin-rich materials (Li et al. 2023; Verma et al. 2023). The production of nanobiochar involves various techniques, notably ball milling and ultrasonication, each offering specific advantages in terms of particle size reduction and structural modification (Chausali et al. 2021). Moreover, modifications such as metal and acid–base treatments are employed to enhance nanobiochar properties, thereby improving its efficiency in pollutant removal processes (Qiu et al. 2022). By incorporating nanomaterials onto biochar surfaces, novel biochar nanohybrids are developed, exhibiting enhanced heavy metal adsorption capabilities and presenting promising avenues for environmental remediation (Vishnu et al. 2022). Furthermore, nanomaterials possess superior functional groups and catalytic degradation abilities, positioning them as valuable tools for mitigating pollutants and advancing nanoremediation strategies (Chaudhary et al. 2023). Zhou et al. (2013) demonstrated that chitosan-modified biochar effectively removed Cd^{2+}, Pb^{2+}, and Cu^{2+} from solutions, with Pb^{2+} exhibiting a high Langmuir sorption capacity despite slow kinetics. Boutsika et al. (2014) utilized biochar from malt spent rootlets (MSR) to achieve complete removal of Hg(II), showing a sorption capacity of 103 mg g^{-1}. Shi et al. (2020) developed $ZnCl_2$–

modified glue residue biochar for Cr(VI) sorption, surpassing previous sorbents with a capacity of 325.5 mg g^{-1}. Liu et al. (2014) enhanced Ni(II) removal using zinc borate-treated lotus stalks biochar, achieving 3–10 times higher sorption. Additionally, biochar-ZnS nanocomposites, synthesized by depositing ZnS nanocrystals onto biochar, exhibited a significantly enhanced sorption capacity of 368 mg/g, while manganese dioxide-loaded biochar demonstrated improved HM ion removal from aqueous solutions (Ahuja et al. 2022). However, the leaching of modified materials raises significant environmental concerns due to the potential release of harmful substances into the surrounding ecosystem. When materials are modified or treated with various chemicals or compounds to enhance their properties or functionality, there is a risk that these added components may leach out over time, especially when exposed to environmental conditions such as water or soil. This leaching process can lead to the contamination of groundwater, soil, and nearby water bodies, posing risks to both human health and the environment. Additionally, the leached substances may bioaccumulate in organisms, leading to further ecological disruptions up the food chain. Therefore, it is crucial to thoroughly assess the leaching potential of modified materials and implement appropriate measures to mitigate any adverse environmental impacts, ensuring the sustainable use and disposal of such materials. Hence, further studies are of utmost importance (Nath et al. 2019) (Table 16.2).

16.6 Modification of Biochar for Wastewater Treatment

Modification of biochar for wastewater treatment has emerged as a dynamic and promising field, offering tailored solutions to address the diverse challenges of pollutant removal. Through various modification strategies such as chemical, physical, and biological methods, biochar's surface properties can be engineered to enhance its adsorption capacity, selectivity, and overall performance in wastewater treatment. For instance, surface functionalization of biochar with amine groups has been shown to significantly improve its affinity for heavy metals like cadmium and copper, through chelation mechanisms (Zhou et al. 2013; Liu et al. 2020b). Furthermore, the introduction of metal nanoparticles onto biochar surfaces, such as silver or iron oxide nanoparticles, has exhibited enhanced antimicrobial activity and pollutant degradation, rendering it effective in tackling organic pollutants and pathogens in wastewater (Huang et al. 2017). Additionally, the incorporation of nanomaterials like graphene oxide or carbon nanotubes into biochar matrices has demonstrated improved conductivity and photocatalytic activity, facilitating the degradation of persistent organic pollutants under light irradiation (Luo et al. 2022). Moreover, biochar can also be tailored through pyrolysis conditions and precursor materials to optimize its physical and chemical properties for specific wastewater treatment applications. Overall, the modification of biochar offers a versatile and sustainable approach to wastewater treatment, presenting a multitude of strategies that can be tailored to meet the demands of different contaminants and treatment scenarios. While biochar exhibits notable

Table 16.2 Recent advances in nanomodified biochar for heavy metal removal from wastewaters

Material	Mechanism involved	Efficiency	References
Electrolytic manganese slag (EMS) based double network hydrogel (EMS/PAA hydrogel)	Activation of EMS was done by NaOH. Mechanism of elimination of heavy metals by EMS/PAA gels chiefly entails inner-sphere complexation, electrostatic attraction, and coordination interactions	The hydrogel demonstrates significant selectivity in adsorbing Pb (II), Cd (II), and Cu (II), with capacities of 153.85, 113.63, and 54.35 $mg.g^{-1}$ respectively	Ma et al. (2024)
Biochar (CHKBC) co-modified with KOH and chitosan	Adsorption mechanisms include ion exchange, π–π interactions, electrostatic attraction, pore filling, and hydrogen bonding	The CHKBC displayed an augmented presence of functional groups like −COOH, $-NH_2$, and −OH, resulting in a significant rise in the maximum adsorption of Methylene blue by the biochar from 8.83 $mg\ g^{-1}$ to 62.04 $mg\ g^{-1}$, marking a 7.03-fold enhancement	Su et al. (2024)
Shell dust from *F. bengalensis* and *P. globosa*	The elimination process involved chemical adsorption, ion exchange and surface precipitation	Biosorbent of Cd^{2+}, Co^{2+}, $Cr_2O_7{}^{2-}$ and Cu^{2+}	Paul and Aditya (2024)
NaOH-modified *Luffa cylindrica*fiber	Common mechanisms involved coagulation, biological treatment, oxidation, and adsorption	89.9 to 99.5% phenol removal from waste water	Ogunniyi et al. (2024)
Grafting of $-NO_2$ and MnO_x loading binary amendment on biochar developed from bagasse	The sorption of Sb(III) was controlled by multi-layered heterogeneous chemisorption, involving complexation among $-NO_2$/Mn–O bond and Sb(III)	Enhancement of antimonite Sb(III)) adsorption (142.86 mg/g)	Zhang et al. (2024)

(continued)

Table 16.2 (continued)

Material	Mechanism involved	Efficiency	References
TiO_2-Nanoparticle from *Cola nitida* pod extract. Modification of the adsorbent developed from groundnut shell impregnated by TiO_2-NP (TiO_2-NPs-GNSAC)	Adsorption mechanism follows physisorption process	81.78% ibuprofen removal from pharmaceutical waste water	Olowonyo et al. (2024)
Thermally converted *Arachis hypogaea* husk into biochar	The adsorbent material contains magnesium (Mg) and chromium (Cr) on its surface	In a continuous flow fixed bed setup, the adsorbent's capacity for adsorbing Cr(VI) was 50 mg g^{-1}	Singh et al. (2023)
Nano ZnO/ZnS modified biochar produced by slow pyrolysis of zinc-contaminated corn stover	The biochar modified with zinc minerals exhibited an improved porous structure, with zinc minerals uniformly attached to the surface of the biochar	Pb(II), Cu(II), and Cr(VI) removal at 135.8, 91.2, and 24.5 mg g^{-1}	Li et al. (2018)

heavy metal ion adsorption capabilities from contaminated solutions, it fares comparatively lower than established biosorbents like activated carbon (Liu et al. 2018). Consequently, recent research has prioritized enhancing biochar's environmental benefits and remediation efficacy through surface property improvements and novel structures. Strategies include enhancing porosity, surface area, and surface functional groups. Modification approaches encompass chemical, physical, magnetic, and impregnation with mineral oxides methods. Incorporating Mg, Al, or Mn oxides onto biochar surfaces results in biochar-based composites, enhancing the sorption of metal cations and oxyanions in water solutions. Again, Pb sorption using biochar derived from cypress sawdust treated with $MgCl_2$, revealing a 7.4-fold increase in sorption compared to the untreated biochar (Sizmur et al. 2017).

16.7 Techniques Used for the Characterization of Biochar

Biochar's physical and chemical properties, including pH, surface area, porosity, charge, functional groups, and mineral content, are pivotal for metal sorption. Despite its carbon and ash composition, biochar's characteristics vary based on biomass type and production conditions. Effective utilization hinges on biomass selection.

Comprehensive biochar characterization is essential for understanding its composition, functional groups, stability, and structure, guiding its application. Characterization techniques play a pivotal role in understanding the physical, chemical, and structural properties of biochar, thereby guiding its optimal utilization in diverse applications. Scanning Electron Microscopy (SEM) and Transmission Electron Microscopy (TEM) unveil the surface morphology and nanoscale structures of biochar, while X-ray Diffraction (XRD) elucidates its crystalline structure and mineral phases. Fourier Transform Infrared Spectroscopy (FTIR) and Raman Spectroscopy provide insights into the functional groups and molecular structure, respectively, offering crucial information on biochar's chemical composition and reactivity (El-Azazy et al. 2021). Brunauer–Emmett–Teller (BET) Surface Area Analysis and Thermogravimetric Analysis (TGA) quantify surface area, pore characteristics, and thermal stability, essential for assessing biochar's adsorption capacity and application suitability (El-Azazy et al. 2021). Elemental analysis techniques and Solid-State Nuclear Magnetic Resonance (NMR) further contribute to understanding biochar's elemental composition and carbon bonding (El-Azazy et al. 2021). These characterization techniques collectively provide a comprehensive understanding of biochar's properties, facilitating its efficient utilization across various environmental and agricultural domains.

16.8 Conclusion

Utilizing biochar for heavy metal removal presents a promising approach. Biochar-based nanocomposites, formed by integrating various nanomaterials onto biochar surfaces, offer enhanced properties. These nanocomposites can be tailored using diverse feedstocks, pyrolysis conditions, and synthesis techniques, impacting biochar's physicochemical attributes significantly. Nanobiochar exhibits superior characteristics compared to pristine biochar, including heightened surface area, increased functionalities, and enhanced sorption capabilities. This review provides a comprehensive overview of biochar's wide-ranging applications in wastewater treatment for heavy metal removal. Exploring sorption mechanisms forms the basis of understanding biochar's sorption behavior, prompting research focus on biochar modification to boost performance. Strategies aim to augment biochar's surface area, porosity, and surface sorption sites, enhancing its efficacy in metal removal.

References

Abdelbasir SM, McCourt KM, Lee CM, Vanegas DC (2020) Waste-derived nanoparticles: synthesis approaches, environmental applications, and sustainability considerations. Front Chem 8:782

Abubackar HN, Veiga MC, Kennes C (2011) Biological conversion of carbon monoxide: rich syngas or waste gases to bioethanol. Biofuels Bioprod Biorefin 5(1):93–114

Ahuja R, Kalia A, Sikka R, P C (2022) Nano modifications of biochar to enhance heavy metal adsorption from wastewaters: a review. ACS omega 7(50):45825–45836

Boutsika LG, Karapanagioti HK, Manariotis ID (2014) Aqueous mercury sorption by biochar from malt spent rootlets. Water Air Soil Pollut 225:1–10

Chakraborty I, Sathe SM, Dubey BK, Ghangrekar MM (2020) Waste-derived biochar: applications and future perspective in microbial fuel cells. Biores Technol 312:123587

Chaturvedi S, Singh SV, Dhyani VC, Govindaraju K, Vinu R, Mandal S (2021) Characterization, bioenergy value, and thermal stability of biochars derived from diverse agriculture and forestry lignocellulosic wastes. Biomass Convers Biorefinery 1–14

Chaudhary P, Ahamad L, Chaudhary A, Kumar G, Chen WJ, Chen S (2023) Nanoparticle-mediated bioremediation as a powerful weapon in the removal of environmental pollutants. J Environ Chem Eng 11(2):109591

Chausali N, Saxena J, Prasad R (2021) Nanobiochar and biochar based nanocomposites: advances and applications. J Agric Food Res 5:100191

Chew JJ, Doshi V (2011) Recent advances in biomass pretreatment–torrefaction fundamentals and technology. Renew Sustain Energy Rev 15(8):4212–4222

Couto N, Rouboa A, Silva V, Monteiro E, Bouziane K (2013) Influence of the biomass gasification processes on the final composition of syngas. Energy Proced 36:596–606

Das AP, Sukla LB, Pradhan N, Nayak S (2011) Manganese biomining: a review. Biores Technol 102(16):7381–7387. https://doi.org/10.1016/J.BIORTECH.2011.05.018

Das AP, Ghosh S, Mohanty S, Sukla LB (2015) Advances in manganese pollution and its bioremediation. In Sukla LB, Pradhan N, Panda S, Mishra BK (eds) Environmental microbe biotechnology 1st ed, pp 313–328. Springer, Cham. https://doi.org/10.1007/978-3-319-19018-1_16

Dey S, Tripathy B, Kumar MS, Das AP (2023) Ecotoxicological consequences of manganese mining pollutants and their biological remediation. Environ Chem Ecotoxicol. https://doi.org/10.1016/J.ENCECO.2023.01.001

El-Azazy M, Nabil I, Hassan SS, El-Shafie AS (2021) Adsorption characteristics of pristine and magnetic olive stones biochar with respect to clofazimine. Nanomaterials 11(4):963

Ercan B, Alper K, Ucar S, Karagoz S (2023) Comparative studies of hydrochars and biochars produced from lignocellulosic biomass via hydrothermal carbonization, torrefaction and pyrolysis. J Energy Inst 109:101298

Gao J, Zhang L, Liu S et al (2022) Enhanced adsorption of copper ions from aqueous solution by two-step DTPA-modified magnetic cellulose hydrogel beads. Int J Biol Macromol. https://doi.org/10.1016/j.ijbiomac.2022.05.073

Gong Z, Wang H, Vayenas DV, Yan Q (2022) Enhanced electrochemical removal of sulfadiazine using stainless steel electrode coated with activated algal biochar. J Environ Manage 306:114535

Guo Y, Zhang J, Zhao Y, Wang S, Jiang C, Zheng C (2017) Chemical agglomeration of fine particles in coal combustion flue gas: experimental evaluation. Fuel 203:557–569

Guo M, Xiao P, Li H (2019) Valorization of agricultural byproducts through conversion to biochar and bio-oil. Byproducts Agric Fish Adding Value Food Feed Pharma Fuels 501–522

Hansen V, Müller-Stöver D, Ahrenfeldt J, Holm JK, Henriksen UB, Hauggaard-Nielsen H (2015) Gasification biochar as a valuable by-product for carbon sequestration and soil amendment. Biomass Bioenerg 72:300–308

Huang JF, Shi QS, Feng J, Chen MJ, Li WR, Li LQ (2017) Facile pyrolysis preparation of rosin-derived biochar for supporting silver nanoparticles with antibacterial activity. Compos Sci Technol 145:89–95

Kambo HS, Dutta A (2015) A comparative review of biochar and hydrochar in terms of production, physico-chemical properties and applications. Renew Sustain Energy Rev 45:359–378

Kapoor RT (2021) Biochar a sustainable solution for wastewater treatment: Current status and perspectives. In: Microbial ecology of wastewater treatment plants, pp 217–236. Elsevier

Li HB, Dong XL, da Silva EB et al (2017) Mechanisms of metal sorption by biochars: biochar characteristics and modifications. Chemosphere 178:466–478. https://doi.org/10.1016/j.chemosphere.2017.03.072

Li C, Zhang L, Gao Y, Li A (2018) Facile synthesis of nano ZnO/ZnS modified biochar by directly pyrolyzing of zinc contaminated corn stover for Pb (II), Cu (II) and Cr (VI) removals. Waste Manage 79:625–637

Li Y, Xu R, Wang H, Xu W, Tian L, Huang J, Liang C, Zhang Y (2022) Recent advances of biochar-based electrochemical sensors and biosensors. Biosensors 12(6):377

Li Q, Zhang X, Mao M, Wang X, Shang J (2023) Carbon content determines the aggregation of biochar colloids from various feedstocks. Sci Total Environ 880:163313

Lima RMAP, Dos Reis GS, Thyrel M, Alcaraz-Espinoza JJ, Larsson SH, de Oliveira HP (2022) Facile synthesis of sustainable biomass-derived porous biochars as promising electrode materials for high-performance supercapacitor applications. Nanomaterials 12(5):866

Liu H, Liang S, Gao J, Ngo HH, Guo W, Guo Z, Li Y (2014) Development of biochars from pyrolysis of lotus stalks for Ni (II) sorption: using zinc borate as flame retardant. J Anal Appl Pyrol 107:336–341

Liu Z, Singer S, Tong Y, Kimbell L, Anderson E, Hughes M, Zitomer D, McNamara P (2018) Characteristics and applications of biochars derived from wastewater solids. Renew Sustain Energy Rev 90:650–664

Liu Y, Xu J, Cao Z, Fu R, Zhou C, Wang Z, Xu X (2020a) Adsorption behavior and mechanism of Pb (II) and complex Cu (II) species by biowaste-derived char with amino functionalization. J Colloid Interface Sci 559:215–225

Liu W, Li Y, Feng Y, Qiao J, Zhao H, Xie J, Fang Y, Shen S, Liang S (2020b) The effectiveness of nanobiochar for reducing phytotoxicity and improving soil remediation in cadmium-contaminated soil. Sci Rep 10(1):858

Lopez G, Aguado R, Olazar M, Arabiourrutia M, Bilbao J (2009) Kinetics of scrap tyre pyrolysis under vacuum conditions. Waste Manage 29(10):2649–2655

Luo H, Yu S, Zhong M, Han Y, Su B, Lei Z (2022) Waste biomass-assisted synthesis of TiO2 and N/O-contained graphene-like biochar composites for enhanced adsorptive and photocatalytic performances. J Alloy Compd 899:163287

Lyu H, Gao B, He F, Zimmerman AR, Ding C, Huang H, Tang J (2018) Effects of ball milling on the physicochemical and sorptive properties of biochar: experimental observations and governing mechanisms. Environ Pollut 233:54–63

Ma M, Ke X, Wang T, Li J, Ye H (2024) A novel double-network hydrogel made from electrolytic manganese slag and polyacrylic acid-polyacrylamide for removal of heavy metals in wastewater. J Hazard Mater 462:132722

Maliutina KM, Omoriyekomwan JE, He C, Fan L, Folli A (2023) Biomass-derived carbon nanostructures and their applications as electrocatalysts for hydrogen evolution and oxygen reduction/evolution. Front Environ Eng 2

Marzbali MH, Kundu S, Halder P, Patel S, Hakeem IG, Paz-Ferreiro J, Madapusi S, Surapaneni A, Shah K (2021) Wet organic waste treatment via hydrothermal processing: a critical review. Chemosphere 279:130557

Mishra S, Das AP (2008) Hexavalent chromium reduction and 16S rDNA identification of bacteria isolated from a Cr (VI) contaminated site. Internet J Microbiol 7(1)

Naghdi M, Taheran M, Pulicharla R, Rouissi T, Brar SK, Verma M, Surampalli RY (2019) Pine-wood derived nanobiochar for removal of carbamazepine from aqueous media: adsorption behavior and influential parameters. Arab J Chem 12(8):5292–5301

Nath BK, Chaliha C, Kalita E (2019) Iron oxide permeated mesoporous rice-husk nanobiochar (IPMN) mediated removal of dissolved arsenic (As): chemometric modelling and adsorption dynamics. J Environ Manage 246:397–409

Ndirangu SM, Liu Y, Xu K, Song S (2019) Risk evaluation of pyrolyzed biochar from multiple wastes. J Chem 2019:1–28

Ogunniyi S, Emenike EC, Iwuozor KO, Ighalo JO, Ezzat AO, Adewoye TL, Egbemhenghe A, Okoro HK, Adeniyi AG (2024) Removal of phenol from wastewater using Luffa cylindrica fibers in a packed bed column: optimization, isotherm and kinetic studies. Heliyon

Okoroanyanwu U, Bhardwaj A, Einck V, Ribbe A, Hu W, Rodriguez JM, Schmidt WR, Watkins JJ (2021) Rapid preparation and electrochemical energy storage applications of silicon carbide and silicon oxycarbide ceramic/carbon nanocomposites derived via flash photothermal pyrolysis of organosilicon preceramic polymers. Chem Mater 33(2):678–694

Olowonyo IA, Salam KK, Aremu MO, Lateef A (2024) Synthesis, characterization, and adsorptive performance of titanium dioxide nanoparticles modified groundnut shell activated carbon on ibuprofen removal from pharmaceutical wastewater. Waste Manag Bullet 1(4):217–233

Othmani A, John J, Rajendran H, Mansouri A, Sillanpää M, Chellam PV (2021) Biochar and activated carbon derivatives of lignocellulosic fibers towards adsorptive removal of pollutants from aqueous systems: Critical study and future insight. Sep Purif Technol 274:119062

Pariyar P, Kumari K, Jain MK, Jadhao PS (2020) Evaluation of change in biochar properties derived from different feedstock and pyrolysis temperature for environmental and agricultural application. Sci Total Environ 713:136433

Paul P, Aditya G (2024) The shells of edible freshwater snails as a biosorbent for multi-metal ions from wastewater: implication in effluent purification and waste valorisation. Sep Purif Technol 331:125623

Piskorz J, Majerski P, Radlein D, Vladars-Usas A, Scott DS (2000) Flash pyrolysis of cellulose for production of anhydro-oligomers. J Anal Appl Pyrol 56(2):145–166

Prabhakar A, Mishra S, Das AP (2019) Isolation and identification of lead (Pb) solubilizing bacteria from automobile waste and its potential for recovery of lead from end of life waste batteries. Geomicrobiol J 36(10):894–903. https://doi.org/10.1080/01490451.2019.1654044

Pratap T, Chaubey AK, Patel M, Mlsna TE, Pittman Jr CU, Mohan D (2022) Nanobiochar for aqueous contaminant removal. In: Sustainable biochar for water and wastewater treatment. Elsevier, pp 667–704

Qian H, Wang J, Yan L (2020) Synthesis of lignin-poly(N-methyl aniline)-reduced graphene oxide hydrogel for organic dye and lead ions removal. J Bioresour Bioprod 5(3):204–210. https://doi.org/10.1016/j.jobab.2020.07.006

Qiu B, Shao Q, Shi J, Yang C, Chu H (2022) Application of biochar for the adsorption of organic pollutants from wastewater: modification strategies, mechanisms and challenges. Sep Purif Technol 300:121925

Raj K, Das AP (2023) Lead pollution: impact on environment and human health and approach for a sustainable solution. Environ Chem Ecotoxicol 5:79–85. https://doi.org/10.1016/J.ENCECO.2023.02.001

Ramanayaka S, Tsang DC, Hou D, Ok YS, Vithanage M (2020) Green synthesis of graphitic nanobiochar for the removal of emerging contaminants in aqueous media. Sci Total Environ 706:135725

Ren X, Tang L, Wang J, Almatrafi E, Feng H, Tang X, Yu J, Yang Y, Li X, Zhou C, Zeng Z (2021) Highly efficient catalytic hydrogenation of nitrophenols by sewage sludge derived biochar. Water Res 201:117360

Robinson J, Dodds C, Stavrinides A, Kingman S, Katrib J, Wu Z, Medrano J, Overend R (2015) Microwave pyrolysis of biomass: control of process parameters for high pyrolysis oil yields and enhanced oil quality. Energy Fuels 29(3):1701–1709

Ru B, Wang S, Dai G, Zhang L (2015) Effect of torrefaction on biomass physicochemical characteristics and the resulting pyrolysis behavior. Energy Fuels 29(9):5865–5874

Sarker TR, Nanda S, Dalai AK, Meda V (2021) A review of torrefaction technology for upgrading lignocellulosic biomass to solid biofuels. BioEnergy Res 14:645–669

Shi Y, Shan R, Lu L, Yuan H, Jiang H, Zhang Y, Chen Y (2020) High-efficiency removal of Cr (VI) by modified biochar derived from glue residue. J Clean Prod 254:119935

Singh K, Azad SK, Dave H, Prasad B, Maurya DM, Kumari M, Dubey D, Rai AK, Sillanpää M, Sah MP, Prasad KS (2023) Effective removal of Cr (VI) ions from the aqueous solution by agro-waste-based biochar: an exploration of batch and column studies. Biomass Convers Biorefinery 1–15

Sizmur T, Fresno T, Akgül G, Frost H, Moreno-Jiménez E (2017) Biochar modification to enhance sorption of inorganics from water. Biores Technol 246:34–47

Su X, Wang X, Ge Z, Bao Z, Lin L, Chen Y, Dai W, Sun Y, Yuan H, Yang W, Meng J (2024) Koh-activated biochar and chitosan composites for efficient adsorption of industrial dye pollutants. Chem Eng J 150387

Tan GC, Sun WL, Xu YR et al (2016) Sorption of mercury (II) and atrazine by biochar, modified biochars and biochar based activated carbon in aqueous solution. Biores Technol 211:727–735. https://doi.org/10.1016/j.biortech.2016.03.147

Thomson R, Kwong P, Ahmad E, Nigam KDP (2020) Clean syngas from small commercial biomass gasifiers; a review of gasifier development, recent advances and performance evaluation. Int J Hydrog Energy 45(41):21087–21111

Tian Y, Chen N, Yang X, Zhang R, Xu J, Li C, Yang W, Ren N, Feng Y, He W (2024) Achieved advanced wastewater phosphate removal via improved ionic interference resistance in a migration-electric-field assisted electrocoagulation (MEAEC) system with modified-biochar pseudocapacitor electrode. Sep Purif Technol 328:124972

Verma R, Dutta S, Kumar A, Dabodiya TS, Kumar N, Karuppasamy KSK, Sangmesh B, Jaiswal A, Jaiswal KK (2023) Biochar-based nanocomposite materials: types, characteristics, physical activation, and diverse application scenarios. Biochar-based nanocomposites for contaminant management: synthesis, contaminants removal, and environmental sustainability. Springer International Publishing, Cham, pp 3–18

Vieira FR, Luna CMR, Arce GL, Ávila I (2020) Optimization of slow pyrolysis process parameters using a fixed bed reactor for biochar yield from rice husk. Biomass Bioenerg 132:105412

Vishnu D, Dhandapani B, Vaishnavi G, Preethi V (2022) Synthesis of tri-metallic surface engineered nanobiochar from cynodondactylon residues in a single step-batch and column studies for the removal of copper and lead ions. Chemosphere 286:131572

Wang T, Zhai Y, Zhu Y, Li C, Zeng G (2018) A review of the hydrothermal carbonization of biomass waste for hydrochar formation: process conditions, fundamentals, and physicochemical properties. Renew Sustain Energy Rev 90:223–247

Wang J, Guo M, Luo Y et al (2021) Production of magnetic sodium alginate polyelectrolyte nanospheres for lead ions removal from wastewater. J Environ Manage. https://doi.org/10.1016/j.jenvman.2021.112506

Yan B, Zheng J, Feng L, Du C, Jian S, Yang W, Wu YA, Jiang S, He S Chen W (2022) Wood-derived biochar as thick electrodes for high-rate performance supercapacitors. Biochar 4(1):50

Yue L, Lian F, Han Y, Bao Q, Wang Z, Xing B (2019) The effect of biochar nanoparticles on rice plant growth and the uptake of heavy metals: implications for agronomic benefits and potential risk. Sci Total Environ 656:9–18

Zhang HG, Zhuo YQ, Zhang XM, Zhang L, Xu PY, Tian HR, Lin SC, Zhang Q, Xie SY, Zheng LS (2021) Synthesis of fullerenes from a nonaromatic chloroform through a newly developed ultrahigh-temperature flash vacuum pyrolysis apparatus. Nanomaterials 11(11):3033

Zhang X, Wells M, Niazi NK, Bolan N, Shaheen S, Hou D, Gao B, Wang H, Rinklebe J, Wang Z (2022) Nanobiochar-rhizosphere interactions: implications for the remediation of heavy-metal contaminated soils. Environ Pollut 299:118810

Zhang L, Yang D, Tao Y, Jiang X, Wang L, Wang H, Liu Y, Meng X, Niu Y, Niu Y, Luo M (2024) Efficiently simultaneous sorption and oxidation of antimonite on nitroso grafting-manganese loading binary-modified biochar from aqueous solution. Sep Purif Technol 337:126398

Zhou Y, Gao B, Zimmerman AR, Fang J, Sun Y, Cao X (2013) Sorption of heavy metals on chitosan-modified biochars and its biological effects. Chem Eng J 231:512–518

Chapter 17
Laccase Enzyme for the Environmental Remediation of Bisphenol A

Ranjit Das, Guiying Li, and Taicheng An

Abstract The environmental presence of endocrine-disrupting compounds, particularly Bisphenol A (BPA), originating from both natural and anthropogenic sources, results in harmful effects on the environment and Earth's biota. Addressing this issue, there is an imperative need for the development of efficient and eco-friendly remediation methods. Laccase offers a great platform for designing appropriate technology to remediate BPA-polluted environments. This chapter aims to discuss the toxicological consequences of BPA contamination and its impacts on the environment. It also provides an overview of the progress made in this field, emphasizing the promising future of laccase as a new biocatalyst for the bioremediation of emerging pollutants such as BPA present in different compartments of the environment. In conclusion, this chapter highlights the potential of laccase enzyme as a sustainable and efficient tool for mitigating BPA contamination in the environment, underscoring the importance of further research to optimize its application in real-world scenarios.

Keywords Bisphenol A · Laccase · Toxicological · Biocatalyst · Bioremediation

17.1 Introduction

Bisphenol A (BPA) poses a significant concern due to its extensive utilization as a plasticizer in both industrial and residential settings, notably in the manufacture of polycarbonate plastics, epoxy resins, and various polymers. The worldwide demand

R. Das (✉)
Department of Microbiology, All India Institute of Medical Sciences (AIIMS), Kalyani, Nadia, West Bengal, India
e-mail: ranmicro_85@yahoo.co.in

G. Li · T. An
Guangdong Key Laboratory of Environmental Catalysis and Health Risk Control, Guangdong-Hong Kong-Macao Joint Laboratory for Contaminants Exposure and Health, Institute of Environmental Health and Pollution Control, Guangdong University of Technology, Guangzhou, China

A. Kuanar et al. (eds.), *Recent Advances in Bioremediation and Phytoremediation*, Environmental Science and Engineering,
https://doi.org/10.1007/978-3-031-77884-1_17

for BPA exceeded approximately 6.5 million tons in 2012 (Huang et al. 2012; Zhou et al. 2015; Im et al. 2016; Larsson et al. 2017). China has exhibited robust usage, with an estimated demand of about 2.25 × 106 tons in 2010 (Huang et al. 2012). Importantly, the United States Environmental Protection Agency (EPA) estimates that over 1 million pounds of BPA leak into the environment on an annual basis, with over 90% of tested human bodies, especially infants and children, containing high levels of BPA (Kabir et al. 2015; Seachrist et al. 2016).

The frequent occurrence of BPA in the environment, coupled with its toxic effects on organisms, necessitates stringent regulations on its production and usage (Chen et al. 2016a, b). BPA functions as an endocrine disruptor and exhibits carcinogenic or genotoxic activity toward various organisms, posing a significant threat to human health (Grignard et al. 2012; Rochester et al. 2015; Bereketoglu et al. 2017; Wu et al. 2017;Tarafdar et al. 2022). Therefore, it is important to protect both the environment and human health from the detrimental effects of BPA and to establish efficient systems utilizing indigenous microorganisms. This approach could represent one of the most significant steps toward the bioremediation of environments exposed to BPA. Using microorganisms to degrade BPA is widely considered a safe and economical alternative to commonly employed chemical processes. In the current scenario of BPA remediation prospects, microbial enzymes offer promising avenues for cost-effective, efficient, and environmentally friendly solutions. Ligninolytic enzymes have been extensively studied as agents in the biodegradation of various xenobiotics, including BPA (Zeng et al. 2017; Zhang et al. 2020; Akram et al. 2022). Ligninolytic enzymes, such as manganese peroxidase (MnP), lignin peroxidase (LiP), versatile peroxidase (VP), and laccases, have been extensively studied as agents in the biodegradation of various xenobiotics, including BPA (Karigar and Rao 2011; Husain and Qayyum 2012; Cajthaml 2015; Ba and Kumar 2017; de Freitas et al. 2017; Das et al. 2018 and 2023).

Laccase (benzenediol: oxygen oxidoreductase; EC 1.10.3.2) is a member of the largest protein superfamily of multi-copper oxidases (MCOs), catalyzing the oxidation of a wide variety of organic and inorganic substrates while reducing molecular oxygen to water (Dwivedi et al. 2011).Laccases are predominantly found in a wide range of organisms, including plants, insects, fungi, bacteria, and archaea (Koschorreck et al. 2008; Uthandi et al. 2010; Chen et al. 2016a, b; Hautphenne et al. 2016). Its exceptional oxidation potential and broad substrate range have significantly elevated theimportance of laccase in various industrial and biotechnological applications. Specifically, laccases play a crucial role in the degradation of aromatic pollutants, pulp delignification, biobleaching, and decolorization of textile dyes (Telke et al. 2009; Zhang et al. 2016; Das et al. 2018; Akram et al. 2022; Naseem et al. 2023).

The catalytic function of laccase depends on protein structure, particularly on amino acid residues coordinate with Cu^{2+}ions (Daroch et al. 2014).Structurally, a typical laccase is multi-domain proteins with a conserved structural core. This core contains multiple copper-binding sites responsible for the catalytic activity of enzyme. It contains four copper atoms that catalytically distributed in three distinct copper binding sites such as Type 1 (T1), Type 2 (T2), and Type 3 (T3) (Chen et al. 2015). The catalytic actions of laccases are mainly involved in the coordination of amino acids (particularly, histidine and cysteine residues) and copper atoms, present in the copper-binding sites (Reiss et al. 2013; Das et al. 2018). The transition of charge molecules from cysteine residue to Cu^{2+} at the T1 site is also engaged to produce the color of the blue laccase plus an absorption peak at around 600 nm (Duraoet al. 2008; Daroch et al. 2014).In contrast, the yellow or white laccase is notably different due to its available amino acid residues present in the copper centers and their synchronization with Cu^{2+} sphere at the T1 site, usually aids the color and spectral properties in the laccases (Mate et al. 2013). As compared to fungal sources, the laccases from bacteria display more pH and temperature stability due to distinct protein structure and interaction with the Cu^{2+} (Hildén et al. 2009).

In this chapter, we explore several common themes that have emerged regarding the important field of enzyme-mediated bioremediation, particularly focusing on the application of laccases for the degradation of BPA. Specifically, this chapter provides detailed insights into the basic structural concepts and physicochemical properties of BPA. Additionally, it discusses the toxicological consequences of BPA contamination and its impacts on the environment. Furthermore, the chapter elucidates the structural and functional concepts of laccase enzymes, followed by an exploration of the degradation mechanisms of BPA facilitated by laccases. Emphasis is placed on the promising future of laccase as a novel biocatalyst for the bioremediation of BPA. Overall, this chapter offers an overview of the utilization of laccase enzymes for the environmental remediation of Bisphenol A, highlighting their potential, challenges, and future directions in this field.

17.2 Physio-Chemical Properties and Applications of Bisphenol A

Bisphenol A [BPA; 2,2-bis(4-hydroxyphenyl) propane], is a synthetic aromatic compound with a molecular structure comprises two phenol rings connected by a methyl bridge, with two methyl functional groups attached to the bridge and two hydroxyl residues are directly bounded to aromatic ring (Kamaraj et al. 2014). It is a white, crystalline solid with the molecular formula $C_{15}H_{16}O_2$ and molecular weight of 228.29 g mol^{-1}(Fig. 17.1). The presence of the hydroxyl group renders BPA slightly acidic and susceptible to oxidation. Like other phenols, BPA has the potential to undergo conversion into ethers, esters, and salts. It boasts a relatively low melting point of around 158 °C (316°F), rendering it suitable for various manufacturing processes, and a boiling point of 220 °C (at a pressure of 5 hPa). BPA exhibits moderate polarity (log Kow 3.3–3.5) with good solubility in organic solvents such as ethanol, acetone, and ether, and moderate solubility in water (approximately 300 mg/L at 25 °C). BPA is reactive and can undergo polymerization reactions to form polycarbonate plastics and epoxy resins. Additionally, it can undergo esterification reactions with carboxylic acids to yield various derivatives (Im et al. 2016; Michałowicz 2014).

Fig. 17.1 Chemical structure of Bisphenol A (BPA)

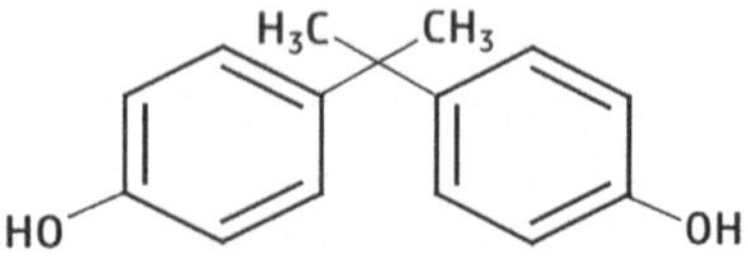

Bisphenol A (BPA)

Indeed, BPA undergoes electrophilic substitutions such as nitration, sulphonation, or alkylation (Flint et al. 2012; Michałowicz 2014; Im et al. 2016). BPA ranks as one of the most widely produced chemicals globally among endocrine disruptor compounds (Huang et al. 2012). It is a synthetic organic compound extensively utilized as a substrate and precursor in the synthesis of polycarbonate plastics, epoxy resins, and other polymers (Mokra et al. 2015; Yang et al. 2014; Badiefar et al. 2015). Due to its favorable mechanical properties, low water adsorption, and thermal stability in synthetic polymers, BPA is commonly used in the manufacture of various products including plastic bottles, thermal papers, water pipes, food containers, toys, medical equipment, dental sealants, and electronic devices (Huang et al. 2012; Michałowicz 2014; Zhou et al. 2015; Im et al. 2016; Larsson et al. 2017). Large quantities of thermal paper are used in various applications such as register receipts, books, faxes, and labels. After recycling, papers are repurposed to produce brochures, tickets, mailing envelopes, newspapers, kitchen rolls, toilet paper, and food cartons (Nam et al. 2010; Liao et al. 2012; Michałowicz 2014; Chen et al. 2016b). Moreover, BPA is extensively applied in the manufacture of polyester, polyacrylates and lacquer coatings for tins. Upon degradation, these materials may significantly contribute to various environmental settings and food sources (Vandenberg et al. 2012a).

17.3 Toxicity of Bisphenol A

Bisphenol A (BPA) is a chemical percussor moleculethat commonly used in the production of plastics and resins, particularly in the manufacture of polycarbonate plastics and epoxy resins. It has been a topic of concern due to its potential adverse effects on human health. Here, we have discussed some points of BPA's toxicity and its impact on human health (Table 17.1):

17.3.1 Toxicity of BPA on Reproductive System

Bisphenol A (BPA) has emerged as a significant concern in reproductive health due to its endocrine-disrupting properties. It can mimic estrogen, leading to hormonal imbalances that may interfere with normal reproductive processes (Vandenberg et al. 2012b). Exposure to BPA has been linked to reproductive health issues,

Table 17.1 Toxic effects of bisphenol A in different systems

Different system	Adverse effects	Possible issues	References
Reproductive system	Disruption of hormonal balance, reduced fertility, reproductive organ abnormalities, and disrupt reproductive function, affecting sperm quality,	Infertility, miscarriage, preterm birth, low birth weight, and gestational diabetes, hormonal disorders, reproductive organ abnormalities	(Vandenberg et al. 2012b; Rochester et al. 2013; Harley et al. 2017; Manikkam et al. 2013)
Developmental System	Interference with fetal development, altered neurobehavioral development, increased risk of birth defects	Birthdefects, neuro-developmental disorders, growth abnormalities, cognitive deficits, increased anxiety-like behaviour, aggression, social deficits, and reduced memory, obesity, insulin resistance, and metabolic disorders	(Angle et al. 2013; Braun and Hauser, 2011)
Cardiovascular System	Metabolic disorders and functions, Increased risk of hypertension,	Metabolic dysfunction, obesity, diabetes, and cardiovascular diseases, hypertension, coronary artery disease and angina	(Shankar and Teppala, 2012; Melzer et al. 2010; Magdalena et al. 2012;Rochester et al. 2018)
Other Parts of the Body	Effect on neurotransmitter systems, disrupt liver function	Autism disorder, liver diseases, cancers in different organs	(Lee et al. 2018; Rahbar et al. 2017; Leung et al. 2017; Williams et al. 2013)

including disruptions in fertility, development of reproductive organs, and reproductive function (Rochester et al. 2013). BPA interrupts hormonal signaling pathways, disrupting reproductive function, affecting sperm quality and quantity in men, and menstrual cycle regularity and ovulation in women. Prenatal exposure to BPA has been connected with an increased risk of pregnancy complications, including miscarriage, preterm birth, low birth weight, and gestational diabetes (Harley et al. 2017). Earlier research suggests that exposure to BPA may adversely affect reproductive processes, leading to various abnormalities and complications (Table 17.1). It is also evident that BPA exposure during gestation may have transgenerational effects, impacting the reproductive health of subsequent generations. Epigenetic modifications induced by BPA exposure can also alter gene expression patterns linked to reproductive function (Manikkam et al. 2013).

17.3.2 Toxic Effects of BPA on Developmental System

Exposure to BPA during critical periods of development has raised concerns due to its potential to disrupt normal growth and development, particularly affecting offspring's height, weight, and neural development (Angle et al. 2013). Studies on animal models have shown that prenatal and neonatal exposure to BPA can disrupt brain development, resulting in cognitive deficits, hyperactivity, increased anxiety-like behavior, aggression, social deficits, and reduced memory (Braun and Hauser 2011; Rochester et al. 2018). BPA exposure during early development may program metabolic pathways, leading to long-term effects on growth, metabolism, and adiposity (Table 17.1). Evidence from studies on animal models has shown that BPA exposure can predispose offspring to obesity, insulin resistance, and metabolic disorders (Angle et al. 2013).

17.3.3 Toxic Effects of BPA on Metabolic and Cardiovascular Systems

Emerging evidence suggests that BPA exposure may also be associated with metabolic disorders such as obesity, diabetes, and cardiovascular diseases. BPA has been implicated in disrupting metabolic pathways and contributing to the development of these conditions. Earlier, Alonso-Magdalena et al. (2012) established that BPA acts as an estrogenic compound, influencing metabolic processes such as glucose metabolism and lipid accumulation, and its ability to modulate gene expression and signaling pathways involved in metabolic regulation contributes to metabolic dysfunction. Moreover, low-dose BPA exposure also impacts cardiovascular health and behavior in offspring, as well as sex-specific epigenetic modifications, leading to alterations in cardiovascular function and behavior (Kundakovic et al. 2013). Previously, Melzer et al. (2010) established a positive correlation between urinary BPA and an increased risk of coronary artery disease and angina, highlighting the potential cardiovascular effects of BPA exposure (Table 17.1).

17.3.4 Other Forms of Toxicity by BPA

BPA exposure has been shown to affect neurotransmitter systems in the brain, including dopamine, serotonin, and glutamate, which play crucial roles in mood regulation, cognition, and behavior (Williams et al. 2013). In one study, Lee et al. (2018) investigated the association between BPA exposure and serum levels of liver enzymes in children. The findings of the study indicated a significant positive correlation between alanine aminotransferase (ALT) levels and BPA exposure in children. The study suggested that BPA exposure may have adverse effects on liver health, as reflected by elevated ALT levels, in pediatric populations (Lee et al. 2018). There is sufficient evidence to prove that neurological disorders like autism spectrum disorder and various cancers in the ovary, lung, and breast (Table 17.1) have been triggered by increased levels of BPA (Rahbar et al. 2017; Leung et al. 2017).

17.4 Environmental Abundance and Impact of BPA

BPA residues present in the naturalenvironmentsis solely attributed to anthropogenic activities. The primary sources of BPA contamination in human surroundings include its production, treatment, and processing, as well as discharges of industrial wastewater, leachates from landfills, manufacturing processes involving BPA, waste from combustion of electronic goods, spraying of paint, and hydrolysis of various polymers including epoxy resins and polycarbonates (Huang et al. 2012; Michałowicz 2014; Chen et al. 2016b). Consequently, BPA is ubiquitously distributed and detected across various environmental compartments such as soil, sediment, air, municipal waste, biota samples, and food (Grignard et al. 2012; Huang et al. 2012; Ji et al. 2013; Peng et al. 2015; Chen et al. 2016b). Understanding the environmental fate of BPA and determining its levels in ecosystems are crucial for revealing potential risks and developing strategies to reduce exposure rates and mitigate toxicological effects on populations.

Various studies have reported BPA concentrations ranging between 0.55 and 147 μg/kg on a dry weight basis (d.w.) in agricultural soils in the United States of America (Kinney et al. 2008), < 0.24 and 492 μg/kg d.w. in sediments of sixteen major rivers in Taiwan (Michalowicz, 2014), 0.001 and 20 μg/L in groundwater from England (Stuart et al. 2011), and < 0.001 and 92 μg/L in surface water in China, Germany, Portugal, Korea, and the UK (Luo et al. 2014; Michalowicz 2014; Careghini et al. 2015; Bereketoglu et al. 2017). However, higher concentrations of BPA up to 17.2 mg/L have been detected in hazardous waste landfill leachate at various sites in Japan (Yamamoto et al. 2001). Additionally, several studies have identified BPA concentrations up to 790 μg/kg fresh weight in food, primarily derived from polymer packaging materials (Noonan et al. 2011). BPA concentrations were also observed to be 26.6 mg/kg in food-contact paper, 42.6 g/kg in thermal receipt paper, and 0.86 μg/L in drinking water/commercial drinks (Suciu et al. 2013; Babu et al. 2015; Careghini et al. 2015). BPA leaching from plastic waste into water has been well documented. With a moderate solubility (120–300 mg/L) at pH 7.0, BPA can also be adsorbed to sediment. Despite its low volatility, BPA exhibits slow biodegradation in acclimated wastewater treatment systems, with half-lives typically ranging from 2.5 to 4 days (Staples et al. 1998).

In toxicological studies performed in vivo, BPA concentrations of 50 mg/kg have been defined as "low" dose, while concentrations up to 200 mg/kg have been used to investigate its effects (Alonso-Magdalena et al. 2010; Wang and Bennett 2012; Xi et al. 2011; Tan et al. 2013). The human population is primarily exposed to BPA through diet, while air, dust, water, and skin contact with thermal paper are other possible sources of exposure (Michałowicz 2014; Seachrist et al. 2016; Testai et al. 2016). BPA in food and beverages accounts for approximately 50% of daily human internal exposure, whereas skin contact with BPA-containing thermal paper constitutes the remaining 50% of internal exposure for the general population (Vandenberg et al. 2012a). Oral exposure to BPA facilitates quick absorption from the gastrointestinal tract due to high pre-systemic metabolism in the gut and liver, resulting in non-toxic conjugation metabolites. BPA exhibits low systemic bioavailability (approximately 1–10% in humans) and a half-life of a few hours. In contrast, parental exposure to BPA results in 100% systemic bioavailability, with the full dose reaching the systemic circulation intact (Ginsberg et al. 2002; Testai et al. 2016).

While BPA is biodegradable in sediments and soil under aerobic conditions, its continuous production and high release rates compared to removal rates contribute to its accumulation in natural environments, posing a potential threat to ecosystems (Careghini et al. 2015; Zhou et al. 2016; Bereketoglu et al. 2017). Moreover, BPA acts as an environmental endocrine disruptor and genotoxic agent towards algae, invertebrates, and fish, and exhibits estrogenic effects on all living organisms, particularly humans, leading to various health issues such as reproductive damage, growth impairment, and the development of serious conditions like cancer (Badiefar et al. 2015; Kabir et al. 2015; Rochester et al. 2015; Seachrist et al. 2016; Wu et al. 2017). Additionally, BPA disrupts the function of multiple hormones including sex hormones, thyroxin, insulin, and leptin, and causes hepatotoxic, immunotoxic, mutagenic, and carcinogenic effects (Huang et al. 2012; Michałowicz 2014; Liu et al. 2016).

The harmful effects of BPA in the environment have become a significant social concern, prompting public prioritization of the development of cost-effective technologies for cleaning up BPA-contaminated environments (Li et al. 2012; Kamaraj et al. 2014; Badiefar et al. 2015). Various physical, chemical, and biological methods have been developed for treating numerous BPA-contaminated sites, offering potential for partial or total restoration of polluted systems. Enzymes, in particular, have gained importance as green catalysts in this regard due to their cost-effectiveness, efficiency, and environmental acceptability.

17.5 Enzymatic Aspect of BPA Degradation

Over the past several years, environmental engineers have been engaged in continuous efforts to develop and enhance innovative technologies. Notably, various conventional methods, including advanced oxidation, membrane separation, solvent extraction, photocatalytic degradation, adsorption, and microbial degradation, have been employed to eliminate toxic phenolic chemicals from environmental matrices (Kang and Kondo 2006; Liu et al. 2016). Recently, the application of biodegradation has proven to be an advanced technique for pollutant removal processes. Microbial degradation plays a crucial role in eliminating BPA pollution from the environment (Zhang et al. 2013; Yang et al. 2014; Li et al. 2012). However, the application of this technique remains under consideration, as it can be easily inhibited by various factors, including the activity of introduced organisms and environmental circumstances (Perelo 2010). Biological degradation of phenolic pollutants has garnered attention from environmentalists due to its potential to target a wide spectrum of pollutants and achieve their complete mineralization into inorganic compounds. Biological or enzymatic treatments present a new approach to oxidize various hazardous or xenobiotic organic pollutants into harmless, stable, and water-insoluble polymers (Kang and Kondo 2006; Husain and Qayyum 2012; Liu et al. 2016).Bacterial strains have played a significant role in the degradation of BPA in the environment. Various bacterial strains have demonstrated the ability to utilize BPA as a unique source of carbon and energy, including gram-negative strains such as *Sphingomonas* sp., *Pseudomonas* sp., *Achromobacter* sp., *Nitrosomonas* sp., *Serratia* sp., *Bordetella* sp., *Alcaligenes* sp., *Pandoraea* sp., *Klebsiella* sp., and *Cupriavidus* sp., as well as gram-positive strains like *Streptomyces* sp. and *Bacillus* sp. (Kang and Kondo 2006; Zhang et al. 2007, 2013; Telke et al. 2009; Li et al. 2012; Kamaraj et al. 2014; Badiefar et al. 2015).In Gram-positive bacteria, BPA is completely degraded, leading to the production of intermediates such as 4-hydroxyacetophenone and 4-hydroxybenzoic acid. Furthermore, these metabolites undergo complete mineralization into CO_2 and H_2O (Li et al. 2012; Zhang et al. 2013).

Recently, the application of enzymes isolated from natural microbes has gained significant recognition for the removal of pollutants. Their metabolic pathways and enzymatic capabilities for converting various toxic organic pollutants into environmentally benign compounds are well documented (van Hamme et al. 2003; Fuentes et al. 2014).Various enzymes, such as cytochrome P450 monooxygenase, ammonia monooxygenase (AMO), and oxidoreductive enzymes like horseradish peroxidase, laccase, and tyrosinase, have been identified to participate in BPA degradation (Huang and Weber 2005; Gassara et al. 2013; Zhang et al. 2013; Ba and Kumar 2017). Both tyrosinase and laccase are copper-containing enzymes that require readily available O_2 for catalytic transformation of numerous substrates (Harms et al. 2011). These two phenol oxidases differ in their oxidation mechanism of phenolic compounds. Laccases oxidize phenolic compounds to generate phenoxy radicals and water, while tyrosinases oxidize phenolics to generate quinones with subsequent release of H_2O (Bollag 1992). Although research on tyrosinases in environmental applications is limited, their potential has been demonstrated in various studies using both free and immobilized forms to eliminate environmental contaminants (Ba and Kumar 2017).

Enzymes produced by ligninolytic fungi, such as lignin peroxidase (LiP), versatile peroxidases (VP), manganese peroxidases (MnP) and laccases, are nonspecific and have gained attention for initiating degradation of many recalcitrant organic pollutants (Karigar and Rao 2011; Husain and Qayyum 2012; Cajthaml 2015; Ba and Kumar 2017; Zeng et al. 2017). The white-rot fungi *Pleurotusostreatus* has been observed to degrade BPA, with in vitro experiments demonstrating the involvement of MnP in BPA degradation and the production of various transformation products (Hirano et al. 2000). Similarly, MnP produced by *Phanerochaetechrysosporium* and *Trametes versicolor* effectively removed BPA from culture medium in laboratory tests, with a decline in estrogenic activity over time, suggesting susceptibility of potential BPA transformation products to MnP-mediated degradation (Tsutsumi et al. 2001).Laccase finds widespread application in various biotechnological, chemical, and industrial systems (Shraddha et al. 2011; Lu et al. 2012; Yang et al. 2014; Zhang et al. 2016).

Previous reports have demonstrated that in the presence of NADH, cytochrome P450 monooxygenase from *Sphingomonas* sp. strain AO1 efficiently degrades BPA into 2,2-bis(4-hydroxyphenyl)-1-propanol and 1,2-bis(4-hydroxyphenyl)-2-propanol via hydroxylation of the methyl group and the quaternary carbon, respectively (Sasaki et al. 2005). P450 monooxygenase is also involved in ipso substitution in other xenobiotic compounds comprisingof phenol at quaternary carbon (Kolvenbach et al. 2007).Ammonia monooxygenase (AMO) is another enzyme implicated in BPA degradation. The degradation efficiency of a *Nitrosomonas europaea* strain was found to be inhibited by the presence of allylthiourea, an inhibitor of AMO (Roh et al. 2009).

However, several important factors remain challenging, including the high toxicity levels of compounds, the large volumes of pollutants, and the need to remediate contaminated water and soil in compliance with environmental standards. Hence, it is essential to devise specific strategies to develop effective technologies that can operate across diverse environments and adverse conditions (Ma et al. 2016).The use of microbial enzymes, particularly laccases, has emerged as a promising alternative to advanced treatment processes for removing organic contaminants. The primary advantages of enzymatic processes include high activity under ambient temperature and pressure conditions, low energy requirements, low toxicity, and simple process control and maintenance (Hilden et al. 2009; Chen et al. 2016a).The strategic utilization of natural biocatalysts, such as laccase from various sources, may offer a viable pathway for bioremediation that can function effectively across diverse environments and operational limitations. It is crucial to elucidate the structural and functional concepts of the laccase enzyme for this purpose.

17.6 Structure and Functional Mechanism of Laccase

Structurally, laccases are typically multi-domain glycoproteins encoded by 500–550 amino acids, with a molecular weight of about50–130 kDa (Das et al. 2018; Koschorreck et al. 2008; Malhotra and Suman 2021). This core harbors multiple copper-binding sites responsible for the catalytic activityof the enzymes. As shown in Fig. 17.2, laccases often house four copper atoms distributed among three distinct copper centers: Type 1 (T1), Type 2 (T2), and Type 3 (T3) (Dwivedi et al. 2011; Shraddha et al. 2011; Malhotra and Suman 2021). These copper atoms are coordinated by amino acid residues, particularly histidine and cysteine residues, forming the active sites where catalysis occurs. Notably, laccases exhibit significant uniqueness in amino acid sequence alignment with other multi-copper oxidases, revealing four conserved histidine-rich copper-binding motifs. Figure 17.3 depicted the strictly conserved copper-binding motifs within the amino acid sequence of laccases encompass Cu I (HLHG), Cu II (HDH), Cu III (HPIHLH), and Cu IV (HCHILEH), as observed in various fungal and bacterial laccases (Das et al. 2018; Dwivedi et al. 2011; Shraddha et al. 2011).

These copper sites are classified as Type-1 (T1) or blue copper center, Type-2 (T2) or normal copper, and Type-3 (T3) or coupled binuclear copper centers. However, yellow or white laccases, primarily found in fungi but recently observed in bacteria as well, display notable differences due to variations in available amino acid residues within the copper centers and their coordination with Cu2 + ions at the T1 site. These differences typically influence the color and spectral properties of laccases (Mate et al. 2013; Das et al. 2020; Malhotra and Suman 2021). In comparison to fungal sources, laccases derived from bacteria demonstrate greater pH and temperature stability owing to distinct protein structures and interactions with Cu^{2+} (Hildén et al. 2009; Chandra and Chowdhary 2015; Das et al. 2020).

To facilitate its catalytic function, laccase relies on copper atoms distributed across its three distinct copper centers (Fig. 17.2). The enzyme initiates the catalytic process by extracting electrons from substrates, converting them into free radicals that can undergo polymerization. Upon receiving four electrons, the enzyme transfers them to molecular oxygen, ultimately yielding a water molecule (Bourbonnais and Paice 1990; Dwivedi et al. 2011). In summary, laccase catalysis encompasses three primary steps.

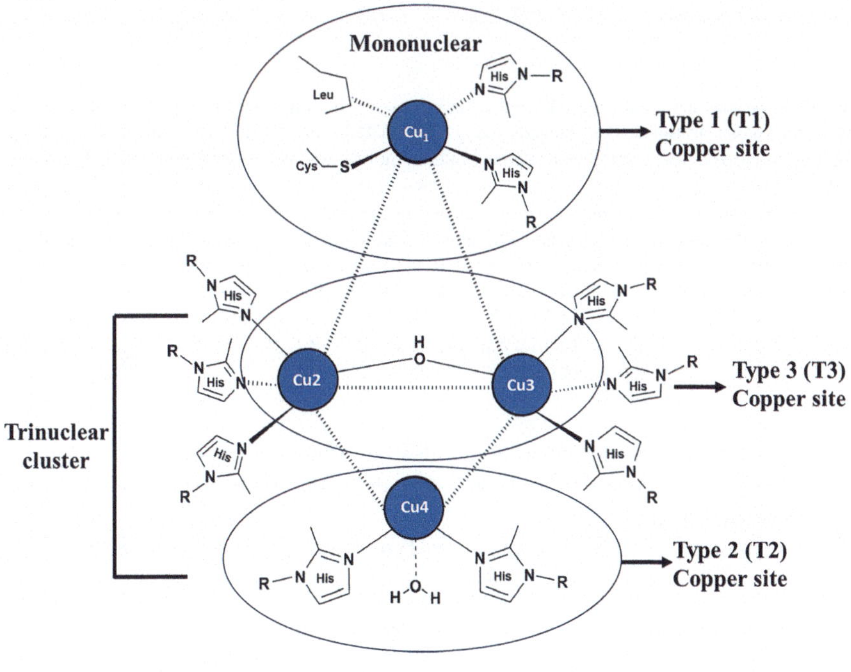

Fig. 17.2 Schematic representation of laccase active siteshowing the relative orientation ofthe copper atoms

```
                                  T2 T3                        T2 T3
Populus trichocarpa       77  VTVHWHGVS 85         217   LLWHAHISW 125
Agaricus bisporus         79  VSIHWHGFF 87         120   FWYHSHLST 128
Trametes versicolor       81  TSIHWHGFF 89         122   FWYHSHLST 130
Cerrena unicolor          82  TSIHWHGFF 90         123   YWYHSHLST 131
S. Maltophilia            115 TSIHWHGIL 123        153   YWYHSHSMF 161
Bacillus sp. GZB          102 TVVHLHGGA 110        150   LWYHDHAMA 158
B. velezensis             102 TVVHLHGGA 110        150   LWYHDHAMA 158
B. amyloliquefaciens      102 TVVHLHGGA 110        150   LWYHDHAMA 158
B. tequilensis            102 TVVHLHGGV 110        150   LWYHDHAMA 158
B. Subtilis               102 TVVHLHGGV 110        150   LWYHDHAMA 158
Pseudomonas stutzeri      102 TVVHLHGGV 110        150   LWYHDHAMA 158
                              .  :* **                   :*  *

                              T1   T2 T3              T3 T3    T1
Populus trichocarpa       455 SHPFHLHGF 463     517 LHCHLEVHTTWGLKM 531
Agaricus bisporus         408 AHPFHLHGH 416     469 LHCHIDWHLEAGLAI 483
Trametes versicolor       406 PHPFHLHGH 414     471 LHCHIDFHLDAGFAI 485
Cerrena unicolor          399 PHPFHLHGH 407     462 LHCHIDWHLEAGFAV 476
S. Maltophilia            524 QHPIHLHGV 532     582 YHCHLLYHMEAGMMR 596
Bacillus sp. GZB          418 THPIHLHLV 426     490 WHCHILEHEDYDMMR 504
B. velezensis             418 THPIHLHLV 426     490 WHCHILEHEDYDMMR 504
B. amyloliquefaciens      418 THPIHLHLV 426     490 WHCHILEHEDYDMMR 504
B. tequilensis            418 THPIHLHLV 426     490 WHCHILEHEDYDMMR 504
B. Subtilis               418 THPIHLHLV 426     490 WHCHILEHEDYDMMR 504
Pseudomonas stutzeri      418 THPIHLHLV 426     490 WHCHILEHEDYDMMR 504
                               *  :*:*              ***:    *      :
```

Fig. 17.3 Amino acid sequence alignment of laccases. The alignment was performed based on the homologous amino acid sequences using CLUSTALW program. The amino acid sequence alignment showed four conserved copper-binding motifs are highlighted with thick bar above of the residues. The histidine residues are marked in red and distinguish as T1, T2 and T3 on above of the residues, where cysteine residue is marked in green. The identical amino acid residues are designated with asterisk. The amino acid sequences of laccase shown in the alignment are from *Populus trichocarpa* (CAA74103.1), *Agaricus bisporus* (AAC18877.1), *Trametes versicolor* (AAC49828.1), *Cerrena unicolor* (AEQ35306.1), *Stenotrophomonas maltophilia* (WP_012481059.1), *Bacillus* sp. GZB (KX777662.1) *Bacillus velezensis* (WP_045206917.1), *Bacillus amyloliquefaciens* (WP_053573639.1), *Bacillus tequilensis* (WP_024715866.1), *Bacillus Subtilis* (PDB ID: 1GSK_A), Pseudomonas stutzeri (AND62506.1)

i. Reduction of Type-1 copper by the substrate's reducing agent.
ii. Transfer of electrons from Type-1 copper to the Type-2 and Type-3 copper trinuclear cluster.
iii. Reduction of O_2 to H_2O occurs at the Type-2 and Type-3 copper centers.

The overall reaction can be summarized as the enzymatic oxidation of substrates coupled with the reduction of molecular oxygen to water:

$$4RH + O2 \rightarrow 4R\cdot + 2H2O$$

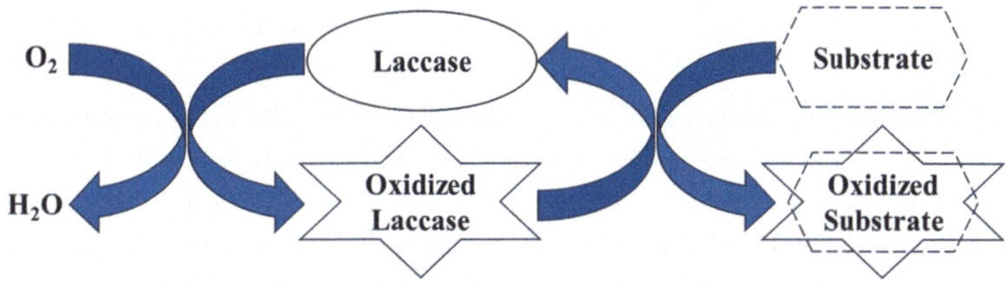

Fig. 17.4 Schematic illustration the reaction mechanism of laccase

In most catalytic reactions involving organic and inorganic substrate catalyzed by laccases, the binding of substances at the T1 site precedes sequential electron transfer from T1 to T2/T3 binuclear sites, leading to the transformation of organic and inorganic substrates while reducing molecular O_2 to H_2O (Fig. 17.4). These mechanisms signify significant advancements in catalytic reactions for remediating numerous organic and inorganic substances facilitated by laccases. Consequently, laccases have emerged as superior biocatalysts in various industrial, biotechnological, and environmental applications due to their broad substrate specificity and high redox potential (Chandra and Chowdhary 2015; Forootanfarand Faramarzi 2015; Akram et al. 2022; Naseem et al. 2023).

17.7 Laccase Mediated BPA Degradation Mechanism

Laccases are enzymes capable of oxidizing a wide range of organic and inorganic substrates, including phenolic compounds like bisphenol A (BPA). As shown in Fig. 17.5, the degradation of BPA by laccase typically involves oxidation reactions that lead to the breakdown of the BPA molecule into smaller, less toxic compounds (Telke et al. 2009; Erkurt 2015; Das et al. 2018 and 2019). It has been well postulated that laccase or ligninolytic enzymes could facilitate the polymerization of BPA, leading to the formation of larger, more complex polymers through coupling reactions between C–C or C–O groups of the phenolic residues. Subsequent cleavage of the oligomers produces phenol and 4-isopropenylphenol (Fukuda et al. 2004; Huang and Weber 2005). Previous reports have corroborated the degradation of BPA by laccases through a condensation pathway, as evidenced by the identification of high molecular weight degradation products and polymeric structures (Torres-Duarte et al. 2012; Cabana et al. 2007).Furthermore, various previous studies have reported the oxidative biodegradation of BPA by laccases, yielding 4-isopropenylphenol and hydroquinone as intermediates (Kolvenbach et al. 2007; Das et al. 2018). The 4-isopropenylphenol undergoes direct conversion to hydroquinone through a •OH attack (Das et al. 2018). Telke et al. (2009) reported hydroquinone as an intermediate in the degradation of BPA by laccase from *Pseudomonas* sp. LBC1. Subsequent oxidation of hydroquinone results in the formation of various organic acids. These acidsinclude 2-hydroxypropanoic acid, 2-methylbutanoic acid, and 3-methylbutanoic acid (Das et al.

2018). Alternatively, BPA oxidation by laccase proceeds through 4,4'-dihydroxy-alpha-methylstilbene to 4-hydroxybenzaldehyde and eventually to benzoic acid, as previously reported (Galliker et al. 2010; Das et al. 2018 and 2019). Besides, direct ring cleavage of benzoic acid leads to the formation of low molecular weight products such as 2-hydroxypropanoic acid, 2-methylbutanoic acid, and 3-methylbutanoic acid (Das et al. 2018 and 2019; Daâssi et al. 2016). Overall, laccase-mediated degradation offers a promising approach for mitigating BPA contamination in various environmental matrices, including soil, water, and wastewater. However, the efficiency of BPA degradation using laccase may depend on factors such as the enzyme source, reaction conditions (pH, temperature, substrate concentration), and the presence of other compounds in the environment. Additionally, the specific degradation pathways and intermediates formed during laccase-mediated BPA degradation may vary depending on the experimental conditions and the properties of the laccase enzyme used.

17.8 Laccase as an Important Biocatalyst for BPA Remediation

Laccase emerges as a significant biocatalyst owing to its broad oxidation–reduction potential, finding applications across various domains including food processing and stabilization, chemical synthesis, biobleaching of paper pulp, biosensing, textile finishing, bioremediation, and the degradation of environmental phenolic pollutants and endocrine disruptors (Forootanfar and Faramarzi 2015; Ashe et al. 2016; Akram et al. 2022).The origins of laccase trace back to 1883 when it was first identified in *Rhus vernicifera*, the Japanese lacquer tree, thus earning its name (Reinhammar 1984). Since the nineteenth century, this enzyme has undergone extensive study due to its proficiency in oxidizing a wide array of phenolic compounds. Laccase offers distinct advantages over other conventional chemical or microbial catalysts due to its high catalytic efficiency and broad substrate specificity. Consequently, its applications have gained significant attention across numerous industrial sectors (Fernández-Fernández et al. 2013). Its capability to degrade phenolic compounds positions it as a valuable catalytic agent in applications ranging from dye decolorization to xenobiotics degradation within various industrial waste treatment processes.

Despite its extensive industrial use in sectors such as textile dyeing, pulp bleaching, food enhancement, polymer synthesis, and biosensor development, the potential of laccase biocatalysis for water and wastewater treatment remains underexplored (Dwivedi et al. 2011; Garcia et al. 2011; Shraddha et al. 2011). Developing a successful enzyme biocatalysts system requires maintaining the enzyme's functional activity and stability while maximizing its catalytic efficiency (Naseem et al. 2023). Chen et al. (2016a) have developed a novel biocatalyst by immobilizing laccase from the white-rot fungi *Trametes versicolor* on the yeast cell surface, demonstrating its efficacy in treating recalcitrant organic contaminants such as Bisphenol

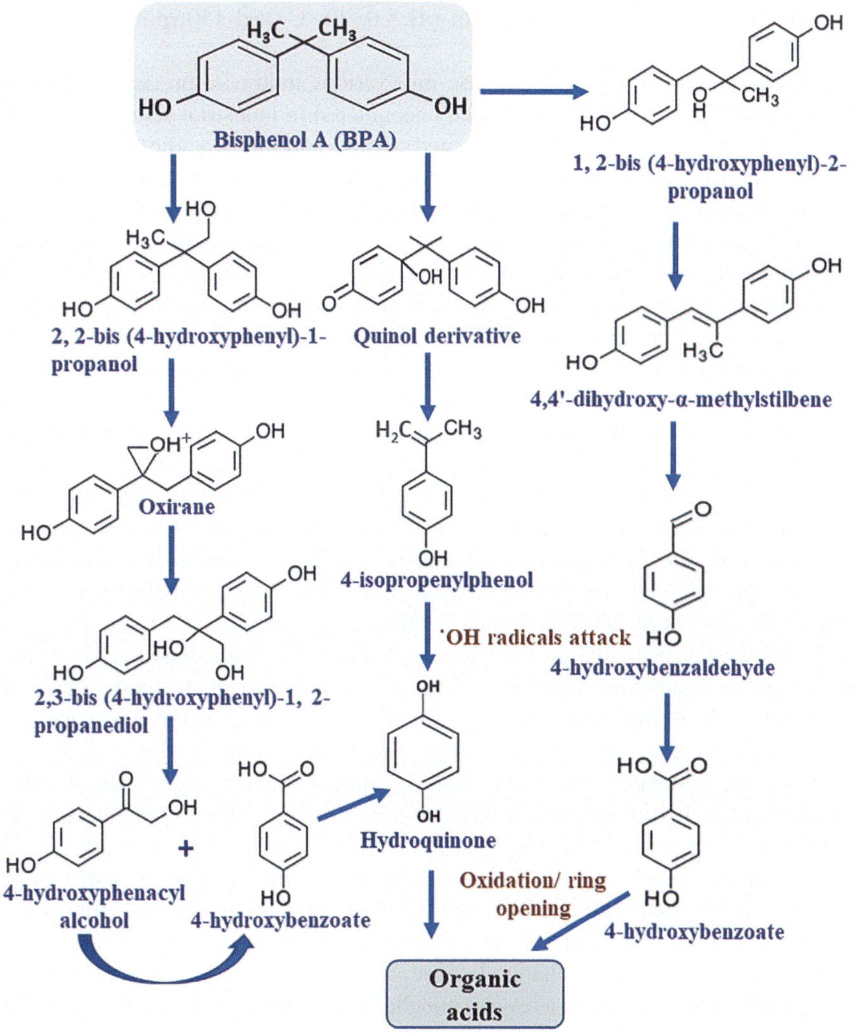

Fig. 17.5 Possible degradation pathways of bisphenol A (BPA) by laccase

A and sulfamethoxazole (an antibiotic). The immobilized laccase retained 74% of its initial activity after eight repeated batch reactions. Previous studies have also observed enhancements in enzyme stability or longevity through immobilization on various matrix materials (Datta et al. 2013; Ba and Kumar 2017).Latif et al. (2022) successfully immobilized laccase from *Trametes versicolor* onto Cu-alginate beads using the entrapment method and investigated its effectiveness in degrading bisphenol A (BPA) in an aqueous medium. Compared to the free enzyme, the immobilized laccase exhibited higher storage stability and achieved a 96.12% degradation of BPA (10 mg/L) under conditions of pH 5.0, 30 °C, and 150 rpm after 1 h of reaction.

However, incorporating free laccase into various industrial processes remains challenging. The harsh conditions often encountered in industrial settings, such as high temperatures and/or pressures, elevated salt concentrations, acidic or alkaline pH levels, oxidative environments, high shear forces, or short processing times, demand enzymes with exceptional resilience. Moreover, there is a pressing need for enzymes with novel catalytic properties to drive innovation and advancement in industrial biotechnology. Additionally, the high cost associated with isolation and purification, coupled with issues related to non-reusability and structural instability, pose practical challenges for laccase applications. These obstacles, however, can be overcome by utilizing natural biocatalysts such as spore-associated laccase (Dwivedi et al. 2011; Das et al. 2018).Spores serve as a protective mechanism enabling survival in environments characterized by starvation, exposure to toxic chemicals, or harsh conditions (Cho et al. 2011). Furthermore, laccase associated with spores as natural biocatalysts exhibits greater stability and resistance to heat, alkaline and acidic conditions, UV light, hydrogen peroxide, and other chemical agents. Recently, thermostable laccases like CotA from *Bacillus* species have found extensive application in the biotransformation of phenols and synthetic dyes (Cho et al. 2011; Zhang et al. 2013; Brander et al. 2014; Chen et al. 2015; Das et al. 2018).

The efficiency of oxidation by a particular laccase is influenced by the nature and position of substituents on the phenolic ring, allowing it to oxidize typical substrates such as monophenols (e.g., sinapinic acid) or guaiacol. Furthermore, laccase exhibits the capability to oxidize polyamines, aminophenol, lignin, aryl diamine, some inorganic ions, and potentially mitigate the toxicity of various polycyclic aromatic hydrocarbons (PAHs), multiple azo dyes, and organophosphorus pesticides (Zeng et al. 2016; Zhang et al. 2016; Chandra and Chowdhary 2015). Certain pseudomonads (e.g., *Pseudomonas* sp. strain LBC1) produce extracellular laccases capable of catalyzing BPA transformation reactions. The laccase from strain LBC1 efficiently degrades BPA even without redox mediators, which typically facilitate electron transfer and enhance laccase activity (Telke et al. 2009). *Fusarium incarnatum* UC-14 laccase in a reverse micelles system has also been applied for the bioremediation of BPA (Chhaya and Gupte 2013). While most studied laccases have been reported from fungal origins, there is increasing interest in bacterial laccases due to their impressive stability across variable pH and temperature, along with well-established bacterial expression systems (Cho et al. 2011; Brander et al. 2014). However, applications of laccases from *Bacillus* spores in environmental pollutants removal, particularly BPA degradation, remain limited. Some studies have employed spore laccases for the decolorization of various synthetic dyes (Lu et al. 2012; Wang et al. 2010; Lončar et al. 2013). Importantly, spore laccases serve as natural biocatalysts and remain stable under diverse environmental conditions for extended periods while maintaining high enzyme activity. These inert characteristics make them highly suitable for various industrial and biotechnological processes.

17.9 Challenges and Future Prospects

Biodegradation, particularly through microbial enzymes like laccase, has emerged as a sustainable, economical, and environmentally friendly alternative for remediating contaminated environments. Consequently, recent efforts have been directed towards developing new technologies utilizing various microbial enzymes to detoxify environmental pollutants and convert them into simpler or less toxic compounds under favorable environmental conditions (Zhang et al. 2020; Akram et al. 2022; de Freitas et al. 2017; Das et al. 2018 and 2023; Chen et al. 2016a, b; Hautphenne et al. 2016). Research now emphasizes cost-effective cleanup techniques with minimal long-term environmental damage. Despite its considerable potential, several challenges hinder the widespread implementation of laccase-based remediation for BPA-contaminated environments (de Freitas et al. 2017; Zhang et al. 2020; Akram et al. 2022). These challenges include the limited stability (pH, temperature, and salt concentration) of laccase enzymes under bioremediation process, substrate inhibition at high BPA concentrations, and the need for efficient enzyme immobilization techniques.

Future research efforts should focus on addressing these challenges through the development of novel enzyme engineering strategies, immobilization techniques, and synergistic approaches involving laccase-mediator systems. Additionally, comprehensive environmental risk assessments and life cycle analyses are necessary to evaluate the long-term sustainability and ecological implications of laccase-based BPA remediation technologies.

17.10 Conclusions

Laccase enzyme offers a promising biotechnological tool for the environmental remediation of bisphenol A contamination. Its ability to degrade BPA under mild reaction conditions offers a sustainable and cost-effective alternative to conventional remediation methods. However, further research is needed to overcome existing challenges and optimize laccase-based remediation strategies for real-world applications.

References

Akram F, Ashraf S, Haq IU, Shah FI, Aqeel A (2022) Eminent industrial and biotechnological applications of laccases from bacterial source: a current overview. Appl Biochem Biotechnol 194:2336–2356

Alonso-Magdalena P, Vieira E, Soriano S, Menes L, Burks D, Quesada I, Nadal A (2010) Bisphenol A exposure during pregnancy disrupts glucose homeostasis in mothers and adult male offspring. Environ Health Perspect 118:1243–1250

Alonso-Magdalena P, Ropero AB, Soriano S, García-Arévalo M, Ripoll C, Fuentes E, Quesada I, Nadal Á (2012) Bisphenol-A acts as a potent estrogen via non-classical estrogen triggered pathways. Mol Cell Endocrinol 355:201–7

Angle BM, Do RP, Ponzi D, Stahlhut RW, Drury BE, Nagel SC, Welshons WV, Besch-Williford CL, Palanza P, Parmigiani S, vom Saal FS, Taylor JA (2013) Metabolic disruption in male mice due to fetal exposure to low but not high doses of bisphenol A (BPA): evidence for effects on body weight, food intake, adipocytes, leptin, adiponectin, insulin and glucose regulation. Reprod Toxicol 42:256–68

Ashe B, Nguyen LN, Hai FI, Lee D-J, van de Merwe JP, Leusch FDL, Price WE, Nghiem LD (2016) Impacts of redox-mediator type on trace organic contaminants degradation by laccase: degradation efficiency, laccase stability and effluent toxicity. Int Biodeterior Biodegrad 113:169–176

Babu S, Uppu SN, Martin B, Agu OA, Uppu RM (2015) Unusually high levels of bisphenol A (BPA) in thermal paper cash register receipts (CRs): development and application of a robust LC-UV method to quantify BPA in CRs. Toxicol Mech Methods 25:410–416

Badiefar L, Yakhchali B, Rodriguez-Couto S, Veloso A, García-Arenzana JM, Matsumura Y, Khodabandeh M (2015) Biodegradation of Bisphenol A b y the newly isolated Enterobacter gergoviae strain BYK-7 enhanced using genetic manipulation. RSC Adv 5:29563–29572

Ba S, Kumar VV (2017) Recent developments in the use of tyrosinase and laccase in environmental applications. Criti Rev Biotechnol. 37:819–832

Bereketoglu C, Arga KY, Eraslan S, Mertoglu B (2017) Analysis of transcriptional profiles of *Saccharomyces cerevisiae* exposed to Bisphenol A. Curr Genetics 63:253–274

Bollag JM (1992) Decontaminating soil with enzymes. Environ Sci Technol 26:1876–1881

Bourbonnais R, Paice MG (1990) Oxidation of non-phenolic substrates. An expanded role for laccase in lignin biodegradation. FEBS Lett 267:99–102

Brander S, Mikkelsen JD, Kepp KP (2014) Characterization of an alkali- and halide-resistant laccase expressed in E. coli: CotA from *Bacillus clausii*. PLoS ONE 9:e99402

Braun JM, Hauser R (2011) Bisphenol A and children's health. Curr Opin Pediatr 23:233–239

Cabana H, Jones JP, Agathos SN (2007) Elimination of endocrine disrupting chemicals using white rot fungi and their lignin modifying enzymes: a review. Eng Life Sci 7:429–456

Cajthaml T (2015) Biodegradation of endocrine-disrupting compounds by ligninolytic fungi: mechanisms involved in the degradation. Environ Microbiol 17:4822–4834

Careghini A, Mastorgio AF, Saponaro S, Sezenna E (2015) Bisphenol A, nonylphenols, benzophenones, and benzotriazoles in soils, groundwater, surface water, sediments, and food: a review. Environ Sci Pollut Res 22:5711–5741

Chandra R, Chowdhary P (2015) Properties of bacterial laccases and their application in bioremediation of industrial wastes. Environ Sci Process Impacts 17:326–42

Chen Y, Stemple B, Kumar M, Wei N (2016) Cell surface display fungal laccase as a renewable biocatalyst for degradation of persistent micropollutants Bisphenol A and sulfamethoxazole. Environ Sci Technol 50:8799–8808

Chen B, Xu WQ, Pan XR, Lu L (2015) A novel non-blue laccase from *Bacillus amyloliquefaciens*: secretory expression and characterization. Int J Biol Macromol. 76 39e44

Chen D, Kannan K, Tan H, Zheng Z, Feng YL, Wu Y et al (2016) Bisphenol analogues other than BPA: environmental occurrence, human exposure, and toxicity-a review. Environ Sci Technol 50:5438–5453

Chhaya U, Gupte A (2013) Possible role of laccase from *Fusarium incarnatum* UC-14 in bioremediation of Bisphenol A using reverse micelles system. J Hazard Mater 254–255:149–156

Cho E-A, Seo J, Lee D-W, Pan J-G (2011) Decolorization of indigo carmine by laccase displayed on Bacillus subtilis spores. Enzyme Microbial Technol. 49:100–104

Daâssi D, Prieto A, Zouari-Mechichi H, Martínez MJ, Nasri M, Mechichi T (2016) Degradation of Bisphenol A by different fungal laccases and identification of its degradation products. Int Biodeterior Biodegrad 110:181–188

Daroch M, Houghton CA, Moore JK, Wilkinson MC, Carnell AJ, Bates AD, Iwanejko LA (2014) Glycosylated yellow laccases of the basidiomycete *Stropharia aeruginosa*. Enzym Microb Technol. 58e59, 1e7

Das R, Li G, Mai B, An T (2018) Spore cells from BPA degrading bacteria Bacillus sp. GZB displaying high laccase activity and stability for BPA degradation. Sci Total Environ 640–641:798–806

Das R, Liang Z, Li G, Mai B, An T (2019) Genome sequence of a spore-laccase forming, BPA-degrading Bacillus sp. GZB isolated from an electronic-waste recycling site reveals insights into BPA degradation pathways. Arch Microbiol 201:623–638

Das R, Liang Z, Li G, An T (2020) A non-blue laccase of Bacillus sp. GZB displays manganese-oxidase activity: a study of laccase characterization, Mn(II) oxidation and prediction of Mn(II) oxidation mechanisms. Chemosphere, 252, 126619

Das R, Yao P, Yin H, Liang Z, Li G, An T (2023) BPA degradation using biogenic manganese oxides produced by an engineered Escherichia coli with a non-blue laccase from Bacillus sp. GZB Chemosphere 326:138407

Datta S, Christena LR, Rajaram YR (2013) Enzyme immobilization: an overview on techniques and support materials. 3 Biotech 3:1–9

de Freitas EN, Bubna GA, Brugnari T, Kato CG, Nolli M, Rauen TG, de Peralta Muniz Moreira RF, Peralta RA, Bracht A, de Souza CGM, Peralta RM (2017) Removal of bisphenol A by laccases from *Pleurotusostreatus* and *Pleurotuspulmonarius* and evaluation of ecotoxicity of degradation products. Chem Eng J 330:1361–1369

Durao P, Chen Z, Fernandes AT, Hildebrandt P, Murgida DH, Todorovic S, Pereira MM, Melo EP, Martins LO (2008) Copper incorporation into recombinant CotA laccase from *Bacillus subtilis*: characterization of fully copper loaded enzymes. J Biol Inorg Chem 13:183–193

Dwivedi UN, Singh P, Pandey VP, Kumar A (2011) Structure–function relationship among bacterial, fungal and plant laccases. J Mol Catal B: Enzyme 68:117–128

Erkurt HA (2015) Biodegradation and Detoxification of BPA: Involving Laccase and a Mediator. Clean—Soil. Air, Water 43:932–939

Fernández-Fernández M, Sanromán MA, Moldes D (2013) Recent developments and applications of immobilized laccase. Biotechnol Avd. 31:1808–1825

Flint S, Markle T, Thomson S, Wallace E (2012) Bisphenol A exposure effects and policy; a wildlife perspective. J Environ Manag 104:19–34

Forootanfar H, Faramarzi MA (2015) Insights into laccase producing organisms, fermentation states, purification strategies, and biotechnological applications. Biotechnol Prog 31:1443–63

Fuentes S, Méndez V, Aguila P, Seeger M (2014) Bioremediation of petroleum hydrocarbons: catabolic genes microbial communities, and applications. Appl Microbiol Biotechnol 98:4781–4794

Fukuda T, Uchida H, Suzuki M, Miyamoto H, Morinaga H, Nawata H, Uwajima T (2004) Transformation products of Bisphenol A by a recombinant *trametesvillosa* laccase and their estrogenic activity. J Chem Technol Biotechnol 79:1212–1218

Galliker P, Hommes G, Schlosser D, Corvini PFX, Shahgaldian P (2010) Laccase modified silica nanoparticles efficiently catalyze the transformation of phenolic compounds. J Colloid Interface Sci 349:98–105

Garcia HA, Hoffman CM, Kinney KA, Lawler DF (2011) Laccase-catalyzed oxidation of oxybenzone in municipal wastewater primary effluent. Water Res 45:1921–1932

Gassara F, Brar SK, Verma M, Tyagi RD (2013) Bisphenol A degradation in water by ligninolytic enzymes. Chemosphere 92:1356–1360

Ginsberg G, Hattis D, Sonawane B, Russ A, Banati P, Kozlak M, Smolenski S, Golbe R (2002) Evaluation of child/adult pharmacokinetic differences from a database derived from the therapeutic drug literature. Toxicol Sci 66:185–200

Grignard E, Lapenna S, Bremer S (2012) Weak estrogenic transcriptional activities of Bisphenol A and Bisphenol S. Toxicol in Vitro 26:727–731

Harley KG, Berger K, Rauch S, Kogut K, Henn BC, Calafat AM, Huen K, Eskenazi B, Holland N (2017) Association of prenatal urinary phthalate metabolite concentrations and childhood BMI and obesity. Pediatr Res 82:405–415

Harms H, Schlosser D, Wick LY (2011) Untapped potential: exploiting fungi in bioremediation of hazardous chemicals. Nat Rev Microbiol 9:177–192

Hautphenne C, Penninckx M, Debaste F (2016) Product formation from phenolic compounds removal by laccases: A review. Environ Technol Innovation 5:250–266

Hilden K, Hakala TK, Lundell T (2009) Thermotolerant and thermostable laccases. Biotechnol Lett 31:1117–1128

Hirano T, Honda Y, Watanabe T, Kuwahara M (2000) Degradation of bisphenol A by the lignin-degrading enzyme, manganese peroxidase, produced the white-rot basidiomycete. Pleurotusostreatus Biosci Biotechnol Biochem. 64:1958–1962

Huang Q, Weber JW (2005) Transformation and removal of Bisphenol A from aqueous phase via peroxidase-mediated oxidative coupling reactions: efficacy, products, and pathways. Environ Sci Technol 39:6029–6036

Huang YQ, Wong CKC, Zheng JS, Bouwman H, Barra R, Wahlström B, Neretin L, Wong MH (2012) Bisphenol A (BPA) in China: a review of sources, environmental levels, and potential human health impacts. Environ Int 42:91–99

Husain Q, Qayyum S (2012) Biological and enzymatic treatment of bisphenol A and other endocrine disrupting compounds: a review. Crit Rev Biotechnol 33:260–292

Im J, Loffler FE (2016) Fate of bisphenol a in terrestrial and aquatic environments. Environ Sci Technol 50:8403–8416

Ji K, Hong S, Kho Y, Choi K (2013) Effects of bisphenol S exposure on endocrine functions and reproduction of zebrafish. Environ Sci Technol 47:8793–8800

Kabir ER, Rahman MS, Rahman I (2015) A review on endocrine disruptors and their possible impacts on human health. Environ Toxicol Pharmacol 40:241–258

Kamaraj M, Sivaraj R, Venckatesh R (2014) Biodegradation of Bisphenol A by the tolerant bacterial species isolated from coastal regions of Chennai, Tamil Nadu. India. Int Biodeterior Biodegrad. 93:216–222

Kang JH, Kondo F (2006) Bisphenol A in the surface water and freshwater snail collected from rivers around a secure landfill. Bull Environ Contam Toxicol 76:113–118

Karigar CS, Rao SS (2011) Role of microbial enzymes in the bioremediation of pollutants: A review. Enzyme Res. Article ID 805187

Kinney CA, Furlong ET, Kolpin DW, Burkhardt MR, Zaugg SD, Werner S, L., Bossio, J, P., Benotti, M.J., (2008) Bioaccumulation of pharmaceuticalsand other anthropogenic waste indicators in earthworms fromagricultural soil amended with biosolid or swine manure. Environ Sci Technol 42:1863–1870

Kolvenbach B, Schlaich N, Raoui Z, Prell J, Zühlke S, Schäffer A, Guengerich FP, Corvini PFX (2007) Degradation pathway of bisphenol A: does ipso substitution apply to phenols containing a quaternary α-carbon structure in the para position? Appl Environ Microbiol 73:4776–4784

Koschorreck K, Richter SM, Ene AB, Roduner E, Schmid RD, Urlacher VB (2008) Cloning and characterization of a new laccase from *Bacillus licheniformis* catalyzing dimerization of phenolic acids. Appl Microbiol Biotechnol 79:217–224

Kundakovic M, Gudsnuk K, Franks B, Madrid J, Miller RL, Perera FP, Champagne FA (2013) Sex-specific epigenetic disruption and behavioral changes following low-dose in utero bisphenol A exposure. Proc Natl Acad Sci U S A 110:9956–9961

Larsson K, Lindh CH, Jönsson BAG, Giovanoulis G, Bibi M, Bottai M, Bergström A, Berglund M (2017) Phthalates, non-phthalate plasticizers and Bisphenols in Swedish preschool dust in relation to children's exposure. Environ Int 102:114–124

Latif A, Maqbool A, Sun K, Si Y (2022) Immobilization of Trametes Versicolor laccase on Cu-alginate beads for biocatalytic degradation of bisphenol A in water: Optimized immobilization, degradation and toxicity assessment. J Environ Chemi Eng. 10:107089

Lee S, Lee HA, Park B, Han H, Park BH, Oh SY, Hong YS, Ha EH, Park H (2018) A prospective cohort study of the association between bisphenol A exposure and the serum levels of liver enzymes in children. Environ Res 161:195–201

Leung YK, Govindarajah V, Cheong A, Veevers J, Song D, Gear R, Zhu X, Ying J, Kendler A, Medvedovic M, Belcher S, Ho S-M (2017) Gestational high fat diet and bisphenol A exposure heightens mammary cancer risk. Endocr Relat Cancer 24:365–378

Liao C, Liu F, Kannan K (2012) Bisphenol S, a new bisphenol analogue in paper products and currency bills and its association with bisphenol A residues. Environ Sci Technol 46:6515–6522

Li GY, Zu L, Wong PK, Hui X, Lu Y, Xiong JK, An TC (2012) Biodegradation and detoxification of Bisphenol A with one newly-isolated strain Bacillus sp. GZB: kinetics, mechanism and estrogenic transition. Bioresour Technol 114:224–230

Liu N, Liang G, Dong X, Qi X, Kim J, Piao Y (2016) Stabilized magnetic enzyme aggregates on graphene oxide for high performance phenol and Bisphenol A removal. Chemical Eng J. 306(2016):1026–1034

Lončar N, Bŏzić N, Lopez-Santin J, Vujčić Z (2013) *Bacillus amyloliquefaciens* laccase–from soil bacteria to recombinant enzyme for waste water decolorization. Bioresour Technol 147:177–183

Lu L, Zhao M, Wang TN, Zhao L-Y, Du M-H, Li T-L, Li D-B (2012) Characterization and dye decolorization ability of an alkaline resistant and organic solvents tolerant laccase from *Bacillus licheniformis* LS04. Bioresour Technol 115:35–40

Luo Y, Guo W, Ngo HH, Nghiem LD, Hai FI, Zhang J, Liang S, Wang XC (2014) A review on the occurrence of micropollutants in the aquatic environment and their fate and removal during wastewater treatment. Sci Total Environ 473–474:619–641

Ma C, Qin D, Sun Q, Zhang F, Liu H, Yu C-P (2016) Removal of environmental estrogens by bacterial cell immobilization technique. Chemosphere 144:607–614

Malhotra M, Suman SK (2021) Laccase-mediated delignification and detoxification of lignocellulosic biomass: removing obstacles in energy generation. Environ Sci Pollut Res 28:58929–58944

Manikkam M, Tracey R, Guerrero-Bosagna C, Skinner MK (2013) Plastics derived endocrine disruptors (BPA, DEHP and DBP) induce epigenetic transgenerational inheritance of obesity, reproductive disease and sperm epimutations. PLoS ONE 8:e55387

Mate DM, Garcia-Ruiz E, Camarero S, Shubin VV, Falk M, Shleev S, Ballesteros AO, Alcalde M (2013) Switching from blue to yellow: altering the spectral properties of a high redox potential laccase by directed evolution. Biocatal Biotransform 31:8–21

Melzer D, Rice NE, Lewis C, Henley WE, Galloway TS (2010) Association of urinary bisphenol a concentration with heart disease: evidence from NHANES 2003/06. PLoS ONE 5:e8673

Mot AC, Parvu M, Damian G, Irimie FD, Darula Z, Medzihradszky KF, Brem B, Silaghi-Dumitrescu R (2012) A "yellow" laccase with "blue" spectroscopic features, from *Sclerotinia sclerotiorum*. Process Biochem 47:968–975

Michałowicz J (2014) Bisphenol A – Sources, toxicity and biotransformation. Environ Toxicol Pharmacol 37:738–758

Mokra K, Kocia M, Michałowicz J (2015) Bisphenol A and its analogs exhibit different apoptotic potential in peripheral blood mononuclear cells (in vitro study). Food Chemical Toxicol. 84:79–88

Nam S, Seo Y, Kim M (2010) Bisphenol A migration frompolycarbonate baby bottle with repeated use. Chemosphere 79:949–952

Naseem S, Rawal RS, Pandey D, Suman SK (2023) Immobilized laccase: an effective biocatalyst for industrial dye degradation from wastewater. Environ Sci Pollut Res 36:84898–84917

Noonan GO, Ackerman LK, Begley TH (2011) Concentration of bisphenol A in highly consumed canned foods on the U.S. market. J Agric Food Chem 59:7178–7185

Palanza P, Morellini F, Parmigiani S, vom Saal FS (1999) Prenatal exposure to endocrine disrupting chemicals: effects on behavioral development. Neurosci Biobehav Rev 23:1011–1027

Peng YH, Chen YJ, Chang YJ, Shih YH (2015) Biodegradation of Bisphenol A with diverse microorganisms from river sediment. J Hazard Mater 286:285–290

Perelo LW (2010) Review: In situ and bioremediation of organic pollutants in aquatic sediments. J Hazard Mater 177:81–89

Rahbar MH, Swingle HM, Christian MA, Hessabi M, Lee M, Pitcher MR, Campbell S, Mitchell A, Krone R, Loveland KA, Patterson DG Jr (2017) Environmental exposure to dioxins, dibenzofurans, bisphenol A, and phthalates in children with and without autism spectrum disorder living near the Gulf of Mexico. Int J Environ Res Public Health 14:1425

Reinhammar B (1984) Laccase. In: Lontie R (ed) Copper proteins and copper enzymes, vol 3. CRC Press, Boca Raton, pp 1–35

Reiss R, Ihssen J, Thony-Meyer L (2011) *Bacillus pumilus* laccase: a heat stable enzyme with a wide substrate spectrum. BMC Biotechnol 11:9

Reiss R, Ihssen J, Richte, M, Eichhorn E, Schilling B, Thony-Meyer L (2013) Laccase versus laccase-like multi-copper oxidase: a comparative study of similar enzymes with diverse substrate spectra. PloS One 8:e65633

Rochester JR, Bolden AL (2015) Bisphenol S and F: A systematic review and comparison of the hormonal activity of Bisphenol A substitutes. Environ Health Perspect 123:643–650

Rochester JR (2013) Bisphenol A and human health: a review of the literature. Reprod Toxicol 42:132–155

Rochester JR, Bolden AL, Kwiatkowski CF (2018) Prenatal exposure to bisphenol A and hyperactivity in children: a systematic review and meta-analysis. Environ Int 114:343–356

Roh H, Subramanya N, Zhao F, Yu CP, Sandt J, Chu KH (2009) Biodegradation potential of wastewater micropollutants by ammonia-oxidizing bacteria. Chemosphere 77:1084–1089

Sasaki M, Akahira A, Tsuchido T, Matsumura Y (2005) Purification of cytochrome P450 and ferredoxin involved in BPA degradation by *Sphingomonas* sp. strain AO1. Appl Environ Microbiol 171:8024–8030

Seachrist DD, Bonk KW, Ho S-M, Prins GS, Soto AM, Keri RA (2016) A review of the carcinogenic potential of Bisphenol A. Reproductive Toxicol 59:167–182

Shankar A, Teppala SD (2012) Urinary bisphenol A and hypertension in a multiethnic sample of US adults. J Environ Public Health 481641:1–5

Shraddha., Shekhar, R., Sehgal, S., Kamthania, M., Kumar, A., (2011) Laccase: microbial sources, production, purification, and potential biotechnological applications. Enzyme Res 2011:217861

Staples CA, Dorn PB, Klecka GM, O'Block ST, Harris LR (1998) A review of the environmental fate, effects, and exposures of bisphenol A. Chemosphere 36:2149–73

Stuart ME, Manamsa K, Talbot JC, Crane EJ (2011) Emerging contaminants in groundwater. OR/11/013. British Geological Survey Open Report. Keyworth, Nottingham (UK), p 123

Suciu NA, Tiberto F, Vasileiadis S, Lamastra L, Trevisan M (2013) Recycled paper–paperboard for food contact materials: contaminants suspected and migration into foods and food simulant. Food Chem 141:4146–4151

Tan W, Huang H, Wang Y, Wong TY, Wang CC, Leung LK (2013) Bisphenol A differentially activates protein kinase C isoforms in murine placental tissue. Toxicol Appl Pharmacol 269:163–168

Tarafdar A, Sirohi R, Balakumaran PA, Reshmy R, Madhavan A, Sindhu R, Binod P, Kumar Y, Kumar D, Sim SJ (2022) The hazardous threat of Bisphenol A: Toxicity, detection and remediation. J Hazard Mater 423:127097

Telke AA, Kalyani DC, Jadhav UU, Parshetti GK, Govindwar SP (2009) Purification and characterization of an extracellular laccase from a *Pseudomonas* sp. LBC1 and its application for the removal of Bisphenol A. J. Mol. Catal. B: Enzyme 61:252–260

Testai E, Hartemann P, Rodríguez-Farre E, Rastogi SC (2016) The safety of the use of bisphenol A in medical devices. Regulatory Toxicol Pharmacol. 79:106–107

Torres-Duarte C, Viana MT, Vazquez-Duhalt R (2012) Laccase-mediated transformations of endocrine disrupting chemicals abolish binding affinities to estrogen receptors and their estrogenic activity in zebrafish. Appl Biochem Biotechnol 168:864–876

Tsutsumi Y, Haneda T, Nishida T (2001) Removal of estrogenic activities of bisphenol A and nonylphenol by oxidative enzymes from lignin-degrading basidiomycetes. Chemosphere 42:271–276

Uthandi S, Saad B, Humbard MA, Maupin-Furlow JA (2010) LccA, an archaeal laccase secreted as a highly stable glycoprotein into the extracellular medium by *Haloferaxvolcanii*. Appl Environ Microbiol 76:733–743

Vandenberg L, Chahoud I, Heindel J, Padmanabhan V, Paumgartten F, Schoenfelder G (2012) Urinary, circulating, and tissue biomonitoring studies indicate widespread exposure to bisphenol A. Cien Saude Colect. 17:407–434

Vandenberg LN, Colborn T, Hayes TB, Heindel JJ, Jacobs DR Jr, Lee DH, Shioda T, Soto AM, vom Saal FS, Welshons WV, Zoeller RT, Myers JP (2012) Hormones and endocrine-disrupting chemicals: low-dose effects and nonmonotonic dose responses. Endocr Rev 33:378–455

van Hamme JD, Singh A, Ward OP (2003) Recent advances in petroleum microbiology. Microbiol Mol Bio Rev. 67:503–549

Wang JC, Bennett M (2012) Aging and atherosclerosis: mechanisms, functional consequences, and potential therapeutics for cellular senescence. Circ Res 111:245–259

Wang CL, Zhao M, Li D, Cui DZ, Lu L, Wei X (2010) Isolation and characterization of a novel Bacillus subtilis WD23 exhibiting laccase activity from forest soil. Afr J Biotechnol 9:5496–5502

Williams SA, Jasarevic E, Vandas GM, Warzak DA, Geary DC, Ellersieck MR, Roberts RM, Rosenfeld CS (2013) Effects of developmental bisphenol A exposure on reproductive-related behaviors in California mice (Peromyscus californicus): a monogamous animal model. PLoS ONE 8:e55698

Wu M, Pan C, Chen Z, Jiang L, Lei P, Yang M (2017) Bioconcentration pattern and induced apoptosis of Bisphenol A in zebrafish embryos at environmentally relevant concentrations. Environ Sci Pollut Res 24:6611–6621

Xi W, Lee CK, Yeung WS, Giesy JP, Wong MH, Zhang X, Hecker M, Wong CKC (2011) Effect of perinatal and postnatal bisphenol A exposure to the regulatory circuits at the hypothalamus–pituitary–gonadal axis of CD-1 mice. Reprod Toxicol 31:409–417

Yamamoto T, Yasuhara A, Shiraishi H, Nakasugi O (2001) Bisphenol A in hazardous waste landfill leachates. Chemosphere 42:415–418

Yang J, Lin Q, Ng TB, Ye X, Lin J (2014) Purification and Characterization of a Novel Laccase from *Cerrena* sp. HYB07 with Dye Decolorizing Ability. PLoS ONE 9, e110834

Zeng J, Zhu Q, Wu Y, Lin X (2016) Oxidation of polycyclic aromatic hydrocarbons using Bacillus subtilis CotA with high laccase activity and copper independence. Chemosphere 148:1–7

Zeng S, Zhao J, Xia L (2017) Simultaneous production of laccase and degradation of bisphenol A with Trametes versicolor cultivated on agricultural wastes. Bioprocess Biosyst Eng 40:1237–1245

Zhang C, Zeng G, Yuan L, Yu J, Li J, Huang G, Xi B, Liu H (2007) Aerobic degradation of Bisphenol A by *Achromobacterxylosoxidans* strain B-16 isolated from compost leachate of municipal solid waste. Chemosphere 68:181–190

Zhang R, Wang L, Han J, Wu J, Li C, Ni L, Wang Y (2020) Improving laccase activity and stability by HKUST-1 with cofactor via one-pot encapsulation and its application for degradation of bisphenol A. J Hazard Mater 383:121130

Zhang W, Yin K, Chen L (2013) Bacteria-mediated BisphenolA degradation. Appl. Microbial. Biotechnol 97:5681–5689

Zhang Y, Li X, Hao Z, Xi R, Cai Y, Liao X (2016) Hydrogen peroxide-resistant CotA and YjqC of *Bacillus altitudinis* spores are a promising biocatalyst for catalyzing reduction of sinapic acid and sinapine in rapeseed meal. PLoS ONE 11:e0158351

Zhou D, Yang J, Li H, Lu Q, Liu YD, Lin KF (2016) Ecotoxicological evaluation of low concentration bisphenol A exposure on the soil nematode Caenorhabditis elegans and intrinsic mechanisms of stress responsive in vivo. Environ Toxicol Chem 9999:1–7

Zhou NA, Kjeldal H, Gough HL, Nielsen JL (2015) Identification of putative genes involved in Bisphenol A degradation using differential protein abundance analysis of Sphingobium sp. BiD32. Environ Sci Technol 49:12231–12241

The manufacturer's authorised representative in the EU is Springer Nature Customer Service Centre GmbH, Europaplatz 3, 69115 Heidelberg, Germany. If you have any concerns regarding our products, please contact ProductSafety@springernature.com

Printed and bound by CPI Group (UK) Ltd, Croydon, CR0 4YY

01/10/2026

02225960-0017